Heat Transfer and Fluids Properties of Nanofluids

Heat Transfer and Fluids Properties of Nanofluids

Editor

S M Sohel Murshed

MDPI • Basel • Beijing • Wuhan • Barcelona • Belgrade • Manchester • Tokyo • Cluj • Tianjin

Editor
S M Sohel Murshed
Instituto Superior Técnico
University of Lisbon
Lisbon
Portugal

Editorial Office
MDPI
St. Alban-Anlage 66
4052 Basel, Switzerland

This is a reprint of articles from the Special Issue published online in the open access journal *Nanomaterials* (ISSN 2079-4991) (available at: www.mdpi.com/journal/nanomaterials/special_issues/heat_transfer_fluids_properties_nanofluids).

For citation purposes, cite each article independently as indicated on the article page online and as indicated below:

LastName, A.A.; LastName, B.B.; LastName, C.C. Article Title. *Journal Name* **Year**, *Volume Number*, Page Range.

ISBN 978-3-0365-7313-7 (Hbk)
ISBN 978-3-0365-7312-0 (PDF)

Contents

About the Editor

S M Sohel Murshed

Prof. S M Sohel Murshed is currently a professor at the Department of Mechanical Engineering of the Instituto Superior Técnico, University of Lisbon, Portugal. He received his PhD in Mechanical and Aerospace Engineering from Nanyang Technological University (NTU) Singapore in 2007. Then, he worked as a postdoctoral researcher at NTU, Singapore, and at the University of Central Florida, USA. He was also a visiting professor at Rochester Institute of Technology (USA), Imperial College London (UK) and IIT (India).

Prof. Murshed is listed in the rankings of the World's Top 2% Scientists (for both the career and single year of 2021 impact lists) that was published by the Stanford University group in October 10, 2022. Notably, he ranked 45th among 763 listed scientists from Portugal in the list for impact in a single year, 2021. Prof. Murshed has received the University of Lisbon's Scientific Award 2022 in the areas of Mechanical, Naval and Aerospace (Aircrafts) Engineering for his exceptional scientific contributions and impacts.

Besides 12 books and 35 book chapters, he has so far authored/co-authored more than 250 scientific publications in leading international journals and conferences. Several of his papers have been classified as Highly Cited Papers by Web of Science. His current citation count is 9330, with an h-index of 39 and an i10-index of 73 (GS).

Prof. Murshed serves on the editorial team and board of several international journals such as an associate editor of the *Journal of Thermal Analysis and Calorimetry* (Springer). He is a management committee member nominated by FCT (Portugal) of the European COST action and has regularly been serving as an expert evaluator for the European Commission and for funding agencies of different countries.

Prof. Murshed's research interests include nanofluids, ionanocolloids, micro- and nanoscale heat transfer, nanomaterial–fluid systems, heat exchangers, microfluidics, renewable energy harvesting and storage, advanced energy and cooling technologies.

Preface to "Heat Transfer and Fluids Properties of Nanofluids"

This reprint comprises twelve high-quality articles covering a wide range of important research topics on nanofluids.

The first article by Ali and co-workers reviews the state-of-the-art research on popular carbon-based nanofluids and their applications in thermal and energy systems.

A forced convection study on thermal and hydraulic performances of nanofluids containing carbon, and Al_2O_3 and SiO_2 nanoparticles in PEG in a fully developed turbulent flow in a square, heated pipe is reported by Afan and co-workers in the second article.

In the third article, Giwa and co-authors performed a characterization of the morphology and stability of dispersed nanoparticles and determined the viscosity and electrical conductivity of DW-based MWCNT-Fe_2O_3 hybrid nanofluids at different temperatures and concentrations.

A numerical investigation to assess the effects of polydispersity and sintering on the effective thermal conductivity of nanoparticles' aggregates conducted by Karagiannakis and co-workers is reported in the fourth article.

The fifth article presents the thermal performances of DW-based Al_2O_3 and TiO_2 nanofluids in a compact plate heat exchanger by comparing experimental and numerical investigations performed by Ajeeb and Murshed.

In the sixth article, Zhang and co-workers determined isotherms, streamlines, and heat transfer rates under various conditions and parameters by employing the Darcy–Brinkman and energy transport equations in a numerical study of natural convection heat transfer in a porous annulus filled with a Cu nanofluid.

Ahmed and co-authors presented a numerical investigation of the impacts of melting on the convective flow of Al_2O_3-based PCM within cylindrical tubes containing cross-shape heated sections in the seventh article.

The eight article reports a numerical study on the natural convection within an inversed T-shaped enclosure filled by NePCM conducted by Abderrahmane and co-workers.

Pereira and co-authors reported a review of the possible mechanisms and characteristics of nanoparticle deposition and its impact on various factors such as surface roughness and wettability, the density of vaporized core points, and thermal resistance in the ninth article.

The tenth article presents an experimental study on the effects of nanoparticle size and concentration on pool boiling heat transfer with TiO2 nanofluids on laser-textured Cu surfaces conducted by Hadžić and co-workers.

A numerical study on the appearance of a solitary wave particle concentration in nanofluids under a light field conducted by Livashvili and co-authors is reported in the eleventh article.

The final article reports on a molecular dynamics simulation of the behaviors of water nanodroplets impinging on moving surfaces performed by Zhang and co-authors.

The Special Issue Editor acknowledges all the authors for their valuable contributions to the book and the Reviewers for assessing the quality of each article.

S M Sohel Murshed
Editor

nanomaterials

MDPI

Review

Carbon-Based Nanofluids and Their Advances towards Heat Transfer Applications—A Review

Naser Ali [1], Ammar M. Bahman [2], Nawaf F. Aljuwayhel [2,*], Shikha A. Ebrahim [2], Sayantan Mukherjee [3] and Ali Alsayegh [4]

1 Nanotechnology and Advanced Materials Program, Energy and Building Research Center, Kuwait Institute for Scientific Research, Safat 13109, Kuwait; nmali@kisr.edu.kw
2 Mechanical Engineering Department, College of Engineering and Petroleum, Kuwait University, P.O. Box 5969, Safat 13060, Kuwait; a.bahman@ku.edu.kw (A.M.B.); shikha.ebrahim@ku.edu.kw (S.A.E.)
3 Thermal Research Laboratory (TRL), School of Mechanical Engineering, Kalinga Institute of Industrial Technology, Bhubaneswar, Odisha 751024, India; 1881148@kiit.ac.in
4 School of Aerospace, Transport and Manufacturing (SATM), Cranfield University, Cranfield MK43 0AL, UK; a.alsayegh@cranfield.ac.uk
* Correspondence: dr.nawaf@ku.edu.kw

Citation: Ali, N.; Bahman, A.M.; Aljuwayhel, N.F.; Ebrahim, S.A.; Mukherjee, S.; Alsayegh, A. Carbon-Based Nanofluids and Their Advances towards Heat Transfer Applications—A Review. *Nanomaterials* **2021**, *11*, 1628. https://doi.org/10.3390/nano11061628

Academic Editor: S M Sohel Murshed

Received: 26 May 2021
Accepted: 17 June 2021
Published: 21 June 2021

Abstract: Nanofluids have opened the doors towards the enhancement of many of today's existing thermal applications performance. This is because these advanced working fluids exhibit exceptional thermophysical properties, and thus making them excellent candidates for replacing conventional working fluids. On the other hand, nanomaterials of carbon-base were proven throughout the literature to have the highest thermal conductivity among all other types of nanoscaled materials. Therefore, when these materials are homogeneously dispersed in a base fluid, the resulting suspension will theoretically attain orders of magnitude higher effective thermal conductivity than its counterpart. Despite this fact, there are still some challenges that are associated with these types of fluids. The main obstacle is the dispersion stability of the nanomaterials, which can lead the attractive properties of the nanofluid to degrade with time, up to the point where they lose their effectiveness. For such reason, this work has been devoted towards providing a systematic review on nanofluids of carbon-base, precisely; carbon nanotubes, graphene, and nanodiamonds, and their employment in thermal systems commonly used in the energy sectors. Firstly, this work reviews the synthesis approaches of the carbon-based feedstock. Then, it explains the different nanofluids fabrication methods. The dispersion stability is also discussed in terms of measuring techniques, enhancement methods, and its effect on the suspension thermophysical properties. The study summarizes the development in the correlations used to predict the thermophysical properties of the dispersion. Furthermore, it assesses the influence of these advanced working fluids on parabolic trough solar collectors, nuclear reactor systems, and air conditioning and refrigeration systems. Lastly, the current gap in scientific knowledge is provided to set up future research directions.

Keywords: carbon nanotubes; graphene; nanodiamond; parabolic trough solar collector; nuclear reactor; air conditioning and refrigeration

1. Introduction

Since the 20th century, scientists have been working with considerable effort to develop fluids that can surpass those conventionally known by the scientific society and industry in terms of thermal and physical performance. The idea of dispersing solid particles of millimeter (mm) and micrometer (μm) in size is the milestone, which was physically initiated by Ahuja [1,2] in 1975, Liu et al. [3] in 1988, and other researchers at Argonne National Laboratory (ANL) [4–6] in 1992 on the bases of Maxwell theoretical work [7]. Such suspensions have shown tremendous improvements in heat transfer characteristics compared to their base fluids. This is due to the dispersed solid particles' significantly higher thermal conductivity compared to their hosting fluid, which would enhance the effective thermal conductivity of the colloidal. The term 'effective' is generally used when referring to the net property of a solid–liquid suspension [8]. However, it was found that in flow areas of low velocities, the particles hosted by the suspension tended to deposit from its carrier liquid. Additionally, hence the fluid starts to lose its tuned properties. Furthermore, clogging of small passages was also experienced due to the significant level of agglomeration between the dispersed particles, and therefore making it extremely challenging to employ in heat transfer devices containing small channels. This is when, in 1993, Masuda et al. [9] conceived the idea of fabricating suspensions with ultrafine particles of silica, alumina, and titanium dioxide, where these dispersions were afterward given the name 'Nanofluids' by Choi and Eastman [10], in 1995, as a result of their extensive research work at ANL. According to the founders, a nanofluid can be generally defined as an advanced category of fluid that is produced by homogeneously dispersing low concentrations (preferably $\leq$1 vol. %) of particles of less than 100 nanometers (nm) in size within a non-dissolving base fluid [11]. Both Masuda et al.'s [9] and Choi and Eastman's [10] primary motivation at that time was to overcome the limitations associated with suspensions made by their counterparts (i.e., colloidal containing millimeter or micrometer sized particles). In addition, Choi and Eastman [10] have theoretically known beforehand that reducing the size of the dispersed particles to the nanoscale would greatly enlarge the particle exposed surface area to the surrounding, and thus increasing the suspension overall thermal conductivity [12]. The significant variation in thermal conductivity between solid particles and liquids can be clearly seen in Figure 1 for some of the most commonly used particles and base fluids, at room temperature and atmospheric pressure, for fabricating nanofluids [13–17]. It is worth noticing that CuO, MgO, Al_2O_3, ZnO, TiO_2, Fe_2O_3, SiO_2, Ag, Cu, Au, Al, Fe, carbon nanotubes (CNTs), and multiwalled carbon nanotubes (MWCNTs) stands for cupric oxide, magnesium oxide, aluminum oxide, zinc oxide, titanium dioxide, iron(III) oxide, silicon dioxide, silver, copper, gold, aluminum, iron, carbon nanotubes, and multiwalled carbon nanotubes, respectively. Furthermore, the thermal conductivity of some of the materials shown in Figure 1 was seen to have a significant scatter of data across the literature, which can be linked to several factors such as the purity, crystallinity, particle size, and the determination approach used to find this thermal property. In addition, the thermal conductivity of graphene after being subjected to oxidization (i.e., having the form of graphene oxide) gets highly reduced, where it can reach values between 1000 and 2 W/m·K [18–20].

Following their success, many researchers started to explore and develop this class of engineered fluid via modifying their production route, enhancing the suspension stability, and improving the colloidal thermal conductivity [13,21,22]. As of today, nanofluids are seen to have potential usage in a wide range of areas, including the energy sector, construction and building, transportation, oil and gas, medical sector, etc. [23–34]. Figure 2a shows the increasing trend in scientific publications in the field of nanofluids from 1995 to 2020, while Figure 2b illustrates the different types of these published documents that are available in the same database. It is worth mentioning that the data in Figure 2 was obtained from Elsevier's abstract and citation database, Scopus, via searching through the word 'Nanofluid' [35].

Despite the promising achievements that nanofluids could deliver to the scientific community, there are still some obstacles that need to be overcome before this category of fluids can be industrially accepted. For example, the colloidal preparation phase is still considered one of the most significant challenges, as this stage can strongly influence the fluid physical stability and effective thermophysical properties [13,36]. Meaning that if the fabrication process used was not well structured before being executed, the chances of an unstable nanofluid being produced is likely to occur. As a result, some of the suspension's thermophysical properties will gradually degrade with time due to the separation of particles from the hosting base fluid. Almurtaji et al. [37] have illustrated in their published work the relationship between the effective thermal conductivity and the physical stability of suspensions. They showed that the effective thermal conductivity of a nanofluid could reach its optimum possible value when the dispersion is physically stable, and vice versa. In addition, the commonly employed two-step fabrication method that relies on an ultrasonic bath type device, was reported to raise the as-prepared nanofluid temperature and that the surrounding atmospheric conditions govern this increase in temperature along with the sonicator working power. Thus, it is highly unlikely that similar nanofluids can be produced through the conventional two-step route without simultaneously fabricating the products at the same preparation conditions. A more convenient two-step method employed for nanofluid production would be the two-step controlled sonicator bath temperature approach, as was reported by Ali et al. [8,11] and Song et al. [38]. The aforementioned approach would eliminate the rise in bath temperature obstacle, and hence will ensure an optimum level of nanofluids reproducibility to the manufacturer at any surrounding atmospheric conditions, and even when using different types of bath sonicators. Furthermore, as the thermal properties of a nanofluid are influenced mainly by the dispersed particles compared to its base fluid, researchers have been focusing more on carbon-based materials. This is because some of these materials, in the nanoscale, have exceptional thermophysical properties compared to other commonly used materials (e.g., metals and oxides) [39–41]. For instance, CNTs and graphene have significantly elevated thermal conductivity [42,43], large aspect ratio [44], lower density [45,46], lower erosion and corrosion surface effects [47], higher stability [43], and lower pressure drop and pumping power requirement in comparison to other types of nanomaterials [48,49]. Figure 3 demonstrates common allotropes of carbon nanomaterials.

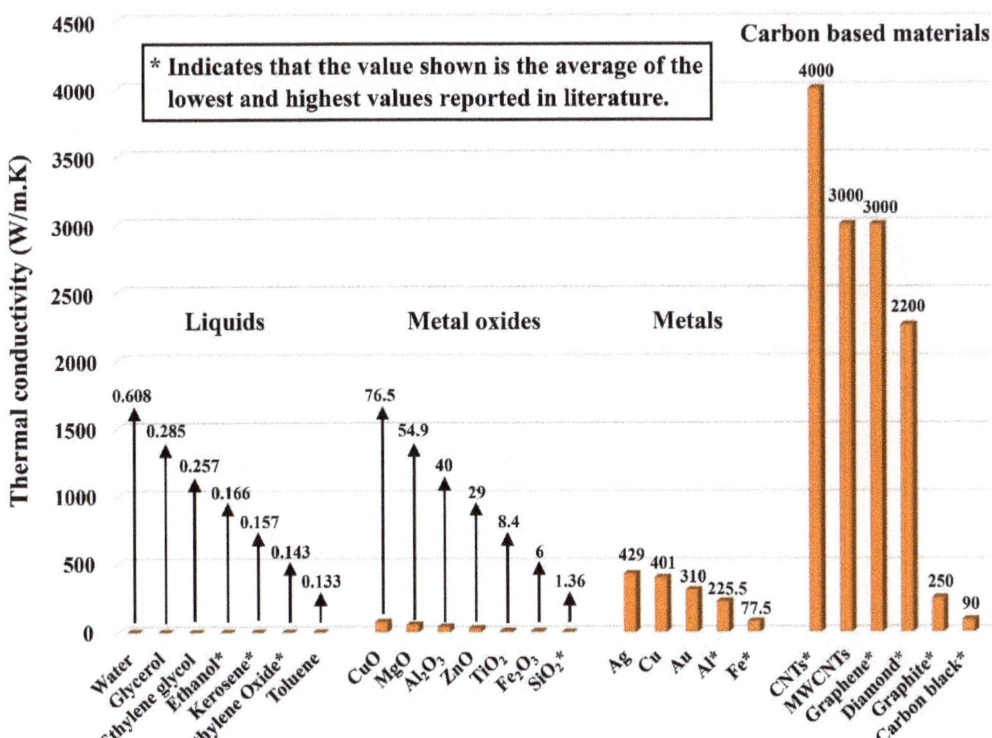

Figure 1. Thermal conductivity of commonly used particles and base fluids for fabricating nanofluids showing an order of magnitude higher in the thermal property for some of the carbon-based materials.

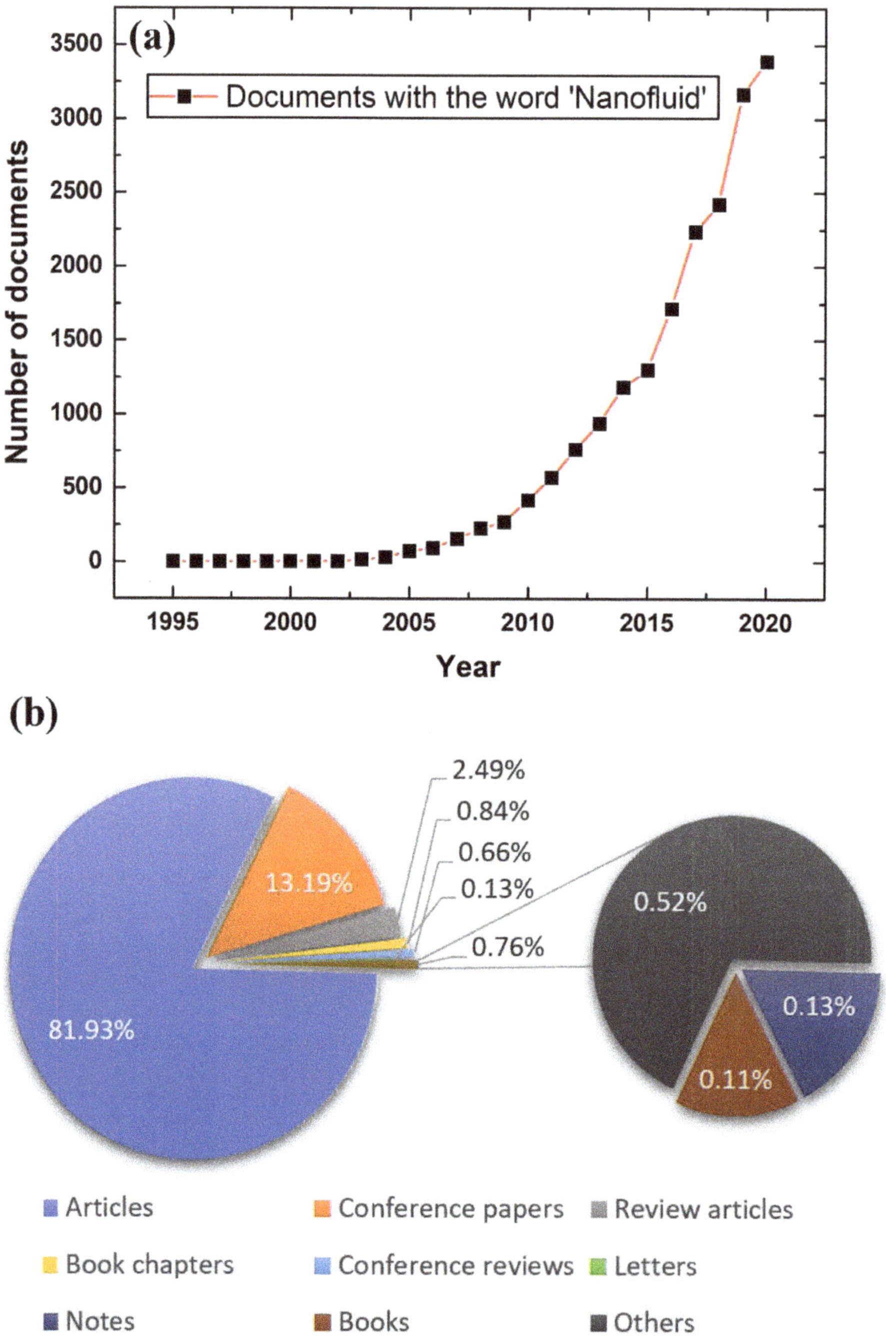

Figure 2. Search result obtained from Scopus database on nanofluids, where (**a**) illustrates the number of published works per year and (**b**) shows the percentage of each type of these documents [35].

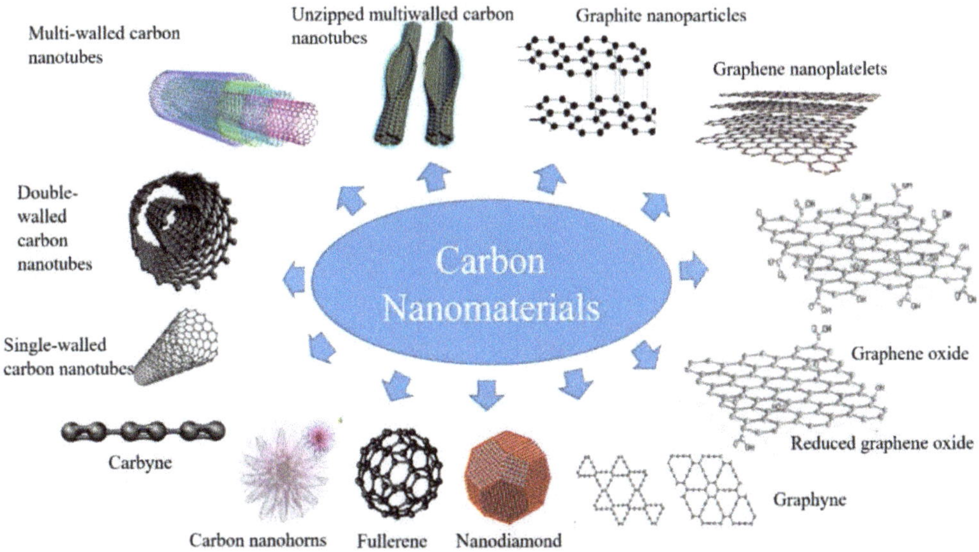

Figure 3. Common allotropes of carbon nanomaterials that grant distinctive thermophysical properties [50].

Many published numerical and experimental studies on nanofluids fabricated with particles of carbon-based materials were found in the literature, which show the continued growth of interest in such materials [35,51–54]. Figure 4 classifies these documents in terms of the number of available publications at the Scopus database for each type of carbon-based material used in nanofluids production. The single-walled carbon nanotube (SWCNT) and double-walled carbon nanotube (DWCNT) abbreviations in Figure 4 refer to the single-walled carbon nanotube and double-walled carbon nanotube, respectively. During the reviewing process, which led to the formation of Figure 4, the authors remarkably recognized that the researchers had used different sonication duration and intensities to fabricate their nanofluids. However, some of the suspensions had the same particles type, size, and hosting base fluid. This shows that, up to today, there is no standard fabrication method for the production of the colloidal. The authors have also found that dispersing carbon-based materials, such as walled carbon nanotubes (MCNTs) and graphene, can tremendously enhance the quality of biofuels blends, in specific biodiesel [55,56]. This includes lowering the brake specific fuel consumption, stabilizing the fuel consumption rate and brake thermal efficiency, and improving the diesel engine performance and the resulting emissions from the combustion process.

This review paper provides an overview of three types of carbon-based nanofluids: CNT, nanodiamond (ND), and graphene. The selection reason for these three carbon-based particles is due to their outstanding thermal properties compared to any other sort of nanoscaled solids. Hence, they can be considered promising candidates for fabricating nanofluids targeted towards heat transfer applications. The main contribution of the present review study is that this work starts from the synthesis stage of these three carbon-based materials, followed by their dispersed form, and up to their employment in selected energy applications. Furthermore, recommendations on the different nanofluids production methods used are shown along with the colloidal stability and its effect on the thermophysical properties. Moreover, the experimental measuring devices and theoretical equations used to determine and predict the thermophysical properties are provided. In addition, the research work done on utilizing these carbon-based suspensions are presented for three thermal applications, namely, parabolic trough solar collectors (PTSCs), nuclear reactors, and air conditioning and refrigeration (AC&R) systems, with a comparison to those of conventional working fluids. Finally, the gaps in present scientific knowledge that scientists need to tackle are highlighted in order to promote these advanced types of heat transfer fluids commercially.

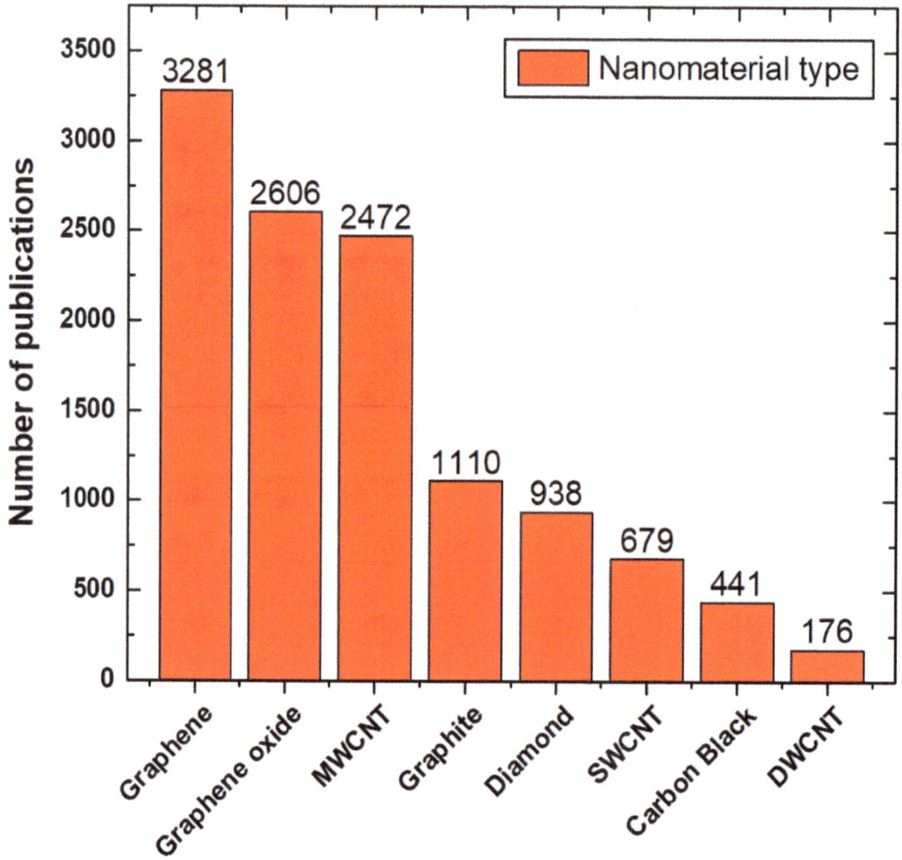

Figure 4. The number of publications available at the Scopus database for common carbon-based material used in nanofluids fabrication [35].

2. Synthesis of Nanoscaled Carbon-Based Materials

Carbon ranks as the 4th most common element after hydrogen, helium, and oxygen in our solar system, and the 17th in the crust of our planet [57]. Remarkably, this element is distinctive so that when the crystal structure of carbon atoms is changed into deferent arrangements, the material properties significantly differ [58–63]. For example, both ND and graphene are made of carbon but of different atomic bounds arrangement. While the first is an electrical isolator and transparent towards visible light waves, the second has excellent electrical conductivity with complete visible light blockage. Such materials that contain various arrangements of carbon atoms are known as 'allotropes of carbon', which means that the material has chemically identical elements but with different atomic arrangements, and hence different physical properties. Due to this fact, many allotropes of carbon exist or have been discovered by scientists, e.g., diamond, graphene, and CNTs. The following Sections 2.1–2.3. will provide a short overview of the fabrication of three allotropes of carbon in the nanoscale, namely ND, graphene, and CNTs. Knowing the production methods of these materials is essential and will, later on, help us understand which nanofluid fabrication route is suitable to conduct.

2.1. Nanodiamonds

NDs have existed for billions of years in nature within meteorites, crude oil, interstellar dust protoplanetary nebulae, and different sediment layers of the Earth's crust. Nevertheless, the synthetization process of this valuable material only started in the second half of the nineteenth century through either exposing graphite to high pressure and high temperature conditions, or by the explosive detonation of bulk graphite [64–66]. The first is known as the high-pressure and high-temperature (HPHT) approach, whereas the second route is known as the detonation technique. In the literature, it was reported that the first study conducted on the preparation of NDs was performed by Bovenkerk et al. [67], in 1959, after which Danilenko [68] used the detonation technique as part of his synthesis approach. Furthermore, many approaches were developed afterward for fabricating ND, such as the microplasma-assisted formation [69], chemical vapor deposition (CVD) method [70], laser ablation [71], high energy ball milling of microdiamonds produced from high pressure and high temperature conditions [72], high energy ball milling of ultra-fine graphite powder [73,74], ultrasound cavitation [75], chlorination of carbides [76], carbon onions irradiated by electron [77], and irradiation of graphite by ion beam [78]. In addition to the previous synthesizing methods, El-Eskandarany has proposed a novel approach for producing superfine NDs from commercial graphite powders and SWCNTs under ambient temperature and atmospheric pressure conditions, using a high-energy ball mill technique [79]. It is important to note that, according to Ali et al. [66] and Mochalin et al. [80], the most common types of NDs seen today are the detonation NDs (DNDs) and the HPHT-NDs. From the aforementioned production routes, it can be concluded that the synthesized NDs can only be produced as independent solid particles, and therefore cannot be grown within liquids through chemical and/or physical approaches. Regardless of the method used, the production of NDs usually involves three major phases, which are 1—synthesis (methods mentioned earlier), 2—processing, and 3—modification. The processing stage, which follows the synthesis phase, enhances the as-produced NDs purity by removing the metals and metals oxides along with the non-diamond carbons that remain attached to the ND surface. Hence, a high level of sp^3 carbon bonded diamond nanoparticles can be obtained. This can be done by using oxidants such as nitric acid (HNO_3), perchloric acid ($HCLO_4$), or hydrochloric acid (HCL) [81]. Furthermore, the modification phase is essential so that the fabricated NDs can meet the requirements of their targeted application. Modification can be performed using either surface functionalization (widely used) or doping of the NDs particles. It is important to note that some researchers have recently started focusing on the doping technique due to the distinct optical properties gained from this NDs modification approach [82,83]. Figure 5 shows the three phases involved in the production of NDs [84].

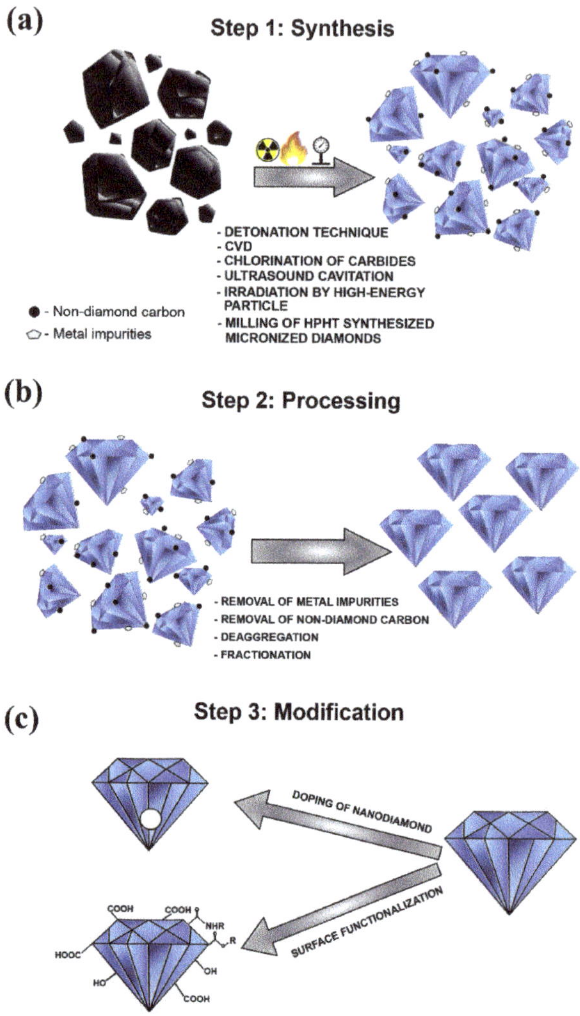

Figure 5. Phases involved in the production process of nanodiamonds, where (**a**) illustrates the synthesis phase, (**b**) demonstrates the processing phase, and (**c**) shows the modification phase. Reproduced with permission from [84]. Elsevier, 2019.

2.2. Graphene

Graphene is a type of carbon material that originates from bulk graphite. It has the shape of a 2-dimensional (2D) (i.e., monolayer) sheet of one-atom thickness and lattice of hexagonally arranged sp^2 bonded carbon atoms [85]. The material itself was successfully synthesized for the first time in 2004 by Novoselov et al. [86], through mechanical exfoliating graphite with Scotch tape. Furthermore, the development in the field has resulted in categorizing graphene by the materials architecture structure, which ranges from zero-dimensional (0D) graphene quantum dots, one-dimensional (1D) graphene fibers and nanoribbons, and 2D graphene nanomesh, rippled/wrinkled and multisheet [87]. Figure 6 shows an illustration of the different categories of graphene based on their dimensionality and bandgap opening. Regarding 2D graphene sheets, few suitable techniques are commonly employed for producing such material, which are mechanical exfoliation [86], sublimation of silicon carbide (SiC) [88], laser-induced graphene [89,90], covalent [91,92] or non-covalent [93] exfoliation of graphite in liquids, and CVD growth [94]. These fabrication methods produce graphene in a solid form except for the liquid-phase exfoliation, which delivers the material as part of a suspension.

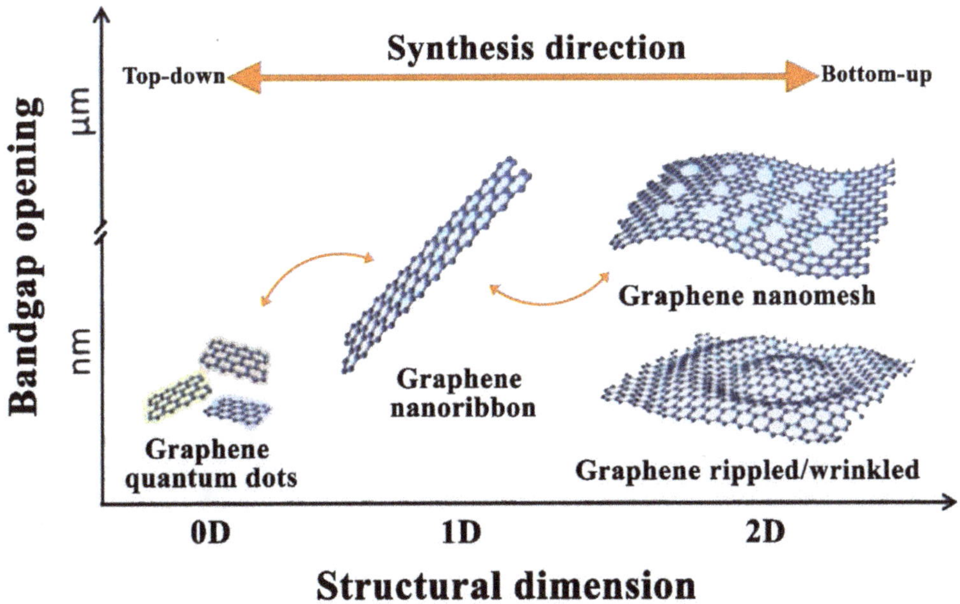

Figure 6. Different forms of graphene based on their dimensionality and bandgap opening, where graphene quantum dots, graphene nanoribbon, and both graphene nanomesh and graphene rippled/wrinkled have a structural dimension of 0D, 1D, and 2D, respectively.

The mechanical exfoliation method was the first approach for obtaining graphene. In this method, the small mesas of highly oriented pyrolytic graphite are repeatedly peeled out with a Scotch tape, and hence the attached thin films on the tape are of monolayer graphene. This production route is highly reliable and allowed the preparation of high-quality graphene sheets of up to 100 μm in thickness [86]. Other less common types of mechanical exfoliation are also available, such as ball milling of graphite nanoparticles [95] and hammering graphite [96]. Furthermore, the high temperature sublimation of SiC, which was developed initially for the electronics industry, relies on the thermal decomposition of a SiC substrate via either an electron beam or resistive heating to epitaxial graphene under ultrahigh vacuum condition. This results in the desorption of the silicon (Si) on the wafer surface, and therefore causing the surface atoms to arrange into forming hexagonal lattice. Moreover, fabricating graphene through laser-inducement is performed under ambient atmosphere by subjecting carbon dioxide (CO_2) pulsed laser to a substrate containing carbon-based materials. This approach combines 3-dimensional (3D) graphene fabrication and patterning into a single step without having to use wet chemical steps. In addition, exfoliation of graphite in liquids or liquid-phase exfoliation depends on the employment of external peeling force, such as an ultrasonic horn sonicator, to separate the graphene sheets from the immersed bulk graphite in a solvent of suitable surface tension. The solvent used in the process is usually a non-aqueous solution, such as N-methyl-2-pyrrolidone (NMP), but aqueous solutions can also be employed if surfactant was added. It is important to note that the yield of the liquid-phase exfoliation process is relatively low, and thus centrifugation is used to gain a significant fraction of monolayer and few-layer graphene flakes in the final dispersion [97]. On the other hand, the CVD production route uses hydrocarbon gases to grow graphene on a targeted substrate by carbon diffusion and segregation of high carbon solubility metallic substrates, such as nickel (Ni), or by surface adsorption of low carbon solubility metals (e.g., Cu) [98,99]. From all of the previous methods, CVD has shown to be the most successful, promising, and feasible approach in the field for producing monolayer graphene of high quality and large area [94]. For deeper insight into the various graphene synthesis methods, the reader is referred to the published work of Rao et al. [100].

2.3. Carbon Nanotubes

Although carbon is known as a ubiquitous material in nature, CNTs are not, where this allotrope material is a human-made seamless cylindrical form of carbon. It is believed that the oldest CNTs existed on damascene swords [101]. Still, their first proof of presence was in 1952 through the transmission electron microscopy (TEM) images published by Radushkevich and Lukyanovich [102], after which Boehm [103] and Oberlin et al. [104] obtained similar images along with describing the currently widely accepted CNTs growth model. Conceptually, CNTs are graphene sheets rolled into cylindrical tubes, of less than 1 nm in diameter, with a half fullerenes caped end. Based on the number of consistent tubes (i.e., rolled-up graphene sheets), CNTs can be classified as SWCNTs, DWCNTs, and MWCNTs. As the terms suggest, the SWCNTs consist of only one tube, whereas DWCNTs and MWCNTs comprise two and three (or more) tubes, respectively [105,106]. Figure 7 shows the mechanism in which CNTs are formed and their three different types. It is important to note that some researchers distinguished between the three tubes form of CNTs and those of a higher number of tubes, where they have categorized the first as the triple-walled carbon nanotubes (TWCNTs) and the second as MWCNTs [107,108].

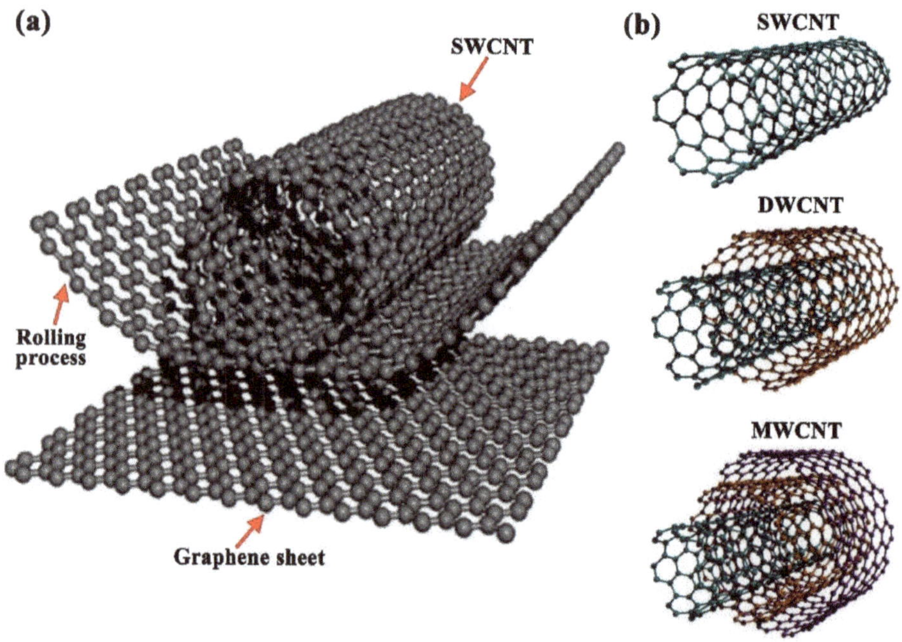

Figure 7. Carbon nanotubes formation and classifications, where (**a**) illustrates the rolling mechanism of graphene sheet into SWCNT and (**b**) demonstrates the three different categories of CNTs, namely SWCNT, DWCNT, and MWCNT.

There are three main synthesis methods for producing CNTs, which are the arc discharge, laser ablation, and CVD [109–111]. Other approaches, such as diffusion and premised flame method, can be used for CNTs fabrication but are less frequently utilized [112]. All three primary production methods depend on the carbon feedstock, either as a solid phased carbon source (arc discharge and laser ablation) or carbonaceous gases (CVD method). An example of the gases employed in the CVD process include carbon monoxide (CO), ethanol, and acetylene. Moreover, the final product is always delivered in a dried form; thus, CNTs cannot be grown within liquids as dispersions. In the arc discharge process, doped graphite rods or two catalysts loaded are vaporized at 4000–5000 K, within a closed chamber, by an electric arc placed between them, after which the resulting deposit is of CNTs. Like the arc discharge method, the laser ablation route relies on the evaporation of a carbon feedstock, usually a graphite rod with a metallic based catalyst, to obtain the CNTs. The difference between this approach and the previous one is that the laser ablation uses high energy laser irradiation to heat the carbon source, and thus causing the phase transformation (i.e., from the solid to gaseous phase). Additionally, the final product gets accumulated in a cold trap located within the chamber. Therefore, this technique is much more efficient than the arc discharge process in terms of the losses in the as-produced CNTs. On the other hand, the CVD, which was mentioned earlier in Section 2.2, decomposes carbonaceous gases on catalytic nanoparticles to produce the CNTs. The catalytic nanoparticles used for this purpose are either grown while conducting the process or are initially fabricated through a separate procedure. Furthermore, the advantage associated with this production technique is the high level of control over the synthesis process parameters such as carbon supply rate, growth temperature, catalyst particles size, and type of substrate used for the CNTs growth.

3. Preparation of Nanofluids

Nanofluids can be formed by dispersion particles made of single elements (e.g., Cu and Fe), single element oxides (e.g., CuO and Al_2O_3), alloys (e.g., stainless steel), metal carbides (e.g., silicon carbide and zirconium carbide), metal nitrides (e.g., silicon nitride and titanium nitride), or carbon-based materials in a none dissolving base fluid such as water, methanol, glycol, ethylene glycol (EG), transformer oil, kerosene, and/or different types of refrigerants with or without the use of surfactant/s [13,113]. The nanosuspension is given the name 'nanofluid' when one type of nanoparticles is used in the fabrication process; in contrast to the previous category, dispersions formed by employing two or more types of nanoparticles are classified as 'hybrid nanofluids' [114,115]. To the best of the authors knowledge, unlike the previous two nanofluids categories that are subjected to the number of different particles used in the process, there does not exist a specific classification for nanofluids made of more than one type of base fluid. However, researchers could have used the terms 'Bi-liquid nanofluid' or 'Tri-liquid nanofluid' to refer to their nanofluid that is made from two or three base fluids, respectively. Figure 8 shows an illustration of the conventional nanofluid and the hybrid nanofluid. In addition, the homogeneity and physical stability of the dispersion depend significantly on the implemented preparation approach, which can substantially influence the effective thermophysical properties of the as-prepared suspension. Knowing the aforementioned is essential when selecting the appropriate type of nanofluid for any targeted application [116]. In general, two known fabrication processes are currently used for producing nanofluids, namely, the one-step (also referred to as the single-step) method and the two-step approach [37]. It is important to note that some researchers prefer to classify the one-step production processes into two categories, which are the one-step physical technique and the one-step chemical approach, resulting in three types of methods of nanofluid fabrication for these groups [117,118]. A summary of the two fabrication schemes (i.e., the one-step and two-step methods) is presented in the following subsections.

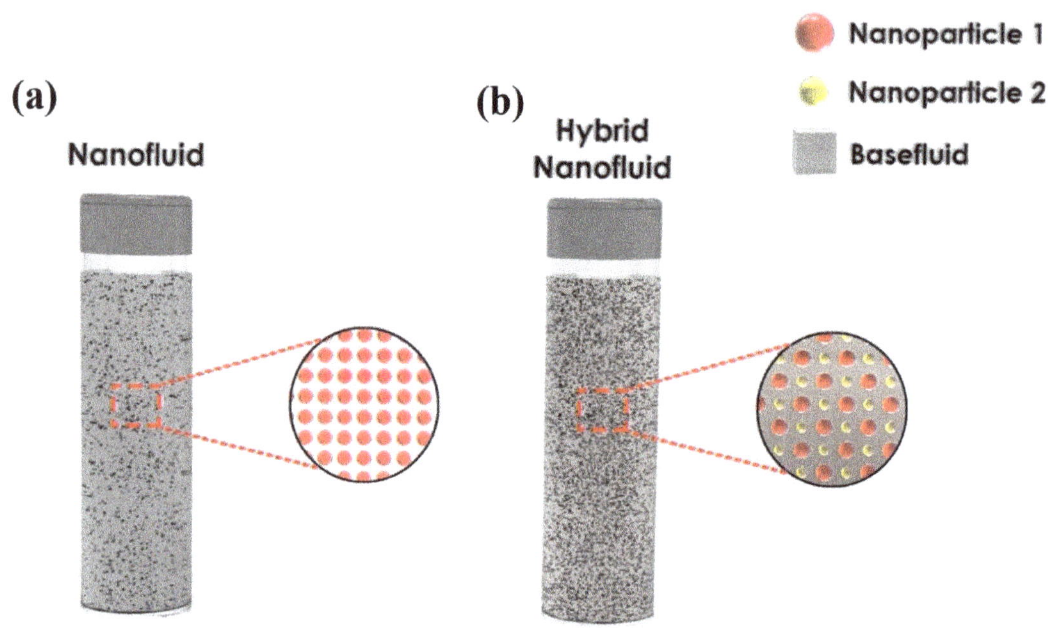

Figure 8. Schematic demonstration to compare between conventional (**a**) and hybrid (**b**) nanofluids that uses the same base fluid.

3.1. One-Step Method

The production of nanofluids by the one-step method is conducted by simultaneous synthesizing and dispersing the nanoparticles in the base fluid. Thus, the storage, drying, and transportation of nanoparticles are unnecessary [119]. Furthermore, the dispersed particles in this bottom-up process avoid oxidization from their surrounding environment. In addition, this technique is well known to highly eliminate clustering and agglomeration of dispersed particles within the hosting fluid, and hence coagulation of nanoparticles in real-life applications that uses microchannels can be minimized with an increase in the level of the physical stability of the colloidal compared to the two-step production approach. Moreover, this method allows greater control over the size and shape of the dispersed nanoparticles during the fabrication process. Nevertheless, the presence of residual reactants as a result of uncompleted reactions has always been a major drawback of such a production route. Other disadvantages can also be experienced when following the single-step synthesis approach, such as the inconsistency of the scale for industrial applications, which can only be used with base liquids of low pressure, high production cost, and limitation in the types of nanofluids that can be fabricated compared to the two-step route [120–123].

One of the most common one-step approaches is the one that was established by Eastman et al. [21]. In this method, nanofluids are synthesized by evaporating a bulk material, after which the evaporated particles get deposited then condensed in a thin film of base fluid attached to a vessel wall due to centrifugation. Figure 9 demonstrates the aforementioned one-step approach. Many researchers have continuously worked on developing the one-step fabrication approach through physical and/or chemical means. Today, different methods have been acknowledged to be in the one-step nanofluid production category [36,120,124]. Figure 10 shows some of the commonly known one-step nanofluid fabrication routes in the field where their method of conduct can be found fully explained in the published work of Ali et al. [13] and Mukherjee et al. [36].

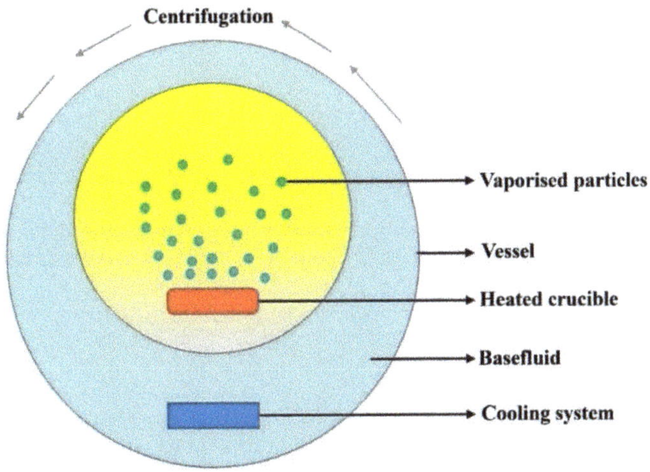

Figure 9. Eastman et al. [21] one-step method of evaporation and centrifugation for nanofluids fabrication [37].

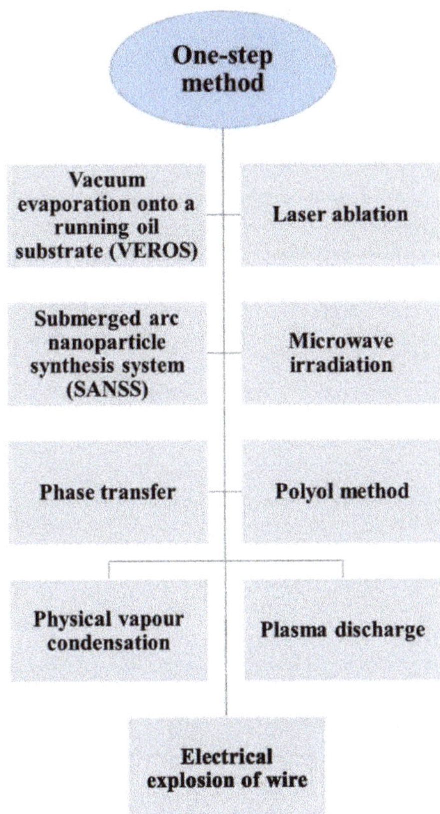

Figure 10. Different nanofluid production approaches that fall under the one-step method.

3.2. Two-Step Method

Unlike the one-step method, the two-step approach is a top-down process that uses dried nanoparticles that were initially prepared, through physical or chemical processes, after which these particles get dispersed in a base fluid through ultrasonic agitation [125–128], magnetic stirring [129–132], homogenizing [131,133,134], or ball milling (least commonly used) [16,135,136] with or without adding surfactant(s) to the mixture. Other less common dispersion routes can also be used, such as dissolver, kneader, three roller mill, stirred media mill, and disc mill [137]. Figure 11 demonstrates an example of the two-step method, where a bath type ultrasonic device is used to form the suspension. In addition to the bath type sonicators, some researchers have employed the probe/horn type sonicators to fabricate their nanofluids. They have reported higher particles dispersion capability and enhanced suspensions thermal properties using this type of device compared to the bath type dispersers [138]. The reason behind the previously achieved improvements in the suspension is that the probe device provides focused and intense ultrasonication effects, reaching up to 20 kW/L, to the mixture in an evenly distributed manner [139]. This is something that the bath type sonicators cannot provide due to its low relative intensity (i.e., 20–40 W/L) and non-uniform distribution of the ultrasonication effect on the fabricated nanofluid. It is important to note that the bath type ultrasonicator is more applicable for commercial scale production of nanofluids. In contrast, the probe type is better suited for synthesis at the lab scale. Regardless of the type of two-step mixing approach used, this method is still considered as a cost-effective process that is appropriate for both small- and large-scale production of any type of nanofluids, which is seen as a favorable approach to many researchers in the field [140]. However, some of the critical issues associated with this method during nanofluids fabrication are the agglomeration of the nanoparticles due to the very high surface energy between the particles, and the notable increase in the process temperature with fabrication time when using some of the mixing devices (e.g., bath type ultrasonic device) [8,13]. The first obstacle causes the suspension to be in a weak physical stability state that results from the nanoparticles undergoing agglomeration, which is followed by separation of the particles from the base fluid in the form of sediments. Thus, the nanofluid thermophysical properties degrade with time. As for the raise in fabrication process temperature problem, the reproducibility of similar nanofluids (i.e., obtaining suspensions with the same thermophysical properties) would be impossible to achieve. This is because different bath type ultrasonic devices and/or surrounding atmospheric conditions lead to varying the thermophysical properties and physical stability of the fabricated colloidal [8,38]. There are several ways to overcome the aforementioned limitations in the two-step method. For example, surfactants can be added to the mixture to reduce the level of particles agglomeration, and the sonicator bath temperature could be controlled throughout the fabrication process by equipping the device with a temperature regulator. Other approaches used to physically stabilize the as-prepared dispersions are mentioned afterward in the nanofluid stability enhancement section (Section 4.2). When preparing nanofluids, the nanoparticles and surfactants (if required) are added to the base fluid with respect to either volume (vol.) or weight (wt.) percentage (%). Most researchers tend to use the vol. % to calculate the added nanopowder to the base fluid, which can be estimated through the appropriate formulae presented in Table 1.

Table 1. Fraction calculation Formulae for different forms of nanofluids.

Type of Particles	Type of Base-Fluid	Fraction (%)	Formulae	Ref.	Eq.
Single type	Single type	vol.	$\frac{V_{np}}{V_{np}+V_{bf}} \times 100$; or $\frac{\left(\frac{m}{\rho}\right)_{np}}{\left(\frac{m}{\rho}\right)_{np}+\left(\frac{m}{\rho}\right)_{bf}} \times 100$	[13,37]	(1)

Table 1. *Cont.*

Type of Particles	Type of Base-Fluid	Fraction (%)	Formulae	Ref.	Eq.
Single type	Two type	vol.	$\frac{\left(\frac{m}{\rho}\right)_{np}}{\left(\frac{m}{\rho}\right)_{np}+\left[\left(\frac{m}{\rho}\right)_{bf1}+\left(\frac{m}{\rho}\right)_{bf2}\right]}\times 100$; where $bf1$ and $bf2$ have equal volume ratio	[141]	(2)
Two type	Single type	vol.	$\frac{\left(\frac{m}{\rho}\right)_{np1}+\left(\frac{m}{\rho}\right)_{np2}}{\left[\left(\frac{m}{\rho}\right)_{np1}+\left(\frac{m}{\rho}\right)_{np2}\right]+\left(\frac{m}{\rho}\right)_{bf}}\times 100$; where $np1$ and $np2$ have equal volume ratio	[142, 143]	(3)
Two type	Two type	vol.	$\frac{\left(\frac{m}{\rho}\right)_{np1}+\left(\frac{m}{\rho}\right)_{np2}}{\left[\left(\frac{m}{\rho}\right)_{np1}+\left(\frac{m}{\rho}\right)_{np2}\right]+\left[\left(\frac{m}{\rho}\right)_{bf1}+\left(\frac{m}{\rho}\right)_{bf2}\right]}\times 100$; where $np1$ and $np2$ have equal volume ratio as well as $bf1$ and $bf2$	[144]	(4)

Where V, m, ρ, np, $np1$, $np2$, bf, $bf1$, and $bf2$ represent the volume, mass, density, single type of nanoparticles, first type of nanoparticles, second type of nanoparticles, single type of base fluid, first type of base fluid, and second type of base fluid, respectively. In addition to the equations shown in Table 1, one can use the following three equations to determine the vol. % for their nanofluids when having two different particles and/or two base fluids concentration ratio(s).

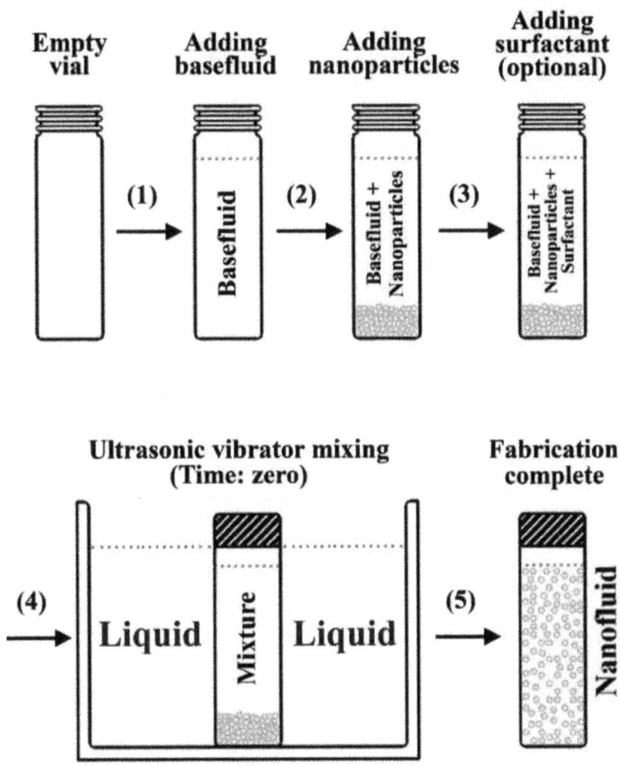

Figure 11. Example of nanofluids two-step preparation using a bath type ultrasonic device.

For single type of nanoparticles and two different types of base fluids:

$$\frac{\left(\frac{m}{\rho}\right)_{np}}{\left(\frac{m}{\rho}\right)_{np} + \left[\left(\frac{m}{\rho}\right)_{bf1} \times \frac{A}{A+B} + \left(\frac{m}{\rho}\right)_{bf2} \times \frac{B}{A+B}\right]} \times 100 \tag{5}$$

where the ratio of $bf1 : bf2$ is equal to $A : B$.

For two different types of nanoparticles and single type of base fluid:

$$\frac{\left(\frac{m}{\rho}\right)_{np1} \times \frac{A}{A+B} + \left(\frac{m}{\rho}\right)_{np2} \times \frac{B}{A+B}}{\left[\left(\frac{m}{\rho}\right)_{np1} \times \frac{C}{C+D} + \left(\frac{m}{\rho}\right)_{np2} \times \frac{D}{C+D}\right] + \left(\frac{m}{\rho}\right)_{bf}} \times 100 \tag{6}$$

where the ratio of $np1 : np2$ is equal to $C : D$.

For two different types of nanoparticles and two types of base fluid:

$$\frac{\left(\frac{m}{\rho}\right)_{np1} + \left(\frac{m}{\rho}\right)_{np2}}{\left[\left(\frac{m}{\rho}\right)_{np1} \times \frac{C}{C+D} + \left(\frac{m}{\rho}\right)_{np2} \times \frac{D}{C+D}\right] + \left[\left(\frac{m}{\rho}\right)_{bf1} \times \frac{A}{A+B} + \left(\frac{m}{\rho}\right)_{bf2} \times \frac{B}{A+B}\right]} \times 100 \tag{7}$$

where the ratio of $np1 : np2$ and $bf1 : bf2$ are equal to $C : D$ and $A : B$, respectively.

3.3. Carbon-Based Nanofluids Fabrication

As was explained previously in Section 2, carbon allotropes, whether ND, graphene, or CNT, have their own production routes and final product form. For instance, it was shown that both NDs and CNTs could only be produced in the form of dried particles, whereas graphene can be fabricated as dried sheets or as part of a dispersion. Therefore, depending on the type of nanoscaled carbon allotrope and base fluid desired for synthesizing the nanofluid, the production process can be constrained by only the two-step method or the manufacturer can be left with the freedom of selecting any of the two approaches. In general, the two-step method is the only approach that can be employed for fabricating dispersions containing NDs or CNTs, while both one- and two-step routes can be used for producing graphene nanofluids. Nevertheless, the majority of the studies have shown the adaptation of the two-step method for producing graphene nanofluids, which can be justified by the difficulties associated with the single-step route of fabrication and the limitations in the type of base fluid that can be used (see Section 3.1) [145,146]. Some of the research work published on fabricating NDs, graphene, and CNTs nanofluids using the two-step method are listed in Table 2. Note that the single-step graphene nanofluid production was excluded from Table 2 because it is precisely the same as liquid-phase exfoliation of graphene; thus, the reader can find further information's within the sources provided previously in Section 2.2 and the work published by Texter [147]. Nevertheless, it is worth mentioning that the common base fluids used in the graphene suspension one-step (or liquid-phase exfoliation) approach are n-methyl-2-pyrrolidone (NMP), γ-butyrolactone (GBL), n,n-dimethylacetamide (DMAC), n,n-dimethylformamide (DMF), dimethylsulfoxide (DMSO), ortho-dichlorobenzene (ODCB), acetonitrile (ACN), and water with the aid of surfactant [148].

Table 2. Published work on nanodiamond, graphene, and carbon nanotubes nanofluids produced using the two-step approach.

Material	Base Fluid	Particles Dimensions (nm)	Particles Concentration	Additional Information	Ref.
ND	EG	30–50	<1.4 vol. %	- Dispersion was performed with an ultrasonic vibration device for 3 h.	[149]
	EG	5–10	0.25–5.0 vol. %	- Purification and surface modification of the particles were done using a mixture of nitric acid, perchloric acid, and hydrochloric acid. - Dispersion was performed via continuous sonication.	[150]
	EG	5–10	0.25–1.0 vol. %	- Purification and surface modification of the particles were done using a mixture of nitric acid and perchloric acid. - Nanofluid pH adjustment: 7–10. - Dispersion was performed by magnetic stirring and ultrasonic sonication for 3 h.	[151]
	EG—water	30–50	0. 5–2.0 vol. %	- Purification and surface modification of the particles were done using a mixture of nitric acid, perchloric acid, and hydrochloric acid. - Base fluid used was a mixture of 55% distilled water and 45% of EG. - Dispersion was performed by sonication for 3 h.	[152]
	EG and mineral oil	5	2.0 g	- NDs were prepared by detonation followed by functionalization. - For the EG base fluid: the particles and 48 g of dimethylsulfoxide (DMSO) were bath sonicated for 30 min then magnetic stirred with 50 mL of glycidol for 24 h. - For the mineral oil base fluid: the particles, 2.0 g of oleic acid, and 63 g of octane were bath sonicated for 1 h	[153]
	Highly refined thermal oil	3–10	0.25–1.0 wt %	- Non-ionic sorbitane trioleate (Span 85) was used as a surfactant in a surfactant to particles ratio of 7:1. - Dispersion was performed by a probe-type sonicator for 1 h.	[154]

Table 2. *Cont.*

Material	Base Fluid	Particles Dimensions (nm)	Particles Concentration	Additional Information	Ref.
	Naphthenic transformer oil (NTO)	10	1.0 g	- The particles, 2.0 g of oleic acid, and 50 g of octane were high energy ultrasonicated for 30 min. - The previous mixture was added to the base fluid then sonicated for an additional 1.0 h.	[155]
	propylene glycol (PG)—water	5–10	0.2–1.0 vol. %	- The particles were initially purified then treated with acid. - The base fluid contained a mixture of PG and water at ratios of 20:80, 40:60, and 60:40, respectively. - Fabrication was performed through a bath type sonicator for 2.0 h.	[156]
Graphene	Water	2–5 *	10 mg/mL	- Graphene powder was produced through a modified hummer method (i.e., mechanical exfoliation) followed by surface treatment. - Nanofluid fabrication was done through mixture centrifugation at 6000 rpm for 10 min.	[157]
	Water	6000–8000 *	0.001–0.01 vol. %	- Graphene powder was initially oxidized using sulfuric acid and nitric acid. - Nanofluid was produced by ultrasonicating the mixture for 2.0 h.	[158]
	Water	2 *	0.025–0.1 wt %	- Graphene powder was initially oxidized using sulfuric acid and nitric acid. - Nanofluid was produced by continuous sonication using a high-power probe type ultrasonicator.	[159]
	EG and water	–	0.005–0.056 vol. %	- Fabricated graphene was treated with acid for better dispersion. - Nanofluid was produced by sonicating the mixture for 30–45 min. - Solution pH value was adjusted to around 6–7.	[160]
	Glycerol	15–50 *	13 wt %	- Graphene was surface functionalized. - Nanofluid was produced by sonicating the mixture for 10 min.	[161]

Table 2. *Cont.*

Material	Base Fluid	Particles Dimensions (nm)	Particles Concentration	Additional Information	Ref.
CNTs	Water	9–15 ^	0.5 wt %	- MWCNTs powder was surface functionalized via nitric and sulfuric acid of 1:3 ratio, respectively. - Nanofluid was produced by probe sonication for 5 min.	[162]
	Vegetable cutting oil	10–20 ^	0.6 vol. %	- Functionalized MWCNTs were used. - Fabrication process consisted of three mixing stages: 1—mechanical mixing for 60 min at 750 rpm, 2—ultrasonic homogenizer for 60 min, and 3—magnetic stirring for 60 min at 1500 rpm.	[163]
	Turbine meter oil	5–16.1 ^	0.05–0.4 wt %	- Triton X100 was added as a surfactant to the base fluid in a ratio of 1:3, respectively. - Fabrication process consisted of: 1—mixing the surfactant with the base fluid using an electric mixer for 20 min at 1500 rpm, 2—adding and dispersing the MWCNTs using the same device for 4 h, 3—additional mixing using a probe sonicator for 2 h.	[164]
	Water	2–4 ^	0.01–0.5 vol. %	- DWCNTs functionalized by carboxylic acid were used. - Nanofluid production was conducted by magnetic stirring for 2.5 h, followed by ultrasonication for 5 h.	[165]
	EG	2–4 ^	0.02–0.6 vol. %	- DWCNTs functionalized by carboxylic acid were used. - Fabrication was performed by magnetic stirring for 2.5 h, then sonication for 6 h.	[166]
	Water	1–2 ^	0.1–0.5 vol. %	- SWCNTs nanofluids were prepared by first adding sodium dodecyl sulfate (SDS) surfactant then mixing with a high-pressure homogenizer for 1 h.	[167]
	Water	0.8–1.6 ^	0.3 vol. %	- Nanofluid production consisted of SWCNTs, sodium deoxycholate surfactant (0.75 vol. %), and the base fluid. - Mixing was conducted by bath sonication for 6 h, followed by probe sonication for 2 h.	[168]

Note: * and ^ refers to graphene sheet thickness and CNTs outer diameter, respectively.

4. Nanofluids Stability

4.1. Stability Mechanism and Evaluation

The stability of nanofluids is of major concern for maintaining the thermophysical properties of the mixture [169]. Specifically, the stability of the suspension combines several aspects such as dispersion stability, kinetic stability, and chemical stability [120,170]. The dispersion stability deals with nanoparticles aggregation within the colloidal, while the kinetic stability describes the Brownian motion of nanoparticles hosted by the base fluid (i.e., sedimentation of randomly agglomerated particles due to gravity). As for the chemical stability, it is associated with the chemical reactions that occur between the nanoparticles themselves and between the nanoparticles and the surrounding base fluid. However, it is essential to note that chemical reactions in a nanofluid are minimized or halted at low temperature conditions (i.e., below the temperature point of a chemical reaction). Hence, agglomeration and sedimentation of nanoparticles would be the primary aspects concerned with suspension stability. When a nanofluid is physically unstable, the formed sedimentation can have one of three behaviors, namely; 1—dispersed sedimentation, 2—flocculated sedimentation, or 3—mixed sedimentation [8]. Figure 12 shows a schematic illustration of the realistic reflection for the three types of sedimentation behaviors. In addition, the speed at which the sediment forms and settles within an unstable suspension can be classified into two main regions. The first is known as the rapid settling region, which occurs at the beginning stage of the separation of the particles from the hosting base fluid; and the following stage is called the slow settling region, where the changes in sediment formation and settling becomes insignificant along the shelving lifetime [171]. Figure 13 demonstrates an example of the two sedimentation speed formation regions from Witharana et al. [171] investigation. Furthermore, there are about eight techniques that can be used to evaluate the stability of nanofluids, such as 1—sedimentation photographical capturing method, 2—dynamic light scattering (DLS) approach, 3—zeta potential analysis, 4—3-ω approach, 5—scanning electron microscopy (SEM) analysis, 6—TEM characterization, 7—spectral analysis, and 8—centrifugation method. From the previous stability evaluation methods, the sedimentation photographical capturing approach is considered as the most reliable route between them all, but at the expense of time (i.e., it takes a very long time to conduct and analyze). The DLS approach usually over-predicts the size of the particles, especially when using a non-ionized base fluid (e.g., deionized water), where the analysis can show larger values (from 2 to 10 nm more) than the actual particle size [172]. Such results are very problematic and misleading when analyzing nanofluids, especially when the dispersed particles are 10 nm or less in size, where the oversized prediction can incorrectly indicate an instability state. On the other hand, the zeta potential analysis should only be used as a supportive characterization tool. This is because if the nanoparticles and/or the base fluid are non-polar or even of low polarity, there may be other mechanisms affecting the suspension stability [172]. Thus, it is highly recommended to use multiple approaches (e.g., three methods) to determine the stability of the nanofluid. A detailed description of each of the experimental stability evaluation approaches, and their advantages and limitations can be found in the work published by Ali et al. [13]. Other than the previous stability evaluation approaches, Carrillo-Berdugo et al. [173] have proposed a novel theory-based design framework for determining the polarity between the solid and liquid interface, which can be used to adjust the interface tension by adding the required number of dispersive components to meet those of the dispersed nanomaterial.

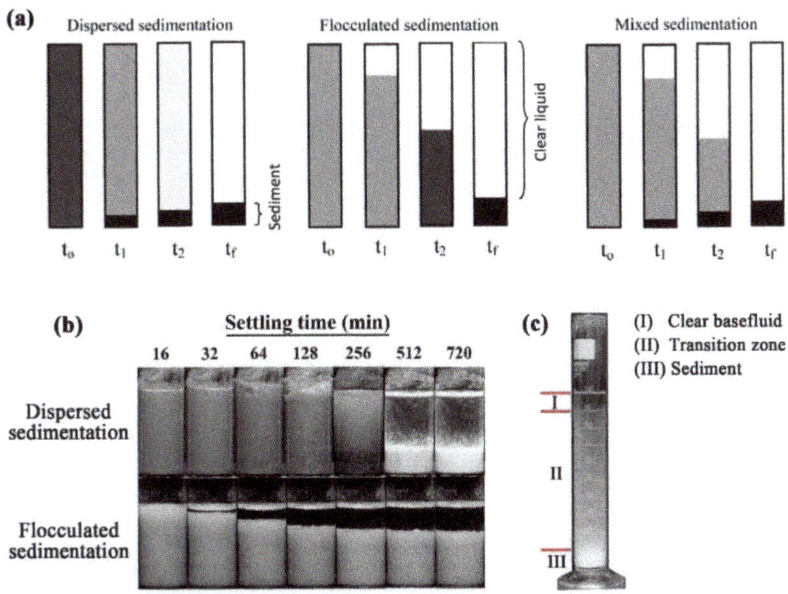

Figure 12. The three types of sedimentation behaviours, where (**a**) shows their schematic mechanism from the starting time (t_o) to the finishing time (t_f) [8], (**b**) demonstrates the dispersed and flocculated sedimentation behaviors from Ali et al. [8] experimental work, and (**c**) represents the mixed sedimentation behavior shown in Ma and Alain [174] investigation.

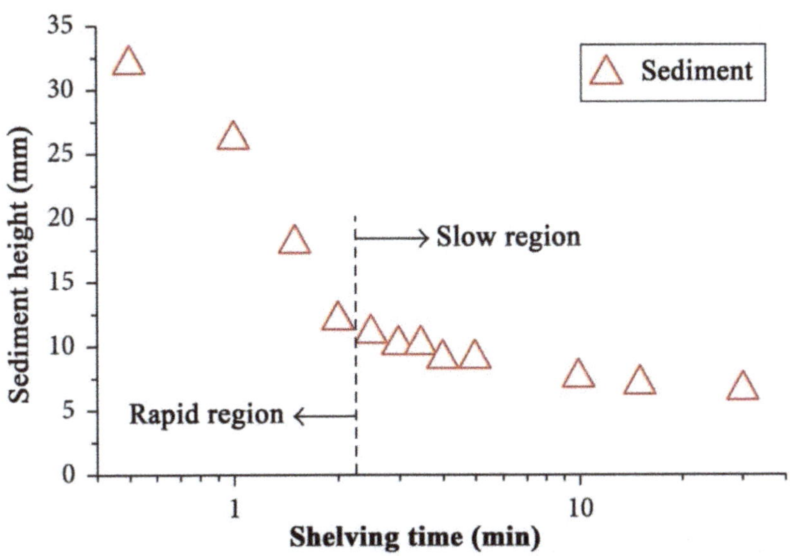

Figure 13. A demonstration of the two sedimentation regions in terms of settling speed, where (the left side) shows the rapid region in which the sediment height changes rapidly, and (the right side) illustrates the slow region, where the changes in the sediment height are very slow to the point where it can be negligible [13].

4.2. Stability Enhancements

Several approaches have been shown to improve the stability of nanofluids successfully. These methods are subdivided into two main categories, which are in the form of physical and chemical routes. The physical approach involves the employment of high energy forces such as ultrasonication, magnetic stirring, homogenizer (or probe sonicator), or even ball milling, which is rarely reported [117,175]. Figure 14 shows the four previous physical stability methods. Unlike the ultrasonication and homogenization methods, the magnetic stirring approach is considered as the most basic route that can be applied to break-down clusters of nanoparticles, within the suspension, with very low performance effectiveness when compared to the other two physical methods [176]. Furthermore, in the literature [177], high pressure homogenization was shown to provide better stability characteristics than ultrasonication to the as-produced nanofluids. In addition, the mixing duration and intensity used in the sonicator device were commonly seen to vary from one research work to another in an attempt to physically stabilize the nanofluid. A good explanation for the aforementioned method is that the mixing power cannot be maintained constant throughout the process due to the voltage fluctuation that the device experienced. Therefore, Yu et al. [178] suggested relying on the relation between the suspension absorption spectra against the total energy supplied to the mixture as a relative solution to the sonication time.

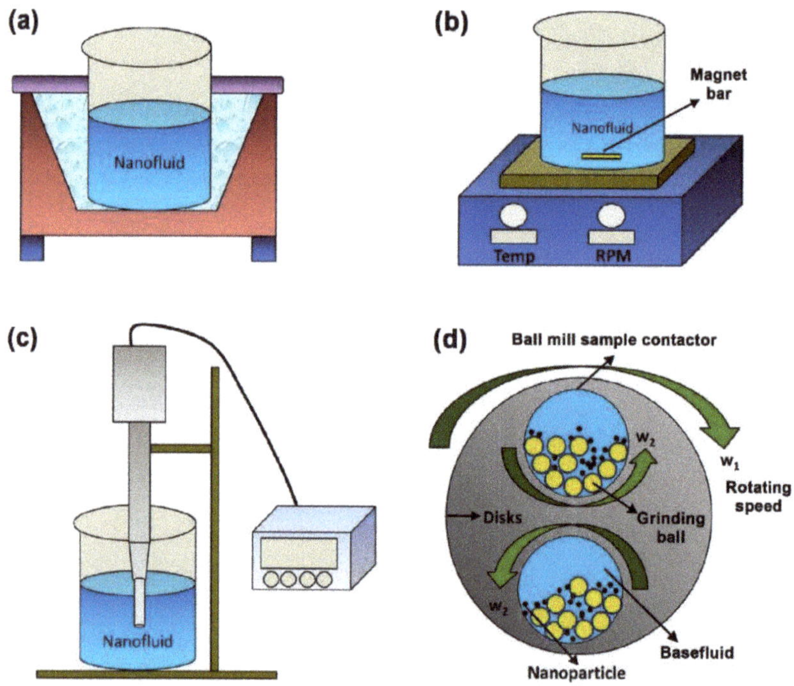

Figure 14. Physical dispersion stability enhancement devices, where (**a**) shows the ultrasonic bath sonicator, (**b**) demonstrate the magnetic stirrer, (**c**) illustrates the homogenizer/prob sonicator, and (**d**) shows the ball milling device. Reproduced with permission from [116]. Elsevier, 2020.

On the other hand, the chemical route stabilizes the suspension by declustering the agglomerated nanoparticles by alternating the pH value of the base fluid or the mixture, adding surfactant(s) to the solid–liquid matrix, or modifying the surface of the nanoparticles. Nanofluids pH alteration affects the level of free cations or anions charges in the media surrounding the dispersed particles, and hence the hydrophilicity or hydrophobicity nature of the particles changes causing the colloidal to either stabilize or destabilize [179,180]. The disadvantage of the previous method is that fabricating suspensions of high or low pH values may be corrosive for high heat flux applications. In addition, surfactants are essential when dispersing nanomaterials of hydrophobic nature (e.g., CNTs and graphene) in a polar base fluid (e.g., water), and vice versa [181,182]. This is because the added surfactant would act as a bridge between the nanoparticles and the hosting fluid, and therefore would improve the dispersion stability of the particles through increasing the repulsive force between the particles themselves and reducing the interfacial tension between the base fluid and the hosted particles. Surfactants are categorized based on their head group charge as cationic, non-ionic, anionic, and amphoteric. Table 3 shows some of the surfactants used in the nanofluids preparation process according to their head group charge [118].

Table 3. Examples of surfactants used in nanofluids fabrication categorized by classifications based on their head group charge.

Surfactant Classification	Head Group Charge	Example(s)
Cationic	+ve	Cetyltrimethyl ammonium bromide (CTAB), distearyl dimethyl ammonium chloride (DSDMAC), and benzalkonium chloride (BAC).
Non-ionic	neutral or uncharged	Oleic acid, polyvinylpyrrolidone (PVP), Arabic gum (AG), Tween 80, and oleylamine.
Anionic	−ve	Sodium dodecyl benzenesulfonate (SDBS), and SDS.
Amphoteric	+ve and −ve	lecithin.

The downside from using surfactants is that the nanofluid becomes more viscous; starts to generate foam when being heated or cooled down; can be lost at high temperatures, and would reduce the overall thermal conductivity of the suspension. As for the nanoparticles surface modification technique, the particles are either initially functionalized (before the dispersion process), or the functionalized materials themselves are added to the colloidal (where they get grafted to the surface of the segregated particles), and therefore forming a new particle surface exposure to the hosting base fluid [183,184]. The drawback of using functionalized materials as stabilizers is that they tend to reduce the overall thermal conductivity of the produced nanofluid due to having a significantly lower thermal conductivity than the dispersed nanoparticles. Figure 15 recaps all of the nanofluid stability improvement methods that were mentioned earlier in this section.

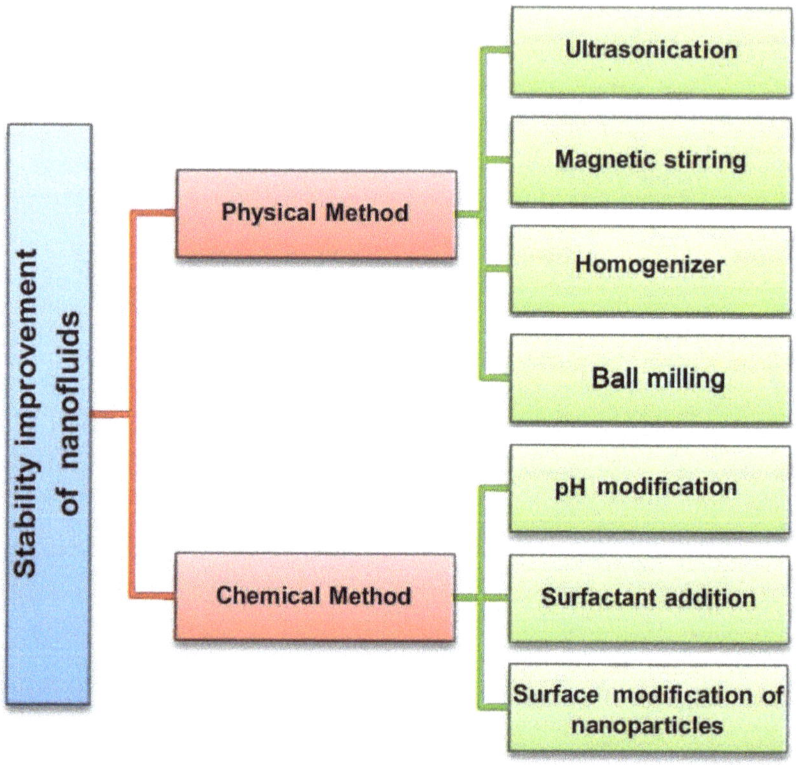

Figure 15. Nanofluids stability improvement methods categorized by their physical and chemical methods.

5. Stability Effect on Thermophysical Properties

The thermophysical properties govern the heat transfer rate that the nanofluid can provide to the system in which it is employed as a working fluid. Nanofluids thermal properties, such as the thermal conductivity, greatly depend on the type of base fluid, nanoparticles material, morphological characteristics of the particles, nanoparticles concentration, and homogeneity of nanoparticles dispersion in the hosting base fluid. The dispersion characteristics of the suspension are subjected to alteration with the change in stability of the particles in their surrounding environment (i.e., base fluid). For such reason, the stability of a nanofluid is considered as a significant factor to maintain the heat transfer rate from and to the colloidal. This section covers the influence of stability on nanofluids effective thermal conductivity and effective viscosity. It is important to highlight that the effect of suspension stability, as a parameter, on the effective density was not reported across the literature, but rather the added surfactants and particles concentration were seen responsible for the changes caused to nanofluids densities [185–187]. This is because nanofluids effective density (ρ_{nf}) is constrained by its overall volume and mass, where it can be directly calculated from extending the rule of mixtures (i.e., Equation (8)):

$$\rho_{nf} = f_V \times \rho_{np} + (1 - f_V) \times \rho_{bf} \quad (8)$$

where f_V is the particles volumetric fraction, ρ_{np} is the density of the nanoparticles, and ρ_{bf} is the density of the base fluid. Similarly, the effective specific heat capacity of the colloidal was not shown to be linked to the dispersion stability. The main parameter that affects nanofluids effective specific heat capacity is the particles concentration included in the mixture. This is because increasing the nanoparticles concentration would result in enhancing the overall thermal performance of the suspension, and hence less heat would be required to raise the temperature of the fabricated nanofluid, and vice versa [188]. In general, nanofluids effective specific heat capacity is lower than their base fluids [126,189]. According to Ali et al. [13] and other researchers [190–193], the most accurate theoretical model for calculating the effective specific heat capacity of a nanofluid ($C_{p_{nf}}$) is the following equation:

$$C_{p_{nf}} = \frac{\rho_{bf} \times (1 - f_V)}{\rho_{nf}} \times C_{p_{bf}} + \frac{\rho_{np} \times f_V}{\rho_{nf}} \times C_{p_{np}} \quad (9)$$

where $C_{p_{bf}}$ and $C_{p_{np}}$ are the specific heat capacities of the base fluid and the nanoparticles, respectively. Experimentally, the $C_{p_{nf}}$ can be determined using the differential scanning calorimetry (DSC) technique, which basically measures the amount of heat required to be delivered to both test sample and reference source, of well-known heat capacity, so that a temperature rise can be achieved [188].

5.1. Effective Thermal Conductivity

Thermal conductivity enhancement of heat transfer fluids has always been the main driving force that motivated researchers into developing nanofluids. This is because the solid particles added to the liquid have tremendously higher thermal conductivity compared to that of the base fluid, and thus cause the effective thermal conductivity of the mixture to improve significantly. At the early stages of their discovery, the claims on the enhancement caused by the dispersed particles on the hosting fluid were seen as a controversial topic because many published works across the literature reported divergence in the level of enhancement and measurement results were difficult to be replicated [194–197]. Nevertheless, a worldwide round-robin, including 33 research institutes, have demonstrated acceptable consistency in measuring the effective thermal conductivity of nanofluids, despite the fact that they unexplored any anomalous improvement in the effective thermal property [198]. Up to today, the effective thermal conductivity of the suspension remains a complicated topic, where it involves many vital elements such as the particles type and morphology, particles concentration, base fluid type and temperature, added surfactants (if any), and dispersion stability [13,199,200]. When constraining the first four parameters in fabricating a dispersion, the optimum effective thermal conductivity is usually reached when the particles are well distributed in the hosting fluid with minimum to no agglomerations/clustering between them. Since a stable state nanofluid reflects that its nanoparticles are homogeneously dispersed within the hosting base fluid, it should theoretically result in a superior overall suspension thermal conductivity to those of an unstable state. The potential influence of nanoparticles agglomeration on the thermal conduction emphasizes that colloid chemistry will play a significant role in enhancing the thermal conductivity of nanofluids. Scientists such as Yu et al. [201], Haghighi et al. [202], and Li et al. [203] have all proven, through their research work, that stabilized nanofluids have greater and steady effective thermal conductivity than their counterparts. Prasher et al. [204] and Wang et al. [205] explained this observation by analyzing the effect of nanoparticles aggregation on the thermal conductivity of nanofluids, where they assumed that solid liner and side chains get formed by particles clustering. Based on the researcher's conclusion, these chains are mainly responsible for enhancing the suspensions thermal conductivity. Still, as more nanoparticles get accumulated, the cluster becomes heavier, and therefore separates from the base fluid due to the gravitational force. The aforementioned causes the thermal conductivity of the colloidal to degrade, with respect to settling time, until it decreased to a minimum possible value when total separation is attained. The previous claim was also supported by the work of Hong et al. [206], where they examined the effective thermal conductivity of SWCNTs—water dispersion with magnetic-field-sensitive nanoparticles (Fe_2O_3) under various magnetic field strengths. In their experiment, the researchers successfully interconnected the dispersed CNTs using Fe_2O_3 nanoparticles and the employed magnetic field, and thus forming a well aligned chains of nanomaterials. This resulted in the effective thermal conductivity to increase by 50% over that of the base fluid. However, as the holding time under the magnetic field increased, the nanomaterials started to form larger clumps that caused the suspension effective thermal conductivity to degrade. Other studies have also proven the enhancement in nanofluids thermal conductivity through the chain concept, such as the work of Wright et al. [207], Wensel et al. [208], and Hong et al. [209]. All three groups of scholars relied on the magnetic field to form the dispersed particles connected networks in the host fluid. However, the first used a novel alignment approach via coating the SWCNTs with Ni, whereas the other two achieved the interconnection with the aid of metal oxide nanoparticles (e.g., Fe_2O_3 and MgO). It is important to note that different types of base fluid and surfactants were used in the three previous studies. Younes et al. [210] have suggested an innovative nanoscale aggregation process that can be adopted to form nanosolids with an interconnect chain capability when dispersed in liquids. In their work, they coated the CNTs through their aggregation process with metal oxide nanoparticles and different types of surfactants. Afterwards, the scholars filtered and dried the aggregate to obtain their as-prepared CNTs-based nanosolids. These newly formed nanomaterial can interconnect when dispersed in a non-aqueous solution by applying a magnetic field. Figure 16 illustrates the effective thermal conductivity degradation theory, which describes the mechanism in which the particles separate from the base fluid due to the formation of both linear and side chains. Other aspects that have less influence on the effective thermal conductivity of nanofluids includes the liquid layering near the outer particles surface [211], Brownian motion of dispersed particles [212,213], thermophoresis [214,215], near-field radiation [216,217], and ballistic transport and nonlocal effects [218,219].

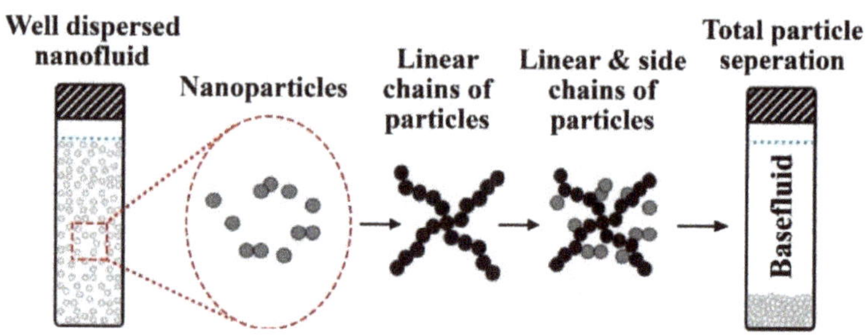

Figure 16. Nanoparticles separation due to the formation of both linear and side chains in the base fluid.

Many different techniques have been adopted for measuring the thermal conductivity of nanofluids, namely; 1—cylindrical cell method, 2—temperature oscillation approach, 3—steady state parallel-plate method, 4—3-ω method, 5—thermal constants analyzer approach, 6—thermal comparator method, 7—flash lamp method, 8—transient hot-wire method, 9—laser flash method, and 10—transient plane source. More details on the usage, advantages, and disadvantages of these thermal conductivity measurement techniques can be found in Mashali et al. [17], Paul et al. [220], Qiu et al. [170], and Tawfik [221] published works. Among the previously mentioned techniques, the transient hot-wire approach was mainly adopted across the literature for nanofluids effective thermal conductivity measurements, although it was the first measuring route for such property [17]. The reasons that attracted researchers into favoring the transient hot-wire method among other methods is due to its capability of eliminating measurements errors caused by the natural convection of the fluid, its minimal amount of time required to perform each measurement (i.e., within seconds), and its simple conceptual design compared to other available devices or apparatuses. One thing that needs to be emphasized here is that the high thermal conductivities of graphene, ND, and CNT found in the literature are based on theoretical calculations for a single particle, and that when attempting to measure this thermal property for a powder sample, the results will show tremendously lower values due to the presence of air along with the limitation associated with the measuring tool itself [222,223].

Furthermore, researchers have published numerous amounts of literature on improving nanofluids effective thermal conductivity over their base fluids [224–228]. For example, Yu et al. [224] compared the thermal conductivity of their stable graphene—EG nanofluids to that of pure EG. The researchers have dispersed 2.0 and 5.0 vol. % of graphene, of 0.7–1.3 nm in size, in EG to fabricate their nanofluids at a set of temperatures from 10 to 60 °C, using the two-step approach. They have found that the as-prepared 5.0 vol. % suspension had 86% enhanced thermal conductivity over its base fluid at 60 °C. Yarmand et al. [225] synthesized water based nanofluids from 0.02 to 0.1 wt % of functionalized graphene nanoplatelets using the two-step method at 20–40 °C. The functionalization process was conducted through an acidic treatment to the graphene powder by dispersing the as-received graphene in a 1:3 mixture of HNO_3 and H_2SO_4, respectively. They found that the formation of sedimentation within their as-fabricated nanofluids was minimal throughout their 245 h test. The heat transfer coefficient improved by 19.68% compared to the base fluid when using the 0.1 wt % nanofluid. Furthermore, Yarmand et al. [225] concluded that the thermal property of the suspension is influenced by the temperature of fabrication and the dispersed solid concentration. Zhang et al. [226] compared the thermal conductivity of three ionic based nanofluids containing graphene sheets, graphite nanoparticles, and SWCNTs. All three types showed enhanced thermal conductivity with a partial increase in viscosity compared to their base fluids. Nevertheless, the nanofluid fabricated from graphene had a higher increase in thermal conductivity compared to the other two types of dispersions. Ghozatloo et al. [227] studied the effect of time, temperature, and concentration on the thermal conductivity of pure and functionalized CVD graphene–water nanofluids. The functionalizing process of graphene was conducted through an alkaline method, and the suspensions were fabricated using sonication (i.e., the two-step approach). Moreover, the concentration used in the production of the suspension was of 0.01–0.05 wt %, and the duration of the dispersion mechanism was 1 h. The authors found that the nanofluids samples containing pure graphene had promptly developed clusters between its solid content, whereas the functionalized suspensions were highly stable. Furthermore, the effective thermal conductivity was seen to reduce to a certain extent for all nanofluids after the time of production. In addition, the enhancement in the effective thermal conductivity using functionalized graphene showed to be 13.5% (0.05 wt %) and 17% (0.03 wt %) over 25 °C and 50 °C water, respectively. Askari et al. [46] experimentally investigated the thermal and rheological properties of 0.1–1.0 wt % CVD nanoporous graphene–water nanofluids along with heat transfer suspension effect on the thermal performance of a counter-flow arranged mechanical wet tower. The base fluid used in the two-step suspension fabrication was taken from one of the working cooling towers located in South Iran to reflect a real-life case scenario. Different types of surfactants were used to stabilize the dispersion of the as-prepared nanofluids, such as AG, Tween 80, CTAB, Triton X-100, and Acumer Terpolymer. The authors found through analyzing the physical stability of their nanofluids, utilizing the sedimentation capturing method and zeta potential measurements, that using Tween80 as a disperser resulted in a stabilized suspension that can last for up to two months. Furthermore, their 1.0 wt % nanofluid showed a 16% increase in the thermal conductivity at a dispersion temperature of 45 °C. At the same time, the low concentration suspensions would be appropriate for industrial applications because of their increasing effect on the effective density and viscosity. Moreover, the as-produced nanofluids enhanced the efficiency, cooling range, and tower characteristic compared to the conventional base fluid. For example, using a 0.1 wt % fabricated nanofluid had resulted in a 67% increase in the cooling range and a 19% decrease in the overall water consumption. Goodarzi et al. [229] studied the effective thermal conductivity, specific heat capacity, and viscosity of their as-prepared nitrogen-doped graphene–water nanofluids along with their convective heat transfer behavior when employed in a double-pipe type heat exchanger. The authors used 0.025 wt % of Triton X-100, as their surfactant, along with 0.01–0.06 wt % of graphene to prepare the suspensions using the two-step method. Their results showed that the examined thermophysical properties where very sensitive to both temperature and concentration. As an example, the effective thermal conductivity of their suspension showed an increase from 0.774 to 0.942 W/m·K with the increase in temperature (from 20 to 60 °C). The maximum effective thermal conductivity achieved by the scholars was 37% higher than that of the base fluid. Furthermore, they found that increasing the concentration of their nanosheets in the base fluid had caused the heat transfer coefficient of their working fluid to improve but at the same time results in increasing the pressure drop in the system and the pumping power requirement. Liu et al. [230] examined the effective thermal conductivity and physical stability of their synthesized graphene oxide–water nanofluids. Moreover, the mass fraction and temperature of the investigated samples were 1.0–4.5 mg/mL and 25–50 °C, respectively. The researchers found that they can achieve a homogeneously stable nanofluid for about 3 months using their preparation process. They also found that the effective thermal conductivity of their as-prepared nanofluid was 25.27% higher than the base fluid, at 4.5 mg/mL mass fraction, and a temperature of 50 °C. Ghozatloo et al. [228] explored the possibility of improving the convection heat transfer behavior of a shell and tube heat exchanger, under laminar flow conditions, using CVD graphene nanofluid of water base. The researchers also investigated the effect of temperature and solid dispersed concentration of the mixture on the thermal conductivity and convective heat transfer coefficients. The dispersions were prepared using 0.05, 0.075, and 0.1 wt % of treated CVD graphene and 15 min sonication in water. According to the authors outcomes, using 0.05, 0.075, and 0.1 wt % suspensions, at 25 °C, enhanced the thermal conductivity over pure water by 15.0%, 29.2%, and 12.6%, respectively. Moreover, the convective heat transfer coefficients of the as-produced mixtures depended on the flow conditions in which the working fluid undergoes but were in all cases higher than the base fluid. From the previously mentioned studies, it can be concluded that carbon-based nanomaterials can form stabilized nanofluids, either by selecting the appropriate base fluid–nanoscaled material matrix or through external physical and/or chemical approaches. Moreover, these suspensions have enhanced thermal properties compared to their conventional base fluids, but the level of enhancement gets affected by parameters such as concentration, temperature, physical stability, etc. Thus, such factors should be carefully considered to obtain the optimum suspension thermophysical condition.

Besides the experimental studies, many researchers have developed theoretical correlations to predict the effective thermal conductivity of nanofluids. Still, most of these formulas have shown conceptual limitations towards their experimental origin. Table 4 demonstrates the developments in the effective thermal conductivity equations.

Table 4. Developments of nanofluids effective thermal conductivity formulas.

Developer/s	Year	Formula	Dependent Parameter	Limitations
Maxwell [231]	1890	$\frac{k_{eff}}{k_{bf}} = \frac{k_{np}+2k_{bf}+2\,f_V\,(k_{np}-k_{bf})}{k_{np}+2k_{bf}-f_V\,(k_{np}-k_{bf})}$; where k_{eff}, k_{bf}, and k_{np} are the effective thermal conductivity of the nanofluid, base fluid thermal conductivity, and nanoparticles thermal conductivity, respectively.	f_V	Suited for spherical shaped particles.

Table 4. *Cont.*

Developer/s	Year	Formula	Dependent Parameter	Limitations
Jefferson et al. [232]	1958	$k_{eff} = k_{bf}\left\{\left(1-1.21\,f_V{}^{2/3}\right) + 0.4875\,f_V{}^{1/3}\left[\frac{\ln\frac{k_{np}}{k_{bf}}-1}{0.25+\left(0.403\,f_V{}^{-1/3}-0.5\right)\left(\ln\frac{k_{np}}{k_{bf}}-1\right)}\right]\right\}$	f_V	The model is used for spherical particles but always underestimate the effective thermal conductivity by 25%.
Hamilton and Crosser [233]	1962	$\frac{k_{eff}}{k_{bf}} = \frac{k_{np}+(n-1)k_{bf}-(n-1)\,f_V\,(k_{bf}-k_{np})}{k_{np}+(n-1)k_{bf}-f_V\,(k_{bf}-k_{np})}$	f_V and n	Preferred for spherical and cylindrical shaped particles with $n = 3/\psi$, where n and ψ are the empirical shape factor and particle sphericity, respectively. For perfectly spherical particles $\psi = 1$.
Wasp et al. [234]	1977	$\frac{k_{eff}}{k_{bf}} = \frac{k_{np}+2k_{bf}-2\,f_V\,(k_{bf}-k_{np})}{k_{np}+2k_{bf}+f_V\,(k_{bf}-k_{np})}$	f_V	Particles should have a sphericity of ≤ 1.
Yu and Choi [235]	2003	$\frac{k_{eff}}{k_{bf}} = \frac{k_{np}+2k_{bf}+2\,f_V\,(k_{np}-k_{bf})(1+\beta)^3}{k_{np}+2k_{bf}-f_V\,(k_{bf}-k_{np})(1+\beta)^3}$; where β is the ratio of the nanolayer thickness to the particle radius.	f_V, interfacial particle layer, and radius	Modified version of the Maxwell [231] model for spherical particles. The main problem is that it is inadequate the non-linear trend of thermal conductivity.
Xuan et al. [236]	2003	$\frac{k_{eff}}{k_{bf}} = \frac{k_{np}+2k_{bf}-2\,f_V\,(k_{bf}-k_{np})}{k_{np}+2k_{bf}+f_V\,(k_{bf}-k_{np})} + \frac{f_V\,\rho_{np}\,C_{np}}{2k_{bf}}\sqrt{\frac{k_B T}{3\pi r_c}}\nu$; where k_B is the Boltzmann constant (1.381×10^{-23} J/K), T is the temperature of the mixture, r_c is the particle apparent radius, and ν is the kinematic viscosity of the liquid.	$f_V, \rho_{np}, C_{np}, T, r_c$, and ν	Hard to predict the thermal conductivity for linear temperatures.
Nan et al. [237]	2003	$\frac{k_{eff}}{k_{bf}} = \frac{3+f_V\left(\frac{k_{np}}{k_{bf}}\right)}{3-2f_V}$	f_V	Can only be used with CNTs nanofluids.
Kumar et al. [218]	2004	$\frac{k_{eff}}{k_{bf}} = 1 + c\frac{2k_B\,T\,f_V\,r_m}{\pi\,\nu\,d_{np}^2\,k_{bf}}\,(1-f_V)\,r_{np}$; - For none-spherical particles: $d_{np} = \frac{6V_{np}}{A_{np}}$; - For CNTs: $d_{np} = \frac{1.5\,ab}{a+\left(\frac{b}{2}\right)}$; where c is a constant value from 2.9 to 3.0, r_m is the radius of the fluid medium particles, r_{np} is the particles radius, d_{np} is the nanoparticles mean diameter, V_{np} is the volume of the particles, A_{np} is the area of the particles, a is the length of the CNT, and b is the outer diameter of the CNT.	f_V dimensions of the particles, T, and ν	The Brownian motion has the dominative effect on the thermal conductivity prediction over all other factors.

Table 4. *Cont.*

Developer/s	Year	Formula	Dependent Parameter	Limitations
Jang and Choi [213]	2004	$k_{eff} = k_{bf}(1 - f_V) + k_{np}\, f_V + 3C_1\left(\frac{d_{bf}}{d_{np}}\right) k_{bf}\, f_V\, Re^2_{d_{np}}\, Pr;$ - $Re_{d_{np}} = \frac{C_{R.M.}\, d_{np}}{v};$ - $C_{R.M.} = \frac{k_B\, T}{3\, \pi \mu d_{np}\, \ell_{bf}};$ where C_1 is a proportional constant, d_{bf} is the diameter of the base fluid molecule, $Re_{d_{np}}$ is the Reynolds number as defined above, Pr is the Prandtl number, $C_{R.M.}$ is the nanoparticle random motion velocity, and ℓ_{bf} is the mean-free path of the base fluid molecule.	f_V dimensions of the particles, T, v, and ℓ_{bf}	Both conduction and convection heat transfer are accounted for, while the heating duration is much higher.
Yu and Choi [238]	2004	$\frac{k_{eff}}{k_{bf}} = 1 + \frac{n f_{V_e} A}{1 - f_{V_e} A};$ - $f_{V_e} = r\, f_V;$ - $A = \frac{1}{3} \sum_{j=a,b,c} \frac{k_{pj} - k_{bf}}{k_{pj} - (n-1) k_{bf}};$ where f_{V_e} is the equivalent volume concentration of complex ellipsoids particles, r is the volume ratio, a, b, and c are the semi-axes of the particle (for sphere $a = b = c$), A is a parameter that reflects the equation shown above, and k_{pj} is the equivalent thermal conductivity of the ellipsoids particle.	f_V, n, and interfacial resistance	This is a renovated Hamilton and Crosser [233] model with $n = 3/\psi^{-\alpha}$, where α is an empirical parameter that depends on both particle sphericity and the particle to liquid thermal conductivity ratio. In addition, this model includes the interface layer between the particles and the surrounding liquid but cannot predict the nonlinear behaviour of the thermal conductivity.
Prasher et al. [239]	2005	$\frac{k_{eff}}{k_{bf}} = \left(1 + A'\, f_V\, Re^{m'}\, Pr^{0.333}\right) \frac{(1+2\alpha) + 2\, f_V (1-\alpha)}{(1+2\alpha) - f_V (1-\alpha)};$ - $\alpha = \frac{2\, R_b\, K_m}{d_{np}};$ where A' is a constant that is independent of the type of base fluid, m' is a constant that depends on the base fluid type, Re is the Reynolds number, α is a parameter that reflects the equation shown above, R_b is the impact of interfacial resistance with a magnitude in the range of 0.77×10^{-8} to 20×10^{-8} $Km^2\, W^{-1}$, and K_m is the matrix conductivity.	f_V, R_b, and d_{np}	Only considers the dispersed particles convection effect.
Xue [240]	2005	$\frac{k_{eff}}{k_{bf}} = \frac{1 - f_V + 2\, f_V \left(\frac{k_{np}}{k_{np} - k_{bf}}\right) \ln\left(\frac{k_{np} + k_{bf}}{2k_{bf}}\right)}{1 - f_V + 2\, f_V \left(\frac{k_{bf}}{k_{np} - k_{bf}}\right) \ln\left(\frac{k_{np} + k_{bf}}{2k_{bf}}\right)}$	f_V	Suitable for nanofluids made of dispersed CNTs.
Murshed et al. [241]	2006	$\frac{k_{eff}}{k_{bf}} = \frac{\left[1 + 0.27\, f_V^{\frac{4}{3}} \left(\frac{k_{np}}{k_{bf}} - 1\right)\right] \left[1 + \frac{0.52\, f_V}{1 - f_V^{\frac{1}{3}}} \left(\frac{k_{np}}{k_{bf}} - 1\right)\right]}{1 + f_V^{\frac{4}{3}} \left(\frac{k_{np}}{k_{bf}} - 1\right) \left(\frac{0.52\, f_V}{1 - f_V^{\frac{1}{3}}} + 0.27\, f_V^{\frac{1}{3}} + 0.27\right)}$	f_V	The particles need to be uniformly dispersed in the suspension for appropriate effective thermal conductivity prediction.

Table 4. *Cont.*

Developer/s	Year	Formula	Dependent Parameter	Limitations
Vajjha et al. [242]	2010	$k_{eff} = \frac{k_{np}+2k_{bf}-2(k_{bf}-k_{np})\,f_V}{k_{np}+2k_{bf}\,(k_{bf}-k_{np})\,f_V}\,k_{bf} + 5 \times 10^4 \beta\, f_V\, \rho_{bf} C_{p_{bf}} \sqrt{\frac{k_B\, T}{\rho_{np}\, d_{np}}}\, f\,(T,\, f_V)$; - $f\,(T,\, f_V) = (2.8217 \times 10^{-2}\, f_V + 3.917 \times 10^{-3})\left(\frac{T}{T_o}\right) + (-3.0669 \times 10^{-2}\, f_V - 3.91123 \times 10^{-3})$; where β is the fraction of the liquid volume that moves with the particle, $f\,(T,\, f_V)$ is a function that depends on the fluid temperature and particles concentration as defined above, and T_o is a reference temperature that equals 273 K	f_V, particles type, and base fluid temperature	Limited to nanofluids of temperatures between 295 and 363 K.
Xing et al. [243]	2016	$k_{eff} = \left(1 + \frac{\eta' f_V}{\frac{3k_{bf}}{\eta' k^c_{33}} + 3H(\eta' P)}\right) k_{bf} + 0.5\, f_V\, \rho_{CNT}\, C_{p_{CNT}} \sqrt{\frac{k_B\, T}{3\,\pi\,\mu\, r_m}}$; - $\eta' = \left[(2 \times 10^8) a^2 - 13.395\, a + 0.2533\right] f_V{}^{-(6988.1\, a + 0.1962)}$; - $H = \frac{1}{P^2-1}\left[\frac{P}{\sqrt{P^2-1}} \ln\left(P + \sqrt{P^2-1}\right) - 1\right]$; - $P = \frac{a}{b}$; - $k^c_{33} = \frac{k_{np}}{1+\frac{2R_k\, k_{np}}{a}}$; where η' is the modified straightness ratio, H is a factor reflected by the equation defined above, P is the CNT length to diameter ratio, k^c_{33} is the equivalent thermal conductivity of the CNT along the longitudinal axes, R_k is the Kaptiza radius and is equal to 8×10^{-8} m^2 K/W, μ is the dynamic viscosity, ρ_{CNT} is the density of the CNT, and is the $C_{p_{CNT}}$ specific heat capacity of the CNT.	f_V, T, and aspect ratio	Can only be used for CNTs suspensions. Furthermore, not all of the parameters are accounted in the correlation, while the effect of the micro-motion is the most significant parameter.
Gao et al. [244]	2018	$\frac{k_{eff}}{k_{bf}} = \frac{3+\eta^2\, f_V}{\left[k_{bf}\left(\frac{2\,R_b}{L} + 13.4\sqrt{t}\right)\right](3-\eta\, f_V)}$; where L is the length of the nanoplatelet, t is the nanoplatelet thickness, and η is the average flatness ratio of the graphene nanoplatelet.	f_V, L, t, R_b, and η.	This model is designed for suspensions of water, as the base fluid, and graphene nanoplatelet.
Li et al. [245]	2019	$\frac{k_{eff}}{k_{bf}} = \frac{k_{pe}+2k_{bf}+2(k_{pe}-k_{bf})\left(1-\frac{t_{nL}}{r_{np}}\right)^3 f_V}{k_{pe}+2k_{bf}-(k_{pe}-k_{bf})\left(1-\frac{t_{nL}}{r_{np}}\right)^3 f_V}$; where k_{pe} is the equivalent particle thermal conductivity, and t_{nl} is the thickness of the nanolayer surrounding the particle.	f_V, t_{nl}, r_{np}, and fluid temperature	This model is a modified form of the Yu and Choi model with the nanolayer constant value changed to quadratic.

Table 4. *Cont.*

Developer/s	Year	Formula	Dependent Parameter	Limitations
Jóźwiak et al. [246]	2020	$\frac{k_{eff}}{k_{bf}} = \frac{\omega f_V(k_{np}-\omega k_{bf})(\gamma_1^2-\gamma^2+1)+(k_{np}+\omega k_{bf})\gamma_1^2[f_V\gamma^2(\omega-1)+1]}{\gamma_1^2(k_{np}+\omega k_{bf})-(k_{np}-\omega k_{bf})f_V(\gamma_1^2+\gamma^2-1)}$; - $\omega = \frac{k_{IN}}{k_{bf}}$; - $\gamma = 1 + \frac{t_{nl}}{r_{CNT}}$; - $\gamma_1 = 1 + \frac{t_{nl}}{2r_{CNT}}$; where ω, γ, γ_1 are factors representing the equations shown above, k_{IN} is the interfacial nanolayer thermal conductivity, and r_{CNT} is the radius of a single CNT.	f_V, and particles morphology	This is a modified version of the Murshed et al. [241] model, which is suitable for ionic liquid nanofluids (also known as ionanofluids) with dispersed CNTs.

5.2. *Effective Viscosity*

The effective viscosity of nanofluids is part of the chain that determines the applicability of using such a category of suspensions in heat transfer applications. Since it is a transport property directly related to the dynamic performance of the heat transfer system, where an increase in colloidal viscosity would lead to an increase in the friction coefficient and thus a raise in the pressure losses in the system. The heat transfer system then compensates for this pressure difference by increasing the pumping power, and accordingly, more electrical power gets consumed. For such a reason, many research studies have been devoted to investigating the link between the nanofluids effective viscosity and the different parameters associated with the suspension, such as nanoparticles shape, size, concentration, dispersion stability, and mixture temperature [247–256]. Mena et al. [257] suggested that nanofluids fabricated with nanoparticles concentration of up to 13 vol. % behaves as Newtonian fluids (i.e., their viscosity is independent of shear strain). In addition, many researchers proved that the stability of nanofluids has an inverse relationship with their effective viscosity. Meaning that well-dispersed suspensions tend to have lower effective viscosity than those of poor stability [202,258–260]. If the viscosity of a shelved nanofluid was to be categorized according to its stability condition, then there would exist one to three different viscosity regions. To be more precise, a well-dispersed suspension would roughly have a uniform viscosity along its column, while three different viscosity regions would exist in the semi-stable case, and two different viscosity regions would form in the unstable separation scenario. Figure 17 demonstrates the three stability cases with their different viscosity regions. As for the nanofluid in its dynamic form, these viscosity regions would most likely still exist within the suspension while flowing in the hosting system. Knowing this, one can explain why the unstable suspension would require higher pumping power compared to the stable form of the same dispersion. To calculate the percentage of viscosity increase that the dispersed particles cause to the base fluids, the following equation can be used [154]:

$$Viscosity\ increase\ (\%) = \left(\frac{\mu_{eff}}{\mu_{bf}} - 1\right) \times 100 \quad (10)$$

where μ_{eff} and μ_{bf} are the effective viscosities of the nanofluid and the base fluid, respectively. Furthermore, the most common and widely used approach for determining nanofluids viscosity is via the rotational viscosity measurement method [261]. In this method, a shaft is inserted in the sample, after which the rotational speed and the torque per unit length of the shaft are used to determine the viscosity of the nanofluid. Other measuring techniques are also used, such as the capillary viscometer, concentric cylinder viscometer, rheonuclear magnetic resonance, and rheoscope [262–264]. To be noted that, according to Prasher et al. [265], in order for a nanofluid to improve the performance of its hosting thermal application, the increase in effective viscosity should not exceed four times the mixture enhanced effective thermal conductivity, otherwise the working fluid would not serve the hosting system in terms of overall performance. This can be theoretically calculated through the following equations [266]:

$$\frac{\left(\frac{\mu_{eff}}{\mu_{bf}}\right)-1}{\left(\frac{k_{eff}}{k_{bf}}\right)-1} < 4 \tag{11}$$

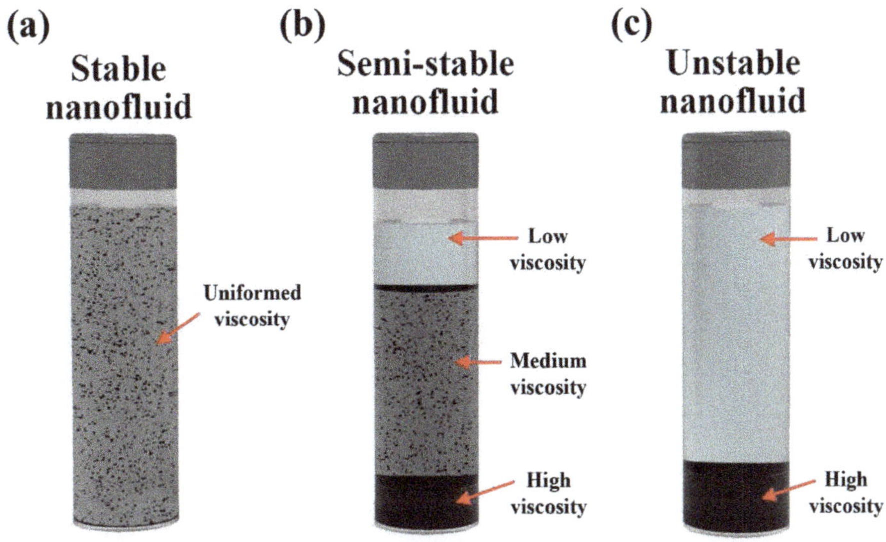

Figure 17. Nanofluids viscosity classification, where (**a**) shows the stable, (**b**) illustrates semistable, and (**c**) demonstrates the unstable cases of the suspension.

In another study, Akhavan-Zanjani et al. [267] investigated the thermal conductivity and viscosity of nanofluids made of graphene, water, and polyvinyl alcohol (PVA) surfactant. The wt % used were of 0.005–0.02% and the mass ratio of the PVA employed was 3:1. The authors found a significant increase in the as-prepared fluid thermal conductivity with a moderate raise in the viscosity. The highest recorded percentages for the thermal conductivity and viscosity were 10.30% and 4.95%, respectively. Iranmanesh et al. [268] analyzed the effect of two preparation parameters, namely; the concentration and temperature, on aqueous graphene nanosheets nanofluids thermal conductivity and viscosity. They used 0.075–0.1 wt % to fabricate their nanofluids using the two-step method at 20–60 °C. The findings indicated that the wt % used had a clear effect on the viscosity and thermal conductivity on the prepared dispersion. Moreover, the temperature, as a parameter, was seen to have a larger influence on the level of the final product viscosity compared to the added solid concentration. Although the study avoided any employment of surfactants, the authors as-prepared suspension was stable for several days. Such an observation is not new and was also reported by other researchers, such as Mehrali et al. [159], where they successfully fabricated stabilized nanofluids made of the graphene–water mixture without the need for surfactants or graphene functionalization, but these are rare cases because of the attraction nature between the head groups of both particles and the base fluid molecules. Ghazatloo et al. [269] have developed a model that can predict the effective viscosity of CVD graphene–water and CVD graphene–EG nanofluids at ambient temperature. They used experimental measurements of the property and employed a commonly used model to form their correlation. Moreover, the researchers used 0.5–1.5 vol % of 60 nm graphene sheets with two separate base fluids (i.e., water and EG), after which the content was subjected to sonication for 45 min. For their water based nanofluids, a volume ratio of 1.5:1 of SDBS surfactant to solid content was used to physical stabilize the dispersion. The outcome of their research indicated that the effective viscosity remarkably increased with the raise of vol. %, and hence the concentration as a parameter had a significant effect on the property. Furthermore, the Batchelor model [270] showed some deviation from the experimental data of the as-prepared suspensions viscosity. This variation in results were reduced by the newly developed model, which, unlike the previous model, considered the solid additive geometry. The comparison between the authors model, Batchelor model, and experimental results is demonstrated in Figure 18.

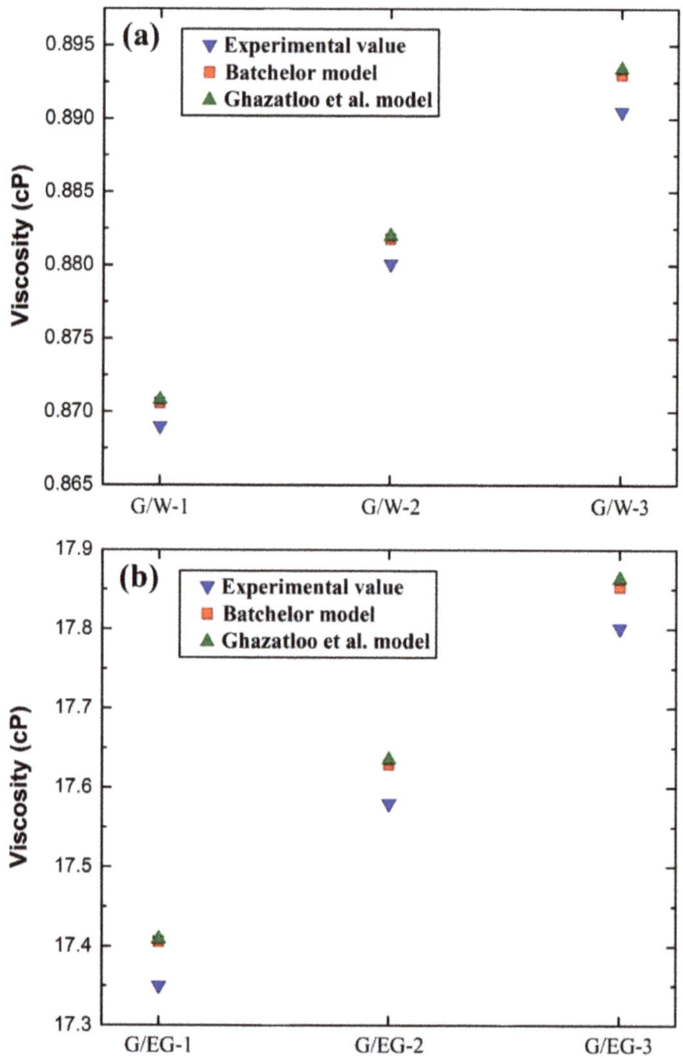

Figure 18. Comparison between the Ghazatloo et al. [269] model, Batchelor [270] model, and experimental effective viscosity, where (**a**) shows the results for graphene–water nanofluid of 0.5 vol. % (G/W-1), 1.0 vol. % (G/W-2), and 1.5 vol. % (G/W-3), and (**b**) illustrates the values for graphene–EG suspensions of 0.5 vol. % (G/EG-1), 1.0 vol. % (G/EG-2), and 1.5 vol. % (G/EG-3).

On the other hand, in terms of the effective viscosity theoretical models developments, Table 5 lists these correlations with their year of development, dependent parameters, and limitations. From the formulas shown in Table 5, it can be concluded that most of the authors have used specific experimental conditions to come up with their correlations, and hence the majority of the models are limited to their operating conditions (i.e., they cannot be accounted as universal models).

Table 5. Developments of nanofluids effective viscosity formulas.

Developer/s	Year	Formula	Dependent Parameter	Limitations
Einstein [271]	1906	$\mu_{eff} = \mu_{bf}(1 + 2.5f_V)$	f_V	Suited for suspensions of <0.02 vol. % and spherical shaped particles.
Hatschek [272]	1913	$\mu_{eff} = \mu_{bf}(1 + 2.5f_V)$	f_V	Designed for suspensions with up to 40 vol. % of spherical particles but does not account for the size of the dispersed particle. The formula also showed very large deviation from the actual viscosity value.
Saitô [273]	1950	$\mu_{eff} = \mu_{bf}\left(1 + \frac{1.25f_V}{1-\frac{f_V}{0.87}}\right)$	f_V	Preferred for dispersions of small spherical particles and is affected by the Brownian motion of the particles.
Mooney [274]	1951	$\mu_{eff} = \mu_{bf}\exp\left(\frac{2.5f_V}{1-CFf_V}\right)$; where CF is the self-crowding factor.	f_V, and CF	This is an extended version of the Einstein's [271] formula that can be used for suspensions of spherical particles with any concentration. The downside is that the modeled suspension needs to meet the functional equation so that the μ_{eff} can be independent of the stepwise sequence of adding further particles concentrations.
Brinkman [275]	1952	$\mu_{eff} = \mu_{bf}(1 - f_V)^{-2.5}$	f_V	Enhanced form of the previous Einstein [271] formula, where it can be used for particles concentrations of up to 4 vol. %.
Roscoe [276]	1952	$\mu_{eff} = \mu_{bf}(1 - Sf_V)^{S'}$; where S is a constant that is equal to 1 (for very diverse particles sizes), -2.5 (for similar particles sizes and <0.05 vol. %), and 1.35 (for higher vol. %); and S' is a constant that is equal to -2.5 (for the very diverse particles sizes case and the >0.05 vol. % suspension) and 1 (for the <0.05 vol. % of similar sized particles).	f_V	Can be used with any dispersion concentration but the particles need to be of spherical shape.
Maron and Pierce [277]	1956	$\mu_{eff} = \mu_{bf}\left(1 - \frac{f_V}{f_P}\right)^{-2}$; where f_P is the packing fraction of the particles.	f_V, and f_P	Suitable for suspensions of small spherical particles and of similar sizes.

Table 5. *Cont.*

Developer/s	Year	Formula	Dependent Parameter	Limitations
Krieger and Dougherty [278]	1959	$\mu_{eff} = \mu_{bf}\left(1 - \frac{f_V}{f_P}\right)^{-2.5\,f_P}$;	f_V, and f_P	For dispersed spherical particles of ≤0.2 vol. %, but the model does not account for the particle's interfacial layers and their aggregation.
Frankel and Acrivos [279]	1967	$\mu_{eff} = \mu_{bf}\left(\frac{9}{8}\right)\left[\frac{\left(\frac{f_V}{f_m}\right)^{\frac{1}{3}}}{1-\left(\frac{f_V}{f_m}\right)^{\frac{1}{3}}}\right]$; where f_m is the maximum attainable concentration.	f_V	Employed for uniform spherical particles and assumes that the rise in viscosity with the increase in particles concentration is due to their hydrodynamic interactions.
Nielson [280]	1970	$\mu_{eff} = \mu_{bf}\,\exp\left(\frac{f_V}{1-f_P}\right)$	f_V, and f_P	This is a modified form of the Einstein's [271] formula but it lacks accurate suspension viscosity prediction.
Brenner and Condiff [281]	1974	$\mu_{eff}= \mu_{bf}\left[1 + f_V\left(2 + \frac{0.312s}{\ln 2s - 1.5} - \frac{0.5}{\ln 2s - 1.5} - \frac{1.872}{s}\right)\right]$; where s is the axis aspect ratio of the dispersed particle.	f_V, aspect ratio, and shear rate	Shows good prediction capability for dispersed particles of rod shape but less effective for other shapes.
Jeffrey and Acrivos [282]	1976	$\mu_{eff} = \mu_{bf}\left[3 + \frac{4}{3}\left(\frac{s^2 f_V}{\ln\frac{\pi}{f_V}}\right)\right]$	f_V, and aspect ratio	Designed for suspensions of rod-shaped particles.
Batchelor [270]	1977	$\mu_{eff} = \mu_{bf}(1 + 2.5 f_V + 6.2\, f_V^2)$	f_V, and Brownian motion	The model includes the interaction between the particles but fails to provide good prediction agreement.
Graham [283]	1981	$\mu_{eff}= \mu_{bf}\left(\frac{9}{4}\right)\left[1 + \left(\frac{h}{d_{np}}\right)\right]^{-1}\left[\frac{1}{\left(\frac{h}{0.5\,d_{np}}\right)} - \frac{1}{1+\left(\frac{h}{0.5\,d_{np}}\right)} - \frac{1}{\left[1+\left(\frac{h}{0.5\,d_{np}}\right)\right]^2}\right] + \left[1+\left(\frac{5}{2}\right) f_V\right]$; where h is the minimum separation distance between the surface of two spherical particles.	f_V, d_{np}, and h	Suitable for spherical particles only and has good prediction agreement with Einstein [271] formula when very low particles concentrations are used or when μ_{eff} is very close to that of μ_{bf}.
Kitano et al. [284]	1981	$\mu_{eff} = \mu_{bf}\left(1 - \frac{f_V}{f_P}\right)^{-2}$	f_V, and f_P	Similar to the Maron and Pierce [277] formula but the f_P value is preliminarily defined numerically and is better suited for two phase mixtures.
Bicerano et al. [285]	1999	$\mu_{eff} = \mu_{bf}(1 + [\eta]\, f_V + k_H\, f_V^2)$; where $[\eta]$ is the intrinsic viscosity, and k_H is the Huggins coefficient.	f_V, $[\eta]$, and k_H	More determined towards analyzing the relation between particles concentration and μ_{eff}.
Wang et al. [286]	1999	$\mu_{eff} = \mu_{bf}(1 + 7.3\, f_V + 123\, f_V^2)$	f_V	Simple model that was formed from a set of experimental results obtained from modifying the suspension particles size and concentration.

Table 5. *Cont.*

Developer/s	Year	Formula	Dependent Parameter	Limitations
Masoumi et al. [248]	2009	$\mu_{eff} = \mu_{bf} + \frac{\rho_{np}}{72\delta\,Fun.}\left(\frac{1}{2r_{np}}\sqrt{\frac{18k_B T}{2\pi\rho_{np}r_{np}}}\right)\left(2{r_{np}}^2\right)$; - $\delta = 2r_{np}\left(\sqrt[3]{\frac{\pi}{6\,f_V}}\right)$; where δ is the distance between the particles, and $Fun.$ is a correction function.	f_V, T, ρ_{np}, particle size, and Brownian motion	The formula is bound by the experimental conditions that were used in its development.
Chevalier et al. [250]	2009	$\mu_{eff} = \mu_{bf}\left[1 - \frac{f_V}{f_P}\left(\frac{D_a}{2r_{np}}\right)^{3-d_f}\right]^{-2}$; where D_a is the average diameter of the aggregates, and d_f is the fractal dimension, which depends on the shape of the dispersed particles, the type of agglomeration, and the shear flow. f_P and D_a are usually set to 0.65, for random packing of spheres, and 1.8, respectively.	f_V, f_P, r_{np}, and d_f	This model depends on the agglomerate size, and thus it is not optimum for determining the μ_{eff} for stabile suspensions.
Chandrasekar et al. [190]	2010	$\mu_{eff} = 1 - Coef._1\left(\frac{f_V}{1-f_V}\right)^{Coef._2}$; where $Coef._1$ and $Coef._2$ are regression coefficients that can be obtained from preliminary experimental results.	Specific area, ρ_{np}, ρ_{nf}, and sphericity of the particles	Depends on preliminary experimental results to set-up the unknown coefficients.
Bobbo et al. [287]	2012	$\mu_{eff} = \mu_{bf}\left(1 + Coef._1\,f_V + Coef._2\,{f_V}^2\right)$	f_V, and r_{np}	Developed for single-walled carbon nanohorn (SWCNH) and TiO_2 nanofluids based on the Batchelor formula and experimental measurements of the μ_{eff} at a range of temperatures from 283.2 to 353.2 K, and concentrations from 0.01 to 1 wt %.
Esfe et al. [288]	2014	$\mu_{eff} = \mu_{bf}(1.1296 + 38.158\,f_V - 0.0017357\,T)$	f_V, and T	Limited for water based MWCNTs nanofluids of 0–1 vol. %.
Aberoumand et al. [289]	2016	$\mu_{eff} = \mu_{bf}(1.15 + 1.061\,f_V - 0.5442\,{f_V}^2 + 0.1181\,{f_V}^3)$	f_V	Used for low temperature oil based suspensions.
Akbari et al. [290]	2017	$\mu_{eff} = \mu_{bf}(-24.81 + 3.23\,T^{0.08014}\exp(1.838\,{f_V}^{0.002334}) - 0.0006779\,T^2 + 0.024{f_V}^3)$	f_V, and T	Suitable for nanofluids of <3 vol. % and of temperature $\leq$50 °C.
Esfe et al. [291]	2019	$\mu_{eff} = 6.35 + 2.56\,f_V - 0.24\,T - 0.068\,Tf_V + 0.905\,{f_V}^2 + 0.0027\,T^2$	f_V, and T	Suitable for MWCNTs and TiO_2 hybrid nanofluids of f_V between 0.05 and 0.85 vol. %.

Table 5. *Cont.*

Developer/s	Year	Formula	Dependent Parameter	Limitations
Ansón-Casaos et al. [292]	2020	$\mu_{eff} = \mu_{bf}\left(1 - \frac{\chi}{2} f_V\right)^{-2}$; where χ is equal to 2.5 for spherical particles or can be replaced by a function, $f(r_{np})$, to determine the suspension property containing 1D and 2D dispersed solids.	f_V, and χ	Suitable for SWCNTs and graphene oxide.
Ilyas et al. [154]	2020	$\mu_{eff} = \mu_{bf}\ \exp\left(\frac{FP_{\cdot 1}}{T - FP_{\cdot 2}}\right) + FP_{\cdot 3} f_V \exp\left(\frac{FP_{\cdot 4}}{T}\right) - FP_{\cdot 5} {f_V}^2$; where $FP_{\cdot 1}$, $FP_{\cdot 2}$, and $FP_{\cdot 4}$ are the temperature fitting parameters in Kelvin, whereas $FP_{\cdot 3}$ and $FP_{\cdot 5}$ are the dynamic viscosity fitting parameters in Pa.s. The values of these parameters (i.e., $FP_{\cdot 1}$ to $FP_{\cdot 5}$) can be found in the published source.	f_V, FP and T	Suitable for ND dispersed in thermal oil and is valid for the range of $0 \leq f_V \leq 1$ and 298.65 $\leq$ T (K) $\leq$ 338.15.

6. Thermal Applications

The previous sections showed how dispersing carbon-based nanomaterials in conventional working fluids could positively affect these liquids properties, especially when it comes to their overall thermal conductivity. On the contrary, this section concentrates on utilizing carbon-based suspensions in three heat and mass transfer systems widely used in the energy sector, namely, PTSCs, nuclear reactors systems, and AC&R systems. This is because the previous attempts that many researchers undertook to enhance the performance of these systems were mainly through design modifications, such as adding turbulators, geometric and construction materials variations, and surface alterations. However, these techniques have reached a point where limited enhancements can be accomplished. Therefore, to break these boundaries for further progress, some scientists have proposed exchanging the working fluids of these thermal applications with nanofluids [293]. This is because employing a working fluid that possesses higher thermal conductivity would eventually improve the heat transfer rate in these systems, as will be demonstrated next.

6.1. Parabolic Trough Solar Collectors

A PTSC is part of the existing energy solar systems that utilizes solar radiation (usually emitted from the sun) to generate thermal energy with high efficiency [294]. This happens when reflecting concentrated incident sunlight from its reflector surface, which consists of a parabolically curved mirror polished metal, to a focal line where the receiver or absorber tube containing the working fluid is located. The lower temperature heat transfer fluid, which is usually water or oil, then absorbs the solar heat flux from the attached inner tube surface, and thus causes its temperature to raise. Figure 19 shows an example of a real life PTSC system and its working mechanism in a schematic diagram. Based on the system configuration and the application used, the working fluid temperature in a PTSC can exceed 500 °C at concentrated solar power plants (CSPP), for steam power cycles; or can be lower than 100 °C, for industrial process heat (IPH) applications, such as domestic and industrial water heating [295]. Examples of low temperature requirements (i.e., temperature starting from ≤100 °C) for different industrial processes are shown in Table 6 [296]. Most of the modern designs of PTSC contain a sunlight tracking system that helps improve the efficiency of these systems [297].

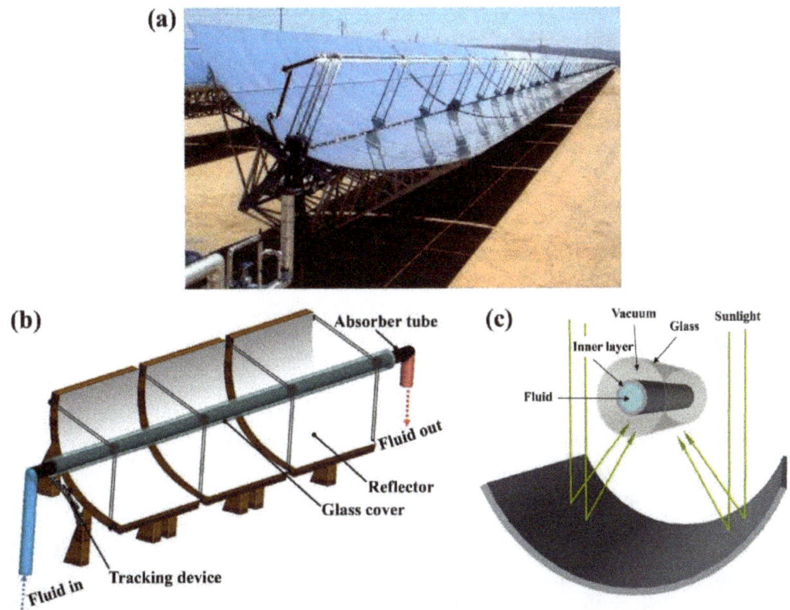

Figure 19. Example of a parabolic trough solar collector system, where (**a**) shows the physical device, (**b**) illustrates its schematic diagram, and (**c**) demonstrates the reflection mechanism of solar radiation on the absorber tube [298].

Table 6. Examples of different industrial processes that utilize parabolic trough solar collector systems and their temperature requirements.

Industry	Process	Required Temperature Range (°C)
Dairy	Boiler feed water	60–90
Agricultural products	Drying	80–200
Textile	Drying	100–130
Chemistry	Petroleum	100–150
Desalinization	Heat transfer fluid	100–250

Since the primary goals in industrial applications are to reduce the processing time, increase the lifetime of the equipment, and decrease the amount of energy consumption, using PTSC systems, these goals can be fulfilled through improving the rate of heat transfer between the absorber tube and the working fluid. One way of achieving this is by utilizing nanofluids as the heat transfer fluid in the PTSC system [299,300]. This is because, as mentioned earlier, nanofluids have higher thermal conductivity than any known conventional heat transfer fluid, which makes them potential candidates for the future of such heat transfer applications. When using carbon-based particles (e.g., MWCNTs, graphene, or NDs), the effective thermal conductivity significantly increases along with the rate of thermal diffusion and effective viscosity of the suspension. Subsequently, this causes the fluid heat capacity, Reynold's number (Re), and Prandtl number (Pr) to decrease. In the case of turbulent flow, the Nusselt number (Nu) depends on both Re and Pr. Thus, a decrease in the two aforementioned parameters would result in fewer or smaller eddy formations within the fluid, and hence the level of turbulently in the flow would reduce. Furthermore, since the effective viscosity of a nanofluid is higher than its base fluid, the pressure drop that will be experienced from using such category of fluids in a PTSC system would be higher than that of the conventional base fluids. To overcome this issue, the PTSC system should take into account the thermophysical properties of the suspension used at its design phase. One important thing to consider is that when using nanofluids, as the PTSC working fluid, the absorber tube needs to be transparent so that the dispersed particles can directly absorb the sunlight throughout their cycle [301].

Although the previous facts showed how promising nanofluids could be when used in PTSC systems, the scientific field is still scarce with the amount of published works that investigate carbon-based nanofluids in such system. Most of the work covered on nanofluids were those involving nanoparticles of Al_2O_3, CuO, TiO_2, Fe_2O_3, SiO_2, Cu, SiC, Fe_3O_4, and limited other literature were found for CNTs, MWCNTs, and SWCNTs [298]. For instance, Kasaeian et al. [302] explored the overall efficiency enhancement of a pilot PTSC system using MWCNTs–mineral oil suspensions of 0.1 wt % and 0.3 wt %. The researchers found that the 0.1 wt % and 0.3 wt % dispersions had improved the system efficiency by 4–5% and 5–7%, respectively, compared to conventional base fluid (i.e., mineral oil). Furthermore, Kasaeian et al. [303] studied the effect of 0.1, 0.2, and 0.3 vol. % of MWCNTs dispersed in EG, as the working fluid, for a direct absorber solar collector attached to a parabolic trough. They found that the optical efficiency reached up to 71.4%, due to the 0.3 vol. % of MWCNTs particles employed in their heat transfer fluid. In addition, the thermal efficiency of their system was found to be 17% higher, when using the 0.3 vol. % nanofluid, than that obtained from pure EG. Moreover, Mwesigye et al. [304] coupled a Monte Carlo ray tracing (MCRT) optical model along with a computational fluid dynamics (CFD) finite volume method (FVM)-based model to analyze a PTSC, hosting a SWCNTs–Therminon VP-1 suspension, thermal performance. The authors found that raising the particles concentration from 0 to 2.5 vol. % caused the entropy generation to reduce by 70%, with the heat transfer rate to increase by 234%, and the thermal efficiency of the system to improve by 4.4%. In addition, Dugaria et al. [305] designed and modeled the optical efficiency of a direct absorber solar collector (DASC) that is connected to a parabolic trough system. In their experiment, they used 0.006, 0.01, 0.02, and 0.05 g/L of SWCNTs to fabricate their aqueous nanofluids. Their results showed that increasing the nanoparticles concentration to more than 0.05 g/L would cause the thermal efficiency to reduce due to the thermal radiation being mostly contained in the surrounding area between the absorber tube inner surface and the nanofluid. In addition, using nanofluids made of 0.05 g/L SWCNTs caused the thermal efficiency of the system, including the optical losses of the concentrating trough, to reach 90.6% at a reduced temperature range (T_m^*) = 0 K·m^2/W and 77.2% at T_m^* = 0.128 K·m^2/W. It is important to note that the thermal efficiency of solar collectors is usually shown in a graph as a function of T_m^*, which is defined for the case of nanofluids as:

$$T_m^* = \frac{\left(T_{m_{nf}} - T_{amb}\right)}{DNI} \quad (12)$$

where $T_{m_{nf}}$, T_{amb}, and DNI are the mean temperature of the nanofluid, ambient air temperature, and direct normal irradiance, respectively. One of the main aspects for the enhancement in the thermal performance of the two aforementioned published works [304,305] was due to the fact that CNTs, along with other carbon-based materials, possessed extremely high solar absorption characteristics (i.e., more than 90%) [306]. Despite the research investigations that were covered in this section on carbon-based nanofluids usage in PTSC's, there are only a few other alternatives [298]. To the best of the authors of this article knowledge, there is still a lack of exploration on utilizing ND's and graphene nanofluids for PTSC's. This shows that further investigation is required from the researchers working in the solar energy field; especially since, for example, nanofluids of ND base showed to contain remarkable optical and thermal properties when studied in other similar applications [307].

6.2. Nuclear Reactors

Nuclear power plants are part of the energy network that has been adopted by many countries across the globe, such as France, USA, UK, Russia, Iran, and UAE, among others to support their growth in energy demands [308]. Unlike most energy sources, the power produced from the fission process of the fuel (i.e., enriched uranium or plutonium) within the nuclear reactor can arguably be considered as one of the solutions for solving the problems associated with climate change and the increasing levels of CO_2 emissions in the atmosphere and its feasibility for none or low oil producing countries [309]. Nuclear technology has seen significant developments throughout the years to enhance the efficiency of these systems, reduce their construction size, and improve their safety standards [310,311]. Historically, the first generation of commercial nuclear reactors were inaugurated in the 1950s, whereas today, the newly introduced fourth generation of reactors are currently being either planned or under construction. In terms of the working fluid, these reactors can be classified into three main groups (i.e., water-cooled reactors (WCRs), gas-cooled reactors (GCRs), and molten solid cooled reactors (MSR)) [312]. The WCRs can be subdivided into further categories, namely, boiling water reactors (BWRs), pressurized water reactors (PWRs), and pressurized heavy water reactors (PHWRs). Furthermore, the thermal transport concept of the BWR and both PWR and PHWR is similar in the sense that the working fluid, in all cases, absorbs the thermal energy from the fuel when it undergoes an excited state. However, the main difference is that PWR and PHWR use pressurizing systems to maintain the working fluid in its liquid phase, and therefore must be separated from the electrical generating cycle for contamination safety concerns. On the other hand, the working fluid in the BWR is boiled to generate steam that is used directly to provide the needed mechanical power to rotate the steam turbine and generate electricity. In addition to being a thermal energy carrier for power generating purposes, the working fluid also takes the role of extracting heat from the nuclear fuel, which is primarily the main concern related to the safe and economic operation and lifespan of the reactor. In some cases where the cooling rate is insufficient or if the control rods fail to operate properly to stabilize or reduce the reaction process, the reactor can experience a loss-of-coolant accident (LOCA) [313]. In such scenarios, the nuclear fuel needs to be rapidly cooled down, using backup water tanks, to avoid a core meltdown crisis and possibly a hydrogen explosion in the chamber. From the aforementioned, one can generalize the modes of heat transfer inside the rector's core based on the driving force of the fluid motion into two main categories; the first is flow boiling, which is a forced convection phenomenon that occurs during normal operating conditions. The second is pool boiling, which is a natural convection heat mechanism that takes place following a reactor LOCA state. Enhancing the heat transfer coefficient (HTC) and critical heat flux (CHF), for flow boiling, or increasing the minimum film boiling temperature (T_{min}) in pool boiling are essential for optimizing these thermal modes outcomes. Whether it comes to improving the energy efficiency or for safety reasons, the aforementioned shows how crucial the role of the working fluid in a nuclear reactor system. Therefore, utilizing working fluids of enhanced thermophysical properties, such as nanofluids, can help in further advancements in the field of nuclear power plants, especially in WCR systems, if properly handled and understood its role in both nuclear flow boiling and pool boiling [314]. This section demonstrates some of the available studies on nanofluids for both thermal modes (i.e., flow and pool boiling), but focuses more on the pool boiling mode due to its important role in designing an emergency core cooling system.

6.2.1. Nanofluids Influence on Flow Boiling

During the normal operating condition, the thermal efficiency of a reactor system depends mainly on the flow boiling of the working fluid, where the working fluid is forced to flow by means of a pump and buoyancy effects. Flow boiling consists of several flow regimes [315] such as liquid single-phase flow, bubbly flow, slug flow, annular flow, mist flow, and vapor single-phase flow, as shown in Figure 20. The existence of each regime is affected by several factors such as the type of working fluid, surface orientation, degrees of liquid subcooling, system pressure, wall temperature, mass flux, surface microstructure (including porosity), wettability, oxidation, and surface roughness [316,317]. Moreover, this mode of thermal transport can remarkably improve the energy efficiency of the system when the HTC and CHF of the working fluid are enhanced [318,319]. Researchers through their intensive studies, have shown that a passive and a safe approach to achieve higher HTC and CHF can be accomplished using nanofluids in the system [320]. The level of enhancement that accompanies the utilization of nanofluids over their other conventional counterparts is tuned by altering nanoparticle morphology (i.e., shape and size), thermophysical properties and different flow parameters (i.e., mass flow rate, channel size, flow direction, and flow regime) [321,322]. Published reports described that nanofluids of enhanced thermal conductivities also showed better convective flow performance [319,323]. Carbon allotropes based nanofluids have higher thermal conductivities compared to the metallic and oxide based nanofluids [324–326]. Therefore, nuclear scientists have already started to investigate the flow boiling performance of various carbon-based nanofluids.

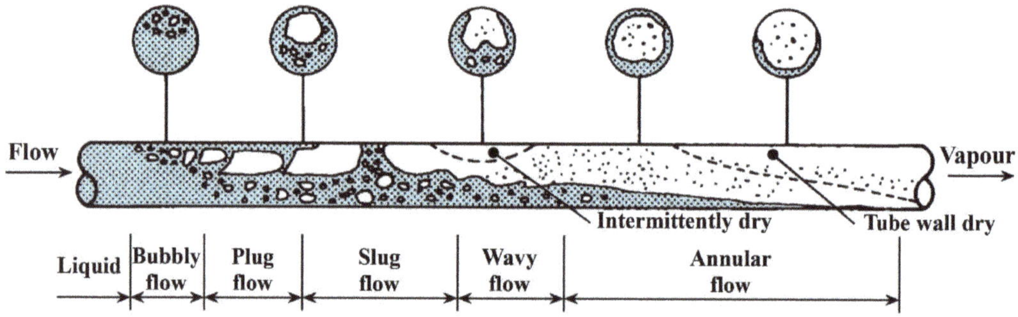

Figure 20. Flow boiling regimes inside a horizontal tube from liquid to vapor phases [327].

Some of the work done on carbon-based nanofluids in flow boiling includes the research conducted by Hashemi and Noie [328]. They experimentally used MWCNTs–water suspensions produced through the two-step fabrication method and stabilized by adding the AG surfactant, of the 1:1 surfactant to the nanomaterial ratio, to the mixture. The MWCNT used had a 10–20 nm outer diameter and 30 µm length, and base fluid concentrations between 0.0005 and 0.005 vol. %. Furthermore, the stability of their dispersions was determined through the zeta potential method, and the testing section consisted of a horizontal stainless steel (SS) tube of 10 mm (diameter) and 200 cm (length). Their findings showed that the HTC of the as-prepared nanofluids was significantly higher than the base fluid, and that the enhancement in the HTC for both types of fluids was influenced by the changes in heat and mass fluxes. In addition, the CHF in the nanofluids case showed a maximum improvement over pure water by approximately 4.3%, when using the 0.005 vol. % suspensions. The same research group [329] also presented the feasibility of AG stabilized MWCTs–water nanofluids at various concentrations (0.001, 0.005, and 0.01 wt %) for flow boiling inside a 2 m long tube placed horizontally under atmospheric condition. The zeta potential analysis of test samples confirmed the well dispersion of nanoparticles in base fluid. Their report describes that CHF of nanoparticles increased significantly due to particle inclusion as well as the increase in mass flux. The CHF enhancement was observed better than water due to nanoparticle deposition and enhancement in wettability of the heating surface. Another flow boiling study under forced convective and nucleate boiling regions using CNTs nanofluid was conducted by Sarafraz and Abad [323]. Their work was performed using statistical, regression and experimental analyses for commercial heat transfer oil based CNTs nanofluids. The nanofluids were prepared by dispersing 0.1 wt % and 0.3 wt % of the dry powder in therminol 66 for a total duration of 25 min, using a two-step procedure (magnetic stirring 15 min then sonication 10 min). The stability of the suspensions was achieved by adding nonylphenol ethoxylates surfactant of 0.1 vol. % to the mixture matrix, after which the zeta potential of the prepared samples was measured. The results showed that the stability of the prepared nanofluids was maintained for 5 days. Additionally, the presence of carbon nanotube within the oil increased the convective

HTC together with a substantial augmentation in the nucleate boiling HTC. The degree of subcooling also increased the HTC in both boiling regions due to enhancements in thermophysical properties of nanofluids. Furthermore, the thermal the thermo-hydraulic performance of the system enhanced by 56% using the carbon nanotube nanofluid compared to the same with pure oil. Sarafraz and Hormozi [330] experimentally investigated flow boiling heat transfer in MWCNT, alumina, and copper oxide flow. Different techniques were implemented for stabilizing based on the type of the dispersed nanomaterial. The results showed that MWCNT–water nanofluids had higher thermal conductivity and boiling thermal performance compared to the other suspensions. It was also found that, as the heat and mass fluxes and the concentration of nanofluids increased, the boiling performance of MWCNT nanofluids intensified. However, the HTC was deteriorated in force convection and nucleate boiling regimes due to the deposition of the nanoparticle on the surface. As a result, the surface roughness decreased over the time since it is affected by the size of nanoparticles, thickness of deposited layer and size of microcavities. This was confirmed by the minimal amount of bubble generation due to reduction in nucleation sites and surface roughness. The researchers finally concluded that the MWCNT–water nanofluids outperformed the other candidates for utilization in cooling applications.

Other studies have found that addition of graphene oxide (GO) into base fluid improve the CHF in flow boiling due to its hydrophilicity feature [331,332]. Lee et al. [331] examined 0.01 vol. % of GO–water nanofluid in round tubes with a length of 0.5 m and an inner diameter of 0.01041 m at two inlet temperatures (25 and 50 °C) and four different mass fluxes (100, 150, 200, and 250 $kg/m^2{\cdot}s$) for low pressure and low flow scenarios. Comparing to other oxide nanofluids from the literature [333], the research group showed maximum CHF enhancements at mass flux of 250 $kg/m^2{\cdot}s$ increased up to 100% and 72% at fixed temperatures of 25 °C and 50 °C, respectively. This significant improvement was due to the liquid film's wettability enhancement caused by the deposition of GO nanoparticles. However, Park and Bang [332] reported limited improvement in CHF of up to 20% when testing GO–water nanofluid in advanced light water reactors (ALWRs) at 50 and 100 $kg/m^2.s$ and subcooling condition of 10 K compared to distilled water. The results showed that GO deposited on the heated surface and changed phase to reduced GO (RGO) during nucleate flow boiling, which might constrain the thermal activity improvement. Zhang et al. [334] examined the deposition of GO in water nanofluids over heating surface with nanoparticle concentration ranging from 0 to 0.05 wt %. They reported that the increase in GO concentration depreciated the heat transfer performances (CHF and HTC) up to 100% and 73% (at 0.05 wt % with 40 mL/min), respectively. In another study, Mohammed [335] varied graphene particle concentration from 0 to 0.5 vol. % in zinc bromide and acetone solution (acetone–$ZnBr_2$).The CHF and HTC on the heated surface increased with GO concentration by up to approximately 52% and 58%, respectively. However, the increase in particle concentration involved a decrease in pressure drop up to 11% approximately.

In terms of nanofluids containing NDs, there were a limited amount of literature covering this topic [336–338]. For instance, DolatiAsl et al. [338] proposed a mathematical model using Kim et al. [336] results to estimate the different parameters that were affecting the CHF when utilizing ND suspensions. Their correlation only required the properties of the nanomaterial and the vol. % employed to predict the CHF. Furthermore, the numerical findings showed that the most effecting parameters on the CHF were the length of the tube (decreasing) and the mass flux (increasing), whereas the particles concentration and thermal conductivity had the lowest influence. They extended the work by developing another correlation that contains the particles property data, and thus only required an input of the type of the particles (i.e., NDs) and the vol. % to perform the prediction. The new model predicted the actual values with a mean absolute error of 9.8% for CHF ranging from 500 to 2500 kW/m^2.

6.2.2. Dispersions Effect on Pool Boiling

Pool boiling heat transfer plays a crucial role in numerous technological and industrial applications. It consists of four different regimes [339]: natural convection (single-phase), nucleate boiling, transition boiling, and film boiling as shown in Figure 21. When a surface is sufficiently heated and plunged in a water pool, film boiling regime occurs where the heated surface is physically separated from the coolant by a stable vapor blanket. This region is denoted by the film boiling regime. In this regime, heat transfers by conduction and radiation only leading to a gradual decrease in the surface temperature. The performance of the pool boiling is evaluated by the enhancement in CHF, HTC, T_{min}, and vapor film thickness. In literature, the effects of the following parameters have been studied: substrate material [340], surface conditions and oxidation [341,342], system pressure [343,344], initial wall temperature [345], shape and dimension of the testing specimen [346,347], degrees of liquid subcooling [348–350], surface wettability and vapor–liquid contact angle [351], surface roughness and wickability [352], and type of quenchant such as water, oil, or nanofluids [353,354]. Recently, researchers have been focused on the effects of the later parameter on pool boiling heat transfer performance.

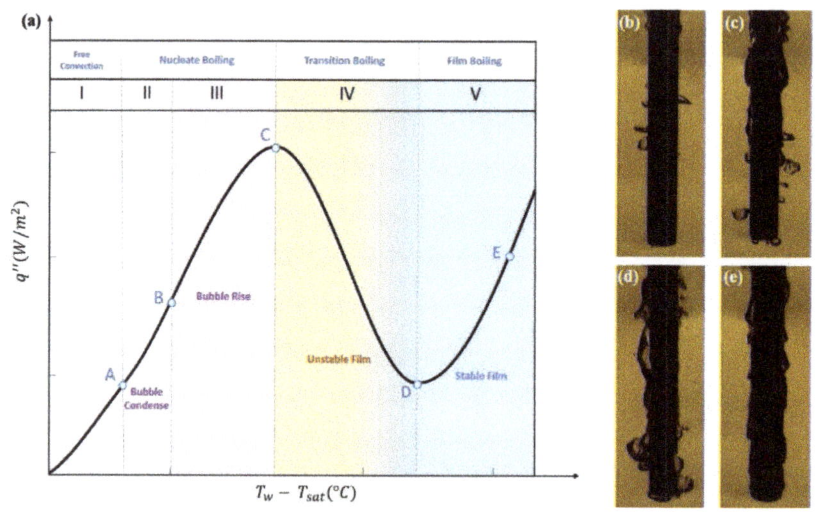

Figure 21. Boiling curve for stagnant water at atmospheric pressure (1 atm), where (**a**) shows the boiling curve and (**b–e**) illustrates the bubble formation within the free convection, nucleate boiling, transition boiling, and film boiling regimes.

Review papers have intensively presented research studies that cover the preparation methods of various nanofluids and test their effect on CHF and HTC [321,355–359]. It was mentioned that carbon-based nanomaterials such as graphene dispersed in water enhanced the heat transfer as compared to any other nanoparticle [355]. Nevertheless, there are many reasons that contribute to the CHF enhancement such as surface roughness [360], deposition of nanoparticles [361], concentration of nanoparticle beyond a certain limit [362,363], increase in surface wettability [364], and capillary effect [365]. Some researchers have explained that the enhancement was a result of the occurrence of cavities on the surface due to the deposition of the dispersed nanomaterial on the surface, especially on surfaces of rough structure. Others mentioned that the increase in surface area of the formed porous layer because the nanoparticle accumulation enhanced heat transfer by disturbing the flow. Active nucleation sites decreased with nanoparticle layers, which significantly increase surface wettability, and therefore enhance CHF. As for HTC enhancement, thermophysical properties has a major role on it. It has been observed that nanoparticle affects the thermal conductivity and surface tension of the quenchant whereas viscosity, density, and heat capacity remain nearly constant [366]. A vapor blanket layer formed during film boiling on the heating surface, which significantly affect the conduction heat transfer, and hence the HTC. Yang and Liu [367] found that as the surface tension decreased, the HTC increase due to the reduction in the formation of bubbles and the active nucleation sites. The effect of the modified surface topologies such as surface roughness was observed to enhance HTC [133,368]. Furthermore, the high particle concentration led to nanoparticle deposition on the surface and thus porous surfaces occur. Depending upon either original surface condition or size of nanoparticle, the surface roughness can be increased [360] or decreased [369]. Generally, HTC was found to be maximum when carbon nanoparticles were used for boiling [355]. Since the current work focuses on the effect of carbon-based nanofluids (i.e., graphene, ND, and CNT) on heat transfer application, various selected studies on CHF and HTC in pool boiling heat transfer have been listed in Tables 7 and 8, respectively.

Table 7. Summary of selected CHF enhancement for various pool boiling studies in water base.

Ref.	Nanofluid	Concentration/Particle Size	Heating Surface	CHF Enhancement%
[370]	CNT	0.1–0.3 wt %	–	Enhanced
[371]	CNT	0.01–0.05 vol. %	Cu block	38.2
[372]	CNT	0.5–4 wt %	Cu plate	60–130
[361]	CNT	0.05 vol. %	SS foil	108 122
[373]	CNT	1.0 vol. %	SS tube	29
[374]	MWCNT	0.01–0.02 wt %	SS cylinder	Enhanced
[371]	MWCNT	0.0001–0.05 vol. %	Cu block	200
[375]	MWCNT	0.1–0.3 wt %	Microfin Cu disk	95
[372]	f-MWCNT	0.5–4 wt %	Cu plate	200
[162]	f-MWCNT	0.25–1 wt %	SS tube	37.5
[376]	f-MWCNT	0.01–0.1 wt %	Cu disk	271.9
[377]	f-MWCNT	0.01 vol. %	Cu block	98.2
[378]	f-SWCNT	2.0 wt	Ni-Cr wire	300
[379]	GO	$\leq$0.001 wt %	Copper plate	Enhanced
[380]	GO	0.0005 wt %	Ni-Cr wire	320
[381]	GO	0.001 vol. %	–	179
[382]	GO	0.0001, 0.0005, 0.0010, and 0.005 wt %	Ni wire	Enhanced
[383]	GO	0.01 vol. %	Ni-Cr wire	–
[366]	ND	1 g/L	Cu plate	Enhanced
[366]	ND	<1 g/L	Cu plate	Deterioration
[384]	ND	0.01–0.1 vol. %	SS plate	Unchanged
[384]	ND	0.01 vol. %	SS plate	11

Note: f-SWCNT, and f-MWCNT refer to functionalized SWCNT, and functionalized MWCNT, respectively.

Table 8. Summary of selected HTC enhancement for various pool boiling studies in water base.

Ref.	Nanofluid	Concentration	Heating Surface	CHF Enhancement%
[373]	MWCNT	1.0 vol. %	Cu block	28.7
[371]	MWCNT	0.0001–0.05 vol. %	Cu block	38.2
[385]	MWCNT	0.25%, 0.5%, and 1.0 vol. %	Ni-Cr wire	320
[375]	MWCNT	0.1–0.3 wt %	Microfin Cu disk	77
[372]	f-MWCNT	0.5–4 wt %	Cu plate	130
[162]	f-MWCNT	0.25–1 wt %	SS tube	66
[376]	f-MWCNT	0.01–0.1 wt %	Cu disk	38.5
[377]	f-MWCNT	0.01 vol. %	Cu block	10.15
[386]	Graphene	0.1 and 0.3 wt %	Cu	96

The third important parameter in pool boiling is T_{min} to be considered in the reactor core under the extreme environment and severe accidents such as LOCA. Understanding this parameter can lead to an improved nuclear cladding performance that provides more efficient and safer future nuclear reactors. Physically, T_{min} is defined as the boundary between film boiling and transition boiling, beyond which temperature liquid loses physical contact with the solid surface and the heat transfer significantly reduces. Fewer studies for the quenching behavior of nanofluids have been conducted in the literature that was focused on T_{min}.

A study by Gerardi [387] showed that T_{min} of a quenched indium-tin-oxide (ITO) rod in a nanofluid pool (ND–water) was found to increase by 30 °C compared to the water pool. Another experimental investigation by Fan et al. [388] was performed in aqueous nanofluids in the presence of four CNTs having various lengths and diameters. It was concluded that the accelerated quenching was clearly related to the enhancement in boiling heat transfer. An increase in T_{min} was exhibited for all cases. The modified quenching and boiling behaviors were elucidated by the accumulative changes in surface properties due to the deposition of CNTs. Given the nearly unvaried contact angles, the consistently increased surface roughness and the formation of porous structure seem to be responsible for quenching and boiling enhancement. In order to achieve better performance, the use of longer and thicker CNTs tends to form a highly porous layer, even upon consecutive quenching, which may induce rewetting by the entrapped liquid in the pores and serve as vapor ventilation channels as well. In another experimental study, Fan et al. [389] examined transient pool boiling heat transfer in aqueous GO nanofluids. They tested various dilute concentrations of the nanofluids up to 0.1 wt %. It was shown that the quenching processes could be accelerated using GO nanofluids as compared to pure water. The boiling behavior during quenching was analyzed in relation to the modified surface properties of the quenched surfaces. T_{min} values were found to increase with raising the concentration of GOs compared to the baseline case of pure water. The results suggested that surface property changes due to the deposition of GOs were responsible for the modified boiling behavior of the nanofluids. In addition, the surface wettability was a nondominant factor in most cases. The surface effects of the deposited layer of GOs were strongly dependent on the material properties, finish, and treatment of the original surfaces. Kim et al. [390] quenched metal spheres made from SS and zircaloy in water-based nanofluid containing low concentration (less than 0.1 vol. %) of ND. They showed that film boiling heat transfer in nanofluids was almost identical to that in pure water. However, subsequent quenches proceeded faster due to the gradual accumulation of nanoparticle deposition on the sphere tended to destabilize the vapor film but, T_{min} remained unchanged. A summary of the previous research studies is listed in Table 9.

Table 9. Summary of selected T_{min} enhancement for various pool boiling studies in water based nanofluids.

Ref.	Nanomaterial(s)	Heating Surface	$T_{min,water}$ (°C)	$T_{min,nanofluid}$ (°C)
[387]	ND (0.01 vol. %)	ITO	230	260
[388]	CNT-1 CNT-2 CNT-3 CNT-4 (0.5 wt.%)	316L SS sphere	215 218 218 219	241, 294, 303, 328, 335 211, 229, 277, 281, 287 228, 246, 254, 262, 264 231, 238, 243, 254, 256

Table 9. *Cont.*

Ref.	Nanomaterial(s)	Heating Surface	$T_{min,water}$ (°C)	$T_{min,nanofluid}$ (°C)
[389]	GO (0.0001 wt %) GO (0.001 wt %) GO (0.005 wt %) GO (0.01 wt %) GO (0.05 wt %) GO (0.1 wt %)	SS sphere	230	236.1 239.6 235.7 235.6 233.1 235.9
[390]	Al_2O_3 SiO_2 ND (0.1 vol. %)	SS	249, 247, 249, 247, 250, 250, 251	244, 343, 345, 394, 348, 399, 389 251, 330, 368, 368, 377, 389, 397 252, 252, 250, 253, 255, 264, 279
	Al_2O_3 SiO_2 ND (0.1 vol. %)	Zr	267, 272, 253, 272, 260, 266, 253	287, 347, 354, 400, 401, 411, 412 282, 323, 362, 372, 415 278, 275, 269, 269, 274, 283, 272

6.3. Air Conditioning and Refrigeration Systems

Air conditioning (AC) is a process used to controls air's thermal and physical properties and then supply it with cooling or heating to an allocated area from its central plant or rooftop units. It also maintains and controls the temperature, humidity, air cleanliness, air movement, and pressure differential in a space within predefined limits so that conditioned space occupants or products enclosed satisfy comfort and health standards [391]. A typical AC or refrigeration system uses a vapor compression cycle to accomplish cooling or heating. The vapor compression cycle consists mainly of a compressor, an evaporator, a condenser, an expansion device, indoor and outdoor fans, and a working fluid. Additionally, secondary heating and cooling loops are implemented to accommodate more extensive systems, as shown in Figure 22. The AC system is potentially used for providing a clean, healthy, and comfortable indoor environment, and saving energy by developing high-efficiency equipment in residential and industrial sectors. However, none of these uses come without associated challenges. The AC&R systems can be operating with a very high temperature lift (different between heat source and heat sink temperatures). For instance, the AC system operates in hot and dry climate countries needs to maintain indoor temperature as low as 18.3 °C (65° F) [392], whereas the refrigeration system needs to run continuously for long hours to sustain freezing chamber temperatures [393]. As a result, the AC&R systems generate a tremendous amount of heat loss to the environment during the compression process, which increases the pressure ratio across the compressor and degrades its efficiency; it increases the compressor discharge temperature and jeopardizes its reliability. Simultaneously, the cooling demand is compromised and the AC or the refrigeration system strives to provide enough cooling in the unit's evaporator (or reject heating in the unit's condenser) and therefore degrades the overall system coefficient of performance (COP).

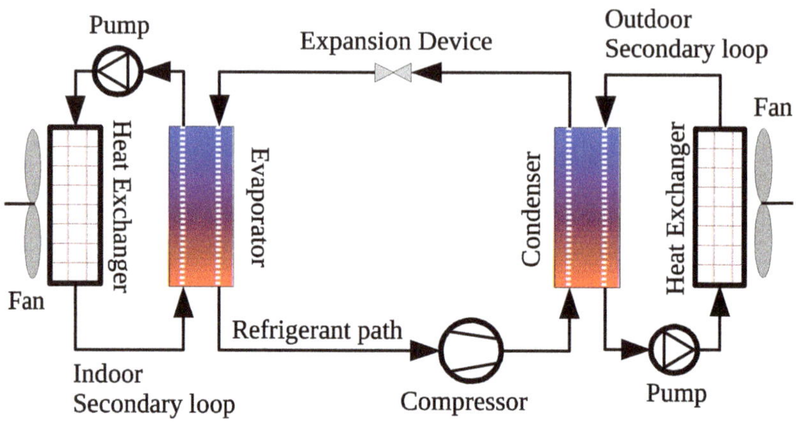

Figure 22. Schematic for a typical vapor compression system with secondary heating and cooling loops.

One of the methods to improve the COP of AC&R systems is to reduce the power consumption in the compressor. Many researchers have already shown that adding nanoparticles to the compressor oil (nanolubricant) reduced its energy consumption because it enhanced the lubricating oil's tribological and thermal properties, which helped improve the compression process, and therefore increased the system COP [394–402]. Lee et al. [403,404] studied the effects of adding nanoparticle to mineral oil. Their results showed that the improvement in the lubricating properties of the mineral oil increases with the addition of the nanoparticle. The authors found that adding the nanoparticle to the compressor oil decreased its friction coefficient by 90%, and thus causing improvement in the compression process and reducing the energy losses in the compressor. Jia et al. [405] investigated the effects of using mineral-based nano-oils in a domestic refrigerator compressor with two different refrigerants, namely, R-134a and R-600. They concluded that the COP values increased by 5.33% when the nano-oil was utilized in the compressor with R-600, whereas no effects were noticed when the same nano-oil was used with R-134a.

Another method to improve the cooling COP is to increase the heat transfer coefficient in the heat exchangers of the AC&R system. Many studies have already shown that mixing nanoparticles with the refrigerant enhanced the heat transfer coefficient of the refrigerant (nanorefrigerant) in the condenser and the evaporator because of the additional nucleate boiling and the higher thermal conductivity of the nanoparticles that enhanced the heat transfer rate, and thus increasing the system COP [286,395,401,406–411]. Since carbon-based nanofluids (i.e., ND, graphene, CNTs, etc.) have better performance due to their superior features compared to other known nanomaterials [17,406,412–418], they could result in significant system performance improvement. Therefore, researchers have further investigated carbon-based nanoparticles for various AC&R applications [419], such as the ones demonstrated in Figure 23. The following sections present a literature review on studies investigating the effect of carbon-based nanoparticles on the thermophysical properties of AC&R refrigerants, followed by a literature review on studies investigating the effect of carbon-based nanoparticle on the AC&R system's performance.

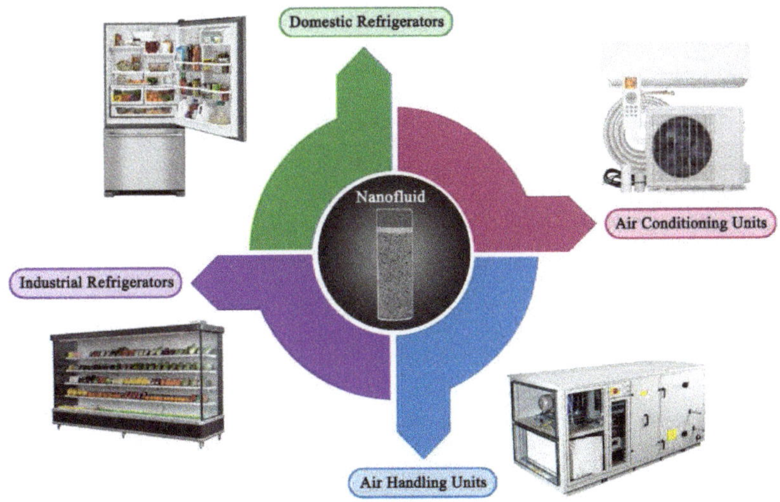

Figure 23. Nanofluids employment in AC&R applications, namely; air conditioning units, air handling units, industrial refrigerators, and domestic refrigerators.

6.3.1. Influence of Carbon-Based Nanoparticles on the Thermophysical Properties of Working Fluid in AC&R Systems

It is evident from the literature that there are many researchers investigated the performance properties (i.e., heat transfer coefficient and viscosity) of nanoparticles applied to the refrigerants in AC&R systems including copper (Cu), aluminum (Al), nickel (Ni), copper oxide (CuO), zinc oxide (ZnO), aluminum oxide (Al_2O_3), titanium oxide (TiO_2), and other metal nanoparticles [325,419–434]. However, only a limited number of research work is available for ND, graphene, and CNTs, which can be summarized in Table 10. Park and Jung [435] investigated the possible contribution of CNT on the nucleate boiling heat transfer coefficients of R-123 and R-134a. They reported an enhancement up to 36.6% in nucleate boiling heat transfer coefficients of the nanorefrigerant at low heat flux compared to the baseline refrigerant. However, as the heat flux increases the enhancement decreased due to robust bubble generation that prevented the CNT from penetrating the thermal boundary layer and touch the surface. The flow boiling heat transfer characteristics and pressure drop were also investigated experimentally by Zhang et al. [436], using MWCNT dispersed in the R-123 refrigerant with SDBS surfactant flowing in a horizontal circular tube heat exchanger. Their results showed that the nanorefrigerant heat transfer coefficient and frictional pressure drop increased with the increase of nanoparticle concentration, mass flux, and vapor quality. Similar conclusions were observed by Sun et al. [437] when they investigated MWCNT with R-141b. Jiang et al. [438] studied the influence of CNT diameters and aspect ratios on CNT–R-113 nanorefrigerant. The study involved four different groups of CNTs with different physical dimensions (diameters, length, and aspect ratio). Their experimental results showed that the thermal conductivities of CNT nanorefrigerant increased proportionally with the increase of CNT's volume fraction and aspect ratio and with the decrease of CNT's diameter. The maximum increase in the thermal conductivity was about 104% for a volume fraction of 1.0 vol. %. Peng et al. [439] studied the influence of CNT physical dimensions such as diameters, length, and aspect ratios for the CNT–R-113–oil mixture. They used the same four different groups of CNTs with different physical dimensions as Jiang et al. [438] and VG68 ester lubricating oil. An enhancement of up to 61% was obtained in the nucleate pool boiling heat transfer coefficient compared to R-311–oil mixture without CNTs. They also showed that the improvement of the nucleate pool boiling heat transfer coefficient increased as the CNTs length increases and as CNTs outside diameter decreases. The heat transfer performance of MWCNT–oil–R-600A nano-refrigerant in horizontal counter-flow double-pipe heat micro-fin tube heat exchanger, was studied by Ahmadpour et al. [440]. Their experiments covered a wide range of parameters, including mass velocity, vapor quality, and condensation pressure. Their results showed that an increase up to 74.8% in the heat transfer coefficient could be achieved with 0.3% nanoparticles concentration at 90 $kg/m^2.s$ mass velocity compared to the pure refrigerant. Kumaresan et al. [441] conducted an experimental study on the convective heat transfer characteristics of secondary refrigerant nanofluids in a tubular heat exchanger. The objective of the secondary refrigerant loop is to reduce the primary refrigerant charge in vapor compression refrigeration systems. The nanofluid used in the study consists of MWCNT dispersed in a water-EG mixture. Their results showed that the maximum enhancement in convective heat transfer coefficient was 160% for the nanofluid containing 0.45 vol. % of MWCNT compared to the base fluid. However, the friction factor was also increased by 8.3 times, which might increase the pumping power and reduce the advantage of the increase in the heat transfer coefficient of the nanofluid [442]. Similar findings were attained by Baskar et al. [443] and Wang et al. [444] when they experimentally tested MWCNT–IPA and graphene–EG in a secondary refrigeration loop, respectively.

The dispersion stability of MWCNT in the R-141b refrigerant with the addition of surfactant was investigated by Lin et al. [445]. Three different types of surfactants, including SDBS, hexadecyl trimethyl ammonium bromide, and nonylphenoxpoly ethanol (NP-10), were tested to prevent the aggregation and sedimentation of MWCNTs during the long-term operation. SDBS was found to have an excellent adsorption ability on the MWCNT surface. It was also shown that the relative concentration increased with decreasing MWCNT length or outer diameter and increasing ultrasonication time. The optimal SDBS concentration for the highest dispersion stability increased proportionally with the increase of the initial MWCNT concentration. However, the SDBS might reduce the nanorefrigerant's thermal conductivity at higher operating temperatures. The thermophysical properties and heat transfer performance of SWCNTs dispersed in the R-134a refrigerant was also investigated by Alawi and Sidik [446]. They found that up to a 43% increase in thermal conductivity can be reached when 5 vol. % of nanoparticle concentration is used in the MWCNT–R-134A nanorefrigerant compared to the pure R-134A refrigerant. Similar to other nanofluids, the thermal conductivity increases with the increase of nanoparticle volume concentrations and with the increase in the temperature of the nano-refrigerant. Moreover, the increase of volume fractions at a constant temperature led to a significant increase in the viscosity and density of the nanorefrigerant.

Dalkilic et al. [447] investigated the stability and viscosity of MWCNTs–polyolester (POE) oil nanolubricants. The study involved using four different refrigeration compressor oil with different values of viscosity (i.e., 32 mm^2/s, 68 mm^2/s, 100 mm^2/s, and 220 mm^2/s) tested at a maximum temperature of 50 °C and a concentration of MWCNTs up to 1 wt %. They reported a substantial augmentation in viscosity up to 90% compared to the viscosity of the base oil. This could reduce the refrigeration efficiency due to the possible increase in the compressor pumping power. Most of the review studies [325,419,420,431,448] have shown that adding nanoparticles always enhances the heat transfer coefficient of the nanofluid mixture due to the higher thermal conductivity of nanorefrigerant and due to the reduction of the thermal boundary layer thickness caused by the presence of nanoparticles. Additionally, nanoparticles increased the viscosity of the nano-refrigerant causing an increase in the frictional pressure drop and therefore might reduce the AC&R system performance. The review studies of references [325,419,420,431,448] covered only CNTs nanomaterial from the carbon family, and therefore further investigations on other types of carbon-based nanoparticles, such as diamonds and graphene, needs to be conducted.

Table 10. List of studies related to carbon-based nanoparticles effect with working fluid in AC&R systems.

Reference	Nanofluid	Test Conditions	Nanoparticle		
			Concentration	Diameter (nm)	Length (μm)
Park and Jung [435]	CNT–R-123 CNT–R-134a	Heat Flux 20–60 kW/m^2	1.0 vol. %	20	1
Zhang et al. [436]	MWCNT–R-123	Heat Flux -	0.02–0.20 vol. %	30–70	2–10
Sun et al. [437]	MWCNT–R-141b	Mass flux 100 to 350 kg/(m^{2}s)	0.059, 0.117 and 0.176 vol. %	8	10–30
Jiang et al. [438]	CNT–R-113	Temperature 303 K	0.2–1.0 vol. %	15–80	1.5–10
Peng et al. [439]	CNT–POE–R-113	Heat Flux 10–80 kW/m^2	0.1–1 wt %	15–80	1.5–10
Ahmadpour et al. [440]	MWCNT–mineral oil–R-600A	Heat Flux -	0.1-.3 wt %	5–15	50
Kumaresan et al. [441]	MWCNT–EG–water	Temperature 273–313K	0.15–0.45 vol. %	30–50	10–20
Baskar et al. [443]	MWCNT–propanol + isopropyl alcohol	Temperature 273–303K	0.15–0.3 vol. %	-	-
Wang et al. [444]	Graphene–EG	Temperature 328–333K	0.01–1 wt %	-	5–15
Lin et al. [445]	MWCNT–R-141b	-	250–750 mg/L	15–80	1.5–10
Alawi and Sidik [446]	SWCNT–R-134a	Temperature 300–320 K	1.0–5.0 vol. %	20	-
Dalkilic et al. [447]	MWCNT–POE	Temperature 288–323 K	0.01–0.1 wt %	10–30	-

6.3.2. Influence of Carbon-Based Nanofluids on the COP and Overall Cooling Performance of AC&R Systems

A limited number of studies are available on how ND, graphene, and CNTs improve system COP and cooling capacity, which can be summarized in Table 11.

Abbas et al. [449] examined CNT mixed with POE oil in an R-134a refrigeration unit. They found that the system COP increased by 4.2% with nanoparticle concentration of 0.1 wt %. The experiment was infeasible beyond this concentration because the main challenge was with the agglomeration due to the strong Van der Waals interactions during the preparation phase. Jalili et al. [450] mixed various concentrations of MWCNT with water to assess the cooling performance of the secondary fluid in the evaporator of the refrigeration system. The transient analysis results showed that the evaporator's inlet temperature increased by 6.5% while the outlet temperature decreased by 14.5% when the water contains 2000 ppm of MWCNT. The significant enhancement in evaporator outlet temperature confirmed the tremendous increase in heat transfer coefficient with MWCNT. According to Kruse and Schroeder [451] and Cremaschi [452], the existence of oil lubricant in heat exchangers acted as insulation and resulted in heat transfer coefficient reduction. However, if the addition of nanoparticles enhanced the oil lubricant, the heat transfer coefficient would be compensated in the heat exchangers due to the improved overall thermophysical properties. Vasconcelos et al. [453] examined MWCNT–water as a secondary fluid in a 4–9 kW refrigeration unit with R-22 as a refrigerant. Due to the high thermal conductivity of the nanofluid, the cooling capacity increased up to 22.2% at the coolant's inlet temperature range of 30–40 °C. Vasconcelos et al. [453] found no significant reduction in the total power consumption. However, the increase in cooling load helped the compressor power consumption to relatively reduced because of the relative increase in evaporation pressure, and therefore the COP increased up to 33.3%. Kamaraj and Manoj Babu [454] replaced the POE oil with POE–mineral oil nano lubricants containing CNT particles with the amount of 0.1 and 0.2 g/L in a R-134a refrigerator. Besides the reduction in cooling time using the new nanofluids by approximately 40%, the COP was improved by 16.7% for 0.2 g/L of CNT using mineral oil. This is mainly due to the enhancement of the heat transfer coefficient in the evaporator without any significant reduction in the compressor power. Yang et al. [455] analyzed graphene nanosheets blended with SUNISO 3GS refrigerant oil in a R-600a refrigerator/freezer. The authors found that the cooling rate freezing rate improved by approximately 5.6% and 4.7%, respectively. The energy analysis yielded that the three concentrations nanolubricants (10 mg/L, 20 mg/L, and 30 mg/L) helped in reducing the compressor discharge temperature by 2.5%, 3.8%, and 4.6%, and dropping the energy consumption by 14.8%, 18.5%, and 20.4%, respectively. Hence, the energy saving was estimated to be up to 20% compared to using pure refrigerant oil. Indeed, the addition of graphene nanosheets with lubricant oil helped to decrease compressor friction losses. However, using graphene nanosheets as nanolubricant required additional surfactants (dispersants), which might increase the compression power because surfactants increase the viscosity and reduce the thermal conductivity at higher operating temperatures. Pico et al. [456] investigated two mass concentrations of ND–POE in a 17.6 kW vapor compression refrigeration system that used variable-speed compressor and refrigerant R-410A. The results showed that the compressor power consumption remained the same due to the type of compressor (i.e., hermetic scroll). On the other hand, the discharge compressor temperature reduced by approximately 3 °C and 4 °C, while the cooling capacity increased by 4.2% and 7%. Therefore, the overall system COP increased by 4% and 8% at 0.1% and 0.5% mass concentrations, respectively. Furthermore, Pico et al. [457] experimentally investigated the same ND–POE nanolubricant with R-32 as a substitute for R-410A. The results showed that for 0.5% mass concentration of diamond nanoparticle added to POE lubricant, the cooling capacity increased by 2.4% and the discharge compressor temperature decreased by approximately 2 °C, and hence the COP enhanced by 3.2%. The reduction of ND–POE performance with R-32 compared to R-410A can be justified with the low mass flow rate, which affected the oil circulation rate of the system operating with R-32.

Table 11. List of studies related to carbon-based nanoparticles effect on AC&R systems performance.

Ref.	Nanofluid	Nanoparticle			Compressor Discharge Temperature	Compressor Power	Cooling Capacity	COP
		Concentration	Diameter (nm)	Length (μm)				
Abbas et al. [449]	CNT–POE–R-134a	0.01–0.1 wt %	–	–	–	Reduced by 2.2%	–	Improved by 4.2%
Jalili et al. [450]	MWCNT–water	0–2000 ppm	10–20	5–15	–	–	–	–
Kamaraj and Manoj Babu [454]	CNT–POE–mineral oil–R-134a	0.1 and 0.2 g/L	13	–	Negligible reduction	Negligible reduction	Improved by 16.7%	Improved by 16.7%
Vasconcelos et al. [453]	MWCNT–water–R-22	0.035–0.212 vol. %	1–2	5–30	–	Negligible reduction	Improved by 22.2%	Improved by 27.3–33.3%
Pico et al. [456]	ND–POE–R-410A	0.1 and 0.5 mass %	3–6	–	Reduced by 3–4 °C	Negligible reduction	Improved by 4.2–7%	Improved by 4–8%
Pico et al. [457]	ND–POE–R-32	0.1 and 0.5 mass %	3–6	–	Reduced by 1.2–2 °C	Negligible reduction	Improved by 1–2.4%	Improved by 1–3.2%
Yang et al. [455]	Graphene–SUNISO 3GS–R-600a	10, 20, and 30 mg/L	100–3000	–	Reduced by 2.5–4.6%	Reduced by 14.8–20.4%	Improved by 5.6%	–
Rahman et al. [458]	SWCNT–R-407c	5 vol. %	–	–	–	Reduced by 4%	–	Improved by 4.3%

7. Environmental Consideration and Potential Health Issues

In addition to the studies that focused on the thermal enhancement that nanofluids can provide to energy systems, researchers have also explored the environmental impact and the potential hazardous towards human health from using these kind of suspensions [459]. In terms of environmental concerns, Meyer et al. [460] proposed an exergoenvironmental analysis method for designing an energy conversion system with an as low as possible environmental impact. Furthermore, an Eco-indicator 99 life cycle impact assessment approach was employed by the scholars to evaluate the environmental impact in a quantitative manner. They obtained the ecoindicator point through undertaken a series of analyses such as the exposure and effect analysis, resource and fate analysis, and damage analysis. One of the main element in determining the fate analysis is the toxicity evaluation of the nanomaterial(s) that form(s) the nanofluid. However, the criteria for measuring the level of toxicity is still not clear, and as such uncertainty always exist [461]. In order to resolve the previous issue, Card and Magnuson [462] came with a two-step approach for evaluating the toxicity of nanomaterials in an objective manner. In the first step, the authors collected the available literature that are related to the toxicity of nanomaterials, after which they started to evaluate and rank these studies according to the suitability of the design, documentation of adopted approach, materials used, and research outcome to produce the reliability 'study score'. As for the next step, the fulfilment of the physicochemical characterization of the nanomaterials is assessed in every study, which is then used to generate a 'nanomaterial score'. In general, the optimum way to reduce the environmental impact is to lower the utilization of resources and the resulting gas emissions during operation condition. For systems the utilize nanofluids, this can be achieved through enhancing their thermal efficiency, reducing their overall system size, and/or reducing the number of nanomaterials used. The reader is guided to the following sources [463–471] for further details on the aforementioned.

On the other hand, users dealing with nanofluids are always under high health risk [472]. This is because in the preparation process of these suspensions there is an inevitable contact between the user and the nanomaterials used, which can therefore enter their blood streams and/or organisms through skin absorption, inhalation, and/or ingestion of these toxic materials. It is important to note that almost all carbon-based nanomaterials, including CNTs, NDs, and graphene, are toxic and that the level of their toxicity increases with the decrease in their size [473]. Moreover, it is important that the reader distinguish between the safety aspect of nanomaterials that are used as medication (or for inner body diagnostic) and those used for other non-medical applications [110,474–477]. The first are safe and non-toxic when transferred to the human body in the appropriate way, whereas the second have high health risks and should be dealt with safety precautions. Some of the studies that were mainly devoted towards evaluating the harm that nanomaterial can cause to human health can be found in the following literature [473,474,478–482], where the material type, shape, size, surface characteristics charge, curvature, free energy, and functionalized groups were taken into account. Although using personal protective equipment and following the safety handling procedures can lower the user health risks, it is believed that part of the problem associated with commercializing nanofluids is due to the increasing concerns from the potential stakeholders and a lack of sufficient research studies [472,481].

8. Discussion and Future Directions

This review article has covered carbon-based nanofluids, from the fabrication stages of the raw nanoparticles materials (i.e., ND, graphene, and CNT) and up to their employment in some of the commonly known thermal applications in the energy industry. In addition, it was shown how dispersions made of carbon allotropes possess the most favorable thermal properties and, when well handled, physical properties compared to any other type of nanofluids or conventional fluids. This is because these carbon-based materials, when dispersed in a base fluid attain unique features such as high thermal conductivity and specific heat capacity, high heat transfer rate, and lower pressure drop in the working system compared to other types of dispersed nanomaterials. Furthermore, the aforementioned suspensions cause the least corrosion and erosion effects on the hosting device [483], all of which are crucial parameters for the operation cycle. Moreover, the influence of the stability of these suspensions on their thermophysical properties was also highlighted along with the development in these properties prediction correlations. Nevertheless, there are still some challenges and gaps in the scientific knowledge that need to be tackled for further advancement in the field and the possible industrial viability of such advanced fluids. Some of these issues are pointed out in this section.

8.1. Challenges in Carbon-Based Nanofluids

Carbon allotropes in the nanoscale have shown to be promising in enhancing the thermal performance of liquids when homogeneously dispersed, and their products are commercially available through a wide range of companies. Nevertheless, these powders are very expansive from an economic perspective compared to other sorts of nanomaterials, which makes their utilization quite questionable in the sense of their feasibility towards the targeted application [484]. Therefore, one of the challenges that need to be focused on is how to fabricate large quantities of these nanopowders at minimal production cost. In the current situation and before even introducing such type of nanofluids to the industry, researchers need to initially evaluate the gained performance enhancement and economic benefits of carbon-based nanofluids for each selected application before hands, and hence more work is needed in this area. Some scholars have proposed combining carbon-based nanomaterials with other cheaper types of nanoparticles (e.g., Cu, Al, and Fe) to form hybrid nanofluids containing carbon allotropes, and thus reducing the suspension cost [485–488]. However, the feasibility of such an approach remains limited, and the consideration of such types of hybrid nanofluids remains in the exploration stages. As it is well-known by now that the favorable thermophysical properties of nanofluids have made such a category of working fluids beneficial when used for enhancing the system performance of many thermal applications. Yet, the stability of the dispersed particles remains a major drawback and thus limiting the widespread of these suspensions. This is because, in an unstable state, the particles tend to cluster into larger forms of agglomerate, and therefore the benefits of the high surface area of the nanoparticles losses its optimum effectiveness on the exposed host (i.e., base fluids). For such a reason, it is essential that any proposed nanofluid to the industry maintains its long-term stability. This is where the preparation phase of the product plays a critical role. In order to overcome this difficulty, scientists have suggested using physical approaches (e.g., sonication) and/or chemical methods, such as surfactants and surface functionalization. Although this can help solve the aggregation problem, the changes caused to the surface of the dispersed particles remain another uncertainty that needs to be understood. In addition, further exploration on combining the two stabilization methods (i.e., physical and chemical routes) need to be conducted. Moreover, a joint international standard database on the thermophysical properties and physical stability of different types of nanofluids, their dispersed nanomaterial(s) concentration, and fabrication approach is strongly needed [198]. This is because even after more than 25 years from the first discovery of nanofluids, scientist are still reporting different thermophysical properties and physical stability for similar synthesized suspensions. In terms of properties prediction, it was shown previously that both effective thermal conductivity and effective viscosity lacks universal correlations and can only be determined through experimental means. However, artificial neural networking that is based on data mining has started to show good accuracy in predicting these properties, but further research is still required in this area [489–492].

8.2. Limitations in Parabolic Trough Solar Collector Systems

In PTSCs, the main challenge is that most of the studies shown in the literature were of pilot-scale tests, and thus further investigations are needed on the real-life application itself. Moreover, systematic studies are required to understand and standardize the influence of operational parameters such as high pressure, high temperature, flow rate, particles concentration, and suspension thermophysical properties on the system performance. Furthermore, there is still a need for a better understanding of the fouling build-up mechanism that is commonly associated with the use of nanofluids in systems of elevated temperatures as this newly introduced thin-film is likely to change the wettability behavior of the surface, and with it the dynamics of the flow [493–495]. Knowing this can open the door for introducing systematic washing routines, whether online, offline, or both, that can help in extending the heat transfer efficiency of the PTSC. Other aspects such as the friction factor and pressure drop should also be considered and explored intensively due to the nature of nanofluids having higher effective viscosity than conventional fluids.

8.3. Limitations in Nuclear Reactor Systems

As with most heat transfer systems, nuclear reactors were shown to have potential performance benefits from replacing their working fluids with nanofluids. Although on some occasions, certain types of nanofluids cannot compile with such application requirements (e.g., gold and platinum), due to the high temperature nature of such systems and the presence of emitted radiations that effects the dispersed particles [496]. Nevertheless, carbon-based materials have shown the capability of being an acceptable candidate for these systems. Despite that, the common challenge with almost every application that uses this category of suspensions remains rounded on the feasibility of such fluids and their particles clustering issue within the hosting system. When focusing on nuclear reactors, these systems design, sizes, and mechanisms can be seen changing rapidly throughout the past 20 years [308,312,497]. While this shows how this area of science is advancing, it also constrains the exploration capability of researchers working in the field of nanofluids. Thus, scientists specialized in nanofluids cannot investigate the performance enhancement caused by their suspensions on pre-existing reactors. Still, at the same time, they need to take into account the operation lifetime of the facility and understand how the isotopes build up and decay within these systems. This is because such changes in these isotopes could cause different behaviors when exposed to the dispersed particles. In addition, the fact that some of the dispersed particles may deposit on the nuclear fuel surface needs to be also considered and evaluated with respect to the possible corrosion development on the outer surface of the fuel. Moreover, studies on the long-term physical stability of nanofluids, when employed in nuclear reactors, remain unknown and need to be investigated. In addition, further work is needed to determine the effect of surfactants, when used as stabilizers, on the heat transfer rate in such application. This is because most (if not all) of these chemicals cannot withstand high temperature operating conditions. In terms of LOCA scenarios, nanofluids can help stop (or reduce) the level of damages that the fuel of elevated temperature may cause to the facility, but the method in which the newly introduced waste can be dealt with remains questionable and needs to be solved. This is because, unlike conventional liquids, the dispersed particles conserve more radiations, and hence remain radioactive for a very long time before they decay and stabilize. When it comes to pool boiling during quenching, to the best of our knowledge, there is still no existing literature that covers the effect of nanofluids on the minimum film boiling temperature (T_{min}) such as what was presented in Section 6.2.2. In general, there is a lack of studies about the impact of nanofluids on T_{min} during quenching. Owing to the importance of T_{min}, various types, concentrations, and sizes of nanofluids should have experimented with to investigate their effects on this parameter in specific. Furthermore, most of the investigations that are concern the effect of nanofluids on the CHF uses block plates, flat plates, or wires. However, research work on other geometries is crucial because it is evidence that the CHF will strongly be influenced by it. In addition, the currently employed models (e.g., Zuber's correlation) fails to accurately predict CHF when using thin wires [498], and therefore scholars need to focus more into developing a universal model that can withstand such limitation.

8.4. Limitations in Air Conditioning and Refrigeration Systems

Based on the research conducted by Hu et al. [499], it was shown that some carbon-based nanomaterials need surfactant(s) to ensure long-term stability and avoid agglomeration when they are mixed with the oil–refrigerant in the AC&R system. Additionally, surfactant might help to increase the performance of the AC&R system because it enhances the nucleate pool boiling heat transfer coefficient. However, this improvement is only attained under limited conditions because it is highly dependent on certain specifications (e.g., nanoparticle type, nanoparticle size, nanoparticle concentration, surfactant type, surfactant concentration, nanolubricant concentration, base fluid type, and heat flux). In addition, Cheng and Liu [500] recommended further investigation on nucleate pool boiling and flow boiling for refrigerant based-nanofluids. Therefore, further investigations must be conducted toward carbon-based nanofluids with refrigerant–oil–surfactants for AC&R applications. According to Bahiraei et al. [501] and Dalkılıç et al. [502], nanofluids helped to improve the heat transfer coefficient in heat exchangers (i.e., spiral-type and double-type heat exchangers). However, due to the increase in the nanofluid's viscosity, the pressure drop can be escalated, especially for low mass flow rates [503,504]. In AC&R systems, the pressure drop is an important factor that needs to be incorporated during the system design phase because any additional increase in the pressure drop in the heat exchangers during the operation of the AC&R system will result in significant degradation in the system COP performance as reported by Sunardi et al. [505] and Tashtoush et al. [506]. Yet, pressure drop due to a carbon-based nanofluids additive has not been studied in AC&R systems and needs to be further investigated. In fact, engineers need robust software tools to design AC&R systems with nanofluid additives. Some studies provided correlations for all thermophysical properties of the nanofluids (heat transfer coefficient, friction factor, thermal conductivity, viscosity, etc.) [507–510]. However, computational models integrating component models (as employed in Bahman et al. [511] and Loaiza et al. [512]) and the influence of carbon-based nanofluids in AC&R systems are still lacking. In addition, those kinds of models might have the potential to predict the overall system performance and ensure optimal operations. Furthermore, experimental investigations were limited to a specific concentration amount of nanofluids. The optimal concentration for maximum AC&R performance can be obtained numerically with the formerly mentioned computational models. Moreover, the determination of the optimal amount of concentration has not been proposed yet in the literature for carbon-based nanofluids in AC&R systems. In the AC&R system, nanofluid additives ultimately enhance the viscosity of the lubricating oil. Conversely, this might relatively increase the compressor power consumption. Therefore, it is vivid to find the relationship between nanofluid viscosity and compressor pumping power for AC&R application [513–515]. A limited number of research employed energy and exergy analysis on carbon-based nanofluids applied in AC&R systems. They have shown that there is a high potential for decreasing the irreversibility with carbon-based nanofluids due to their higher thermal conductivity compared to oxide materials [516]. Therefore, more compressive studies similar to Bahman and Groll [517] are required to identify the AC&R components with major irreversibility when employing nanofluids. Moreover, the literature is scarce in techno-economic analysis for nanofluids in AC&R applications. Although Bhattad et al. [518] showed that nanofluids could result in a higher payback period than the AC&R's components (i.e., heat exchangers), however, further studies need to be conducted on carbon-based nanofluids because by optimizing the number of nanoparticles with respect to operating condition and stability, it can be more cost-effective. Finally, AC&R systems combine several components, as nanofluid pass through these components, it would be compressed, expanded, or changed phases. All these processes may lead to the possibility of getting nanoparticles to be separated from the carrier fluid during long-term operation and probably degrade the system performance. Therefore, the long-term operating performance of the nanorefrigerant (and nanolubricant) must be investigated.

9. Conclusions

This paper provides a comprehensive state-of-the-art review on carbon-based nanofluids, including the initial synthesis methods used for producing carbon nanotubes, graphene, and nanodiamonds, and up to the employment of their dispersions into thermal energy applications, namely; parabolic trough solar collectors, quenching systems, and air conditioning and refrigeration systems. It was shown how some of these nanomaterials could only be fabricated in a dry form, such as high purity nanodiamonds, whereas graphene, for example, can be produced as a dry powder or a suspension. Thus, when selecting the nanofluids' preparation approach containing these nanomaterials, the available options can be narrowed from two routes to only a single process. Furthermore, the main equations used in calculating the volume concentration that are commonly required for the nanofluid two-step production method were provided. Moreover, the physical stability of the suspension, which is considered as one of the most influential aspect that can dramatically affect the thermophysical properties of any nanofluid, was discussed in terms of its formation mechanism and evaluation approaches. Although there are many advanced ways to characterize the dispersion stability, it was concluded that the photographical capturing method remains the most reliable approach due to its capability of determining both short- and long-term dispersion stability of the mixture in real-time and at high accuracy. Nevertheless, this method is very time-consuming to conduct, especially when the characterized sample is of high state of stability. In addition, chemical methods, such as surfactants and surface functionalization; and physical approaches, namely, ultrasonication, magnetic stirring, homogenizer, and ball milling, were also discussed and shown in how they can be employed to improve the level of particles dispersion within an instable suspension. It was concluded that, unlike the chemical approaches, using physical methods for enhancing the dispersion stability is a better option when it comes to conserving the optimum possible effective thermal conductivity of the nanofluid and that between the available physical routes, the homogenizer can provide the best outcomes. In general, the stability of the suspension does not affect the mixture density nor its specific heat capacity but rather influences both the effective thermal conductivity and effective viscosity of the nanofluid. These two properties were seen to degrade gradually with time due to the nanomaterial's agglomerations and their sediment formation. Many methods were shown to determine these two properties (i.e., effective thermal conductivity and effective viscosity), either by experimental means or through pre-existing correlations. Still, up to today, the scientific field has failed to provide a universal formula for both of these two properties, and hence the only reliable approach is through experimental analysis. When it comes to replacing conventional working fluids with carbon-based nanofluid in thermal applications (i.e., parabolic trough solar collectors, nuclear reactor systems, and air conditioning and refrigeration systems), it was proven, at least at the lab and pilot-scale, that such advanced fluids are very beneficial in terms of enhancing the overall performance of these systems, and can therefore be seen as strong candidates for such industries when their associated challenges are solved and fully understood.

Author Contributions: N.A., A.M.B., N.F.A. and A.A. conducted the Introduction Section. N.A. carried out the Synthesis of Nanoscaled Carbon-Based Materials Section, the Preparation of Nanofluids Section, and the Stability Effect on Thermophysical Properties Section. N.A. conducted the Parabolic trough solar collectors Section. S.A.E. and S.M. worked on the Nuclear reactors Section. A.M.B. and N.F.A. conducted the Air conditioning and refrigeration systems Section. N.A., N.F.A., A.M.B. and S.A.E. have worked on the Discussion and Future Directions Section along with the Conclusion Section. All authors have read and agreed to the published version of the manuscript.

Funding: This research received no external funding.

Acknowledgments: The authors of this article are grateful and acknowledge the help provided by their institutes. They are also grateful to the journals that have waved their copy rights fees for some of the figures that were used in this article. This includes Materials Science and Engineering: C, Elsevier (License Number: 5054231133553); and Applied Thermal Engineering, Elsevier (License Number: 5054240415839).

Conflicts of Interest: The authors declare no conflict of interest.

Nomenclature

A	Area (nm^2)
AC	Air conditioning
AC&R	Air conditioning and refrigeration
AG	Arabic gum
ALWR	Advanced light water reactor
ANL	Argonne National Laboratory
BAC	Benzalkonium chloride
BWR	Boiling water reactor
$C_{R.M.}$	Nanoparticle random motion velocity (nm/s)
CF	Self-crowding factor
CFD	Computational fluid dynamics
CHF	Critical heat flux (W/m^2)
CNT	Carbon nanotube
COP	Coefficient of performance
C_p	Specific heat capacity (J/kg·K)
CSPP	Concentrated solar power plant
CTAB	Cetyltrimethyl ammonium bromide
CVD	Chemical vapour deposition
DASC	Direct absorber solar collector
d_{bf}	Diameter of the base fluid molecule (nm)
DND	Detonation nanodiamond
DNI	Direct normal irradiance
d_{np}	Nanoparticles mean diameter
DSC	Differential scanning calorimetry
DSDMAC	Distearyl dimethyl ammonium chloride
DWCNT	Double-walled carbon nanotube
DX	Direct expansion
EG	Ethylene glycol
f_{m}	Maximum attainable concentration
f_{P}	Packing fraction of the particles
f_{V}	Particles volumetric fraction
FVM	Finite volume method
GCR	Gas-cooled reactor
GO	Graphene oxide
HFC	Hydrofluorocarbon
HPHT	High-pressure and high-temperature
HTC	Heat transfer coefficient ($W/m^2 \cdot K$)
IPH	Industrial process heat
k_{pj}	Equivalent thermal conductivity of the ellipsoids particle (W/m·K)
k_B	Boltzmann constant (1.381×10^{-23} J/K)
k_H	Huggins coefficient
K_m	Matrix conductivity (W/m·K)
k_{pe}	Equivalent particle thermal conductivity (W/m·K)
ℓ_{bf}	Mean-free path of the base fluid molecule (nm)
LOCA	Loss-of-coolant accident
m	Mass (Kg)
MCRT	Monte Carlo ray tracing
MSR	Molten solid cooled reactor
MWCNT	Multiwalled carbon nanotube
ND	Nanodiamond
Nu	Nusselt number
PHWR	Pressurized heavy water reactor
POE	Polyolester oil

Pr	Prandtl number
PTSC	Parabolic trough solar collector
PVA	Polyvinyl alcohol
PVP	Polyvinylpyrrolidone
PWR	Pressurized water reactor
r	Volume ratio
R_b	Impact of interfacial resistance (Km^2/W)
r_c	Particle apparent radius (nm)
RGO	Reduced graphene oxide
Re	Reynolds number
R_k	Kaptiza radius (8×10^{-8} m^2 K/W)
r_m	Radius of the fluid medium particles (nm)
SANSS	Submerged arc nanoparticle synthesis system
SDBS	Sodium dodecyl benzenesulfonate
SDS	Sodium dodecyl sulfate
SEM	Scanning electron microscopy
SWCNH	Single-walled carbon nanohorn
SWCNT	Single-walled carbon nanotube
T	Temperature (K or °C)
T_o	Reference temperature (273 K)
T_m	Mean temperature (K or °C)
T_{min}	Minimum film boiling temperature (K or °C)
t_{nl}	Thickness of the nanolayer surrounding the particle (nm)
t_o	Starting time (s)
t_f	Finishing time (s)
TEM	Transmission electron microscopy
TWCNT	Triple-walled carbon nanotube
V	Volume (m^3)
VERSO	Vacuum evaporation onto a running oil substrate
vol. %	Volume percentage
WCR	Water-cooled reactor
wt %	Weight percentage
Greek letters	
β	Ratio of the nanolayer thickness to the particle radius
Δ	Difference
η	Average flatness ratio of the graphene nanoplatelet
$[\eta]$	Intrinsic viscosity
μ	Dynamic viscosity (kg/m·s)
n	Empirical shape factor
ν	Kinematic viscosity (m^2/s)
ψ	Particle sphericity
ρ	Density (kg/m^3)
k	Thermal conductivity (W/m·K)
Subscripts	
amb	Ambient
bf	Base fluid
CNT	Carbon nanotube
eff	Effective
min	Minimum
nf	Nanofluid
np	Nanoparticles
sat	Saturated
sup	Super-heated
w	Water

References

1. Ahuja, A.S. Augmentation of heat transport in laminar flow of polystyrene suspensions. I. Experiments and results. *J. Appl. Phys.* **1975**, *46*, 3408–3416. [CrossRef]
2. Ahuja, A.S. Augmentation of heat transport in laminar flow of polystyrene suspensions. II. Analysis of the data. *J. Appl. Phys.* **1975**, *46*, 3417–3425. [CrossRef]
3. Liu, K.V.; Choi, U.S.; Kasza, K.E. *Measurements of Pressure Drop and Heat Transfer in Turbulent Pipe Flows of Particulate Slurries*; Argonne National Lab: Lemont, IL, USA, 1988.
4. Choi, S.U.; Cho, Y.I.; Kasza, K.E. Degradation effects of dilute polymer solutions on turbulent friction and heat transfer behavior. *J. Non-Newton Fluid Mech.* **1992**, *41*, 289–307. [CrossRef]
5. Choi, U.; France, D.M.; Knodel, B.D. *Impact of Advanced Fluids on Costs of District Cooling Systems*; Argonne National Lab: Lemont, IL, USA, 1992.
6. Choi, U.; Tran, T. Experimental Studies of the Effects of Non-Newtonian Surfactant Solutions on the Performance of a Shell-and-Tube Heat Exchanger. In *Recent Developments in Non-Newtonian Flows and Industrial Applications*; The American Society of Mechanical Engineers New York: New York, NY, USA; FED: Atlanta, GA, USA, 1991; pp. 47–52.
7. Maxwell, J.C. *A Treatise on Electricity and Magnetism*, 2nd ed.; Clarendon Press: Oxford, UK, 1881.
8. Ali, N.; Teixeira, J.A.; Addali, A. Aluminium Nanofluids Stability: A Comparison between the Conventional Two-Step Fabrication Approach and the Controlled Sonication Bath Temperature Method. *J. Nanomater.* **2019**, *2019*, 1–9. [CrossRef]
9. Masuda, H.; Ebata, A.; Teramae, K. *Alteration of Thermal Conductivity and Viscosity of Liquid by Dispersing Ultra-Fine Particles. Dispersion of Al2o3, Sio2 and Tio2 Ultra-Fine Particles*; Netsu Bussei: Tokyo, Japan, 1993; Volume 7, pp. 227–233. [CrossRef]
10. Choi, S.U.S.; Eastman, J.A. Enhancing thermal conductivity of fluids with nanoparticles. In *Proceedings of the 1995 International Mechanical Engineering Congress and Exhibition, San Francisco, CA, USA, 12–17 November 1995*; PBD: Washington, DC, USA; Argonne National Lab: Lemont, IL, USA, 1995; 8p.
11. Naser, A.; Teixeira, J.A.; Ali, N. New pH Correlations for Stainless Steel 316L, Alumina, and Copper(I) Oxide Nanofluids Fabricated at Controlled Sonication Temperatures. *J. Nano Res.* **2019**, *58*, 125–138. [CrossRef]
12. Lee, S.; Choi, S.U.S. *Application of Metallic Nanoparticle Suspensions in Advanced Cooling Systems*; American Society of Mechanical Engineers, Materials Division: Atlanta, GA, USA, 1996; Volume 72, pp. 227–234.
13. Ali, N.; Teixeira, J.A.; Addali, A. A Review on Nanofluids: Fabrication, Stability, and Thermophysical Properties. *J. Nanomater.* **2018**, *2018*, 1–33. [CrossRef]
14. Pop, E.; Varshney, V.; Roy, A.K. Thermal properties of graphene: Fundamentals and applications. *MRS Bull.* **2012**, *37*, 1273–1281. [CrossRef]
15. Han, Z.; Fina, A. Thermal conductivity of carbon nanotubes and their polymer nanocomposites: A review. *Prog. Polym. Sci.* **2011**, *36*, 914–944. [CrossRef]
16. Sezer, N.; Atieh, M.; Koc, M. A comprehensive review on synthesis, stability, thermophysical properties, and characterization of nanofluids. *Powder Technol.* **2019**, *344*, 404–431. [CrossRef]
17. Mashali, F.; Languri, E.M.; Davidson, J.; Kerns, D.; Johnson, W.; Nawaz, K.; Cunningham, G. Thermo-physical properties of diamond nanofluids: A review. *Int. J. Heat Mass Transf.* **2019**, *129*, 1123–1135. [CrossRef]
18. Schwamb, T.; Burg, B.R.; Schirmer, N.C.; Poulikakos, D. An electrical method for the measurement of the thermal and electrical conductivity of reduced graphene oxide nanostructures. *Nanotechnology* **2009**, *20*, 405704. [CrossRef]
19. Mahanta, N.K.; Abramson, A.R. Thermal conductivity of graphene and graphene oxide nanoplatelets. In Proceedings of the 13th InterSociety Conference on Thermal and Thermomechanical Phenomena in Electronic Systems, San Diego, CA, USA, 30 May–1 June 2012; IEEE: New York, NY, USA, 2012; pp. 1–6.
20. Zhang, H.; Fonseca, A.F.; Cho, K. Tailoring Thermal Transport Property of Graphene through Oxygen Functionalization. *J. Phys. Chem. C* **2014**, *118*, 1436–1442. [CrossRef]
21. Eastman, J.A.; Choi, U.S.; Li, S.; Thompson, L.J.; Lee, S. Enhanced thermal conductivity through the development of nanofluids. In *Proceedings of the 1996 MRS Fall Symposium*; George, E.P., Gotthardt, R., Otsuka, K., Trolier-McKinstry, S., Wun-Fogle, M., Eds.; Materials Research Society: Pittsburgh, PA, USA; Boston, MA, USA, 1997; pp. 3–11.
22. Li, Z.X.; Khaled, U.; Al-Rashed, A.A.A.A.; Goodarzi, M.; Sarafraz, M.M.; Meer, R. Heat transfer evaluation of a micro heat exchanger cooling with spherical carbon-acetone nanofluid. *Int. J. Heat Mass Transf.* **2020**, *149*, 119124. [CrossRef]
23. Ilyas, S.U.; Pendyala, R.; Shuib, A.; Marneni, N. A review on the viscous and thermal transport properties of nanofluids. In *International Conference on Process Engineering and Advanced Materials, ICPEAM 2012*; Trans Tech Publications Ltd.: Kuala Lumpur, Malaysia, 2014; pp. 18–27.
24. Shanthi, R.; Anandan, S.; Ramalingam, V. Heat transfer enhancement using nanofluids: An overview. *Therm. Sci.* **2012**, *16*, 423–444. [CrossRef]
25. Wen, D.; Lin, G.; Vafaei, S.; Zhang, K. Review of nanofluids for heat transfer applications. *Particuology* **2009**, *7*, 141–150. [CrossRef]
26. Vékás, L.; Bica, D.; Avdeev, M.V. Magnetic nanoparticles and concentrated magnetic nanofluids: Synthesis, properties and some applications. *China Particuol.* **2007**, *5*, 43–49. [CrossRef]
27. Reddy, K.S.; Kamnapure, N.R.; Srivastava, S. Nanofluid and nanocomposite applications in solar energy conversion systems for performance enhancement: A review. *Int. J. Low Carbon Technol.* **2016**, *12*, 1–23. [CrossRef]

28. Sheikholeslami, M.; Ganji, D.D. Application of Nanofluids. In *Applications of Semi Analytical Methods for Nanofluid Flow and Heat Transfer*; Sheikholeslami, M., Ganji, D.D., Eds.; Elsevier: Amsterdam, The Netherlands, 2018; pp. 1–44.
29. Mansoury, D.; Doshmanziari, F.I.; Kiani, A.; Chamkha, A.J.; Sharifpur, M. Heat Transfer and Flow Characteristics of Al2O3/Water Nanofluid in Various Heat Exchangers: Experiments on Counter Flow. *Heat Transf. Eng.* **2018**, *41*, 1–36. [CrossRef]
30. Chamkha, A.J.; Molana, M.; Rahnama, A.; Ghadami, F. On the nanofluids applications in microchannels: A comprehensive review. *Powder Technol.* **2018**, *332*, 287–322. [CrossRef]
31. Alsayegh, A.; Ali, N. Gas Turbine Intercoolers: Introducing Nanofluids—A Mini-Review. *Processes* **2020**, *8*, 1572. [CrossRef]
32. Mannu, R.; Karthikeyan, V.; Velu, N.; Arumugam, C.; Roy, V.A.L.; Gopalan, A.-I.; Saianand, G.; Sonar, P.; Lee, K.-P.; Kim, W.-J.; et al. Polyethylene Glycol Coated Magnetic Nanoparticles: Hybrid Nanofluid Formulation, Properties and Drug Delivery Prospects. *Nanomaterials* **2021**, *11*, 440. [CrossRef]
33. Martínez-Merino, P.; Sánchez-Coronilla, A.; Alcántara, R.; Martín, E.I.; Carrillo-Berdugo, I.; Gómez-Villarejo, R.; Navas, J. The Role of the Interactions at the Tungsten Disulphide Surface in the Stability and Enhanced Thermal Properties of Nanofluids with Application in Solar Thermal Energy. *Nanomaterials* **2020**, *10*, 970. [CrossRef] [PubMed]
34. Rostami, S.; Aghakhani, S.; Pordanjani, A.H.; Afrand, M.; Cheraghian, G.; Oztop, H.F.; Shadloo, M.S. A Review on the Control Parameters of Natural Convection in Different Shaped Cavities with and Without Nanofluid. *Processes* **2020**, *8*, 1011. [CrossRef]
35. Scopus-Database, Nanofluid Analyze Search Results for Documents Published from 1995 to 2020 with the Word 'Nanofluid'; Elsevier: Amsterdam, The Netherlands. 2021. Available online: www.scopus.com (accessed on 1 April 2021).
36. Mukherjee, S.; Mishra, P.C.; Chaudhuri, P. Stability of Heat Transfer Nanofluids—A Review. *ChemBioEng Rev.* **2018**, *5*, 312–333. [CrossRef]
37. Almurtaji, S.; Ali, N.; Teixeira, J.A.; Addali, A. On the Role of Nanofluids in Thermal-hydraulic Performance of Heat Exchangers-A Review. *Nanomaterials* **2020**, *10*, 734. [CrossRef] [PubMed]
38. Song, Y.Y.; Bhadeshia, H.; Suh, D.-W. Stability of stainless-steel nanoparticle and water mixtures. *Powder Technol.* **2015**, *272*, 34–44. [CrossRef]
39. Ebrahimnia-Bajestan, E.; Niazmand, H.; Duangthongsuk, W.; Wongwises, S. Numerical investigation of effective parameters in convective heat transfer of nanofluids flowing under a laminar flow regime. *Int. J. Heat Mass Transf.* **2011**, *54*, 4376–4388. [CrossRef]
40. Martínez-Cuenca, R.; Mondragón, R.; Hernández, L.; Segarra, C.; Jarque, J.C.; Hibiki, T.; Juliá, J.E. Forced-convective heat-transfer coefficient and pressure drop of water-based nanofluids in a horizontal pipe. *Appl. Therm. Eng.* **2016**, *98*, 841–849. [CrossRef]
41. Benedict, L.X.; Louie, S.G.; Cohen, M.L. Heat capacity of carbon nanotubes. *Solid State Commun.* **1996**, *100*, 177–180. [CrossRef]
42. Yazid, M.N.A.W.M.; Sidik, N.A.C.; Mamat, R.; Najafi, G. A review of the impact of preparation on stability of carbon nanotube nanofluids. *Int. Commun. Heat Mass Transf.* **2016**, *78*, 253–263. [CrossRef]
43. Shah, K.A.; Tali, B.A. Synthesis of carbon nanotubes by catalytic chemical vapour deposition: A review on carbon sources, catalysts and substrates. *Mater. Sci. Semicond. Process.* **2016**, *41*, 67–82. [CrossRef]
44. Wang, H.; Xu, Z.; Eres, G. Order in vertically aligned carbon nanotube arrays. *Appl. Phys. Lett.* **2006**, *88*, 213111. [CrossRef]
45. Hong, P.N.; Minh, D.N.; Van Hung, N.; Minh, P.N.; Khoi, P.H. Carbon Nanotube and Graphene Aerogels—The World's 3D Lightest Materials for Environment Applications: A Review. *Int. J. Mater. Sci. Appl.* **2017**, *6*, 277. [CrossRef]
46. Askari, S.; Lotfi, R.; Seifkordi, A.; Rashidi, A.; Koolivand, H. A novel approach for energy and water conservation in wet cooling towers by using MWNTs and nanoporous graphene nanofluids. *Energy Convers. Manag.* **2016**, *109*, 10–18. [CrossRef]
47. Neuberger, N.; Adidharma, H.; Fan, M. Graphene: A review of applications in the petroleum industry. *J. Pet. Sci. Eng.* **2018**, *167*, 152–159. [CrossRef]
48. Shanbedi, M.; Heris, S.Z.; Amiri, A.; Hosseinipour, E.; Eshghi, H.; Kazi, S. Synthesis of aspartic acid-treated multi-walled carbon nanotubes based water coolant and experimental investigation of thermal and hydrodynamic properties in circular tube. *Energy Convers. Manag.* **2015**, *105*, 1366–1376. [CrossRef]
49. Sadeghinezhad, E.; Mehrali, M.; Saidur, R.; Mehrali, M.; Latibari, S.T.; Akhiani, A.R.; Metselaar, H.S.C. A comprehensive review on graphene nanofluids: Recent research, development and applications. *Energy Convers. Manag.* **2016**, *111*, 466–487. [CrossRef]
50. Tam, N.T.; Phuong, N.V.; Khoi, P.H.; Minh, P.N.; Afrand, M.; Van Trinh, P.; Thang, B.H.; Żyła, G.; Estellé, P. Carbon Nanomaterial-Based Nanofluids for Direct Thermal Solar Absorption. *Nanomaterials* **2020**, *10*, 1199. [CrossRef]
51. Ambreen, T.; Saleem, A.; Park, C. Homogeneous and Multiphase Analysis of Nanofluids Containing Nonspherical MWCNT and GNP Nanoparticles Considering the Influence of Interfacial Layering. *Nanomaterials* **2021**, *11*, 277. [CrossRef] [PubMed]
52. Giwa, S.O.; Sharifpur, M.; Ahmadi, M.H.; Murshed, S.M.S.; Meyer, J.P. Experimental Investigation on Stability, Viscosity, and Electrical Conductivity of Water-Based Hybrid Nanofluid of MWCNT-Fe_2O_3. *Nanomaterials* **2021**, *11*, 136. [CrossRef] [PubMed]
53. Freitas, E.; Pontes, P.; Cautela, R.; Bahadur, V.; Miranda, J.; Ribeiro, A.P.C.; Souza, R.R.; Oliveira, J.D.; Copetti, J.B.; Lima, R.; et al. Article pool boiling of nanofluids on biphilic surfaces: An experimental and numerical study. *Nanomaterials* **2021**, *11*, 125. [CrossRef]
54. Karagiannakis, N.P.; Skouras, E.D.; Burganos, V.N. Modelling Thermal Conduction in Nanoparticle Aggregates in the Presence of Surfactants. *Nanomaterials* **2020**, *10*, 2288. [CrossRef]
55. Khan, H.; Soudagar, M.E.M.; Kumar, R.H.; Safaei, M.R.; Farooq, M.; Khidmatgar, A.; Banapurmath, N.R.; Farade, R.A.; Abbas, M.M.; Afzal, A.; et al. Effect of Nano-Graphene Oxide and n-Butanol Fuel Additives Blended with Diesel—Nigella sativa Biodiesel Fuel Emulsion on Diesel Engine Characteristics. *Symmetry* **2020**, *12*, 961. [CrossRef]

56. Soudagar, M.E.M.; Afzal, A.; Safaei, M.R.; Manokar, A.M.; El-Seesy, A.I.; Mujtaba, M.A.; Samuel, O.D.; Badruddin, I.A.; Ahmed, W.; Shahapurkar, K.; et al. Investigation on the effect of cottonseed oil blended with different percentages of octanol and suspended MWCNT nanoparticles on diesel engine characteristics. *J. Therm. Anal. Calorim.* **2020**. [CrossRef]
57. Zhang, Y.; Yin, Q.Z. Carbon and other light element contents in the Earth's core based on first-principles molecular dynamics. *Proc. Natl. Acad. Sci. USA* **2012**, *109*, 19579–19583. [CrossRef]
58. Ferrari, A.C.; Robertson, J.; Ferrari, A.C.; Robertson, J. Interpretation of Raman spectra of disordered and amorphous carbon. *Phys. Rev. B* **2000**, *61*, 14095–14107. [CrossRef]
59. Wei, L.; Kuo, P.K.; Thomas, R.L.; Anthony, T.R.; Banholzer, W.F. Thermal conductivity of isotopically modified single crystal diamond. *Phys. Rev. Lett.* **1993**, *70*, 3764–3767. [CrossRef]
60. Hodkiewicz, J.; Scientific, T. *Characterizing Carbon Materials with Raman Spectroscopy*; Thermo Scientific Application Note; Thermo Fisher Scientific: Madison, WI, USA, 2010.
61. Dai, L.; Chang, D.W.; Baek, J.-B.; Lu, W. Carbon Nanomaterials for Advanced Energy Conversion and Storage. *Small* **2012**, *8*, 1130–1166. [CrossRef] [PubMed]
62. Kaneko, K.; Ishii, C.; Ruike, M.; Kuwabara, H. Origin of superhigh surface area and microcrystalline graphitic structures of activated carbons. *Carbon* **1992**, *30*, 1075–1088. [CrossRef]
63. Pang, J.; Bachmatiuk, A.; Ibrahim, I.; Fu, L.; Placha, D.; Martynková, G.S.; Trzebicka, B.; Gemming, T.; Eckert, J.; Rümmeli, M.H. CVD growth of 1D and 2D sp2 carbon nanomaterials. *J. Mater. Sci.* **2015**, *51*, 640–667. [CrossRef]
64. van Thiel, M.; Ree, F.H. Properties of carbon clusters in TNT detonation products: Graphite-diamond transition. *J. Appl. Phys.* **1987**, *62*, 1761–1767. [CrossRef]
65. Morita, Y.; Takimoto, T.; Yamanaka, H.; Kumekawa, K.; Morino, S.; Aonuma, S.; Kimura, T.; Komatsu, N. A Facile and Scalable Process for Size-Controllable Separation of Nanodiamond Particles as Small as 4 nm. *Small* **2008**, *4*, 2154–2157. [CrossRef]
66. Ali, M.S.; Metwally, A.; Fahmy, R.H.; Osman, R. Nanodiamonds: Minuscule gems that ferry antineoplastic drugs to resistant tumors. *Int. J. Pharm.* **2019**, *558*, 165–176. [CrossRef] [PubMed]
67. Bovenkerk, H.P.; Bundy, F.P.; Hall, H.T.; Strong, H.M.; Wentorf, R.H. Preparation of Diamond. *Nat. Cell Biol.* **1959**, *184*, 1094–1098. [CrossRef]
68. Danilenko, V.V. On the history of the discovery of nanodiamond synthesis. *Phys. Solid State* **2004**, *46*, 595–599. [CrossRef]
69. Kumar, A.; Lin, P.A.; Xue, A.; Hao, B.; Yap, Y.K.; Sankaran, R.M. Formation of nanodiamonds at near-ambient conditions via microplasma dissociation of ethanol vapour. *Nat. Commun.* **2013**, *4*, 2618. [CrossRef]
70. Frenklach, M.; Howard, W.; Huang, D.; Yuan, J.; Spear, K.E.; Koba, R. Induced nucleation of diamond powder. *Appl. Phys. Lett.* **1991**, *59*, 546–548. [CrossRef]
71. Yang, G.-W.; Wang, J.-B.; Liu, Q.-X. Preparation of nano-crystalline diamonds using pulsed laser induced reactive quenching. *J. Phys. Condens. Matter* **1998**, *10*, 7923–7927. [CrossRef]
72. Boudou, J.P.; Curmi, P.A.; Jelezko, F.; Wrachtrup, J.; Aubert, P.; Sennour, M.; Balasubramanian, G.; Reuter, R.; Thorel, A.; Gaffet, E. High yield fabrication of fluorescent nanodiamonds. *Nanotechnology* **2009**, *20*, 235602. [CrossRef]
73. El-Eskandarany, M.S. Mechanically Induced Graphite-Nanodiamonds-Phase Transformations During High-Energy Ball Milling. *J. Mater. Eng. Perform.* **2017**, *26*, 2974–2982. [CrossRef]
74. Lin, C.R.; Wei, D.H.; Dao, M.K.; Chung, R.J.; Chang, M.H. Nanocrystalline diamond particles prepared by high-energy ball milling method. In *Applied Mechanics and Materials*; Trans Tech Publications Ltd.: Schwyz, Switzerland, 2013; Volume 284, pp. 168–172. [CrossRef]
75. Galimov, A.É.M.; Kudin, A.M.; Skorobogatskii, V.N.; Plotnichenko, V.G.; Bondarev, O.L.; Zarubin, B.G.; Strazdovskii, V.V.; Aronin, A.S.; Fisenko, A.V.; Bykov, I.; et al. Experimental corroboration of the synthesis of diamond in the cavitation process. *Dokl. Phys.* **2004**, *49*, 150–153. [CrossRef]
76. Welz, S.; Gogotsi, Y.; McNallan, M.J. Nucleation, growth, and graphitization of diamond nanocrystals during chlorination of carbides. *J. Appl. Phys.* **2003**, *93*, 4207–4214. [CrossRef]
77. Banhart, F.; Ajayan, P.M. Carbon onions as nanoscopic pressure cells for diamond formation. *Nature* **1996**, *382*, 433–435. [CrossRef]
78. Daulton, T.; Kirk, M.; Lewis, R.; Rehn, L. Production of nanodiamonds by high-energy ion irradiation of graphite at room temperature. *Nucl. Instrum. Methods Phys. Res. Sect. B Beam Interact. Mater. Atoms* **2001**, *175–177*, 12–20. [CrossRef]
79. El-Eskandarany, M.S. Method for Synthesizing Nanodiamonds. U.S. Patent 9,540,245 B1, 10 January 2017.
80. Mochalin, V.N.; Shenderova, O.; Ho, D.; Gogotsi, Y. The properties and applications of nanodiamonds. *Nat. Nanotechnol.* **2012**, *7*, 11–23. [CrossRef]
81. Mashali, F.; Languri, E.M.; Davidson, J.; Kerns, D. Diamond Nanofluids: Microstructural Analysis and Heat Transfer Study. *Heat Transf. Eng.* **2020**, *42*, 479–491. [CrossRef]
82. Crane, M.J.; Petrone, A.; Beck, R.A.; Lim, M.B.; Zhou, X.; Li, X.; Stroud, R.M.; Pauzauskie, P.J. High-pressure, high-temperature molecular doping of nanodiamond. *Sci. Adv.* **2019**, *5*, eaau6073. [CrossRef] [PubMed]
83. Afandi, A.; Howkins, A.; Boyd, I.W.; Jackman, R.B. Nanodiamonds for device applications: An investigation of the properties of boron-doped detonation nanodiamonds. *Sci. Rep.* **2018**, *8*, 1–10. [CrossRef]
84. Tinwala, H.; Wairkar, S. Production, surface modification and biomedical applications of nanodiamonds: A sparkling tool for theranostics. *Mater. Sci. Eng. C* **2019**, *97*, 913–931. [CrossRef]

85. Compton, O.C.; Nguyen, S. Graphene Oxide, Highly Reduced Graphene Oxide, and Graphene: Versatile Building Blocks for Carbon-Based Materials. *Small* **2010**, *6*, 711–723. [CrossRef]
86. Novoselov, K.S.; Geim, A.K.; Morozov, S.V.; Jiang, D.; Zhang, Y.; Dubonos, S.V.; Grigorieva, I.V.; Firsov, A.A. Electric field effect in atomically thin carbon films. *Science* **2004**, *306*, 666–669. [CrossRef]
87. Ma, R.; Zhou, Y.; Bi, H.; Yang, M.; Wang, J.; Liu, Q.; Huang, F. Multidimensional graphene structures and beyond: Unique properties, syntheses and applications. *Prog. Mater. Sci.* **2020**, *113*, 100665. [CrossRef]
88. Moreau, E.; Ferrer, F.J.; Vignaud, D.; Godey, S.; Wallart, X. Graphene growth by molecular beam epitaxy using a solid carbon source. *Phys. Status Solidi (a)* **2010**, *207*, 300–303. [CrossRef]
89. Chyan, Y.; Ye, R.; Li, Y.; Singh, S.P.; Arnusch, C.J.; Tour, J.M. Laser-Induced Graphene by Multiple Lasing: Toward Electronics on Cloth, Paper, and Food. *ACS Nano* **2018**, *12*, 2176–2183. [CrossRef]
90. Ye, R.; James, D.K.; Tour, J.M. Laser-Induced Graphene: From Discovery to Translation. *Adv. Mater.* **2019**, *31*, e1803621. [CrossRef]
91. Stankovich, S.; Dikin, D.A.; Dommett, G.H.B.; Kohlhaas, K.M.; Zimney, E.J.; Stach, E.A.; Piner, R.D.; Nguyen, S.; Ruoff, R.S. Graphene-based composite materials. *Nature* **2006**, *442*, 282–286. [CrossRef] [PubMed]
92. Eda, G.; Fanchini, G.; Chhowalla, M. Large-area ultrathin films of reduced graphene oxide as a transparent and flexible electronic material. *Nat. Nanotechnol.* **2008**, *3*, 270–274. [CrossRef]
93. Hernandez, Y.; Lotya, M.; Rickard, D.; Bergin, S.D.; Coleman, J.N. Measurement of Multicomponent Solubility Parameters for Graphene Facilitates Solvent Discovery. *Langmuir* **2010**, *26*, 3208–3213. [CrossRef] [PubMed]
94. Hussain, A.; Mehdi, S.M.; Abbas, N.; Hussain, M.; Naqvi, R.A. Synthesis of graphene from solid carbon sources: A focused review. *Mater. Chem. Phys.* **2020**, *248*, 122924. [CrossRef]
95. Tetsuka, H.; Nagoya, A.; Fukusumi, T.; Matsui, T. Molecularly Designed, Nitrogen-Functionalized Graphene Quantum Dots for Optoelectronic Devices. *Adv. Mater.* **2016**, *28*, 4632–4638. [CrossRef]
96. Shen, J.; Zhu, Y.; Yang, X.; Li, C. Graphene quantum dots: Emergent nanolights for bioimaging, sensors, catalysis and photovoltaic devices. *Chem. Commun.* **2012**, *48*, 3686–3699. [CrossRef]
97. Novoselov, K.S.; Fal'ko, V.I.; Colombo, L.; Gellert, P.R.; Schwab, M.G.; Kim, K. A roadmap for graphene. *Nature* **2012**, *490*, 192–200. [CrossRef]
98. Tochihara, H.; Pomprasit, A.; Kadowaki, T.; Mizuno, S.; Minagawa, H.; Hayakawa, K.; Toyoshima, I. Decomposition of the surface carbide on Ni(001) induced by copper adsorption and surface segregation of carbon. *Surf. Sci. Lett.* **1991**, *257*, L623–L627.
99. Li, X.; Colombo, L.; Ruoff, R.S. Synthesis of Graphene Films on Copper Foils by Chemical Vapor Deposition. *Adv. Mater.* **2016**, *28*, 6247–6252. [CrossRef] [PubMed]
100. Rao, C.N.R.; Subrahmanyam, K.S.; Matte, H.S.; Abdulhakeem, B.; Govindaraj, A.; Das, B.; Kumar, P.; Ghosh, A.; Late, D.J. A study of the synthetic methods and properties of graphenes. *Sci. Technol. Adv. Mater.* **2010**, *11*, 054502. [CrossRef]
101. Reibold, M.; Paufler, P.; Levin, A.A.; Kochmann, W.; Patzke, N.; Meyer, D.C. Materials: Carbon nanotubes in an ancient Damascus sabre. *Nature* **2006**, *444*, 286. [CrossRef]
102. Radushkevich, L.; Lukyanovich, V.Á. O strukture ugleroda, obrazujucegosja pri termiceskom razlozenii okisi ugleroda na zeleznom kontakte. *Zurn Fisic Chim* **1952**, *26*, 88–95.
103. Boehm, H. Carbon from carbon monoxide disproportionation on nickel and iron catalysts: Morphological studies and possible growth mechanisms. *Carbon* **1973**, *11*, 583–590. [CrossRef]
104. Oberlin, A.; Endo, M.; Koyama, T. Filamentous growth of carbon through benzene decomposition. *J. Cryst. Growth* **1976**, *32*, 335–349. [CrossRef]
105. Costa, S.; Borowiak-Palen, E.; Kruszynska, M.; Bachmatiuk, A.; Kalenczuk, R. Characterization of carbon nanotubes by Raman spectroscopy. *Mater. Sci. Pol.* **2008**, *26*, 433–441.
106. Rafique, I.; Kausar, A.; Anwar, Z.; Muhammad, B. Exploration of Epoxy Resins, Hardening Systems, and Epoxy/Carbon Nanotube Composite Designed for High Performance Materials: A Review. *Polym. Technol. Eng.* **2015**, *55*, 312–333. [CrossRef]
107. Da Cunha, T.H.; De Oliveira, S.; Martins, I.L.; Geraldo, V.; Miquita, D.; Ramos, S.L.; Lacerda, R.G.; Ladeira, L.O.; Ferlauto, A.S. High-yield synthesis of bundles of double- and triple-walled carbon nanotubes on aluminum flakes. *Carbon* **2018**, *133*, 53–61. [CrossRef]
108. Xia, D.; Luo, Y.; Li, Q.; Xue, Q.; Zhang, X.; Liang, C.; Dong, M. Extracting the inner wall from nested double-walled carbon nanotube by platinum nanowire: Molecular dynamics simulations. *RSC Adv.* **2017**, *7*, 39480–39489. [CrossRef]
109. Dresselhaus, G.; Riichiro, S. *Physical Properties of Carbon Nanotubes*; World Scientific: Singapore, 1998.
110. Ibrahim, K.S. Carbon nanotubes-properties and applications: A review. *Carbon Lett.* **2013**, *14*, 131–144. [CrossRef]
111. Mubarak, N.; Abdullah, E.C.; Jayakumar, N.; Sahu, J. An overview on methods for the production of carbon nanotubes. *J. Ind. Eng. Chem.* **2014**, *20*, 1186–1197. [CrossRef]
112. Mittal, G.; Dhand, V.; Rhee, K.Y.; Kim, H.-J.; Jung, D.H. Carbon nanotubes synthesis using diffusion and premixed flame methods: A review. *Carbon Lett.* **2015**, *16*, 1–10. [CrossRef]
113. Ibrahim, M.; Saeed, T.; Chu, Y.-M.; Ali, H.M.; Cheraghian, G.; Kalbasi, R. Comprehensive study concerned graphene nano-sheets dispersed in ethylene glycol: Experimental study and theoretical prediction of thermal conductivity. *Powder Technol.* **2021**, *386*, 51–59. [CrossRef]

114. Yarmand, H.; Gharehkhani, S.; Shirazi, S.F.S.; Goodarzi, M.; Amiri, A.; Sarsam, W.; Alehashem, M.; Dahari, M.; Kazi, S. Study of synthesis, stability and thermo-physical properties of graphene nanoplatelet/platinum hybrid nanofluid. *Int. Commun. Heat Mass Transf.* **2016**, *77*, 15–21. [CrossRef]
115. You, X.; Li, S. Fully Developed Opposing Mixed Convection Flow in the Inclined Channel Filled with a Hybrid Nanofluid. *Nanomaterials* **2021**, *11*, 1107. [CrossRef] [PubMed]
116. Chakraborty, S.; Panigrahi, P.K. Stability of nanofluid: A review. *Appl. Therm. Eng.* **2020**, *174*, 115259. [CrossRef]
117. Yu, W.; Xie, H. A Review on Nanofluids: Preparation, Stability Mechanisms, and Applications. *J. Nanomater.* **2012**, *2012*, 1–17. [CrossRef]
118. Okonkwo, E.C.; Wole-Osho, I.; Almanassra, I.W.; Abdullatif, Y.M.; Al-Ansari, T. An updated review of nanofluids in various heat transfer devices. *J. Therm. Anal. Calorim.* **2020**, 1–56. [CrossRef]
119. Li, Y.; Zhou, J.; Tung, S.; Schneider, E.; Xi, S. A review on development of nanofluid preparation and characterization. *Powder Technol.* **2009**, *196*, 89–101. [CrossRef]
120. Bakthavatchalam, B.; Habib, K.; Saidur, R.; Saha, B.B.; Irshad, K. Comprehensive study on nanofluid and ionanofluid for heat transfer enhancement: A review on current and future perspective. *J. Mol. Liq.* **2020**, *305*, 112787. [CrossRef]
121. Toghraie, D.; Chaharsoghi, V.A.; Afrand, M. Measurement of thermal conductivity of ZnO–TiO2/EG hybrid nanofluid. *J. Therm. Anal. Calorim.* **2016**, *125*, 527–535. [CrossRef]
122. Abbasi, S.M.; Rashidi, A.; Nemati, A.; Arzani, K. The effect of functionalisation method on the stability and the thermal conductivity of nanofluid hybrids of carbon nanotubes/gamma alumina. *Ceram. Int.* **2013**, *39*, 3885–3891. [CrossRef]
123. Yang, L.; Ji, W.; Mao, M.; Huang, J.-N. An updated review on the properties, fabrication and application of hybrid-nanofluids along with their environmental effects. *J. Clean. Prod.* **2020**, *257*, 120408. [CrossRef]
124. Shenoy, U.S.; Shetty, A.N. A simple single-step approach towards synthesis of nanofluids containing cuboctahedral cuprous oxide particles using glucose reduction. *Front. Mater. Sci.* **2018**, *12*, 74–82. [CrossRef]
125. Hwang, Y.; Lee, J.; Lee, C.; Jung, Y.; Cheong, S.; Ku, B.; Jang, S. Stability and thermal conductivity characteristics of nanofluids. *Thermochim. Acta* **2007**, *455*, 70–74. [CrossRef]
126. Sabiha, M.; Mostafizur, R.; Saidur, R.; Mekhilef, S. Experimental investigation on thermo physical properties of single walled carbon nanotube nanofluids. *Int. J. Heat Mass Transf.* **2016**, *93*, 862–871. [CrossRef]
127. Alam Khairul, M.; Saidur, R.; Hossain, A.; Alim, M.A.; Mahbubul, I.M. Heat Transfer Performance of Different Nanofluids Flows in a Helically Coiled Heat Exchanger. *Adv. Mater. Res.* **2014**, *832*, 160–165. [CrossRef]
128. Lee, J.-H.; Choi, S.U.S.; Jang, S.P.; Lee, S.Y. Production of aqueous spherical gold nanoparticles using conventional ultrasonic bath. *Nanoscale Res. Lett.* **2012**, *7*, 420. [CrossRef] [PubMed]
129. Duangthongsuk, W.; Wongwises, S. Measurement of temperature-dependent thermal conductivity and viscosity of TiO2-water nanofluids. *Exp. Therm. Fluid Sci.* **2009**, *33*, 706–714. [CrossRef]
130. Sandhu, H.; Gangacharyulu, D. An experimental study on stability and some thermophysical properties of multiwalled carbon nanotubes with water–ethylene glycol mixtures. *Part. Sci. Technol.* **2016**, *35*, 547–554. [CrossRef]
131. Drzazga, M.; Dzido, G.; Lemanowicz, M.; Gierczycki, A. Influence of nonionic surfactant on nanofluid properties. In Proceedings of the 14th European Conference on Mixing, Warszawa, Poland, 10–13 September 2012; pp. 89–94.
132. Ilyas, S.U.; Pendyala, R.; Narahari, M.; Susin, L. Stability, rheology and thermal analysis of functionalized alumina- thermal oil-based nanofluids for advanced cooling systems. *Energy Convers. Manag.* **2017**, *142*, 215–229. [CrossRef]
133. Wen, D.; Ding, Y. Experimental investigation into the pool boiling heat transfer of aqueous based γ-alumina nanofluids. *J. Nanoparticle Res.* **2005**, *7*, 265–274. [CrossRef]
134. Fontes, D.H.; Ribatski, G.; Filho, E.P.B. Experimental evaluation of thermal conductivity, viscosity and breakdown voltage AC of nanofluids of carbon nanotubes and diamond in transformer oil. *Diam. Relat. Mater.* **2015**, *58*, 115–121. [CrossRef]
135. Farbod, M.; Asl, R.K.; Abadi, A.R.N. Morphology dependence of thermal and rheological properties of oil-based nanofluids of CuO nanostructures. *Colloids Surfaces A Physicochem. Eng. Asp.* **2015**, *474*, 71–75. [CrossRef]
136. Ettefaghi, E.-O.-L.; Mohtasebi, S.S.; Alaei, M.; Ahmadi, H.; Rashidi, A. Experimental evaluation of engine oil properties containing copper oxide nanoparticles as a nanoadditive. *Int. J. Ind. Chem.* **2013**, *4*, 28. [CrossRef]
137. Khan, A.I.; Valan Arasu, A. A review of influence of nanoparticle synthesis and geometrical parameters on thermophysical properties and stability of nanofluids. *Therm. Sci. Eng. Prog.* **2019**, *11*, 334–364. [CrossRef]
138. Noroozi, M.; Radiman, S.; Zakaria, A. Influence of Sonication on the Stability and Thermal Properties of Al2O3Nanofluids. *J. Nanomater.* **2014**, *2014*, 1–10. [CrossRef]
139. Asadi, A.; Pourfattah, F.; Szilágyi, I.M.; Afrand, M.; Żyła, G.; Ahn, H.S.; Wongwises, S.; Nguyen, H.M.; Arabkoohsar, A.; Mahian, O. Effect of sonication characteristics on stability, thermophysical properties, and heat transfer of nanofluids: A comprehensive review. *Ultrason. Sonochemistry* **2019**, *58*, 104701. [CrossRef]
140. Wciślik, S. Efficient Stabilization of Mono and Hybrid Nanofluids. *Energies* **2020**, *13*, 3793. [CrossRef]
141. Hamid, K.A.; Azmi, W.; Mamat, R.; Usri, N.; Najafi, G. Investigation of Al2O3 Nanofluid Viscosity for Different Water/EG Mixture Based. *Energy Procedia* **2015**, *79*, 354–359. [CrossRef]
142. Asadi, A.; Alarifi, I.M.; Foong, L.K. An experimental study on characterization, stability and dynamic viscosity of CuO-TiO2/water hybrid nanofluid. *J. Mol. Liq.* **2020**, *307*, 112987. [CrossRef]

143. Aghahadi, M.H.; Niknejadi, M.; Toghraie, D. An experimental study on the rheological behavior of hybrid Tungsten oxide (WO3)-MWCNTs/engine oil Newtonian nanofluids. *J. Mol. Struct.* **2019**, *1197*, 497–507. [CrossRef]
144. Kakavandi, A.; Akbari, M. Experimental investigation of thermal conductivity of nanofluids containing of hybrid nanoparticles suspended in binary base fluids and propose a new correlation. *Int. J. Heat Mass Transf.* **2018**, *124*, 742–751. [CrossRef]
145. Bahiraei, M.; Heshmatian, S. Graphene family nanofluids: A critical review and future research directions. *Energy Convers. Manag.* **2019**, *196*, 1222–1256. [CrossRef]
146. Le Ba, T.; Mahian, O.; Wongwises, S.; Szilágyi, I.M. Review on the recent progress in the preparation and stability of graphene-based nanofluids. *J. Therm. Anal. Calorim.* **2020**, *142*, 1–28. [CrossRef]
147. Texter, J. Graphene dispersions. *Curr. Opin. Colloid Interface Sci.* **2014**, *19*, 163–174. [CrossRef]
148. Xu, Y.; Cao, H.; Xue, Y.; Li, B.; Cai, W. Liquid-Phase Exfoliation of Graphene: An Overview on Exfoliation Media, Techniques, and Challenges. *Nanomaterials* **2018**, *8*, 942. [CrossRef] [PubMed]
149. Kang, H.U.; Kim, S.H.; Oh, J.M. Estimation of Thermal Conductivity of Nanofluid Using Experimental Effective Particle Volume. *Exp. Heat Transf.* **2006**, *19*, 181–191. [CrossRef]
150. Xie, H.; Yu, W.; Li, Y.; Chen, L. Discussion on the thermal conductivity enhancement of nanofluids. *Nanoscale Res. Lett.* **2011**, *6*, 124. [CrossRef]
151. Yu, W.; Xie, H.; Li, Y.; Chen, L.; Wang, Q. Experimental investigation on the thermal transport properties of ethylene glycol based nanofluids containing low volume concentration diamond nanoparticles. *Colloids Surfaces A: Physicochem. Eng. Asp.* **2011**, *380*, 1–5. [CrossRef]
152. Xie, H.; Yu, W.; Li, Y. Thermal performance enhancement in nanofluids containing diamond nanoparticles. *J. Phys. D Appl. Phys.* **2009**, *42*, 095413. [CrossRef]
153. Branson, B.T.; Beauchamp, P.S.; Beam, J.C.; Lukehart, C.M.; Davidson, J.L. Nanodiamond Nanofluids for Enhanced Thermal Conductivity. *ACS Nano* **2013**, *7*, 3183–3189. [CrossRef]
154. Ilyas, S.U.; Narahari, M.; Pendyala, R. Rheological characteristics of ultrastable diamond-thermal oil nanofluids. *J. Mol. Liq.* **2020**, *309*, 113098. [CrossRef]
155. Shukla, G.; Aiyer, H. Thermal conductivity enhancement of transformer oil using functionalized nanodiamonds. *IEEE Trans. Dielectr. Electr. Insul.* **2015**, *22*, 2185–2190. [CrossRef]
156. Sundar, L.S.; Singh, M.K.; Sousa, A.C.M. Experimental thermal conductivity and viscosity of nanodiamond-based propylene glycol and water mixtures. *Diam. Relat. Mater.* **2016**, *69*, 49–60. [CrossRef]
157. Li, P.; Zheng, Y.; Wu, Y.; Qu, P.; Yang, R.; Zhang, A. Nanoscale ionic graphene material with liquid-like behavior in the absence of solvent. *Appl. Surf. Sci.* **2014**, *314*, 983–990. [CrossRef]
158. Park, S.S.; Kim, N.J. Influence of the oxidation treatment and the average particle diameter of graphene for thermal conductivity enhancement. *J. Ind. Eng. Chem.* **2014**, *20*, 1911–1915. [CrossRef]
159. Mehrali, M.; Sadeghinezhad, E.; Latibari, S.T.; Kazi, S.N.; Mehrali, M.; Zubir, M.N.B.M.; Metselaar, H.S.C. Investigation of thermal conductivity and rheological properties of nanofluids containing graphene nanoplatelets. *Nanoscale Res. Lett.* **2014**, *9*, 15. [CrossRef] [PubMed]
160. Baby, T.T.; Ramaprabhu, S. Investigation of thermal and electrical conductivity of graphene based nanofluids. *J. Appl. Phys.* **2010**, *108*, 124308. [CrossRef]
161. Moghaddam, M.B.; Goharshadi, E.K.; Entezari, M.H.; Nancarrow, P. Preparation, characterization, and rheological properties of graphene–glycerol nanofluids. *Chem. Eng. J.* **2013**, *231*, 365–372. [CrossRef]
162. Babu, K.; Kumar, T.P. Effect of CNT concentration and agitation on surface heat flux during quenching in CNT nanofluids. *Int. J. Heat Mass Transf.* **2011**, *54*, 106–117. [CrossRef]
163. Öndin, O.; Kıvak, T.; Sarıkaya, M.; Yıldırım, Ç.V. Investigation of the influence of MWCNTs mixed nanofluid on the machinability characteristics of PH 13-8 Mo stainless steel. *Tribol. Int.* **2020**, *148*, 106323. [CrossRef]
164. Pourpasha, H.; Heris, S.Z.; Mahian, O.; Wongwises, S. The effect of multi-wall carbon nanotubes/turbine meter oil nanofluid concentration on the thermophysical properties of lubricants. *Powder Technol.* **2020**, *367*, 133–142. [CrossRef]
165. Rahimi, A.; Kasaeipoor, A.; Malekshah, E.H.; Kolsi, L. Experimental and numerical study on heat transfer performance of three-dimensional natural convection in an enclosure filled with DWCNTs-water nanofluid. *Powder Technol.* **2017**, *322*, 340–352. [CrossRef]
166. Shamaeil, M.; Firouzi, M.; Fakhar, A. The effects of temperature and volume fraction on the thermal conductivity of functionalized DWCNTs/ethylene glycol nanofluid. *J. Therm. Anal. Calorim.* **2016**, *126*, 1455–1462. [CrossRef]
167. Said, Z. Thermophysical and optical properties of SWCNTs nanofluids. *Int. Commun. Heat Mass Transf.* **2016**, *78*, 207–213. [CrossRef]
168. Harish, S.; Ishikawa, K.; Einarsson, E.; Aikawa, S.; Inoue, T.; Zhao, P.; Watanabe, M.; Chiashi, S.; Shiomi, J.; Maruyama, S. Temperature Dependent Thermal Conductivity Increase of Aqueous Nanofluid with Single Walled Carbon Nanotube Inclusion. *Mater. Express* **2012**, *2*, 213–223. [CrossRef]
169. Keblinski, P.; Eastman, J.A.; Cahill, D.G. Nanofluids for thermal transport. *Mater. Today* **2005**, *8*, 36–44. [CrossRef]

170. Qiu, L.; Zhu, N.; Feng, Y.; Michaelides, E.E.; Żyła, G.; Jing, D.; Zhang, X.; Norris, P.M.; Markides, C.N.; Mahian, O. A review of recent advances in thermophysical properties at the nanoscale: From solid state to colloids. *Phys. Rep.* **2020**, *843*, 1–81. [CrossRef]
171. Witharana, S.; Hodges, C.; Xu, D.; Lai, X.; Ding, Y. Aggregation and settling in aqueous polydisperse alumina nanoparticle suspensions. *J. Nanoparticle Res.* **2012**, *14*, 851. [CrossRef]
172. Bhattacharjee, S. DLS and zeta potential—What they are and what they are not? *J. Control. Release* **2016**, *235*, 337–351. [CrossRef]
173. Carrillo-Berdugo, I.; Zorrilla, D.; Sánchez-Márquez, J.; Aguilar, T.; Gallardo, J.J.; Gómez-Villarejo, R.; Alcántara, R.; Fernández-Lorenzo, C.; Navas, J. Interface-inspired formulation and molecular-level perspectives on heat conduction and energy storage of nanofluids. *Sci. Rep.* **2019**, *9*, 7595. [CrossRef] [PubMed]
174. Kunsong, M. Sedimentation Behavior of a Fine Kaolinite in the Presence of Fresh Fe Electrolyte. *Clays Clay Miner.* **1992**, *40*, 586–592. [CrossRef]
175. Joni, I.M.; Purwanto, A.; Iskandar, F.; Okuyama, K. Dispersion Stability Enhancement of Titania Nanoparticles in Organic Solvent Using a Bead Mill Process. *Ind. Eng. Chem. Res.* **2009**, *48*, 6916–6922. [CrossRef]
176. Hwang, Y.; Lee, J.-K.; Lee, J.-K.; Jeong, Y.-M.; Cheong, S.-I.; Ahn, Y.-C.; Kim, S.H. Production and dispersion stability of nanoparticles in nanofluids. *Powder Technol.* **2008**, *186*, 145–153. [CrossRef]
177. Fedele, L.; Colla, L.; Bobbo, S.; Barison, S.; Agresti, F. Experimental stability analysis of different water-based nanofluids. *Nanoscale Res. Lett.* **2011**, *6*, 300. [CrossRef]
178. Yu, J.; Grossiord, N.; Koning, C.E.; Loos, J. Controlling the dispersion of multi-wall carbon nanotubes in aqueous surfactant solution. *Carbon* **2007**, *45*, 618–623. [CrossRef]
179. Xie, H.; Lee, H.; Youn, W.; Choi, M. Nanofluids containing multiwalled carbon nanotubes and their enhanced thermal conductivities. *J. Appl. Phys.* **2003**, *94*, 4967. [CrossRef]
180. Xian-Ju, W.; Xin-Fang, L. Influence of pH on Nanofluids' Viscosity and Thermal Conductivity. *Chin. Phys. Lett.* **2009**, *26*, 056626. [CrossRef]
181. Yu, H.; Hermann, S.; Schulz, S.E.; Gessner, T.; Dong, Z.; Li, W.J. Optimizing sonication parameters for dispersion of single-walled carbon nanotubes. *Chem. Phys.* **2012**, *408*, 11–16. [CrossRef]
182. Xia, G.; Jiang, H.; Liu, R.; Zhai, Y. Effects of surfactant on the stability and thermal conductivity of Al2O3/de-ionized water nanofluids. *Int. J. Therm. Sci.* **2014**, *84*, 118–124. [CrossRef]
183. Tang, E.; Cheng, G.; Ma, X.; Pang, X.; Zhao, Q. Surface modification of zinc oxide nanoparticle by PMAA and its dispersion in aqueous system. *Appl. Surf. Sci.* **2006**, *252*, 5227–5232. [CrossRef]
184. Zhang, S.; Han, X. Effect of different surface modified nanoparticles on viscosity of nanofluids. *Adv. Mech. Eng.* **2018**, *10*. [CrossRef]
185. Halelfadl, S.; Maré, T.; Estellé, P. Efficiency of carbon nanotubes water based nanofluids as coolants. *Exp. Therm. Fluid Sci.* **2014**, *53*, 104–110. [CrossRef]
186. Pastoriza-Gallego, M.J.; Casanova, C.; Páramo, R.; Barbés, B.; Legido, J.L.; Piñeiro, M.M. A study on stability and thermophysical properties (density and viscosity) of Al2O3 in water nanofluid. *J. Appl. Phys.* **2009**, *106*, 064301. [CrossRef]
187. Saini, H.; Sandhu, A.; Sharma, S.; Dasaroju, G. Nanofluids: A Review Preparation, Stability, Properties and Applications. *Int. J. Eng. Res. Technol.* **2016**, *5*, 11–16.
188. Zhou, S.; Ni, R. Measurement of the specific heat capacity of water-based Al2O3 nanofluid. *Appl. Phys. Lett.* **2008**, *92*, 093123. [CrossRef]
189. Cabaleiro, D.; Gracia-Fernández, C.; Legido, J.; Lugo, L. Specific heat of metal oxide nanofluids at high concentrations for heat transfer. *Int. J. Heat Mass Transf.* **2015**, *88*, 872–879. [CrossRef]
190. Chandrasekar, M.; Suresh, S.; Bose, A.C. Experimental investigations and theoretical determination of thermal conductivity and viscosity of Al2O3/water nanofluid. *Exp. Therm. Fluid Sci.* **2010**, *34*, 210–216. [CrossRef]
191. Utomo, A.T.; Poth, H.; Robbins, P.T.; Pacek, A.W. Experimental and theoretical studies of thermal conductivity, viscosity and heat transfer coefficient of titania and alumina nanofluids. *Int. J. Heat Mass Transf.* **2012**, *55*, 7772–7781. [CrossRef]
192. Shin, D.; Banerjee, D. Enhancement of specific heat capacity of high-temperature silica-nanofluids synthesized in alkali chloride salt eutectics for solar thermal-energy storage applications. *Int. J. Heat Mass Transf.* **2011**, *54*, 1064–1070. [CrossRef]
193. Starace, A.K.; Gomez, J.C.; Pradhan, S.; Glatzmaier, G.C.; Wang, J. Nanofluid heat capacities. *J. Appl. Phys.* **2011**, *110*, 124323. [CrossRef]
194. Putnam, S.A.; Cahill, D.G.; Braun, P.V.; Ge, Z.; Shimmin, R.G. Thermal conductivity of nanoparticle suspensions. *J. Appl. Phys.* **2006**, *99*, 084308. [CrossRef]
195. Zhang, X.; Gu, H.; Fujii, M. Effective thermal conductivity and thermal diffusivity of nanofluids containing spherical and cylindrical nanoparticles. *J. Appl. Phys.* **2006**, *100*, 44325. [CrossRef]
196. Eapen, J.; Williams, W.C.; Buongiorno, J.; Hu, L.-W.; Yip, S.; Rusconi, R.; Piazza, R. Mean-Field Versus Microconvection Effects in Nanofluid Thermal Conduction. *Phys. Rev. Lett.* **2007**, *99*, 095901. [CrossRef] [PubMed]
197. Timofeeva, E.V.; Gavrilov, A.N.; McCloskey, J.M.; Tolmachev, Y.; Sprunt, S.; Lopatina, L.M.; Selinger, J. Thermal conductivity and particle agglomeration in alumina nanofluids: Experiment and theory. *Phys. Rev. E* **2007**, *76*, 061203. [CrossRef]

198. Buongiorno, J.; Venerus, D.C.; Prabhat, N.; McKrell, T.J.; Townsend, J.; Christianson, R.J.; Tolmachev, Y.; Keblinski, P.; Hu, L.-W.; Alvarado, J.L.; et al. A benchmark study on the thermal conductivity of nanofluids. *J. Appl. Phys.* **2009**, *106*, 094312. [CrossRef]
199. Barai, D.P.; Bhanvase, B.A.; Sonawane, S.H. A Review on Graphene Derivatives-Based Nanofluids: Investigation on Properties and Heat Transfer Characteristics. *Ind. Eng. Chem. Res.* **2020**, *59*, 10231–10277. [CrossRef]
200. Ambreen, T.; Kim, M.-H. Influence of particle size on the effective thermal conductivity of nanofluids: A critical review. *Appl. Energy* **2020**, *264*, 114684. [CrossRef]
201. Yu, W.; Xie, H.; Wang, X. Enhanced Thermal Conductivity of Liquid Paraffin Based Nanofluids Containing Copper Nanoparticles. *J. Dispers. Sci. Technol.* **2011**, *32*, 948–951. [CrossRef]
202. Haghighi, E.B.; Nikkam, N.; Saleemi, M.; Behi, M.; Mirmohammadi, S.A.; Poth, H.; Khodabandeh, R.; Toprak, M.; Muhammed, M.; Palm, B. Shelf stability of nanofluids and its effect on thermal conductivity and viscosity. *Meas. Sci. Technol.* **2013**, *24*. [CrossRef]
203. Li, X.; Zhu, D.; Wang, X.; Wang, N.; Gao, J.; Li, H. Thermal conductivity enhancement dependent pH and chemical surfactant for Cu-H2O nanofluids. *Thermochim. Acta* **2008**, *469*, 98–103. [CrossRef]
204. Prasher, R.; Evans, W.; Meakin, P.; Fish, J.; Phelan, P.; Keblinskia, P. Effect of aggregation on thermal conduction in colloidal nanofluids. *Appl. Phys. Lett.* **2006**, *89*, 143119. [CrossRef]
205. Wang, J.; Zheng, R.; Gao, J.; Chen, G. Heat conduction mechanisms in nanofluids and suspensions. *Nano Today* **2012**, *7*, 124–136. [CrossRef]
206. Hong, H.; Wright, B.; Wensel, J.; Jin, S.; Ye, X.R.; Roy, W. Enhanced thermal conductivity by the magnetic field in heat transfer nanofluids containing carbon nanotube. *Synth. Met.* **2007**, *157*, 437–440. [CrossRef]
207. Wright, B.; Thomas, D.; Hong, H.; Groven, L.; Puszynski, J.; Duke, E.; Ye, X.; Jin, S. Magnetic field enhanced thermal conductivity in heat transfer nanofluids containing Ni coated single wall carbon nanotubes. *Appl. Phys. Lett.* **2007**, *91*, 173116. [CrossRef]
208. Wensel, J.; Wright, B.; Thomas, D.; Douglas, W.; Mannhalter, B.; Cross, W.; Hong, H.; Kellar, J.J.; Smith, P.; Roy, W. Enhanced thermal conductivity by aggregation in heat transfer nanofluids containing metal oxide nanoparticles and carbon nanotubes. *Appl. Phys. Lett.* **2008**, *92*, 023110. [CrossRef]
209. Hong, H.; Luan, X.; Horton, M.; Li, C.; Peterson, G. Alignment of carbon nanotubes comprising magnetically sensitive metal oxides in heat transfer nanofluids. *Thermochim. Acta* **2011**, *525*, 87–92. [CrossRef]
210. Younes, H.; Hong, H.; Peterson, G.P. A Novel Approach to Fabricate Carbon Nanomaterials–Nanoparticle Solids through Aqueous Solutions and Their Applications. *Nanomanufacturing Metrol.* **2021**, 1–11. [CrossRef]
211. Xue, L.; Keblinski, P.; Phillpot, S.; Choi, S.-S.; Eastman, J.; Xue, L.; Keblinski, P.; Phillpot, S.; Choi, S.-S.; Eastman, J. Effect of liquid layering at the liquid–solid interface on thermal transport. *Int. J. Heat Mass Transf.* **2004**, *47*, 4277–4284. [CrossRef]
212. Keblinski, P.; Phillpot, S.; Choi, S.; Eastman, J. Mechanisms of heat flow in suspensions of nano-sized particles (nanofluids). *Int. J. Heat Mass Transf.* **2002**, *45*, 855–863. [CrossRef]
213. Jang, S.P.; Choi, S.U.S. Role of Brownian motion in the enhanced thermal conductivity of nanofluids. *Appl. Phys. Lett.* **2004**, *84*, 4316–4318. [CrossRef]
214. Buongiorno, J. Convective Transport in Nanofluids. *J. Heat Transf.* **2006**, *128*, 240–250. [CrossRef]
215. Koo, J.; Kleinstreuer, C. Impact analysis of nanoparticle motion mechanisms on the thermal conductivity of nanofluids. *Int. Commun. Heat Mass Transf.* **2005**, *32*, 1111–1118. [CrossRef]
216. Domingues, G.; Voltz, S.; Joulain, K.; Greffet, J.-J. Heat Transfer between Two Nanoparticles Through Near Field Interaction. *Phys. Rev. Lett.* **2005**, *94*, 085901. [CrossRef] [PubMed]
217. Shen, S.; Narayanaswamy, A.; Chen, G. Surface Phonon Polaritons Mediated Energy Transfer between Nanoscale Gaps. *Nano Lett.* **2009**, *9*, 2909–2913. [CrossRef] [PubMed]
218. Kumar, D.H.; Patel, H.E.; Kumar, V.R.R.; Sundararajan, T.; Pradeep, T.; Das, S.K. Model for Heat Conduction in Nanofluids. *Phys. Rev. Lett.* **2004**, *93*, 144301. [CrossRef]
219. Keblinski, P.; Cahill, D.G. Comment on "Model for heat conduction in nanofluids". *Phys. Rev. Lett.* **2005**, *95*, 209401. [CrossRef]
220. Paul, G.; Chopkar, M.; Manna, I.; Das, P. Techniques for measuring the thermal conductivity of nanofluids: A review. *Renew. Sustain. Energy Rev.* **2010**, *14*, 1913–1924. [CrossRef]
221. Tawfik, M.M. Experimental studies of nanofluid thermal conductivity enhancement and applications: A review. *Renew. Sustain. Energy Rev.* **2017**, *75*, 1239–1253. [CrossRef]
222. [Pradhan, N.R.; Duan, H.; Liang, J.; Iannacchione, G.S. The specific heat and effective thermal conductivity of composites containing single-wall and multi-wall carbon nanotubes. *Nanotechnology* **2009**, *20*, 245705. [CrossRef]
223. Che, J.; Çagin, T.; Goddard, W.A. Thermal conductivity of carbon nanotubes. *Nanotechnology* **2000**, *11*, 65–69. [CrossRef]
224. Yu, W.; Xie, H.; Wang, X. Significant thermal conductivity enhancement for nanofluids containing graphene nanosheets. *Phys. Lett. A* **2011**, *375*, 1323–1328. [CrossRef]
225. Yarmand, H.; Gharehkhani, S.; Shirazi, S.F.S.; Amiri, A.; Alehashem, M.S.; Dahari, M.; Kazi, S. Experimental investigation of thermo-physical properties, convective heat transfer and pressure drop of functionalized graphene nanoplatelets aqueous nanofluid in a square heated pipe. *Energy Convers. Manag.* **2016**, *114*, 38–49. [CrossRef]
226. Zhang, L.; Chen, L.; Liu, J.; Fang, X.; Zhang, Z. Effect of morphology of carbon nanomaterials on thermo-physical characteristics, optical properties and photo-thermal conversion performance of nanofluids. *Renew. Energy* **2016**, *99*, 888–897. [CrossRef]

227. Ghozatloo, A.; Shariaty-Niasar, M.; Rashidi, A.M. Preparation of nanofluids from functionalized Graphene by new alkaline method and study on the thermal conductivity and stability. *Int. Commun. Heat Mass Transf.* **2013**, *42*, 89–94. [CrossRef]
228. Ghozatloo, A.; Rashidi, A.; Shariaty-Niassar, M. Convective heat transfer enhancement of graphene nanofluids in shell and tube heat exchanger. *Exp. Therm. Fluid Sci.* **2014**, *53*, 136–141. [CrossRef]
229. Goodarzi, M.; Kherbeet, A.; Afrand, M.; Sadeghinezhad, E.; Mehrali, M.; Zahedi, P.; Wongwises, S.; Dahari, M. Investigation of heat transfer performance and friction factor of a counter-flow double-pipe heat exchanger using nitrogen-doped, graphene-based nanofluids. *Int. Commun. Heat Mass Transf.* **2016**, *76*, 16–23. [CrossRef]
230. Liu, W.; Malekahmadi, O.; Bagherzadeh, S.A.; Ghashang, M.; Karimipour, A.; Hasani, S.; Tlili, I.; Goodarzi, M. A novel comprehensive experimental study concerned graphene oxide nanoparticles dispersed in water: Synthesise, characterisation, thermal conductivity measurement and present a new approach of RLSF neural network. *Int. Commun. Heat Mass Transf.* **2019**, *109*, 104333. [CrossRef]
231. Maxwell, J.C. *The Scientific Papers of James Clerk Maxwell*; Cambridge University Press: London, UK, 1890.
232. Jefferson, T.B.; Witzell, O.W.; Sibbitt, W.L. Thermal Conductivity of Graphite—Silicone Oil and Graphite-Water Suspensions. *Ind. Eng. Chem.* **1958**, *50*, 1589–1592. [CrossRef]
233. Hamilton, R.L.; Crosser, O.K. Thermal Conductivity of Heterogeneous Two-Component Systems. *Ind. Eng. Chem. Fundam.* **1962**, *1*, 187–191. [CrossRef]
234. Wasp, E.J.; Kenny, J.P.; Gandhi, R.L. Solid-liquid flow: Slurry pipeline transportation. Pumps, valves, mechanical equipment, economics. *Ser. Bulk Mater. Handl.* **1977**, *1*, 216–219.
235. Yu, W.; Choi, S. The Role of Interfacial Layers in the Enhanced Thermal Conductivity of Nanofluids: A Renovated Maxwell Model. *J. Nanoparticle Res.* **2003**, *5*, 167–171. [CrossRef]
236. Xuan, Y.; Li, Q.; Hu, W. Aggregation structure and thermal conductivity of nanofluids. *AIChE J.* **2003**, *49*, 1038–1043. [CrossRef]
237. Nan, C.-W.; Shi, Z.; Lin, Y. A simple model for thermal conductivity of carbon nanotube-based composites. *Chem. Phys. Lett.* **2003**, *375*, 666–669. [CrossRef]
238. Yu, W.; Choi, S. The role of interfacial layers in the enhanced thermal conductivity of nanofluids: A renovated Hamilton?Crosser model. *J. Nanopart. Res.* **2004**, *6*, 355–361. [CrossRef]
239. Prasher, R.; Bhattacharya, P.; Phelan, P.E. Thermal Conductivity of Nanoscale Colloidal Solutions (Nanofluids). *Phys. Rev. Lett.* **2005**, *94*, 025901. [CrossRef] [PubMed]
240. Xue, Q. Model for thermal conductivity of carbon nanotube-based composites. *Phys. B Condens. Matter* **2005**, *368*, 302–307. [CrossRef]
241. Murshed, S.M.S.; Leong, K.C.; Yang, C. A Model for Predicting the Effective Thermal Conductivity of Nanoparticle-Fluid Suspensions. *Int. J. Nanosci.* **2011**, *5*, 23–33. [CrossRef]
242. Vajjha, R.S.; Das, D.K.; Kulkarni, D.P. Development of new correlations for convective heat transfer and friction factor in turbulent regime for nanofluids. *Int. J. Heat Mass Transf.* **2010**, *53*, 4607–4618. [CrossRef]
243. Xing, M.; Yu, J.; Wang, R. Experimental investigation and modelling on the thermal conductivity of CNTs based nanofluids. *Int. J. Therm. Sci.* **2016**, *104*, 404–411. [CrossRef]
244. Gao, Y.; Wang, H.; Sasmito, A.P.; Mujumdar, A.S. Measurement and modeling of thermal conductivity of graphene nanoplatelet water and ethylene glycol base nanofluids. *Int. J. Heat Mass Transf.* **2018**, *123*, 97–109. [CrossRef]
245. Li, M.-J.; He, Y.-L.; Tao, W.-Q. A novel semi-empirical model on predicting the thermal conductivity of diathermic oil-based nanofluid for solar thermal application. *Int. J. Heat Mass Transf.* **2019**, *138*, 1002–1013. [CrossRef]
246. Jóźwiak, B.; Dzido, G.; Zorębski, E.; Kolanowska, A.; Jędrysiak, R.; Dziadosz, J.; Libera, M.; Boncel, S.; Dzida, M. Remarkable Thermal Conductivity Enhancement in Carbon-Based Ionanofluids: Effect of Nanoparticle Morphology. *ACS Appl. Mater. Interfaces* **2020**, *12*, 38113–38123. [CrossRef]
247. Pantzali, M.; Kanaris, A.; Antoniadis, K.; Mouza, A.; Paras, S. Effect of nanofluids on the performance of a miniature plate heat exchanger with modulated surface. *Int. J. Heat Fluid Flow* **2009**, *30*, 691–699. [CrossRef]
248. Masoumi, N.; Sohrabi, N.; Behzadmehr, A. A new model for calculating the effective viscosity of nanofluids. *J. Phys. D: Appl. Phys.* **2009**, *42*, 055501. [CrossRef]
249. Hosseini, S.M.; Moghadassi, A.R.; Henneke, D.E. A new dimensionless group model for determining the viscosity of nanofluids. *J. Therm. Anal. Calorim.* **2010**, *100*, 873–877. [CrossRef]
250. Chevalier, J.; Tillement, O.; Ayela, F. Structure and rheology of SiO2 nanoparticle suspensions under very high shear rates. *Phys. Rev. E Stat. Nonlin Soft Matter Phys.* **2009**, *80*, 051403. [CrossRef] [PubMed]
251. Nguyen, C.; Desgranges, F.; Galanis, N.; Roy, G.; Maré, T.; Boucher, S.; Mintsa, H.A. Viscosity data for Al2O3–water nanofluid—hysteresis: Is heat transfer enhancement using nanofluids reliable? *Int. J. Therm. Sci.* **2008**, *47*, 103–111. [CrossRef]
252. Jarahnejad, M.; Haghighi, E.B.; Saleemi, M.; Nikkam, N.; Khodabandeh, R.; Palm, B.; Toprak, M.S.; Muhammed, M. Experimental investigation on viscosity of water-based Al2O3 and TiO2 nanofluids. *Rheol. Acta* **2015**, *54*, 411–422. [CrossRef]
253. Dalkilic, A.; Küçükyıldırım, B.; Eker, A.A.; Çebi, A.; Tapan, S.; Jumpholkul, C.; Wongwises, S. Experimental investigation on the viscosity of Water-CNT and Antifreeze-CNT nanofluids. *Int. Commun. Heat Mass Transf.* **2017**, *80*, 47–59. [CrossRef]
254. Contreras, E.M.C.; Oliveira, G.A.; Filho, E.P.B. Experimental analysis of the thermohydraulic performance of graphene and silver nanofluids in automotive cooling systems. *Int. J. Heat Mass Transf.* **2019**, *132*, 375–387. [CrossRef]

255. Asadi, M.; Asadi, A. Dynamic viscosity of MWCNT/ZnO–engine oil hybrid nanofluid: An experimental investigation and new correlation in different temperatures and solid concentrations. *Int. Commun. Heat Mass Transf.* **2016**, *76*, 41–45. [CrossRef]
256. Żyła, G.; Fal, J.; Estellé, P. The influence of ash content on thermophysical properties of ethylene glycol based graphite/diamonds mixture nanofluids. *Diam. Relat. Mater.* **2017**, *74*, 81–89. [CrossRef]
257. Mena, J.B.; de Moraes, A.A.U.; Benito, Y.R.; Ribatski, G.; Parise, J.A.R. Extrapolation of Al2O3–water nanofluid viscosity for temperatures and volume concentrations beyond the range of validity of existing correlations. *Appl. Therm. Eng.* **2013**, *51*, 1092–1097. [CrossRef]
258. Srivastava, S. Effect of aggregation on thermal conductivity and viscosity of nanofluids. *Appl. Nanosci.* **2012**, *2*, 325–331.
259. Sadri, R.; Ahmadi, G.; Togun, H.; Dahari, M.; Kazi, S.N.; Sadeghinezhad, E.; Zubir, N. An experimental study on thermal conductivity and viscosity of nanofluids containing carbon nanotubes. *Nanoscale Res. Lett.* **2014**, *9*, 151. [CrossRef] [PubMed]
260. Asadi, A.; Alarifi, I.M. Effects of ultrasonication time on stability, dynamic viscosity, and pumping power management of MWCNT-water nanofluid: An experimental study. *Sci. Rep.* **2020**, *10*, 1–10. [CrossRef]
261. Doganay, S.; Alsangur, R.; Turgut, A. Effect of external magnetic field on thermal conductivity and viscosity of magnetic nanofluids: A review. *Mater. Res. Express* **2019**, *6*, 112003. [CrossRef]
262. Yiamsawas, T.; Mahian, O.; Dalkilic, A.S.; Kaewnai, S.; Wongwises, S. Experimental studies on the viscosity of TiO2 and Al2O3 nanoparticles suspended in a mixture of ethylene glycol and water for high temperature applications. *Appl. Energy* **2013**, *111*, 40–45. [CrossRef]
263. Walters, K.; Jones, W. Measurement of Viscosity. In *Instrumentation Reference Book*; Boyes, W., Ed.; Butterworth-Heinemann: Boston, MA, USA, 2010; pp. 69–75.
264. Żyła, G.; Cholewa, M. On unexpected behavior of viscosity of diethylene glycol-based MgAl2O4 nanofluids. *RSC Adv.* **2014**, *4*, 26057–26062. [CrossRef]
265. Prasher, R.; Song, D.; Wang, J.; Phelan, P. Measurements of nanofluid viscosity and its implications for thermal applications. *Appl. Phys. Lett.* **2006**, *89*, 133108. [CrossRef]
266. Lyu, Z.; Asadi, A.; Alarifi, I.M.; Ali, V.; Foong, L.K. Thermal and Fluid Dynamics Performance of MWCNT-Water Nanofluid Based on Thermophysical Properties: An Experimental and Theoretical Study. *Sci. Rep.* **2020**, *10*, 1–14. [CrossRef]
267. Akhavan-Zanjani, H.; Saffar-Avval, M.; Mansourkiaei, M.; Ahadi, M.; Sharif, F. Turbulent Convective Heat Transfer and Pressure Drop of Graphene–Water Nanofluid Flowing Inside a Horizontal Circular Tube. *J. Dispers. Sci. Technol.* **2014**, *35*, 1230–1240. [CrossRef]
268. Iranmanesh, S.; Mehrali, M.; Sadeghinezhad, E.; Ang, B.C.; Ong, H.C.; Esmaeilzadeh, A. Evaluation of viscosity and thermal conductivity of graphene nanoplatelets nanofluids through a combined experimental–statistical approach using respond surface methodology method. *Int. Commun. Heat Mass Transf.* **2016**, *79*, 74–80. [CrossRef]
269. Ghozatloo, A.; Azimi, M.S.; Shariaty, N.M.; Morad, R.A. Investigation of nanoparticles morphology on viscosity of nanofluids and new correlation for prediction. *J. Nanostructures* **2015**, *5*, 161–168.
270. Batchelor, G.K. The effect of Brownian motion on the bulk stress in a suspension of spherical particles. *J. Fluid Mech.* **1977**, *83*, 97–117. [CrossRef]
271. Einstein, A. Eine neue Bestimmung der Moleküldimensionen. *Ann. Phys. Leipzig* **1906**, *324*, 289–306. [CrossRef]
272. Hatschek, E. The general theory of viscosity of two-phase systems. *Trans. Faraday Soc.* **1913**, *9*, 80–92. [CrossRef]
273. Saitô, N. Concentration Dependence of the Viscosity of High Polymer Solutions. I. *J. Phys. Soc. Jpn.* **1950**, *5*, 4–8. [CrossRef]
274. Mooney, M. The viscosity of a concentrated suspension of spherical particles. *J. Colloid Sci.* **1951**, *6*, 162–170. [CrossRef]
275. Brinkman, H.C. The Viscosity of Concentrated Suspensions and Solutions. *J. Chem. Phys.* **1952**, *20*, 571. [CrossRef]
276. Roscoe, R. The viscosity of suspensions of rigid spheres. *Br. J. Appl. Phys.* **1952**, *3*, 267–269. [CrossRef]
277. Maron, S.H.; Pierce, P.E. Application of ree-eyring generalized flow theory to suspensions of spherical particles. *J. Colloid Sci.* **1956**, *11*, 80–95. [CrossRef]
278. Krieger, I.M.; Dougherty, T.J. A Mechanism for Non-Newtonian Flow in Suspensions of Rigid Spheres. *Trans. Soc. Rheol.* **1959**, *3*, 137–152. [CrossRef]
279. Frankel, N.; Acrivos, A. On the viscosity of a concentrated suspension of solid spheres. *Chem. Eng. Sci.* **1967**, *22*, 847–853. [CrossRef]
280. Nielsen, L.E. Generalized Equation for the Elastic Moduli of Composite Materials. *J. Appl. Phys.* **1970**, *41*, 4626–4627. [CrossRef]
281. Brenner, H.; Condiff, D.W. Transport mechanics in systems of orientable particles. IV. convective transport. *J. Colloid Interface Sci.* **1974**, *47*, 199–264. [CrossRef]
282. Jeffrey, D.J.; Acrivos, A. The rheological properties of suspensions of rigid particles. *AIChE J.* **1976**, *22*, 417–432. [CrossRef]
283. Graham, A.L. On the viscosity of suspensions of solid spheres. *Flow Turbul. Combust.* **1981**, *37*, 275–286. [CrossRef]
284. Kitano, T.; Kataoka, T.; Shirota, T. An empirical equation of the relative viscosity of polymer melts filled with various inorganic fillers. *Rheol. Acta* **1981**, *20*, 207–209. [CrossRef]
285. Bicerano, J.; Douglas, J.F.; Brune, D.A. Model for the Viscosity of Particle Dispersions. *J. Macromol. Sci. Part C* **1999**, *39*, 561–642. [CrossRef]
286. Wang, X.; Xu, X.; Choi, S.U.S. Thermal Conductivity of Nanoparticle—Fluid Mixture. *J. Thermophys. Heat Transf.* **1999**, *13*, 474–480. [CrossRef]

287. Bobbo, S.; Fedele, L.; Benetti, A.; Colla, L.; Fabrizio, M.; Pagura, C.; Barison, S. Viscosity of water based SWCNH and TiO2 nanofluids. *Exp. Therm. Fluid Sci.* **2012**, *36*, 65–71. [CrossRef]
288. Esfe, M.H.; Saedodin, S.; Mahian, O.; Wongwises, S. Thermophysical properties, heat transfer and pressure drop of COOH-functionalized multi walled carbon nanotubes/water nanofluids. *Int. Commun. Heat Mass Transf.* **2014**, *58*, 176–183. [CrossRef]
289. Aberoumand, S.; Jafarimoghaddam, A.; Moravej, M.; Aberoumand, H.; Javaherdeh, K. Experimental study on the rheological behavior of silver-heat transfer oil nanofluid and suggesting two empirical based correlations for thermal conductivity and viscosity of oil based nanofluids. *Appl. Therm. Eng.* **2016**, *101*, 362–372. [CrossRef]
290. Akbari, M.; Afrand, M.; Arshi, A.; Karimipour, A. An experimental study on rheological behavior of ethylene glycol based nanofluid: Proposing a new correlation as a function of silica concentration and temperature. *J. Mol. Liq.* **2017**, *233*, 352–357. [CrossRef]
291. Esfe, M.H.; Raki, H.R.; Emami, M.R.S.; Afrand, M. Viscosity and rheological properties of antifreeze based nanofluid containing hybrid nano-powders of MWCNTs and TiO2 under different temperature conditions. *Powder Technol.* **2019**, *342*, 808–816. [CrossRef]
292. Ansón-Casaos, A.; Ciria, J.C.; Sanahuja-Parejo, O.; Víctor-Román, S.; González-Domínguez, J.M.; García-Bordejé, E.; Benito, A.M.; Maser, W.K. The viscosity of dilute carbon nanotube (1D) and graphene oxide (2D) nanofluids. *Phys. Chem. Chem. Phys.* **2020**, *22*, 11474–11484. [CrossRef] [PubMed]
293. Kumar, V.; Tiwari, A.K.; Ghosh, S.K. Application of nanofluids in plate heat exchanger: A review. *Energy Convers. Manag.* **2015**, *105*, 1017–1036. [CrossRef]
294. Duffie, J.A.; Beckman, W.A. Concentrating Collectors. In *Solar Engineering of Thermal Processes*, 4th ed.; John Wiley & Sons: Hoboken, NJ, USA, 2013; pp. 322–370.
295. Fernández-García, A.; Zarza, E.; Valenzuela, L.; Pérez, M. Parabolic-trough solar collectors and their applications. *Renew. Sustain. Energy Rev.* **2010**, *14*, 1695–1721. [CrossRef]
296. Lillo, I.; Pérez, E.; Moreno, S.; Silva, M. Process Heat Generation Potential from Solar Concentration Technologies in Latin America: The Case of Argentina. *Energies* **2017**, *10*, 383. [CrossRef]
297. Kalogirou, S. Solar thermal collectors and applications. *Prog. Energy Combust. Sci.* **2004**, *30*, 231–295. [CrossRef]
298. Olia, H.; Torabi, M.; Bahiraei, M.; Ahmadi, M.H.; Goodarzi, M.; Safaei, M.R. Application of Nanofluids in Thermal Performance Enhancement of Parabolic Trough Solar Collector: State-of-the-Art. *Appl. Sci.* **2019**, *9*, 463. [CrossRef]
299. Khan, M.S.; Abid, M.; Ratlamwala, T.A.H. Energy, Exergy and Economic Feasibility Analyses of a 60 MW Conventional Steam Power Plant Integrated with Parabolic Trough Solar Collectors Using Nanofluids. *Iran. J. Sci. Technol. Trans. Mech. Eng.* **2018**, *43*, 193–209. [CrossRef]
300. Abed, N.; Afgan, I.; Iacovides, H.; Cioncolini, A.; Khurshid, I.; Nasser, A. Thermal-Hydraulic Analysis of Parabolic Trough Collectors Using Straight Conical Strip Inserts with Nanofluids. *Nanomaterials* **2021**, *11*, 853. [CrossRef]
301. Ebrazeh, S.; Sheikholeslami, M. Applications of nanomaterial for parabolic trough collector. *Powder Technol.* **2020**, *375*, 472–492. [CrossRef]
302. Kasaeian, A.; Daviran, S.; Azarian, R.D.; Rashidi, A. Performance evaluation and nanofluid using capability study of a solar parabolic trough collector. *Energy Convers. Manag.* **2015**, *89*, 368–375. [CrossRef]
303. Kasaeian, A.; Daneshazarian, R.; Rezaei, R.; Pourfayaz, F.; Kasaeian, G. Experimental investigation on the thermal behavior of nanofluid direct absorption in a trough collector. *J. Clean. Prod.* **2017**, *158*, 276–284. [CrossRef]
304. Mwesigye, A.; Yılmaz, İ.H.; Meyer, J.P. Numerical analysis of the thermal and thermodynamic performance of a parabolic trough solar collector using SWCNTs-Therminol® VP-1 nanofluid. *Renew. Energy* **2018**, *119*, 844–862. [CrossRef]
305. Dugaria, S.; Bortolato, M.; Del Col, D. Modelling of a direct absorption solar receiver using carbon based nanofluids under concentrated solar radiation. *Renew. Energy* **2018**, *128*, 495–508. [CrossRef]
306. Muhammad, M.; Che Sidik, N.A.; Umar, U.; Hamisu, M.; Sa'ad, A.; Malaysia, T.; Lumpur, K.; Sultan, J.; Petra, Y.; Semarak, J. Carbon Nanotube for Solar Energy Applications: A Review. *J. Adv. Res. Fluid Mech. Therm. Sci.* **2019**, *56*, 233–247.
307. Di Rosa, D.; Wanic, M.; Fal, J.; Żyła, G.; Mercatelli, L.; Sani, E. Optical and dielectric properties of ethylene glycol-based nanofluids containing nanodiamonds with various purities. *Powder Technol.* **2019**, *356*, 508–516. [CrossRef]
308. Ho, M.; Obbard, E.; A Burr, P.; Yeoh, G. A review on the development of nuclear power reactors. *Energy Procedia* **2019**, *160*, 459–466. [CrossRef]
309. Fernández-Arias, P.; Vergara, D.; Orosa, J.A. A Global Review of PWR Nuclear Power Plants. *Appl. Sci.* **2020**, *10*, 4434. [CrossRef]
310. Rowinski, M.K.; White, T.; Zhao, J. Small and Medium sized Reactors (SMR): A review of technology. *Renew. Sustain. Energy Rev.* **2015**, *44*, 643–656. [CrossRef]
311. Estrada-Domínguez, L.-A.; Espinosa-Paredes, G.; Nuñez-Carrera, A.; del Valle-Gallegos, E.; Vázquez-Rodriguez, R. Progress and utilization of small nuclear reactors. *Energy Sources, Part A: Recover. Util. Environ. Eff.* **2016**, *38*, 2362–2369. [CrossRef]
312. Abu-Khader, M.M. Recent advances in nuclear power: A review. *Prog. Nucl. Energy* **2009**, *51*, 225–235. [CrossRef]
313. Gu, Z. History review of nuclear reactor safety. *Ann. Nucl. Energy* **2018**, *120*, 682–690. [CrossRef]
314. Buongiorno, J.; Hu, L.-W. Nanofluid heat transfer enhancement for nuclear reactor applications. In *International Conference on Micro/Nanoscale Heat Transfer*; ASME: Shanghai, China, 2009; pp. 517–522.
315. Martin Callizo, C. Flow Boiling Heat Transfer in Single Vertical Channels of Small Diameter. Ph.D. Thesis, KTH, Brinellvägen, Stockholm, 2010.

316. Beck, F.R.; Jin, Y.; Garrett, G.; Cheung, F.-B.; Bajorek, S.M.; Tien, K. Effect of Fouling on Quenching of Simulated Fuel Rods. *Transactions* **2019**, *120*, 1123–1125.
317. Garrett, G.; Beck, F.R.; Jin, Y.; Cheung, F.B.; Bajorek, S.M.; Tien, K.; Hoxie, C. Effects of system parameters on the two-phase flow and heat transfer behavior in a rod bundle. In *4th Thermal and Fluids Engineering Conference*; Begel House Inc.: Las Vegas, NV, USA, 2019; pp. 1383–1396.
318. Mesquita, A.Z.; Rodrigues, R.R. Detection of the Departure from Nucleate Boiling in Nuclear Fuel Rod Simulators. *Int. J. Nucl. Energy* **2013**, *2013*, 1–7. [CrossRef]
319. Kamel, M.S.; Lezsovits, F.; Hussein, A.K. Experimental studies of flow boiling heat transfer by using nanofluids. *J. Therm. Anal. Calorim.* **2019**, *138*, 4019–4043. [CrossRef]
320. Ahn, H.S.; Kim, M.H. A Review on Critical Heat Flux Enhancement with Nanofluids and Surface Modification. *J. Heat Transf.* **2011**, *134*, 024001. [CrossRef]
321. Dadhich, M.; Prajapati, O.S. A brief review on factors affecting flow and pool boiling. *Renew. Sustain. Energy Rev.* **2019**, *112*, 607–625. [CrossRef]
322. Mukherjee, S.; Jana, S.; Mishra, P.C.; Chaudhuri, P.; Chakrabarty, S. Experimental investigation on thermo-physical properties and subcooled flow boiling performance of Al2O3/water nanofluids in a horizontal tube. *Int. J. Therm. Sci.* **2021**, *159*, 106581. [CrossRef]
323. Sarafraz, M.M.; Abad, A.T.K. Statistical and experimental investigation on flow boiling heat transfer to carbon nanotube-therminol nanofluid. *Phys. A Stat. Mech. Appl.* **2019**, *536*, 122505. [CrossRef]
324. Le Ba, T.; Alkurdi, A.Q.; Lukács, I.E.; Molnár, J.; Wongwises, S.; Gróf, G.; Szilágyi, I.M. A Novel Experimental Study on the Rheological Properties and Thermal Conductivity of Halloysite Nanofluids. *Nanomaterials* **2020**, *10*, 1834. [CrossRef] [PubMed]
325. Murshed, S.M.S.; Nieto de Castro, C.A. Superior thermal features of carbon nanotubes-based nanofluids—A review. *Renew. Sustain. Energy Rev.* **2014**, *37*, 155–167. [CrossRef]
326. Younes, H.; Al Ghaferi, A.; Saadat, I.; Hong, H. Nanofluids based on carbon nanostructures. In *Advances in Carbon Nanostructures*; IntechOpen: London, UK, 2016.
327. Manthey, R.; Viereckl, F.; Shabestary, A.M.; Zhang, Y.; Ding, W.; Lucas, D.; Schuster, C.; Leyer, S.; Hurtado, A.; Hampel, U. Modelling of Passive Heat Removal Systems: A Review with Reference to the Framatome BWR Reactor KERENA: Part II. *Energies* **2020**, *13*, 109. [CrossRef]
328. Hashemi, M.; Noie, S.H. Study of flow boiling heat transfer characteristics of critical heat flux using carbon nanotubes and water nanofluid. *J. Therm. Anal. Calorim.* **2017**, *130*, 2199–2209. [CrossRef]
329. Hashemi, M.; Noie, S.H. Investigation of critical heat flux (CHF) enhancement in flow boiling using carbon nanotubes/water nanofluid. *Transp. Phenom. Nano Micro. Scales* **2020**, *8*, 81–88.
330. Sarafraz, M.; Hormozi, F. Comparatively experimental study on the boiling thermal performance of metal oxide and multi-walled carbon nanotube nanofluids. *Powder Technol.* **2016**, *287*, 412–430. [CrossRef]
331. Lee, S.W.; Kim, K.M.; Bang, I.C. Study on flow boiling critical heat flux enhancement of graphene oxide/water nanofluid. *Int. J. Heat Mass Transf.* **2013**, *65*, 348–356. [CrossRef]
332. Park, S.D.; Bang, I.C. Flow boiling CHF enhancement in an external reactor vessel cooling (ERVC) channel using graphene oxide nanofluid. *Nucl. Eng. Des.* **2013**, *265*, 310–318. [CrossRef]
333. Groeneveld, D.; Shan, J.; Vasić, A.; Leung, L.; Durmayaz, A.; Yang, J.; Cheng, S.; Tanase, A. The 2006 CHF look-up table. *Nucl. Eng. Des.* **2007**, *237*, 1909–1922. [CrossRef]
334. Zhang, C.; Zhang, L.; Xu, H.; Wang, D.; Ye, B. Investigation of flow boiling performance and the resulting surface deposition of graphene oxide nanofluid in microchannels. *Exp. Therm. Fluid Sci.* **2017**, *86*, 1–10. [CrossRef]
335. Mohammed, H.I.; Giddings, D.; Walker, G.S. Experimental investigation of nanoparticles concentration, boiler temperature and flow rate on flow boiling of zinc bromide and acetone solution in a rectangular duct. *Int. J. Heat Mass Transf.* **2019**, *130*, 710–721. [CrossRef]
336. Kim, S.J.; McKrell, T.; Buongiorno, J.; Hu, L.-W. Experimental Study of Flow Critical Heat Flux in Alumina-Water, Zinc-Oxide-Water, and Diamond-Water Nanofluids. *J. Heat Transf.* **2009**, *131*, 043204. [CrossRef]
337. Kim, S.J.; McKrell, T.; Buongiorno, J.; Hu, L.-W. Subcooled flow boiling heat transfer of dilute alumina, zinc oxide, and diamond nanofluids at atmospheric pressure. *Nucl. Eng. Des.* **2010**, *240*, 1186–1194. [CrossRef]
338. DolatiAsl, K.; Bakhshan, Y.; Abedini, E.; Niazi, S. Correlations for estimating critical heat flux (CHF) of nanofluid flow boiling. *Int. J. Heat Mass Transf.* **2019**, *139*, 69–76. [CrossRef]
339. Incropera, F.P.; Dewitt, D.P.; Bergman, T.; Lavine, A. *Fundamentals of Mass and Heat Transfer*; John Wiley and Sons: Hoboken, NJ, USA, 2002.
340. Peterson, L.J.; Bajorek, S.M. Experimental Investigation of Minimum Film Boiling Temperature for Vertical Cylinders at Elevated Pressure. In *10th International Conference on Nuclear Engineering*; ASME: Arlington, VA, USA, 2002; pp. 883–892.
341. Sinha, J. Effects of surface roughness, oxidation level, and liquid subcooling on the minimum film boiling temperature. *Exp. Heat Transf.* **2003**, *16*, 45–60. [CrossRef]
342. Lee, C.Y.; Chun, T.H.; In, W.K. Effect of change in surface condition induced by oxidation on transient pool boiling heat transfer of vertical stainless steel and copper rodlets. *Int. J. Heat Mass Transf.* **2014**, *79*, 397–407. [CrossRef]

343. Henry, R. A correlation for the minimum film boiling temperature. *AIChE Symp. Ser.* **1974**, *70*, 81–90.
344. Bahman, A.M.; Ebrahim, S.A. Prediction of the minimum film boiling temperature using artificial neural network. *Int. J. Heat Mass Transf.* **2020**, *155*, 119834. [CrossRef]
345. Kang, J.-Y.; Kim, T.K.; Lee, G.C.; Jo, H.; Kim, M.H.; Park, H.S. Impact of system parameters on quenching heat transfer in the candidate materials for accident tolerant fuel-cladding in LWRs. *Ann. Nucl. Energy* **2019**, *129*, 375–389. [CrossRef]
346. Sakurai, A.; Shiotsu, M.; Hata, K. Transient boiling caused by vapor film collapse at minimum heat flux in film boiling. *Nucl. Eng. Des.* **1987**, *99*, 167–175. [CrossRef]
347. Jun-Young, K.; Cheol, L.G.; Kaviany, M.; Sun, P.H.; Moriyama, K.; Hwan, K.M. Control of minimum film-boiling quench temperature of small spheres with micro-structured surface. *Int. J. Multiph. Flow* **2018**, *103*, 30–42. [CrossRef]
348. Adler, M.R. *The Influence of Water Purity and Subcooling on the Minimum Film Boiling Temperature*; University of Illinois at Urbana-Champaign: Champaign, IL, USA, 1979.
349. Freud, R.; Harari, R.; Sher, E. Collapsing criteria for vapor film around solid spheres as a fundamental stage leading to vapor explosion. *Nucl. Eng. Des.* **2009**, *239*, 722–727. [CrossRef]
350. Ebrahim, S.A.; Alat, E.; Sohag, F.A.; Fudurich, V.; Chang, S.; Cheung, F.-B.; Bajorek, S.M.; Tien, K.; Hoxie, C.L. Effects of Substrate Materials and Surface Conditions on the Minimum Film-Boiling Temperature. *Nucl. Technol.* **2018**, *205*, 226–238. [CrossRef]
351. Ebrahim, S.A.; Chang, S.; Cheung, F.-B.; Bajorek, S.M. Parametric investigation of film boiling heat transfer on the quenching of vertical rods in water pool. *Appl. Therm. Eng.* **2018**, *140*, 139–146. [CrossRef]
352. Carey, V.P. *Liquid-Vapor Phase-Change Phenomena: An Introduction to the Thermophysics of Vaporization and Condensation Processes in Heat Transfer Equipment*; CRC Press: Boca Raton, FL, USA, 2020.
353. Shoji, M.; Witte, L.C.; Sankaran, S. The influence of surface conditions and subcooling on film-transition boiling. *Exp. Therm. Fluid Sci.* **1990**, *3*, 280–290. [CrossRef]
354. Lee, C.Y.; Kim, S. Parametric investigation on transient boiling heat transfer of metal rod cooled rapidly in water pool. *Nucl. Eng. Des.* **2017**, *313*, 118–128. [CrossRef]
355. Khan, A.; Ali, H.M. A comprehensive review on pool boiling heat transfer using nanofluids. *Therm. Sci.* **2019**, *23*, 3209–3237. [CrossRef]
356. Liang, G.; Mudawar, I. Review of pool boiling enhancement with additives and nanofluids. *Int. J. Heat Mass Transf.* **2018**, *124*, 423–453. [CrossRef]
357. Kamatchi, R.; Venkatachalapathy, S. Parametric study of pool boiling heat transfer with nanofluids for the enhancement of critical heat flux: A review. *Int. J. Therm. Sci.* **2015**, *87*, 228–240. [CrossRef]
358. Yazid, M.N.A.W.M.; Sidik, N.A.C.; Yahya, W.J. Heat and mass transfer characteristics of carbon nanotube nanofluids: A review. *Renew. Sustain. Energy Rev.* **2017**, *80*, 914–941. [CrossRef]
359. Barber, J.; Brutin, D.; Tadrist, L. A review on boiling heat transfer enhancement with nanofluids. *Nanoscale Res. Lett.* **2011**, *6*, 280. [CrossRef]
360. Bang, I.C.; Chang, S.H. Boiling heat transfer performance and phenomena of Al2O3–water nano-fluids from a plain surface in a pool. *Int. J. Heat Mass Transf.* **2005**, *48*, 2407–2419. [CrossRef]
361. Pham, Q.; Kim, T.; Lee, S.; Chang, S. Enhancement of critical heat flux using nano-fluids for Invessel Retention–External Vessel Cooling. *Appl. Therm. Eng.* **2012**, *35*, 157–165. [CrossRef]
362. Sakashita, H. CHF and near-wall boiling behaviors in pool boiling of water on a heating surface coated with nanoparticles. *Int. J. Heat Mass Transf.* **2012**, *55*, 7312–7320. [CrossRef]
363. Kathiravan, R.; Kumar, R.; Gupta, A.; Chandra, R. Characterization and Pool Boiling Heat Transfer Studies of Nanofluids. *J. Heat Transf.* **2009**, *131*, 081902. [CrossRef]
364. Coursey, J.S.; Kim, J. Nanofluid boiling: The effect of surface wettability. *Int. J. Heat Fluid Flow* **2008**, *29*, 1577–1585. [CrossRef]
365. Kim, S.; Bang, I.; Buongiorno, J.; Hu, L. Surface wettability change during pool boiling of nanofluids and its effect on critical heat flux. *Int. J. Heat Mass Transf.* **2007**, *50*, 4105–4116. [CrossRef]
366. Kwark, S.M.; Kumar, R.; Moreno, G.; Yoo, J.; You, S.M. Pool boiling characteristics of low concentration nanofluids. *Int. J. Heat Mass Transf.* **2010**, *53*, 972–981. [CrossRef]
367. Yang, X.-F.; Liu, Z.-H. Pool boiling heat transfer of functionalized nanofluid under sub-atmospheric pressures. *Int. J. Therm. Sci.* **2011**, *50*, 2402–2412. [CrossRef]
368. Okawa, T.; Takamura, M.; Kamiya, T. Boiling time effect on CHF enhancement in pool boiling of nanofluids. *Int. J. Heat Mass Transf.* **2012**, *55*, 2719–2725. [CrossRef]
369. Liu, Z.-H.; Xiong, J.-G.; Bao, R. Boiling heat transfer characteristics of nanofluids in a flat heat pipe evaporator with micro-grooved heating surface. *Int. J. Multiph. Flow* **2007**, *33*, 1284–1295. [CrossRef]
370. Sarafraz, M.; Hormozi, F.; Silakhori, M.; Peyghambarzadeh, S. On the fouling formation of functionalized and non-functionalized carbon nanotube nano-fluids under pool boiling condition. *Appl. Therm. Eng.* **2016**, *95*, 433–444. [CrossRef]
371. Park, K.-J.; Jung, D.; Shim, S.E. Nucleate boiling heat transfer in aqueous solutions with carbon nanotubes up to critical heat fluxes. *Int. J. Multiph. Flow* **2009**, *35*, 525–532. [CrossRef]
372. Liu, Z.-H.; Yang, X.-F.; Xiong, J.-G. Boiling characteristics of carbon nanotube suspensions under sub-atmospheric pressures. *Int. J. Therm. Sci.* **2010**, *49*, 1156–1164. [CrossRef]

373. Park, K.-J.; Jung, D. Enhancement of nucleate boiling heat transfer using carbon nanotubes. *Int. J. Heat Mass Transf.* **2007**, *50*, 4499–4502. [CrossRef]
374. Shoghl, S.N.; Bahrami, M.; Moraveji, M.K. Experimental investigation and CFD modeling of the dynamics of bubbles in nanofluid pool boiling. *Int. Commun. Heat Mass Transf.* **2014**, *58*, 12–24. [CrossRef]
375. Sarafraz, M.; Hormozi, F. Experimental investigation on the pool boiling heat transfer to aqueous multi-walled carbon nanotube nanofluids on the micro-finned surfaces. *Int. J. Therm. Sci.* **2016**, *100*, 255–266. [CrossRef]
376. Amiri, A.; Shanbedi, M.; Amiri, H.; Heris, S.Z.; Kazi, S.; Chew, B.; Eshghi, H. Pool boiling heat transfer of CNT/water nanofluids. *Appl. Therm. Eng.* **2014**, *71*, 450–459. [CrossRef]
377. Park, S.-S.; Kim, N.-J. Critical heat flux enhancement in pool-boiling heat transfer using oxidized multi-wall carbon nanotubes. *Int. J. Energy Res.* **2015**, *39*, 1391–1401. [CrossRef]
378. Kumar, R.; Milanova, D. Effect of surface tension on nanotube nanofluids. *Appl. Phys. Lett.* **2009**, *94*, 073107. [CrossRef]
379. Zhang, L.; Fan, L.; Yu, Z.; Cen, K. An experimental investigation of transient pool boiling of aqueous nanofluids with graphene oxide nanosheets as characterized by the quenching method. *Int. J. Heat Mass Transf.* **2014**, *73*, 410–414. [CrossRef]
380. Ahn, H.S.; Kim, J.M.; Kim, M.H. Experimental study of the effect of a reduced graphene oxide coating on critical heat flux enhancement. *Int. J. Heat Mass Transf.* **2013**, *60*, 763–771. [CrossRef]
381. Park, S.D.; Won Lee, S.; Kang, S.; Bang, I.C.; Kim, J.H.; Shin, H.S.; Lee, D.W.; Won Lee, D. Effects of nanofluids containing graphene/graphene-oxide nanosheets on critical heat flux. *Appl. Phys. Lett.* **2010**, *97*, 023103. [CrossRef]
382. Kim, J.M.; Kim, T.; Kim, J.; Kim, M.H.; Ahn, H.S. Effect of a graphene oxide coating layer on critical heat flux enhancement under pool boiling. *Int. J. Heat Mass Transf.* **2014**, *77*, 919–927. [CrossRef]
383. Park, S.D.; Bang, I.C. Experimental study of a universal CHF enhancement mechanism in nanofluids using hydrodynamic instability. *Int. J. Heat Mass Transf.* **2014**, *70*, 844–850. [CrossRef]
384. Truong, B.; Hu, L.-W.; Buongiorno, J.; McKrell, T. Modification of sandblasted plate heaters using nanofluids to enhance pool boiling critical heat flux. *Int. J. Heat Mass Transf.* **2010**, *53*, 85–94. [CrossRef]
385. Kathiravan, R.; Kumar, R.; Gupta, A.; Chandra, R.; Jain, P.K. Pool Boiling Characteristics of Carbon Nanotube Based Nanofluids Over a Horizontal Tube. *J. Therm. Sci. Eng. Appl.* **2009**, *1*, 022001. [CrossRef]
386. Jaikumar, A.; Kandlikar, S.G.; Gupta, A. Pool Boiling Enhancement through Graphene and Graphene Oxide Coatings. *Heat Transf. Eng.* **2016**, *38*, 1274–1284. [CrossRef]
387. Gerardi, C.D. Investigation of the Pool Boiling Heat Transfer Enhancement of Nano-Engineered Fluids by Means of High-Speed Infrared Thermography. Ph.D. Thesis, Massachusetts Institute of Technology, Cambridge, MA, USA, 2009.
388. Fan, L.-W.; Li, J.-Q.; Wu, Y.-Z.; Zhang, L.; Yu, Z.-T. Pool boiling heat transfer during quenching in carbon nanotube (CNT)-based aqueous nanofluids: Effects of length and diameter of the CNTs. *Appl. Therm. Eng.* **2017**, *122*, 555–565. [CrossRef]
389. Fan, L.-W.; Li, J.-Q.; Li, D.-Y.; Zhang, L.; Yu, Z.-T.; Cen, K.-F. The effect of concentration on transient pool boiling heat transfer of graphene-based aqueous nanofluids. *Int. J. Therm. Sci.* **2015**, *91*, 83–95. [CrossRef]
390. Kim, H.; DeWitt, G.; McKrell, T.; Buongiorno, J.; Hu, L.-W. On the quenching of steel and zircaloy spheres in water-based nanofluids with alumina, silica and diamond nanoparticles. *Int. J. Multiph. Flow* **2009**, *35*, 427–438. [CrossRef]
391. Wang, S.K. *Handbook of Air Conditioning and Refrigeration*, 2nd ed.; McGraw-Hill: New York, NY, USA, 2001.
392. Pérez-Tello, C.; Campbell-Ramírez, H.; Suástegui-Macías, J.A.; Reinhardt, M.S. *Methodology of Energy Management in Housing and Buildings of Regions with Hot and Dry Climates*; IntechOpen: London, UK, 2018; pp. 13–29.
393. Said, S.A.; El-Shaarawi, M.A.; Siddiqui, M.U. Alternative designs for a 24-h operating solar-powered absorption refrigeration technology. *Int. J. Refrig.* **2012**, *35*, 1967–1977. [CrossRef]
394. Veera Raghavalu, K.; Govindha Rasu, N. Review on Applications of NanoFluids used in Vapour Compression Refrigeration System for Cop Enhancement. *IOP Conf. Series: Mater. Sci. Eng.* **2018**, *330*, 012112. [CrossRef]
395. Nair, V.; Tailor, P.; Parekh, A. Nanorefrigerants: A comprehensive review on its past, present and future. *Int. J. Refrig.* **2016**, *67*, 290–307. [CrossRef]
396. Bandgar, M.; Ragit, S.; Kolhe, K.; Biradar, N. Effect of Nano Lubricant on the Performance of Vapour Compression Refrigeration System: A Review. *J. Emerg. Technol. Innov. Res.* **2016**, *3*, 56–59.
397. Azmi, W.; Sharif, M.; Yusof, T.; Mamat, R.; Redhwan, A. Potential of nanorefrigerant and nanolubricant on energy saving in refrigeration system—A review. *Renew. Sustain. Energy Rev.* **2017**, *69*, 415–428. [CrossRef]
398. Alawi, O.A.; Salih, J.M.; Mallah, A. Thermo-physical properties effectiveness on the coefficient of performance of Al2O3/R141b nano-refrigerant. *Int. Commun. Heat Mass Transf.* **2019**, *103*, 54–61. [CrossRef]
399. Jwo, C.-S.; Jeng, L.-Y.; Teng, T.-P.; Chang, H. Effects of nanolubricant on performance of hydrocarbon refrigerant system. *J. Vac. Sci. Technol. B: Microelectron. Nanometer Struct.* **2009**, *27*, 1473. [CrossRef]
400. Lee, J.; Cho, S.; Hwang, Y.; Lee, C.; Kim, S.H. Enhancement of Lubrication Properties of Nano-oil by Controlling the Amount of Fullerene Nanoparticle Additives. *Tribol. Lett.* **2007**, *28*, 203–208. [CrossRef]
401. Sheikholeslami, M.; Rezaeianjouybari, B.; Darzi, M.; Shafee, A.; Li, Z.; Nguyen, T.K. Application of nano-refrigerant for boiling heat transfer enhancement employing an experimental study. *Int. J. Heat Mass Transf.* **2019**, *141*, 974–980. [CrossRef]
402. Krishnan, R.S.; Arulprakasajothi, M.; Logesh, K.; Raja, N.D.; Rajendra, M. Analysis and Feasibilty of Nano-Lubricant in Vapour Compression Refrigeration System. *Mater. Today Proc.* **2018**, *5*, 20580–20587. [CrossRef]

403. Lee, K.; Hwang, Y.; Cheong, S.; Kwon, L.; Kim, S.; Lee, J. Performance evaluation of nano-lubricants of fullerene nanoparticles in refrigeration mineral oil. *Curr. Appl. Phys.* **2009**, *9*, e128–e131. [CrossRef]
404. Lee, K.; Hwang, Y.; Cheong, S.; Choi, Y.; Kwon, L.; Lee, J.; Kim, S.H. Understanding the Role of Nanoparticles in Nano-oil Lubrication. *Tribol. Lett.* **2009**, *35*, 127–131. [CrossRef]
405. Jia, T.; Wang, R.; Xu, R. Performance of MoFe2O4–NiFe2O4/Fullerene-added nano-oil applied in the domestic refrigerator compressors. *Int. J. Refrig.* **2014**, *45*, 120–127. [CrossRef]
406. Saidur, R.; Kazi, S.; Hossain, M.; Rahman, M.; Mohammed, H. A review on the performance of nanoparticles suspended with refrigerants and lubricating oils in refrigeration systems. *Renew. Sustain. Energy Rev.* **2011**, *15*, 310–323. [CrossRef]
407. Kasaeian, A.; Hosseini, S.M.; Sheikhpour, M.; Mahian, O.; Yan, W.-M.; Wongwises, S. Applications of eco-friendly refrigerants and nanorefrigerants: A review. *Renew. Sustain. Energy Rev.* **2018**, *96*, 91–99. [CrossRef]
408. Alawi, O.A.; Sidik, N.A.C.; Mohammed, H. A comprehensive review of fundamentals, preparation and applications of nanorefrigerants. *Int. Commun. Heat Mass Transf.* **2014**, *54*, 81–95. [CrossRef]
409. Sun, B.; Yang, D. Flow boiling heat transfer characteristics of nano-refrigerants in a horizontal tube. *Int. J. Refrig.* **2014**, *38*, 206–214. [CrossRef]
410. Zhang, S.; Yu, Y.; Xu, Z.; Huang, H.; Liu, Z.; Liu, C.; Long, X.; Ge, Z. Measurement and modeling of the thermal conductivity of nanorefrigerants with low volume concentrations. *Thermochim. Acta* **2020**, *688*, 178603. [CrossRef]
411. Selimefendigil, F. Experimental investigation of nano compressor oil effect on the cooling performance of a vapor-compression refrigeration system. *J. Therm. Eng.* **2019**, *5*, 100–104. [CrossRef]
412. Celen, A.; Çebi, A.; Aktas, M.; Mahian, O.; Dalkilic, A.S.; Wongwises, S. A review of nanorefrigerants: Flow characteristics and applications. *Int. J. Refrig.* **2014**, *44*, 125–140. [CrossRef]
413. Sundar, L.S.; Singh, M.K.; Ramana, E.V.; Singh, B.K.; Grácio, J.; Sousa, A. Enhanced Thermal Conductivity and Viscosity of Nanodiamond-Nickel Nanocomposite Nanofluids. *Sci. Rep.* **2014**, *4*, 4039. [CrossRef]
414. Sundar, L.S.; Ramana, E.V.; Graca, M.P.; Singh, M.K.; Sousa, A. Nanodiamond-Fe 3 O 4 nanofluids: Preparation and measurement of viscosity, electrical and thermal conductivities. *Int. Commun. Heat Mass Transf.* **2016**, *73*, 62–74. [CrossRef]
415. Bhattad, A.; Sarkar, J.; Ghosh, P. Improving the performance of refrigeration systems by using nanofluids: A comprehensive review. *Renew. Sustain. Energy Rev.* **2018**, *82*, 3656–3669. [CrossRef]
416. Liu, M.; Lin, M.C.; Wang, C. Enhancements of thermal conductivities with Cu, CuO, and carbon nanotube nanofluids and application of MWNT/water nanofluid on a water chiller system. *Nanoscale Res. Lett.* **2011**, *6*, 297. [CrossRef]
417. Maheshwary, P.B.; Handa, C.C.; Nemade, K.R. Preparation of nanorefrigerants using mono-, bi- and tri-layer graphene nanosheets in R134a refrigerant. *AIP Conf. Proc.* **2019**, *2104*, 020017. [CrossRef]
418. Cantuario, T.E.; Fonseca, A.F. High Performance of Carbon Nanotube Refrigerators. *Ann. der Phys.* **2019**, *531*, 1800502. [CrossRef]
419. Yang, L.; Jiang, W.; Ji, W.; Mahian, O.; Bazri, S.; Sadri, R.; Badruddin, I.A.; Wongwises, S. A review of heating/cooling processes using nanomaterials suspended in refrigerants and lubricants. *Int. J. Heat Mass Transf.* **2020**, *153*, 119611. [CrossRef]
420. Molana, M.; Wang, H. A critical review on numerical study of nanorefrigerant heat transfer enhancement. *Powder Technol.* **2020**, *368*, 18–31. [CrossRef]
421. Hamisa, A.H.; Yusof, T.M.; Azmi, W.H.; Mamat, R.; Sharif, M. The stability of TiO2/POE nanolubricant for automotive air-conditioning system of hybrid electric vehicles. *IOP Conf. Series: Mater. Sci. Eng.* **2020**, *863*, 012050. [CrossRef]
422. Nair, V.; Parekh, A.D.; Tailor, P.R. Performance analysis of Al2O3–R718 nanorefrigerant turbulent flow through a flooded chiller tube: A numerical investigation. *J. Braz. Soc. Mech. Sci. Eng.* **2020**, *42*, 1–16. [CrossRef]
423. Safaei, M.R.; Ranjbarzadeh, R.; Hajizadeh, A.; Bahiraei, M.; Afrand, M.; Karimipour, A. Effects of cobalt ferrite coated with silica nanocomposite on the thermal conductivity of an antifreeze: New nanofluid for refrigeration condensers. *Int. J. Refrig.* **2019**, *102*, 86–95. [CrossRef]
424. Aktas, M.; Dalkilic, A.S.; Celen, A.; Cebi, A.; Mahian, O.; Wongwises, S. A Theoretical Comparative Study on Nanorefrigerant Performance in a Single-Stage Vapor-Compression Refrigeration Cycle. *Adv. Mech. Eng.* **2014**, *7*, 138725. [CrossRef]
425. Adelekan, D.S.; Ohunakin, O.; Babarinde, T.O.; Odunfa, M.K.; Leramo, R.O.; Oyedepo, S.O.; Badejo, D.C. Experimental performance of LPG refrigerant charges with varied concentration of TiO 2 nano-lubricants in a domestic refrigerator. *Case Stud. Therm. Eng.* **2017**, *9*, 55–61. [CrossRef]
426. Kumar, R.; Singh, D.K.; Chander, S. An experimental approach to study thermal and tribology behavior of LPG refrigerant and MO lubricant appended with ZnO nanoparticles in domestic refrigeration cycle. *Heat Mass Transf.* **2020**, *56*, 2303–2311. [CrossRef]
427. Krishna Sabareesh, R.; Gobinath, N.; Sajith, V.; Das, S.; Sobhan, C.B. Application of TiO2 nanoparticles as a lubricant-additive for vapor compression refrigeration systems—An experimental investigation. *Int. J. Refrig.* **2012**, *35*, 1989–1996. [CrossRef]
428. Wang, R.; Wu, Q.; Wu, Y. Use of nanoparticles to make mineral oil lubricants feasible for use in a residential air conditioner employing hydro-fluorocarbons refrigerants. *Energy Build.* **2010**, *42*, 2111–2117. [CrossRef]
429. Coumaressin, T.; Palaniradja, K. Performance analysis of a refrigeration system using nano fluid. *Int. J. Adv. Mech. Eng.* **2014**, *4*, 459–470.
430. Xuan, Y.; Li, Q. Heat transfer enhancement of nanofluids. *Int. J. Heat Fluid Flow* **2000**, *21*, 58–64. [CrossRef]
431. Babarinde, T.O.; Akinlabi, S.A.; Madyira, D.M. Enhancing the Performance of Vapour Compression Refrigeration System using Nano Refrigerants: A review. *IOP Conf. Series: Mater. Sci. Eng.* **2018**, *413*, 012068. [CrossRef]

432. Akhavan-Behabadi, M.A.; Sadoughi, M.K.; Darzi, M.; Fakoor-Pakdaman, M. Experimental study on heat transfer characteristics of R600a/POE/CuO nano-refrigerant flow condensation. *Exp. Therm. Fluid Sci.* **2015**, *66*, 46–52. [CrossRef]
433. Ajayi, O.O.; Useh, O.O.; Banjo, S.O.; Oweoye, F.T.; Attabo, A.; Ogbonnaya, M.; Okokpujie, I.P.; Salawu, E.Y. Investigation of the heat transfer effect of Ni/R134a nanorefrigerant in a mobile hybrid powered vapour compression refrigerator. *IOP Conf. Series: Mater. Sci. Eng.* **2018**, *391*, 012001. [CrossRef]
434. Bhutta, M.U.; Khan, Z.A. Wear and friction performance evaluation of nickel based nanocomposite coatings under refrigerant lubrication. *Tribol. Int.* **2020**, *148*, 106312. [CrossRef]
435. Park, K.-J.; Jung, D. Boiling heat transfer enhancement with carbon nanotubes for refrigerants used in building air-conditioning. *Energy Build.* **2007**, *39*, 1061–1064. [CrossRef]
436. Zhang, S.Y.; Ge, Z.; Wang, H.T. Characteristics of Flow Boiling Heat Transfer and Pressure Drop of MWCNT–R123 Nanorefrigerant: Experimental Investigations and Correlations. *Nanoscale Microscale Thermophys. Eng.* **2016**, *20*, 97–120. [CrossRef]
437. Sun, B.; Wang, H.; Yang, D. Effects of surface functionalization on the flow boiling heat transfer characteristics of MWCNT/R141b nanorefrigerants in smooth tube. *Exp. Therm. Fluid Sci.* **2018**, *92*, 162–173. [CrossRef]
438. Jiang, W.; Ding, G.; Peng, H. Measurement and model on thermal conductivities of carbon nanotube nanorefrigerants. *Int. J. Therm. Sci.* **2009**, *48*, 1108–1115. [CrossRef]
439. Peng, H.; Ding, G.; Hu, H.; Jiang, W. Influence of carbon nanotubes on nucleate pool boiling heat transfer characteristics of refrigerant–oil mixture. *Int. J. Therm. Sci.* **2010**, *49*, 2428–2438. [CrossRef]
440. Ahmadpour, M.; Akhavan-Behabadi, M.A.; Sajadi, B.; Salehi-Kohestani, A. Experimental study of R600a/oil/MWCNT nano-refrigerant condensing flow inside micro-fin tubes. *Heat Mass Transf.* **2020**, *56*, 749–757. [CrossRef]
441. Kumaresan, V.; Velraj, R.; Das, S.K. Convective heat transfer characteristics of secondary refrigerant based CNT nanofluids in a tubular heat exchanger. *Int. J. Refrig.* **2012**, *35*, 2287–2296. [CrossRef]
442. Asadi, M.; Asadi, A.; Aberoumand, S. An experimental and theoretical investigation on the effects of adding hybrid nanoparticles on heat transfer efficiency and pumping power of an oil-based nanofluid as a coolant fluid. *Int. J. Refrig.* **2018**, *89*, 83–92. [CrossRef]
443. Baskar, S.; Chandrasekaran, M.; Kumar, T.V.; Vivek, P.; Ramasubramanian, S. Experimental studies on flow and heat transfer characteristics of secondary refrigerant-based CNT nanofluids for cooling applications. *Int. J. Ambient. Energy* **2018**, *41*, 285–288. [CrossRef]
444. Wang, Z.; Han, F.; Ji, Y.; Li, W. Performance and Exergy Transfer Analysis of Heat Exchangers with Graphene Nanofluids in Seawater Source Marine Heat Pump System. *Energies* **2020**, *13*, 1762. [CrossRef]
445. Lin, L.; Peng, H.; Ding, G. Dispersion stability of multi-walled carbon nanotubes in refrigerant with addition of surfactant. *Appl. Therm. Eng.* **2015**, *91*, 163–171. [CrossRef]
446. Alawi, O.A.; Sidik, N.A.C. The effect of temperature and particles concentration on the determination of thermo and physical properties of SWCNT-nanorefrigerant. *Int. Commun. Heat Mass Transf.* **2015**, *67*, 8–13. [CrossRef]
447. Dalkilic, A.S.; Mahian, O.; Kucukyildirim, B.O.; Eker, A.A.; Ozturk, T.H.; Jumpholkul, C.; Wongwises, S. Experimental Study on the Stability and Viscosity for the Blends of Functionalized MWCNTs with Refrigeration Compressor Oils. *Curr. Nanosci.* **2018**, *14*, 216–226. [CrossRef]
448. Patil, M.S.; Kim, S.C.; Seo, J.-H.; Lee, M.-Y. Review of the Thermo-Physical Properties and Performance Characteristics of a Refrigeration System Using Refrigerant-Based Nanofluids. *Energies* **2016**, *9*, 22. [CrossRef]
449. Abbas, M.; Walvekar, R.G.; Hajibeigy, M.T.; Javadi, F.S. Efficient air-condition unit by using nano-refrigerant. In Proceedings of the 1st Engineering Undergraduate Research Catalyst Conference, Selangor, Malaysia, 1–2 July 2013.
450. Jalili, B.; Ghafoori, H.; Jalili, P. Investigation of carbon nano-tube (CNT) particles effect on the performance of a refrigeration cycle. *Int. J. Mater. Sci. Innov.* **2014**, *2*, 8–17.
451. Kruse, H.; Schroeder, M. Fundamentals of lubrication in refrigerating systems and heat pumps. *Int. J. Refrig.* **1985**, *8*, 347–355. [CrossRef]
452. Cremaschi, L. A fundamental view of the flow boiling heat transfer characteristics of nano-refrigerants. In *ASME International Mechanical Engineering Congress and Exposition*; American Society of Mechanical Engineers: Houston, TX, USA, 2012; pp. 2779–2792.
453. Vasconcelos, A.A.; Gómez, A.O.C.; Filho, E.P.B.; Parise, J.A.R. Experimental evaluation of SWCNT-water nanofluid as a secondary fluid in a refrigeration system. *Appl. Therm. Eng.* **2017**, *111*, 1487–1492. [CrossRef]
454. Kamaraj, N.; Manoj Babu, A. Experimental analysis of Vapour Compression Refrigeration System using the refrigerant with Nano particles. In *International Conference on Engineering Innovations and Solutions (ICEIS) 2016*; Seventh Sense Research Group: Tamilnadu, India, 2016; pp. 16–25.
455. Yang, S.; Cui, X.; Zhou, Y.; Chen, C. Study on the effect of graphene nanosheets refrigerant oil on domestic refrigerator performance. *Int. J. Refrig.* **2020**, *110*, 187–195. [CrossRef]
456. Pico, D.F.M.; da Silva, L.R.R.; Schneider, P.; Filho, E.P.B. Performance evaluation of diamond nanolubricants applied to a refrigeration system. *Int. J. Refrig.* **2019**, *100*, 104–112. [CrossRef]
457. Pico, D.F.M.; da Silva, L.R.R.; Mendoza, O.S.H.; Filho, E.P.B. Experimental study on thermal and tribological performance of diamond nanolubricants applied to a refrigeration system using R32. *Int. J. Heat Mass Transf.* **2020**, *152*, 119493. [CrossRef]

458. Rahman, S.; Issa, S.; Said, Z.; El Haj Assad, M.; Zadeh, R.; Barani, Y. Performance enhancement of a solar powered air conditioning system using passive techniques and SWCNT /R-407c nano refrigerant. *Case Stud. Therm. Eng.* **2019**, *16*, 100565. [CrossRef]
459. Mahian, O.; Bellos, E.; Markides, C.N.; Taylor, R.A.; Alagumalai, A.; Yang, L.; Qin, C.; Lee, B.J.; Ahmadi, G.; Safaei, M.R.; et al. Recent advances in using nanofluids in renewable energy systems and the environmental implications of their uptake. *Nano Energy* **2021**, *86*, 106069. [CrossRef]
460. Meyer, L.; Tsatsaronis, G.; Buchgeister, J.; Schebek, L. Exergoenvironmental analysis for evaluation of the environmental impact of energy conversion systems. *Energy* **2009**, *34*, 75–89. [CrossRef]
461. Bertoldo, R.; Mays, C.; Poumadère, M.; Schneider, N.; Svendsen, C. Great deeds or great risks? Scientists' social representations of nanotechnology. *J. Risk Res.* **2016**, *19*, 760–779.
462. Card, J.W.; Magnuson, B.A. A method to assess the quality of studies that examine the toxicity of engineered nanomaterials. *Int. J. Toxicol.* **2010**, *29*, 402–410. [CrossRef]
463. Faizal, M.; Saidur, R.; Mekhilef, S.; Alim, M. Energy, economic and environmental analysis of metal oxides nanofluid for flat-plate solar collector. *Energy Convers. Manag.* **2013**, *76*, 162–168. [CrossRef]
464. Michael Joseph Stalin, P.; Arjunan, T.V.; Matheswaran, M.M.; Dolli, H.; Sadanandam, N. Energy, economic and environmental investigation of a flat plate solar collector with CeO2/water nanofluid. *J. Therm. Anal. Calorim.* **2020**, *139*, 3219–3233. [CrossRef]
465. Hassani, S.; Saidur, R.; Mekhilef, S.; Taylor, R.A. Environmental and exergy benefit of nanofluid-based hybrid PV/T systems. *Energy Convers. Manag.* **2016**, *123*, 431–444. [CrossRef]
466. Sharafeldin, M.; Gróf, G.; Abu-Nada, E.; Mahian, O. Evacuated tube solar collector performance using copper nanofluid: Energy and environmental analysis. *Appl. Therm. Eng.* **2019**, *162*, 114205. [CrossRef]
467. Boyaghchi, F.A.; Chavoshi, M.; Sabeti, V. Optimization of a novel combined cooling, heating and power cycle driven by geothermal and solar energies using the water/CuO (copper oxide) nanofluid. *Energy* **2015**, *91*, 685–699. [CrossRef]
468. Boyaghchi, F.A.; Chavoshi, M. Multi-criteria optimization of a micro solar-geothermal CCHP system applying water/CuO nanofluid based on exergy, exergoeconomic and exergoenvironmental concepts. *Appl. Therm. Eng.* **2017**, *112*, 660–675. [CrossRef]
469. Sahota, L.; Shyam; Tiwari, G. Energy matrices, enviroeconomic and exergoeconomic analysis of passive double slope solar still with water based nanofluids. *Desalination* **2017**, *409*, 66–79. [CrossRef]
470. Grosu, Y.; Anagnostopoulos, A.; Balakin, B.; Krupanek, J.; Navarro, M.E.; González-Fernández, L.; Ding, Y.; Faik, A. Nanofluids based on molten carbonate salts for high-temperature thermal energy storage: Thermophysical properties, stability, compatibility and life cycle analysis. *Sol. Energy Mater. Sol. Cells* **2021**, *220*, 110838. [CrossRef]
471. Sahota, L.; Tiwari, G. Exergoeconomic and enviroeconomic analyses of hybrid double slope solar still loaded with nanofluids. *Energy Convers. Manag.* **2017**, *148*, 413–430. [CrossRef]
472. Khan, I.; Saeed, K.; Khan, I. Nanoparticles: Properties, applications and toxicities. *Arab. J. Chem.* **2019**, *12*, 908–931. [CrossRef]
473. Magrez, A.; Kasas, S.; Salicio, V.; Pasquier, N.; Seo, J.W.; Celio, M.; Catsicas, S.; Schwaller, B.; Forró, L. Cellular Toxicity of Carbon-Based Nanomaterials. *Nano Lett.* **2006**, *6*, 1121–1125. [CrossRef]
474. Bahadar, H.; Maqbool, F.; Niaz, K.; Abdollahi, M. Toxicity of Nanoparticles and an Overview of Current Experimental Models. *Iran. Biomed. J.* **2016**, *20*, 1–11.
475. Khlebtsov, N.; Dykman, L. Biodistribution and toxicity of engineered gold nanoparticles: A review of in vitro and in vivo studies. *Chem. Soc. Rev.* **2011**, *40*, 1647–1671. [CrossRef]
476. Khlebtsov, N.G.; Dykman, L.A. Optical properties and biomedical applications of plasmonic nanoparticles. *J. Quant. Spectrosc. Radiat. Transf.* **2010**, *111*, 1–35. [CrossRef]
477. Ali, N. Assessment of Using 99Mo and 99mTc Isotopes in Kuwait Medical Sector. *Heal. Phys.* **2016**, *110*, 387–390. [CrossRef] [PubMed]
478. Vishwakarma, V.; Samal, S.S.; Manoharan, N. Safety and Risk Associated with Nanoparticles—A Review. *J. Miner. Mater. Charact. Eng.* **2010**, *9*, 455–459. [CrossRef]
479. Pan, Y.; Neuss, S.; Leifert, A.; Fischler, M.; Wen, F.; Simon, U.; Schmid, G.; Brandau, W.; Jahnen-Dechent, W. Size-Dependent Cytotoxicity of Gold Nanoparticles. *Small* **2007**, *3*, 1941–1949. [CrossRef] [PubMed]
480. Schaeublin, N.M.; Braydich-Stolle, L.K.; Maurer, E.I.; Park, K.; MacCuspie, R.; Afrooz, A.; Vaia, R.A.; Saleh, N.B.; Hussain, S.M. Does Shape Matter? Bioeffects of Gold Nanomaterials in a Human Skin Cell Model. *Langmuir* **2012**, *28*, 3248–3258. [CrossRef]
481. Cancino-Bernardi, J.; Paino, I.; Souza, J.; Marangoni, V.; Nogueira, P.; Zucolotto, V. Current Challenges in the Commercialization of Nanocolloids: Toxicology and Environmental Issues. In *Nanocolloids*; Elsevier: Amsterdam, The Netherlands, 2016; pp. 427–463.
482. Ferreira, A.; Cemlyn-Jones, J.; Cordeiro, C.R. Nanoparticles, Nanotechnology and Pulmonary Nanotoxicology. *Rev. Port. de Pneumol. (Engl. Ed.)* **2013**, *19*, 28–37. [CrossRef]
483. Arshad, A.; Jabbal, M.; Yan, Y.; Reay, D. A review on graphene based nanofluids: Preparation, characterization and applications. *J. Mol. Liq.* **2019**, *279*, 444–484. [CrossRef]
484. Zhang, Q.; Huang, J.-Q.; Zhao, M.-Q.; Qian, W.-Z.; Wei, F. Carbon Nanotube Mass Production: Principles and Processes. *ChemSusChem* **2011**, *4*, 864–889. [CrossRef]
485. Syam Sundar, L.; Sousa, A.; Singh, M.K. Heat transfer enhancement of low volume concentration of carbon nanotube-Fe3O4/water hybrid nanofluids in a tube with twisted tape inserts under turbulent flow. *J. Therm. Sci. Eng. Appl.* **2015**, *7*. [CrossRef]
486. Sundar, L.S.; Singh, M.K.; Sousa, A. Enhanced heat transfer and friction factor of MWCNT–Fe3O4/water hybrid nanofluids. *Int. Commun. Heat Mass Transf.* **2014**, *52*, 73–83. [CrossRef]

487. Huang, D.; Wu, Z.; Sunden, B. Effects of hybrid nanofluid mixture in plate heat exchangers. *Exp. Therm. Fluid Sci.* **2016**, *72*, 190–196. [CrossRef]
488. Leong, K.; Ahmad, K.Z.K.; Ong, H.C.; Ghazali, M.; Baharum, A. Synthesis and thermal conductivity characteristic of hybrid nanofluids—A review. *Renew. Sustain. Energy Rev.* **2017**, *75*, 868–878. [CrossRef]
489. Maddah, H.; Aghayari, R.; Ahmadi, M.H.; Rahimzadeh, M.; Ghasemi, N. Prediction and modeling of MWCNT/Carbon (60/40)/SAE 10 W 40/SAE 85 W 90(50/50) nanofluid viscosity using artificial neural network (ANN) and self-organizing map (SOM). *J. Therm. Anal. Calorim.* **2018**, *134*, 2275–2286. [CrossRef]
490. Bakthavatchalam, B.; Shaik, N.B.; Hussain, P.B. An Artificial Intelligence Approach to Predict the Thermophysical Properties of MWCNT Nanofluids. *Processes* **2020**, *8*, 693. [CrossRef]
491. Afrand, M.; Nadooshan, A.A.; Hassani, M.; Yarmand, H.; Dahari, M. Predicting the viscosity of multi-walled carbon nanotubes/water nanofluid by developing an optimal artificial neural network based on experimental data. *Int. Commun. Heat Mass Transf.* **2016**, *77*, 49–53. [CrossRef]
492. Alrashed, A.A.A.A.; Gharibdousti, M.S.; Goodarzi, M.; de Oliveira, L.R.; Safaei, M.R.; Bandarra Filho, E.P. Effects on thermophysical properties of carbon based nanofluids: Experimental data, modelling using regression, ANFIS and ANN. *Int. J. Heat Mass Transf.* **2018**, *125*, 920–932. [CrossRef]
493. Ali, N.; Teixeira, J.A.; Addali, A.; Saeed, M.; Al-Zubi, F.; Sedaghat, A.; Bahzad, H. Deposition of Stainless Steel Thin Films: An Electron Beam Physical Vapour Deposition Approach. *Materials* **2019**, *12*, 571. [CrossRef]
494. Ali, N.; Teixeira, J.A.; Addali, A. Effect of Water Temperature, pH Value, and Film Thickness on the Wettability Behaviour of Copper Surfaces Coated with Copper Using EB-PVD Technique. *J. Nano Res.* **2019**, *60*, 124–141. [CrossRef]
495. Ali, N.; Teixeira, J.A.; Addali, A.; Al-Zubi, F.; Shaban, E.; Behbehani, I. The effect of aluminium nanocoating and water pH value on the wettability behavior of an aluminium surface. *Appl. Surf. Sci.* **2018**, *443*, 24–30. [CrossRef]
496. Buongiorno, J.; Hu, L.-W.; Kim, S.J.; Hannink, R.; Truong, B.; Forrest, E. Nanofluids for Enhanced Economics and Safety of Nuclear Reactors: An Evaluation of the Potential Features, Issues, and Research Gaps. *Nucl. Technol.* **2008**, *162*, 80–91. [CrossRef]
497. Mignacca, B.; Locatelli, G. Economics and finance of Small Modular Reactors: A systematic review and research agenda. *Renew. Sustain. Energy Rev.* **2020**, *118*, 109519. [CrossRef]
498. Hernandez, R.; Folsom, C.P.; Woolstenhulme, N.E.; Jensen, C.B.; Bess, J.D.; Gorton, J.P.; Brown, N.R. Review of pool boiling critical heat flux (CHF) and heater rod design for CHF experiments in TREAT. *Prog. Nucl. Energy* **2020**, *123*, 103303. [CrossRef]
499. Hu, H.; Peng, H.; Ding, G. Nucleate pool boiling heat transfer characteristics of refrigerant/nanolubricant mixture with surfactant. *Int. J. Refrig.* **2013**, *36*, 1045–1055. [CrossRef]
500. Cheng, L.; Liu, L. Boiling and two-phase flow phenomena of refrigerant-based nanofluids: Fundamentals, applications and challenges. *Int. J. Refrig.* **2013**, *36*, 421–446. [CrossRef]
501. Bahiraei, M.; Salmi, H.K.; Safaei, M.R. Effect of employing a new biological nanofluid containing functionalized graphene nanoplatelets on thermal and hydraulic characteristics of a spiral heat exchanger. *Energy Convers. Manag.* **2019**, *180*, 72–82. [CrossRef]
502. Dalkılıç, A.S.; Mercan, H.; Özçelik, G.; Wongwises, S. Optimization of the finned double-pipe heat exchanger using nanofluids as working fluids. *J. Therm. Anal. Calorim.* **2020**, *143*, 859–878. [CrossRef]
503. Darzi, M.; Sadoughi, M.; Sheikholeslami, M. Condensation of nano-refrigerant inside a horizontal tube. *Phys. B Condens. Matter* **2018**, *537*, 33–39. [CrossRef]
504. Sheikholeslami, M.; Darzi, M.; Sadoughi, M. Heat transfer improvement and pressure drop during condensation of refrigerant-based nanofluid; an experimental procedure. *Int. J. Heat Mass Transf.* **2018**, *122*, 643–650. [CrossRef]
505. Sunardi, C.; Markus; Setyawan, A. The effects of the condenser pressure drop on the cooling performance of an air conditioning unit using R-410A. *AIP Conf. Proc.* **2001**, *2018*, 020006.
506. Tashtoush, B.M.; Al-Nimr, M.d.A.; Khasawneh, M.A. Investigation of the use of nano-refrigerants to enhance the performance of an ejector refrigeration system. *Appl. Energy* **2017**, *206*, 1446–1463. [CrossRef]
507. Ambreen, T.; Kim, M.-H. Heat transfer and pressure drop correlations of nanofluids: A state of art review. *Renew. Sustain. Energy Rev.* **2018**, *91*, 564–583. [CrossRef]
508. Jiang, W.; Ding, G.; Peng, H.; Gao, Y.; Wang, K. Experimental and Model Research on Nanorefrigerant Thermal Conductivity. *HVACR Res.* **2009**, *15*, 651–669. [CrossRef]
509. Sidik, N.C.; Alawi, O.A. Computational investigations on heat transfer enhancement using nanorefrigerants. *J. Adv. Res. Des.* **2014**, *1*, 35–41.
510. Ndoye, F.T.; Schalbart, P.; Leducq, D.; Alvarez, G. Numerical study of energy performance of nanofluids used in secondary loops of refrigeration systems. *Int. J. Refrig.* **2015**, *52*, 122–132. [CrossRef]
511. Bahman, A.M.; Ziviani, D.; Groll, E.A. Vapor injected compression with economizing in packaged air conditioning systems for high temperature climate. *Int. J. Refrig.* **2018**, *94*, 136–150. [CrossRef]
512. Loaiza, J.C.V.; Pruzaesky, F.C.; Parise, J.A.R. A numerical study on the application of nanofluids in refrigeration systems. In *International Refrigeration and Air Conditioning Conference at Purdue*; Purdue University: Purdue, IN, USA, 2010; p. 1495.
513. Mahbubul, I.M.; Fadhilah, S.A.; Saidur, R.; Leong, K.Y.; Amalina, M.A. Thermophysical properties and heat transfer performance of Al2O3/R-134a nanorefrigerants. *Int. J. Heat Mass Transf.* **2013**, *57*, 100–108. [CrossRef]

514. Mahbubul, I.; Saidur, R.; Amalina, M. Influence of particle concentration and temperature on thermal conductivity and viscosity of Al2O3/R141b nanorefrigerant. *Int. Commun. Heat Mass Transf.* **2013**, *43*, 100–104. [CrossRef]
515. Mahbubul, I.; Saadah, A.; Saidur, R.; Khairul, M.; Kamyar, A. Thermal performance analysis of Al2O3/R-134a nanorefrigerant. *Int. J. Heat Mass Transf.* **2015**, *85*, 1034–1040. [CrossRef]
516. Li, Z.; Renault, F.L.; Gómez, A.O.C.; Sarafraz, M.; Khan, H.; Safaei, M.R.; Filho, E.P.B. Nanofluids as secondary fluid in the refrigeration system: Experimental data, regression, ANFIS, and NN modeling. *Int. J. Heat Mass Transf.* **2019**, *144*, 118635. [CrossRef]
517. Bahman, A.M.; Groll, E.A. Application of Second-Law Analysis for the Environmental Control Unit at High Ambient Temperature. *Energies* **2020**, *13*, 3274. [CrossRef]
518. Bhattad, A.; Sarkar, J.; Ghosh, P. Energy-Economic Analysis of Plate Evaporator using Brine-based Hybrid Nanofluids as Secondary Refrigerant. *Int. J. Air. Cond. Refrig.* **2018**, *26*, 1850003. [CrossRef]

nanomaterials

Article

Thermal and Hydraulic Performances of Carbon and Metallic Oxides-Based Nanomaterials

Haitham Abdulmohsin Afan [1], Mohammed Suleman Aldlemy [2,3], Ali M. Ahmed [4], Ali H. Jawad [5], Maryam H. Naser [6], Raad Z. Homod [7], Zainab Haider Mussa [8], Adnan Hashim Abdulkadhim [9], Miklas Scholz [10,11,12,13,*] and Zaher Mundher Yaseen [14,15,16,*]

1 Department of Civil Engineering, Al-Maarif University College, Ramadi 31001, Iraq; haitham.afan@uoa.edu.iq
2 Department of Mechanical Engineering, College of Mechanical Engineering Technology, Benghazi 11199, Libya; maldlemy@ceb.edu.ly
3 Center for Solar Energy Research and Studies (CSERS), Benghazi 11199, Libya
4 Engineering Department, Al-Esraa University College, Baghdad 10011, Iraq; ali.majeed@esraa.edu.iq
5 Faculty of Applied Sciences, Universiti Teknologi MARA, Shah Alam 40450, Selangor, Malaysia; ali288@uitm.edu.my
6 Building and Construction Techniques Engineering Department, AL-Mustaqbal University College, Hillah 51001, Iraq; maryamhameed@mustaqbal-college.edu.iq
7 Department of Oil and Gas Engineering, Basrah University for Oil and Gas, Al Basrah 61004, Iraq; raadahmood@yahoo.com
8 College of Pharmacy, University of Al-Ameed, Karbala 56001, Iraq; za.alaady@alameed.edu.iq
9 Department of Computer Engineering, Technical Engineering College, Al-Ayen University, Thi-Qar 64006, Iraq; adnan.hashin@alayen.edu.iq
10 Division of Water Resources Engineering, Faculty of Engineering, Lund University, 221 00 Lund, Sweden
11 Department of Civil Engineering Science, School of Civil Engineering and the Built Environment, University of Johannesburg, Kingsway Campus, Johannesburg 2092, South Africa
12 Institute of Environmental Engineering, Wroclaw University of Environmental and Life Sciences, 50375 Wrocław, Poland
13 Department of Town Planning, Engineering Networks and Systems, South Ural State University, 76, Lenin Prospekt, 454080 Chelyabinsk, Russia
14 Department of Earth Sciences and Environment, Faculty of Science and Technology, Universiti Kebangsaan Malaysia, Bangi 43600, Selangor, Malaysia
15 Adjunct Research Fellow, USQ's Advanced Data Analytics Research Group, School of Mathematics Physics and Computing, University of Southern Queensland, Queensland, QLD 4350, Australia
16 New Era and Development in Civil Engineering Research Group, Scientific Research Center, Al-Ayen University, Nasiriyah 64001, Iraq
* Correspondence: miklas.scholz@tvrl.lth.se (M.S.); yaseen@ukm.edu.my (Z.M.Y.)

Citation: Afan, H.A.; Aldlemy, M.S.; Ahmed, A.M.; Jawad, A.H.; Naser, M.H.; Homod, R.Z.; Mussa, Z.H.; Abdulkadhim, A.H.; Scholz, M.; Yaseen, Z.M. Thermal and Hydraulic Performances of Carbon and Metallic Oxides-Based Nanomaterials. *Nanomaterials* **2022**, *12*, 1545. https://doi.org/10.3390/nano12091545

Academic Editor: S M Sohel Murshed

Received: 30 March 2022
Accepted: 24 April 2022
Published: 3 May 2022

Abstract: For companies, notably in the realms of energy and power supply, the essential requirement for highly efficient thermal transport solutions has become a serious concern. Current research highlighted the use of metallic oxides and carbon-based nanofluids as heat transfer fluids. This work examined two carbon forms (PEG@GNPs & PEG@TGr) and two types of metallic oxides (Al_2O_3 & SiO_2) in a square heated pipe in the mass fraction of 0.1 wt.%. Laboratory conditions were as follows: $6401 \leq Re \leq 11{,}907$ and wall heat flux = 11,205 W/m^2. The effective thermal–physical and heat transfer properties were assessed for fully developed turbulent fluid flow at 20–60 °C. The thermal and hydraulic performances of nanofluids were rated in terms of pumping power, performance index (PI), and performance evaluation criteria (PEC). The heat transfer coefficients of the nanofluids improved the most: PEG@GNPs = 44.4%, PEG@TGr = 41.2%, Al_2O_3 = 22.5%, and SiO_2 = 24%. Meanwhile, the highest augmentation in the Nu of the nanofluids was as follows: PEG@GNPs = 35%, PEG@TGr = 30.1%, Al_2O_3 = 20.6%, and SiO_2 = 21.9%. The pressure loss and friction factor increased the highest, by 20.8–23.7% and 3.57–3.85%, respectively. In the end, the general performance of nanofluids has shown that they would be a good alternative to the traditional working fluids in heat transfer requests.

Keywords: carbon nanostructures; metallic oxides; thermophysical properties; convective heat transfer; turbulent flow

1. Introduction

There was a lot of consideration in the production of working fluids with superior thermal characteristics to enhance heat transfer efficiency in heated pipes [1,2]. The latest research on nanofluids indicates that suspending extremely thermally conductive nanomaterials into the base fluid (e.g., water (DW) or Ethylene glycol (EG)) increases thermal conductivity, an increase in the base fluid's convective heat transfer rate [3,4]. Reducing thermal boundary layer thickness generated by the existence of nanomaterials and their random movement within the base fluid might have a significant impact on such convective heat transfer coefficient augmentation [5,6]. An increase in the nanoparticle mass/volume concentration frequently improves the heat transfer rate of the base fluid. Adding more nanomaterials to the base fluid enhances the Brownian motion-driven variations in the fluid, which leads to a fast heat transfer from the wall to the nanofluid [7,8].

In the heat transfer and hydrodynamic applications, thermal conductivity shows a crucial role in evaluating nanofluid thermal efficiency, which has been calculated based on different factors, involving inlet temperature, nanomaterial types and mass percentage, the nanostructure of particles, base fluid properties, pH values, and types of surfactants/additives [9–11]. Additionally, dynamic viscosity influences the determination of heat and momentum transfer and the device's pumping amount [12]. Meanwhile, less effort was dedicated to density, thermal expansion, and specific heat capacity [13]. The values of density and specific capacity of different nanofluids have been estimated through empirical correlations and equations based on the volume fraction of nanoparticles [14,15].

Numerous experimental and numerical investigation analyses evaluated forced convection heat transfer using different metal and metallic oxides such as Al, Cu, CuO, SiO_2, Al_2O_3, TiO_2, and MWCNTs during various flow regimes. Several mechanical or thermal equipment used constant wall heat flux (WHF) for heat transfer applications. Numerical and experimental efforts studied the effects of thermal and momentum diffusivity on the heat conductivity of various nano-powders under turbulent forced convective heat transfer [11]. The study examined various nanofluid samples (Al_2O_3-H_2O, SiO_2-H_2O, and Cu-H_2O) and various volume percentages (1–3 vol.%) at 30 °C. Improved thermal conductivity had no impact on heat transfer efficiency; but, the Prandtl number (Pr) of nanofluids significantly influenced the Nusselt number value at constant Re. Numerical and experimental analyses observed forced convective heat transfer by flowing graphene nanoplatelet nanofluids through a fully turbulent system inside a horizontally smooth heated pipe [16]. GNPs@H_2O nanofluids increased from 7.96% to 25% in thermal efficiency. Furthermore, Nu_{avg} at 0.1 wt.% revealed enhancements of 75%, 79%, and 83% at 8231, 10,351, and 12,320 W/m^2, respectively. Zubir and his group [17] produced reduced graphene oxide (R-GO) and its influence on a heat exchanger's turbulent convective heat transfer performance. Furthermore, the study noticed substantial improvement in the Nusselt number up to 144% and 63% at the upstream and downstream of the test section, respectively. Laboratory work of PGGNPs with SSA-750 m^2/g was carried out to assess nanofluid flow and heat transfer enhancement [18]. The test section was heated with two rates such as 23,870 and 18,565 W/m^2. Meanwhile, the Re-number during the investigation ranged from 3900 to 11,700. From the results, the heat transfer coefficient improved significantly (around 119% & 84%) at the two heat rates. The performance index of all samples was larger than one, indicating that the synthesized PGGNP@DW nanofluids were effective for convective heat transfer. Yarmand et al. [19] reported the effect of pressure loss, thermophysical characteristics, and convective heat transfer on the stable-doped GNPs nanofluids. Their results showed positive improvements in both Nu-number and heat transfer coefficients by about 26.5% and 19.68%, respectively, at 0.1 wt.%. Lastly, Sadri et al. [20] prepared graphene nanoplatelets using green synthetization. They prepared three samples of C-GNPs nanofluids in 0.025, 0.075, and 0.1 wt.%. The results showed optimum improvements in the Nu-number (18.69%) and convective heat transfer coefficient (37.54%) at Re = 15,927 and 0.1 wt.%. It was clear that the performance index for all CGNP-DW nanofluids was larger than one. This demonstrated the advantage of employing environmentally friendly nanofluids in heat transfer systems. The overall thermal performance of using TiO_2-DW nanofluids reached up to 1.519 as the best value, then reduced by increasing the nanofluid flow [21]. The thermal efficiency of SiO_2-DW in a triangular tube with various turbulators was consistently greater than one. The index increased first with the increase in Re number and then decreased with it. This index reached its maximum at the Reynolds number Re = 6000 [22]. Additionally, in corrugated tubes, several working fluids (DW, GNP-SDBS@DW, Al_2O_3@DW, and SiO_2@DW) and tube shapes (rectangular, triangular, trapezoidal, and curved ribs) were investigated [23]. The overall performance can be enhanced by up to 37% by combining the approaches (GNP-SDBS@DW nanofluids and curved pipe).

A closer look at the literature reveals several gaps and shortcomings of overall thermal performance using carbon and metal-oxide nanofluids within heated pipes. The main purpose was to compare the performance of functionalized carbon nanostructured nanofluids and commercial metallic oxides-based nanofluids. The prepared nanomaterials were characterized via different examinations to show successful chemical reactions. Meanwhile, the nanofluids' thermo-physical properties of PEG@GNPs, PEG@TGr, Al_2O_3, and SiO_2 were measured in the range of temperatures (20–60 °C). The heat transfer and nanofluids flow were evaluated based on several parameters such as the average Nu-number, relative pumping power, and different performance indicators under fully developed turbulent forced convective flow.

2. Materials and Methods

2.1. Functionalization and Preparation Process

Since the raw materials of GNPs and Gr are hydrophobic and cannot dissolve in polar solvents like H_2O, a suitable way to make PEG@GNPs and PEG@TGr hydrophilic is to present the covalent functionalization via acid treatment. The process will dope the surface of GNPs and Gr with -OH- and -COOH. In a typical experiment [24], the chemical reactions were performed by dispersing GNPs (1 g) and Gr (1 g) in the acid medium of $AlCl_3$ (18.54 g) and HCl (10 mL), followed by 1 h microwave radiation. Then, the solution was separated at 11,500 rpm and filtered through a polycarbonate filter (0.45 μm) before sequential washing with DMF, THF, diluted HCl, and enough DI-water to eliminate unreacted $AlCl_3$ and PEG overnight at 60 °C. Furthermore, the dry aluminum oxide (Al_2O_3-NPs = 50 nm) and silicon dioxide (SiO_2-NPs = 50 nm) were ultrasonicated for 1 h to avoid agglomeration/settlement. Nanomaterials were mixed with DW by an ultrasonic probe (Sonics Vibra-Cell, VC 750, Sonics & Materials Inc., Newtown, CT, USA) with an output power of (750 W) and a power supply of (20 kHz) frequency. The production process and nanofluid preparation method were shown in Figure 1 [25].

2.2. Experimental Methodologies

This study was carried out at an inlet temperature of 30 °C; the basic thermophysical properties such as dynamic viscosity and thermal conductivity should be determined initially. The tools of KD2 Pro and Anton Paar Rheometer were used to evaluate thermal conductivity and dynamic viscosity, respectively [26]. In the meantime, for the density readings, a density meter was used at an accuracy level of $\pm 10^{-4}$ g/cm^3. Lastly, a Differential Scanning Calorimeter (data accuracy = ±1.0%) was used to capture the specific heat of the samples. SEM-EDX analysis was conducted to study morphology and elemental structures of the prepared nanomaterials using VEGA3 tool (Tescan, Brno, Czechia).

Experimental model is schematically depicted in Figure 2. The flow loop parts include a magnetic flow meter, a storage tank, a pump, a test section, and a differential pressure transmitter. Each working fluid is driven from a 12 L capacity stainless steel by a magnetic drive pump at the flow rate range of 0–10 LPM. Uncertainties of the flow rate and pressure loss measures were ±0.5% and ±0.075%, respectively.

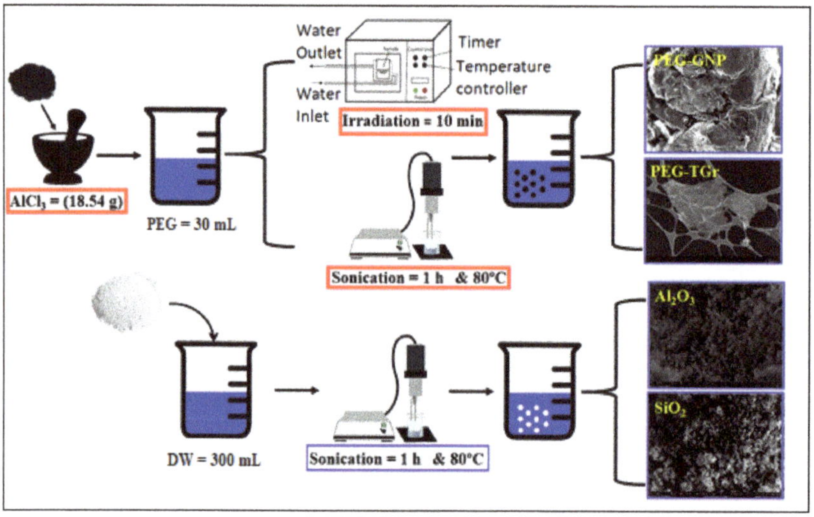

Figure 1. Schematic illustration for the different nanofluids preparation process.

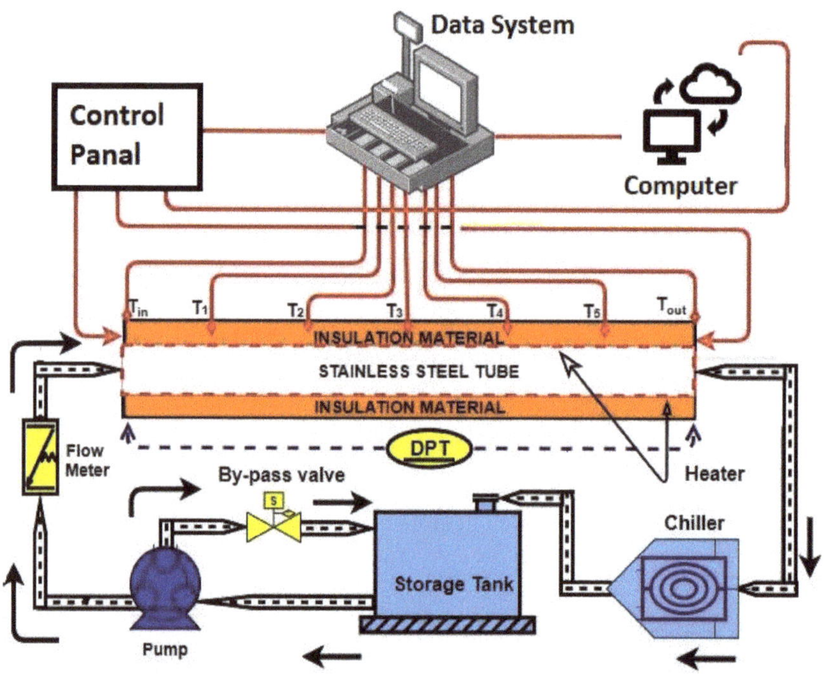

Figure 2. Diagram of the adopted experimental set-up.

The test section is a square heated pipe (length = 1.4 m, inner width = 10 mm, outer width = 12.8 mm). It was heated by a 900 W flexible tape heater attached to a transformer and a power meter. Then, a high-temperature epoxy glue was used for installing 5 T-type thermocouples (uncertainty = ± 0.1 °C) to measure the surface temperature.

Two RTD (PT-100) sensors (uncertainty = ± 0.1 °C) were immersed into the pipe to measure the inlet and outlet temperatures. All temperature measurements were collected by Graphtec (LOGGER GL240). After using the formula ($Q = VI = \dot{m}C_p[T_{out} - T_{in}]$), the maximum heat loss was about 7.2%. This low heat loss rate was thought to have no significant impact on the entire process of heat transfer estimation.

2.3. Data Processing

In the current study, primary data were collected from an experimental setup and handled using very well-known procedures, as described in earlier studies [27]. The present laboratory analysis focused on evaluation of the heat transfer enhancement and hydrodynamic effectiveness under the condition of fully developed turbulent flow. The approximate heat flux, heat transfer coefficient, average Nusselt number, friction factor, Reynolds number, and Prandtl number; are presented as follows:

Heat flux (q'')	$\frac{V \times I}{4D_h L}$	(1)
Heat transfer coefficient (h)	$\frac{q''}{T_w - T_b}$	(2)
Nusselt number (Nu)	$\frac{hD_h}{k}$	(3)
Friction factor (f)	$\frac{\Delta P}{\left(\frac{L}{D}\right)\left(\frac{\rho v^2}{2}\right)}$	(4)
Reynolds Number (Re)	$\frac{4\,\dot{m}}{\pi D_h \mu}$	(5)
Prandtl number (Pr)	$\frac{\mu C_p}{k}$	(6)

In this regard, $T_w = \frac{\sum T}{5}$. (T_w = average wall surface temp.), $T_b = \frac{T_o - T_i}{2}$. $D_h = \frac{4A_c}{P}$, A_c = cross-section area of square pipe, while P is the wetted perimeter.

The Gnielinski [28] relationship is justifiable, especially for the single-phase fluids flowing:

$$Nu = \frac{\left(\frac{f}{8}\right)(Re - 1000)Pr}{1 + 12.7\left(\frac{f}{8}\right)^{0.5}\left(Pr^{2/3} - 1\right)}\left[1 + \left(\frac{d}{L}\right)^{2/3}\right]\left(\frac{Pr_m}{Pr_w}\right)^{0.11} \tag{7}$$

where, Pr_m = the bulk temperature-related Prandtl number and Pr_w = wall temperature-related Prandtl number. The Gnielinski correlation remains valid in the range of 3000 < Re < 5×10^6 and $0.5 < Pr < 2000$.

The Colebrook equation [29] is applicable, based upon Re-number, in order to identify the friction factor of a fully developed turbulent flow using Equation (8).

$$\frac{1}{\sqrt{f}} = -2.0 \log\left(\frac{\varepsilon/D}{3.7} + \frac{2.51}{Re\sqrt{f}}\right) \tag{8}$$

Petukhov's equation [30] of a fully developed turbulent flow is as shown in Equation (9):

$$Nu = \frac{\left(\frac{f}{8}\right)RePr}{1.07 + 12.7\left(\frac{f}{8}\right)^{0.5}\left(Pr^{2/3} - 1\right)} \tag{9}$$

Here, the formula is applicable for the requirements of $3000 < Re < 5 \times 10^6$ and $0.5 < Pr < 2000$.

The values of the Darcy friction factor were determined from the approximate pressure loss along the heated square pipe. The Blasius and Petukhov correlations were employed for the validation of the results obtained for the base fluid [31]:

Petukhov [30]:

$$f = (0.79\ln(Re) - 1.64)^{-2} \tag{10}$$

Blasius [32]:

$$f = \frac{0.316}{Re^{0.25}} \tag{11}$$

A performance index (*PI*) indicates an appropriate parameter to define various velocity and temperature ranges usable by various nanofluids [33]:

$$PI = \frac{h_{nf}/h_{bf}}{\Delta P_{nf}/\Delta P_{bf}} = \frac{R_h}{R_{\Delta P}} \tag{12}$$

where (R_h) is the ratio between nanofluids heat transfer and DW heat transfer, while ($R_{\Delta P}$) is the ratio between nanofluids pressure drop and DW pressure drop. An energy-saving indicator within the turbulent flow region calculated the pumping power using Equation (13).

$$\frac{W_{nf}}{W_{bf}} = \left(\frac{\mu_{nf}}{\mu_{bf}}\right)^{0.25}\left(\frac{\rho_{bf}}{\rho_{nf}}\right)^{2} \tag{13}$$

where $\left(W_{nf}\right)$ is the nanofluids' pumping power and (W_{bf}) is the DW pumping power.

The overall performance was evaluated (in terms of the thermal and hydraulic performances) using a performance evaluation criterion (PEC), which depicts the ratio of the heat performance to the nanofluids compared to DW. The formula of the PEC was expressed as [34]:

$$PEC = \frac{Nu_{nf}/Nu_{bf}}{\left(f_{nf}/f_{bf}\right)^{1/3}} \tag{14}$$

Table 1 presents and outlines the range of uncertainties [35].

Table 1. Uncertainty ranges for heat transfer and fluid flow variables.

Variable	Uncertainty Equations	Uncertainty Values
Reynolds number, *Re*	$\frac{U_{Re}}{Re} = \sqrt{\left(\frac{U_\rho}{\rho}\right)^2 + \left(\frac{U_{\overline{V}}}{\overline{V}}\right)^2 + \left(\frac{U_\mu}{\mu}\right)^2}$	±1.73%
Constant Heat flux, *q*	$\frac{U_q}{q} = \sqrt{\left(\frac{U_V}{V}\right)^2 + \left(\frac{U_I}{I}\right)^2}$	±1.51%
Heat transfer coefficient, *h*	$\frac{U_h}{h} = \sqrt{\left(\frac{U_q}{q}\right)^2 + \left(\frac{U_{(T_w-T_b)}}{(T_w-T_b)}\right)^2}$	±1.52%
Nusselt number, *Nu*	$\frac{U_{Nu}}{Nu} = \sqrt{\left(\frac{U_h}{h}\right)^2 + \left(\frac{U_k}{k}\right)^2}$	±5.23%
Friction factor, *f*	$\frac{U_f}{f} = \sqrt{\left(\frac{U_{\Delta p}}{\Delta p}\right)^2 + \left(\frac{U_\rho}{\rho}\right)^2 + \left(\frac{U_{\overline{V}}}{\overline{V}}\right)^2}$	±1.60%

3. Results and Discussion

3.1. Characterization of Nanofluids

SEM procedures are used to determine the content of particles based on the spectrum of the transmitted beam from the samples; they allow the size of irregularly sized particles or impurities to be determined. Moreover, SEM helps determine the distribution of nanoscale particles onto the surface of any sample. Figure 3a displays the SEM micrograph of the prepared PEG@GNPs; it is evident that the PEG@GNPs included several different-sized GNP-flakes, implying the samples' high-purity level. Examination via the electron beam demonstrated that most flakes were transparent due to the limited number of their layers, though difficulties in determining precise flakes and defects diameter through SEM manifested themselves in sharper planar morphology of the GNP layers on the obtained SEM micrographs. Figure 3b displays the high-resolution SEM image of the PEG@TGr prior to any kind of pre-treatment. Also, consistency and intactness of the grains, curves, and wrinkling were observed on some of the transparent SEM images because of the strict production process. The existence of new functional groups in the PEG@TGr was revealed in observed functionalization-induced wrinkles in the images.

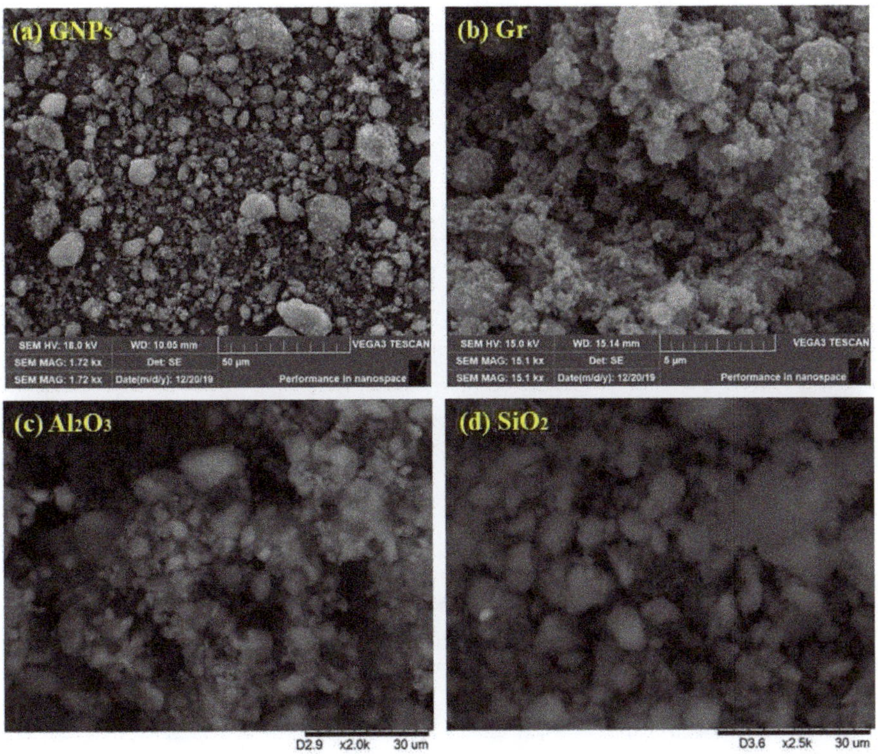

Figure 3. Visualization of different SEM nanoparticles; (**a**) Graphene nanoplatelets, (**b**) Graphene, (**c**) Alumina, (**d**) Silica.

Figure 3c introduces the SEM image of the 0.1 wt.%-Al_2O_3@DW nanofluid; the image illustrates rod-like and rectangular-shaped alumina nanoparticles with a low tendency towards agglomeration of the excellence of the prepared suspension. Furthermore, the image demonstrated exceptional dispersal of the sample after 60 min of ultrasonication. Additionally, the nanoparticles demonstrated a homogeneous grain size (<50 nm), suggesting the prepared nanoparticles were spherical and showed a treatment-dependent size distribution. In the current study, it was observed from Figure 3c that the major bulk of the sample was Al_2O_3. This confirms the high purity of the sample and the suitability of the applied synthesizing methodology. Furthermore, Figure 3d shows the SEM image of 0.1 wt.%-SiO_2@DW after 60 min ultrasonication nanofluid; the image shows the silica nanoparticles showing rod-to-round-like morphological, but with minor clusters and a better suspension. The size of nanoparticles was also found to be uniform, with <50 nm.

Figure 4 shows the EDX analysis for GNPs, Gr, Al_2O_3, and SiO_2 nanomaterials. As shown in Figure 4a,b, the carbon nanostructures show five elements (C, O, Si, S, and Zr). Figure 4c shows two elements only (Al and O). At the same time, Figure 4d presents three different elements (Si, O, and Br). The various elements refer to different synthesizing approaches used in this study.

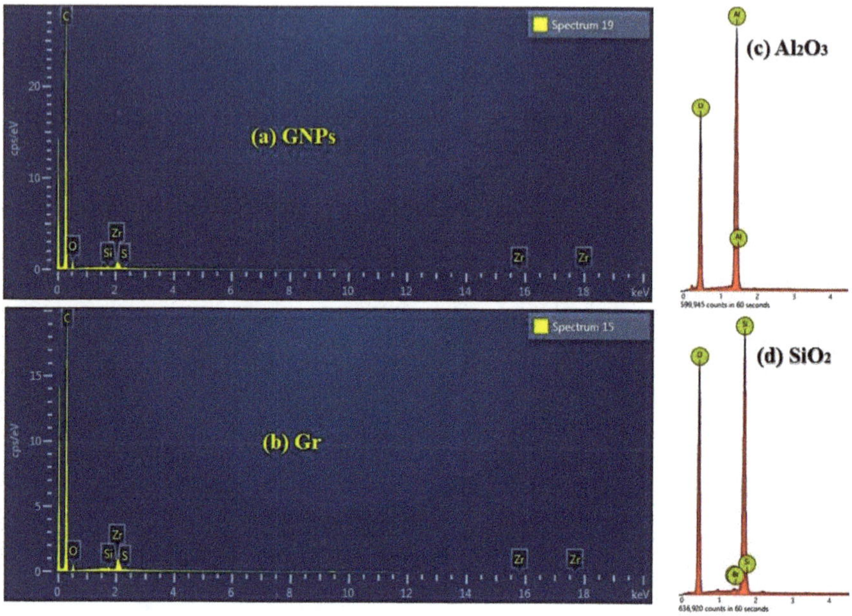

Figure 4. EDX images of different nanoparticles; (**a**) Graphene nanoplatelets, (**b**) Graphene, (**c**) Alumina, (**d**) Silica.

3.2. Thermophysical Properties Measurements

In comparison to distilled water, different nanofluids were described from the perspective of thermophysical properties as a function of mass fractions and temperature, as illustrated in Figure 5. The thermal conductivity of the working fluids plays a critical role in increasing heat removal efficiency from the heat exchangers to the environment. Current findings closely followed existing correlations offered by the National Institute of Science and Technology (NIST) [36], with a maximum standard error of 2%. As shown in Figure 5a, the nanofluids showed considerably higher thermal conductivity than DW; increases in temperature also rose thermal conductivity. The nano-coolants demonstrated a perfect, effective thermal conductivity increase rate at higher mass percentages. The temperature improved thermal conductivity significantly as a result of the increase in the nanoparticles' Brownian motion upon DW. The increases in thermal conductivity were for PEG@GNP = 31.6%, PEG@TGr = 29.74%, SiO_2 = 11.4%, and Al_2O_3 = 8.04% at 0.1 wt.% and 60 °C. Table 2 summarizes the thermal conductivity study by the previous investigators.

Figure 5b compared different nano-coolants and the base fluids in terms of their effective dynamic viscosity at the testing conditions of 0.1 wt.%, the temperature range of 20–60 °C, and a shear rate of 200 s^{-1}. Figure 5b showed a minor increase in the nanofluids' dynamic viscosity following that for DW, and the main reason for this increase is using low concentrations. It is assumed that fluid viscosity increases can result in pumping fluid penalty in the thermal applications; the nanofluids and DW also exhibited reduced dynamic viscosity due to the intermolecular forces degradation at increased temperatures [37]. The dynamic viscosity of all the samples showed a similar decreasing tendency, but the results evidenced increases in the base fluids' dynamic viscosity. This validates the reliability of the proposed synthesis method for nanofluids in this study. Table 3 summarizes the dynamic viscosity study by previous researchers.

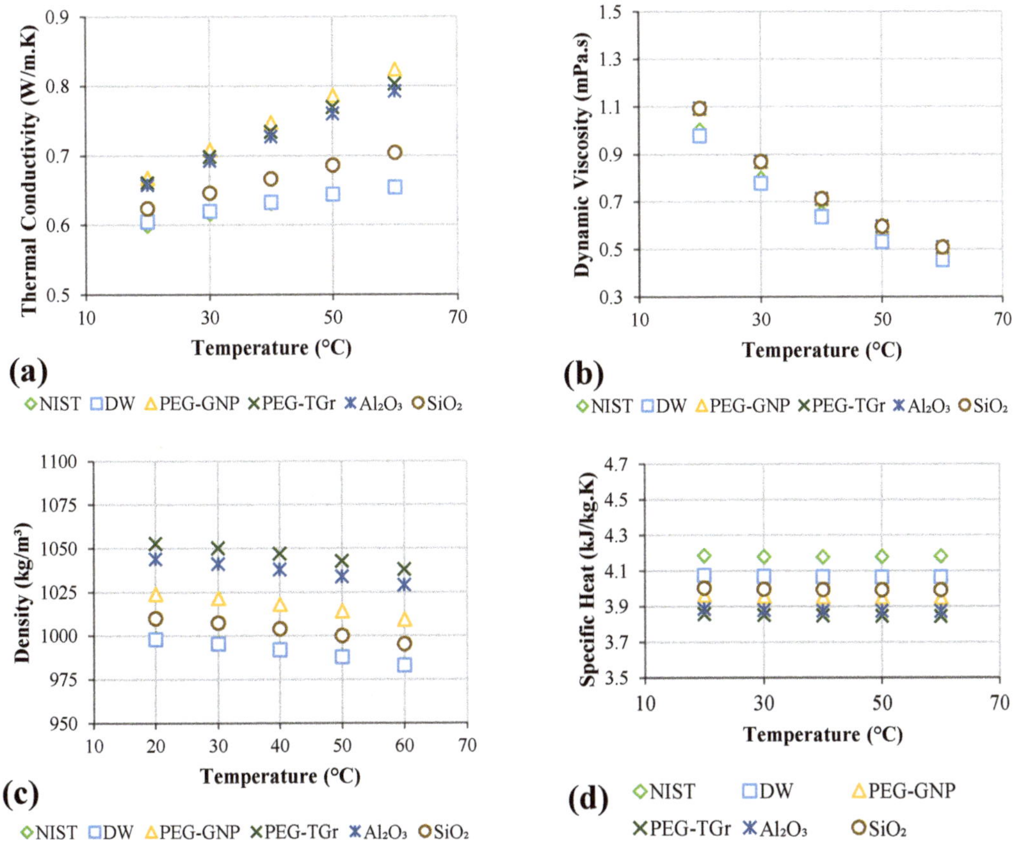

Figure 5. The thermophysical properties of base fluid and nanofluids; (**a**) Thermal conductivity, (**b**) Dynamic viscosity, (**c**) Density, (**d**) Specific heat capacity.

The density of the different working fluids was tested at a temperature range of 20 to 60 °C (see Figure 5c). The data showed a remarkable decrease in density with temperature and a slight increase in density with the nanofluid type. The nanoparticle's density contributed to the improved density of the nano-coolants as it was higher than that of the base fluid. The observed improvement in the nanofluid density was as follows: PEG@GNP = 5.3%, PEG@TGr = 4.5%, Al_2O_3 = 2.6%, and SiO_2 = 1.2% for 0.1 wt.% and 60 °C. However, the density reduced as follows: PEG@GNP = 1.7%, PEG@TGr = 1.8%, Al_2O_3 = 2.1%, and SiO_2 = 2.7% after raising the temperature of the nanofluid from 20 to 60°C, thereby demonstrating the significant role of temperature.

Also, the specific heat capacities are measured in this study (see Figure 5d). The specific heat showed insignificant reductions with temperature increases, but the observed gradient concurred with the specific heat plots reported in the earlier studies [38]. Figure 5d evidenced the average specific heat decreases as follows: PEG@GNP = 5.4%, PEG@TGr = 4.8%, Al_2O_3 = 2.9%, and SiO_2 = 1.8% compared to that of DW. This reduction was the lower specific heat of the solid nanoparticles relative to the base fluid.

Table 2. Summary of thermal conductivity in previous experimental studies.

Study	Nanofluid	Mass/ Volume %	Base Fluid	Temp. Range	Tool	Remarks
[39]	Graphene (Gr)	0.005 and 0.01	Ionic Liquid	From 20 °C to 145 °C	Hot Disc-thermal constant analyzer	Thermal conductivity enhanced by 9.4% at 0.01 wt.%
[40]	Al_2O_3	0.2–1	DW+EG	From 10 °C to 50 °C	KD2pro	Thermal conductivity was enhanced by 8.3% at 1 vol.%
[41]	Graphene (Gr)	0.02–0.2	DW+EG	From 25–65 °C	KD2pro	Thermal conductivity enhanced by 64% at 0.2 wt.%
[42]	Graphene (Gr)	0.5–0.45	DW+EG	30 °C	KD2pro	Thermal conductivity enhanced by 18% at 0.45 vol.%
Current study	PEG@GNP, PEG@TGr, Al_2O_3, and SiO_2	0.1	DW	From 20 °C to 60 °C	KD2pro	PEG@GNP = 31.6%, PEG@TGr = 29.74%, SiO_2 = 11.4%, & Al_2O_3 = 8.04% at 60 °C

Table 3. Summary of viscosity in previous experimental studies.

Study	Nanofluid	Mass/ Volume%	Base Fluid	Temp. Range	Tool	Remarks
[39]	Graphene (Gr)	0.005 and 0.01	Ionic Liquid	From 25 °C to 150 °C	Viscometer	Viscosity enhanced by 29.1% and 13.4% raised for 0.005 and 0.01 wt.%
[40]	Al_2O_3	0.2–1	DW+EG	From 10 °C to 50 °C	Brookfield Viscometer	Viscosity and temperature were in opposite correlation
[41]	Graphene (Gr)	0.02–0.2	DW+EG	From 25–65 °C	Brookfield Viscometer	Viscosity decreases as temperature rises and increases as nanoparticle concentration rises.
Current study	PEG@GNP, PEG@TGr, Al_2O_3, and SiO_2	0.1	DW	From 20 °C to 60 °C	Anton Paar Rheometer	A minor increase in the nanofluids' dynamic viscosity relative to DW.

3.3. Validation Test for Distilled Water

The Nu_{avg} and heat transfer coefficients (h) obtained with the data from Equations (7)–(9) are disclosed in Figure 6a–c. The data demonstrated outstanding agreement between the present findings and equations such as <8% with the Petukhov formula. The Gnielinski equation is better at low-range Re than the Petukhov equation at higher Re-values [43]. Figure 6b–d demonstrated the relative errors between the collected and equations data for average heat transfer coefficients and Nu_{avg}.

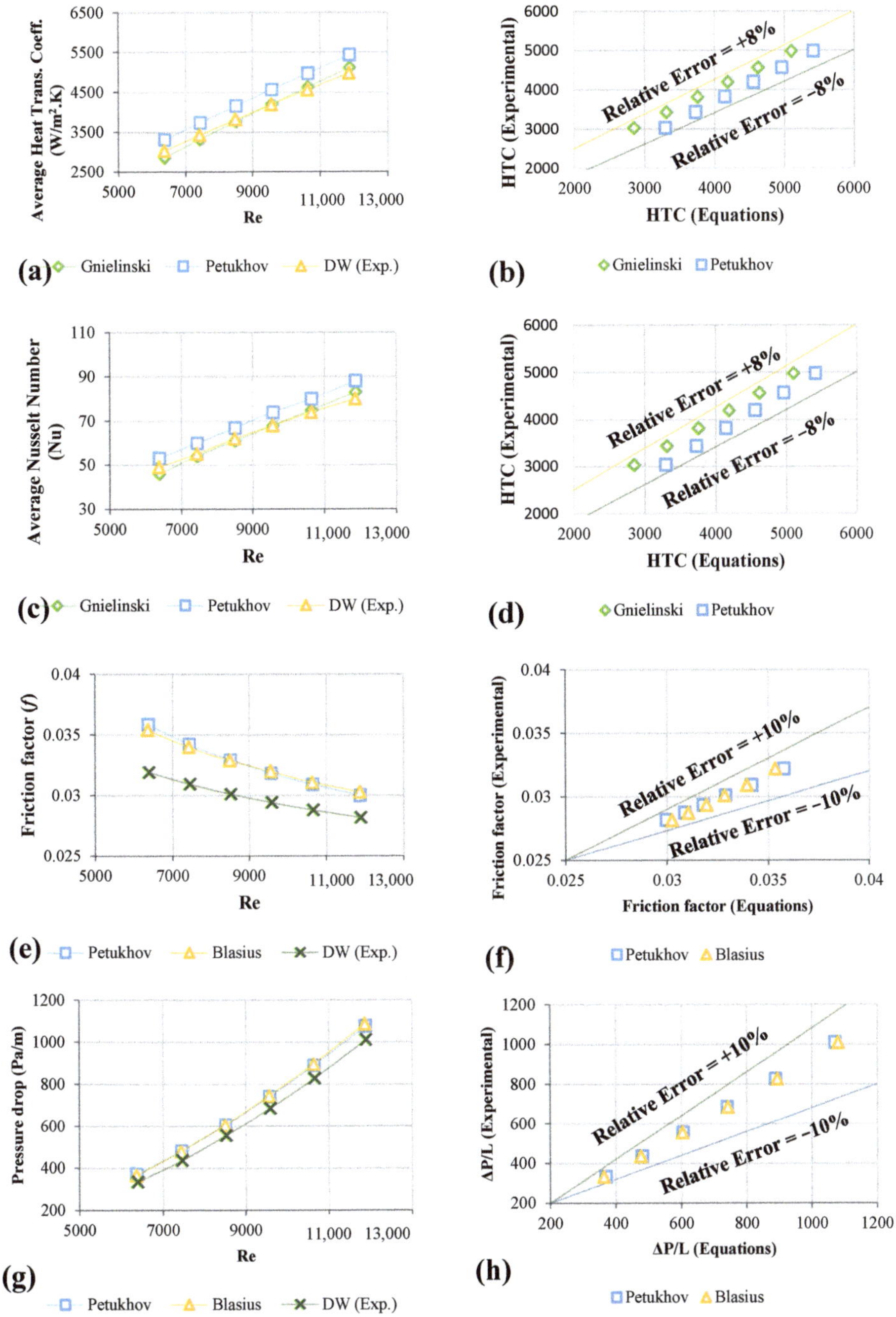

Figure 6. The verification assessment; (**a**) heat transfer coefficients measurement and prediction for 11,205 W/m^2, (**b**) the magnitude of the relative error metric, (**c**) The value of the average Nusselt number at 11,205 W/m^2, (**d**) the magnitude of the relative error metric, (**e**) Frictional head loss, (**f**) the magnitude of the relative error metric, (**g**) Pressure loss, (**h**) the magnitude of the relative error metric.

Assessment of the experimental friction factor was calculated based on the measurement of the pressure loss in the entire applied heating pipe. The validation and verification procedures have been carried out using the Blasius and Petukhov equations for smooth pipes [44,45]. The validation of the experimental data for pressure loss and friction factor is shown in Figure 6e–g, while that of the data from the equations and the current study is shown in Figure 6f–h.

3.4. Convective Heat Transfer of Functionalized Nanofluids

The prepared samples in this study were made without adding surfactant due to their long-term stability [46]. The present study analyzed functionalized and commercial metallic oxide nanofluids to enhance convective heat transfer inside a square heat exchanger. Essentially, turbulent forced convective flow is typically conducted under heat transfer demands.

The convective heat transfer coefficients of the functionalized and metal oxides-based nanofluids are shown in Figure 7a versus multiple nanofluids and Re-numbers. Increased nanofluids convective heat transfer coefficient as the velocity of the working fluid increased. This improvement resulted from solid nanoparticles' Brownian forces, thermal diffusion, and thermophoresis [47]. In the meantime, the increase in heat transfer might also result from the thin thermal boundary layer, which caused the higher velocities that caused thermal conductivity and decreased thermal resistance between the flowing nanofluid and the temperature of the internal wall surface of the heated pipe. Compared to DW, the maximum increase in heat transfer coefficients was as follows: PEG@GNPs = 44.4%, PEG@TGr = 41.2%, Al_2O_3 = 22.5%, and SiO_2 = 24% at 0.1 wt.%. As per experimental data, the increase in heat transfer, as per test data, may be due either to the delay of the thermal boundary layers or due to increased thermal conductivity of the nanofluids.

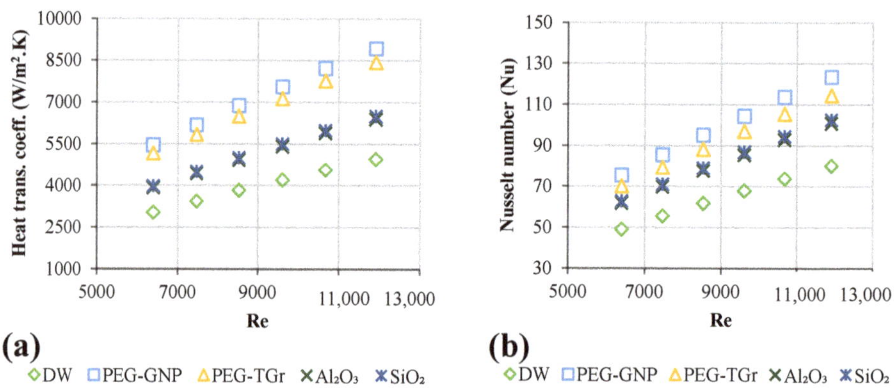

Figure 7. Heat transfer properties of different nanofluid types versus Reynolds numbers; (**a**) Heat transfer coefficients, (**b**) Average Nusselt number.

Figure 7b introduced Nu_{avg} at 11,205 W/m^2 and the Re-number function. The Nu_{avg} revealed an increase for each tested nanofluid. Observable higher Nu_{avg} of nanofluids reflected the decline in the circulation temperature after the working fluid had risen thermal conductivity; this subsequently reduced the temperature gradient between the wall of the tube and bulk fluid contained in the test-section. The maximum rise in Nu_{avg} was noted as follows: PEG@GNPs = 54%, PEG@TGr = 43%, SiO_2 = 28%, and Al_2O_3 = 26% associated with DW.

3.5. Friction Factor of Nanofluids

The nanofluids were evaluated for pressure loss and friction factor when flowing in a square heat exchanger at different Re-numbers. Figure 8a,b showed the measured pressure loss and friction factor for all samples versus the Re. The highest-pressure loss and friction factor increases were 20.8–23.7% and 3.57–3.85%, at a weight percentage of 0.1 wt. % and a velocity of 0.93 m/s, respectively.

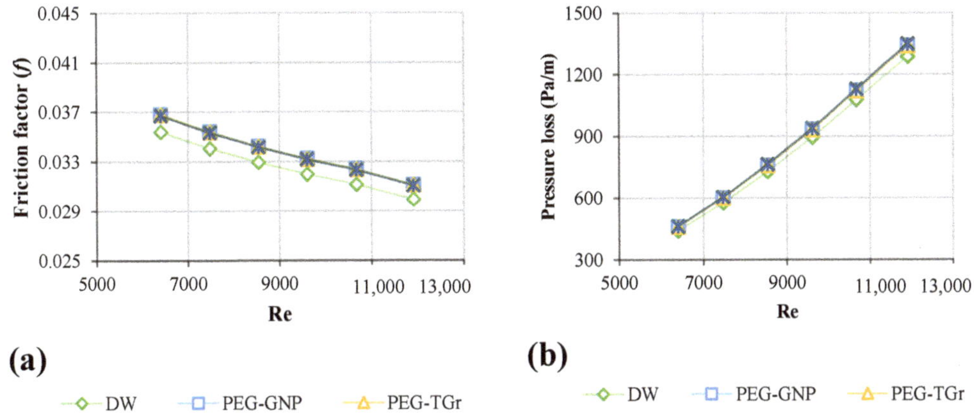

Figure 8. Hydrodynamic properties of different nanofluid types against Reynolds number; (**a**) Friction factor, (**b**) Pressure drop.

Brownian motion significantly affects the momentum transfer between solid nanoparticles and base fluid molecules at a low range of Re-numbers. The friction factor of samples increases slightly due to the Brownian motion [48]. However, this is an inactive mechanism in the high Re range. Mainly, the velocity of the nanofluids can be considered the most significant factor for the development of friction factor at a high Re-number range. The considerable differences between the observed friction factors of functionalized carbon nanostructures, metallic oxides, and distilled water at multiple Re numbers are due to the minor improvement in the viscosities of distilled water and their nanofluids. The variations in the friction factor are based on the nanofluid-related viscous drag. Typically, the density of nanoparticles is a crucial factor in enhancing the nano-coolant friction factor. The combination of dynamic and kinematic viscosities dramatically affects the pressure drop of different nanofluids. The excessive pumping capacity increases with increased dynamic viscosity.

3.6. Performance Index and Performance Evaluation Criterion

Figures 9 and 10 show changes in the performance characteristics of PI and PEC for different types of nanofluid vs. Re-numbers. The average PI, generally with PEC of the analyzed nanofluid, was noted to be >1 [34], indicating the effectiveness of the well-prepared nano-coolants for heated pipe flows. In addition, the carbon-based nanofluids presented higher augmentation of the metallic oxides due to a better rise in heat transport than the increased pressure loss. The PI of carbon and metallic oxides-based nanofluids improved with the Re-number; the maximum thermal efficiency of the nanofluids increased as follows: PEG@GNPs = 2.14, PEG@TGr = 2.05, Al_2O_3 = 1.23, and SiO_2 = 1.19 at Re = 11,907, 0.1 wt.%, and 11,205 W/m^2. This phenomenon was caused by the increased viscosity and thermal conductivity of nanofluids. The dynamic viscosity of a nanofluid can be increased to reduce the thickness of the boundary layer, resulting in an increase in heat transfer, while enhancing thermal conductivity enhances the thermal performance factor [18]. These results also confirm that the positive effects of heat transfer compensate for the negative impacts of pressure loss for carbon and metal-oxide nanofluids within a wide range of inlet temperatures, mass concentrations, and constant flow rates, stating that prepared nanofluids have excellent convective heat transfer capabilities.

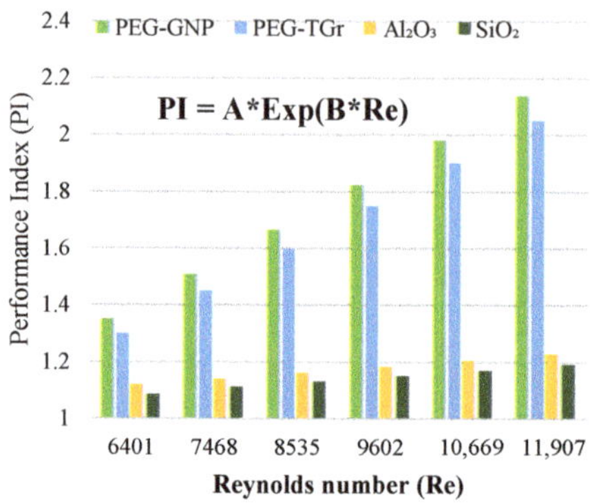

PEG-GNPs	**PEG-TGr**	**Al_2O_3**	**SiO_2**
A = 0.804	A = 0.7776	A = 1.0042	A = 0.9799
B = 8 × 10^{-5}	B = 8 × 10^{-5}	B = 2 × 10^{-5}	B = 2 × 10^{-5}
R^2 = 0.9917	R^2 = 0.9918	R^2 = 0.9998	R^2 = 0.9956

Figure 9. PI against different nanofluids and Reynolds number.

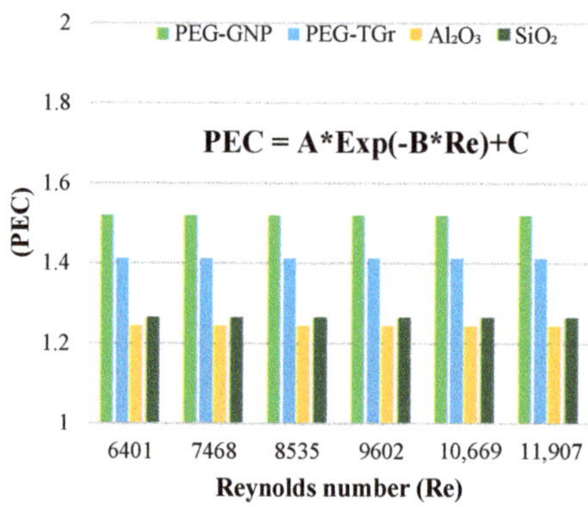

PEG-GNP	**PEG-TGr**	**Al_2O_3**	**SiO_2**
Coefficients (with 95% confidence bounds):	Coefficients (with 95% confidence bounds):	Coefficients (with 95% confidence bounds):	Coefficients (with 95% confidence bounds):
A = 0.4898	A = 0.7094	A = 0.6797	A = 0.119
B = 0.4456	B = 0.7547	B = 0.6551	B = 0.4984
C = 1.52 (1.52, 1.52)	C = 1.412 (1.41, 1.413)	C = 1.243 (1.24, 1.247)	C = 1.264 (1.26, 1.268)

Figure 10. PEC of different nanofluids against Reynolds number.

Additionally, the findings of the PEC have shown a slight reduction in Re numbers. The maximum performance assessment of the nanofluids was as follows: PEG@GNPs = 1.52, PEG@TGr = 1.41, Al_2O_3 = 1.24, and SiO_2 = 1.26.

3.7. Pumping Power of Different Nanofluids

When choosing a heat exchanger, several criteria are the heat transfer rate, pumping power, cost, size and weight, type, and material. Friction effects in nanofluids cause pressure loss, and pressure loss calculations influence pumping power needs. Increased pumping power will result in more extraordinary capital expenses because larger pumps are more expensive and have higher operational costs due to the higher pumping power required. Pumping power measures a system's financial ability to increase industrial and electrical energy. During the design of heat exchangers, it is essential to ensure low pumping power but effective heat transfer to ensure energy conservation. Figure 11 presents the pumping power for prepared nano-coolants at various Re with the working fluids. As the pumping power is dependent on the dynamic viscosity and density of both base fluid and the nanofluids (Equation (13)), the relative pumping power for all the tested samples is less than 1.

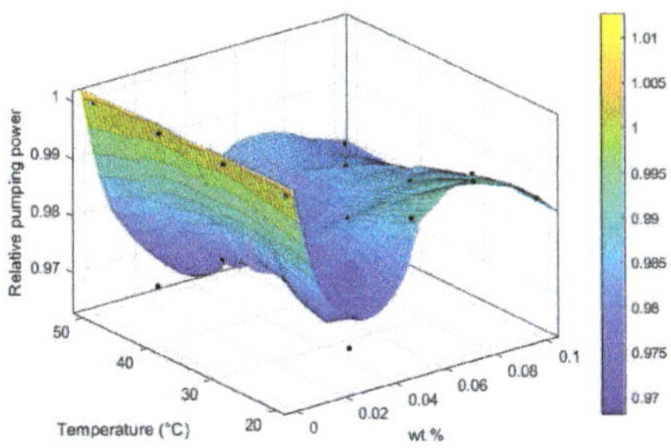

(**a**)
Coefficients:
p = coefficient structure
Goodness of fit:
SSE: 7.37×10^{-5}
R^2: 0.9612
Adjusted R^2: 0.9433
RMSE: 0.002381

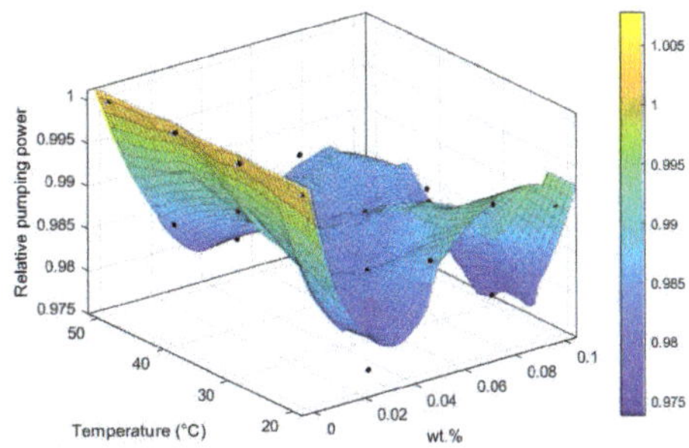

(**b**)
Coefficients:
p = coefficient structure
Goodness of fit:
SSE: 3.253×10^{-5}
R^2: 0.9681
Adjusted R^2: 0.9534
RMSE: 0.001582

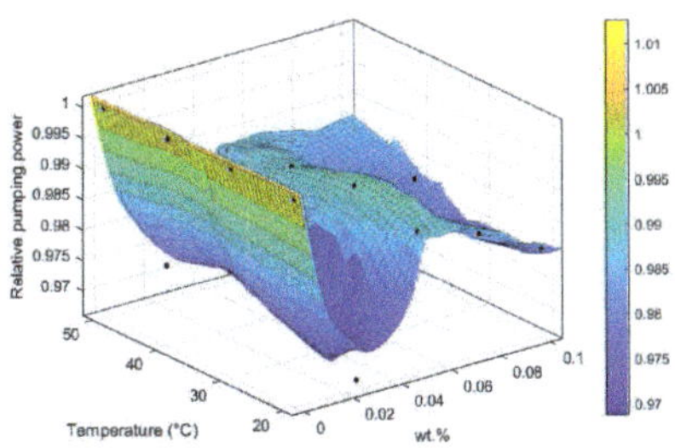

(**c**)
Locally weighted smoothing quadratic regression:
f(x,y) = loess (quadratic) smoothing regression computed from p
where x is normalized by mean 0.05 and std 0.03627
and where y is normalized by mean 35 and std 11.47
Coefficients:
p = coefficient structure
Goodness of fit:
SSE: 4.742×10^{-5}
R^2: 0.9746
Adjusted R^2: 0.9629
RMSE: 0.00191

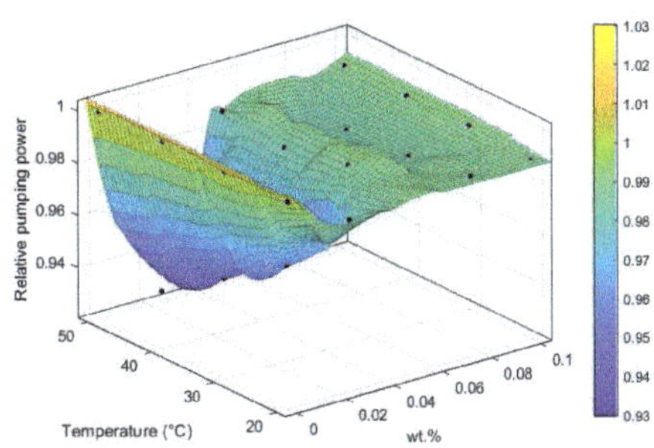

(**d**)
Locally weighted smoothing quadratic regression:
f(x,y) = loess (quadratic) smoothing regression computed from p
where x is normalized by mean 0.05 and std 0.03627
and where y is normalized by mean 35 and std 11.47
Coefficients:
p = coefficient structure
Goodness of fit:
SSE: 2.506×10^{-4}
R^2: 0.9662
Adjusted R^2: 0.9505
RMSE: 0.004391

Figure 11. Relative pumping power of different nanofluids against different temperatures and various wt.%, (**a**) PEG@GNP, (**b**) PEG@TGr, (**c**) SiO_2, and (**d**) Al_2O_3.

4. Conclusions

To improve the thermal performance of a square heat exchanger, four samples of nanofluids were covalently synthesized and produced. In order to start, all nano-coolants' thermophysical properties were tracked versus the temperature to perform studies on the effects of heat and momentum transfer in fully developed turbulent forced convective flow. Different conditions were implemented, such as varying temperatures, different nanofluids, and different Re numbers.

From the results of this work, the following was concluded:

- The nanofluids exhibited the greatest thermal conductivity improvements as follows: PEG@GNPs = 31.6%, PEG@TGr = 29.74%, Al_2O_3 = 10.44%, and SiO_2 = 9.32% at 60 °C and 0.1 wt.%.
- The highest improvement in heat transfer coefficients of the nanofluids was as follows: PEG@GNPs = 44.4%, PEG@TGr = 41.2%, Al_2O_3 = 22.5%, and SiO_2 = 24 % at 0.1 wt.%. Meanwhile, the maximum enhancement in the Nu of the nanofluids was as follows: PEG@GNPs = 35%, PEG@TGr = 30.1%, Al_2O_3 = 20.6%, and SiO_2 = 21.9% at 11,205 W/m^2.
- The most significant pressure loss and friction factor increases were 20.8–23.7% and 3.57–3.85%, respectively. The effective dynamic viscosity significantly impacts the pressure drop for different nanofluids.
- The PI and PEC values of the tested samples were >1 and increased with the Reynolds number.
- Although the required pumping power was slightly increased, this was advantageous for the industrial application of these new working fluids.
- The nonlinear regression was developed for a relative pumping power of different nanofluids against temperature at different mass fractions.

Author Contributions: Conceptualization, M.S.A., A.H.J. and R.Z.H.; data curation, H.A.A., M.H.N., Z.H.M., M.S. and Z.M.Y.; formal analysis, H.A.A., A.M.A., A.H.J., M.H.N., R.Z.H., Z.H.M., A.H.A., M.S. and Z.M.Y.; investigation, M.H.N., A.M.A., R.Z.H., Z.H.M., A.H.A. and Z.M.Y.; methodology, M.S.A., A.H.J. and A.H.A.; resources, H.A.A., M.H.N. and R.Z.H.; software, M.S.A.; supervision, A.H.J. and M.S.; validation, H.A.A., A.M.A., M.S.A., Z.H.M., A.H.A., M.S. and Z.M.Y.; visualization, H.A.A., A.H.J., A.M.A., Z.H.M., A.H.A. and Z.M.Y.; writing–original draft, H.A.A., M.S.A., A.H.J., A.M.A., M.H.N., R.Z.H., Z.H.M., A.H.A., M.S. and Z.M.Y. All authors have read and agreed to the published version of the manuscript.

Funding: This work was financially supported by RainSolutions (Water JPI 2018 Joint Call project).

Institutional Review Board Statement: Not applicable.

Informed Consent Statement: Not applicable.

Data Availability Statement: All the data are presented in the manuscript.

Acknowledgments: The authors would like to thank Al-Mustaqbal University College for providing technical support for this research.

Conflicts of Interest: The authors declare no conflict of interest.

Nomenclature

A_c	Square Cross-section area of pipe, m^2
Al_2O_3	Alumina
$AlCl_3$	Aluminum chloride
CNTs	Carbon nanotubes
C_p	Specific heat capacity, kJ/kg K
Cu	Copper nanomaterials
CuO	Copper oxide
D_h	Hydraulic Diameter of pipe, m
DMF	Dimethylformamide

DPT	Differential Pressure Transmitter
DSC	Differential Scanning Calorimeter
DW	Distilled water
f	Friction factor
GNP	Graphene Nanoplatelets
GO	Graphite oxide
h	Heat transfer coefficient, W/m^2-K
HCl	Hydrochloric acid
I	Electric current, A
k	Thermal conductivity, W/m-K
L	Tube total-length, m
$\dot{m}$	Mass flow rate, kg/s
MWCNT	Multi-Walled Carbon Nanotube
NIST	National Institute of Standards and Technology
Nu	Average Nusselt number
P	Power supply, W
P	Perimeter of square pipe, m
PEC	Performance Evaluation Criterion
PEG	Pentaethylene Glycol
PG	Propylene Glycol
Pr	Prandtl number
q''	Constant Heat flux, W/m^2
Q	Heat transfer amount, W
RGO	Reduced Graphene Oxide
Re	Reynolds number
RTD	Resistance Temperature Detector
SiO_2	Silica nanomaterials
T	Temperature, °C
T-Gr	Thermally Treated Graphene
THF	Tetrahydrofuran
TiO_2	Titanium dioxide
U	Velocity vector, m/s
V	Voltage, V
v	Working fluid velocity, m/s
W	Pumping power, W
Greek symbols	
ρ	Density, kg/m^3
μ	Dynamic Viscosity, Pa·s
ε	Performance index
ΔP	Pressure loss
φ	Mass fraction, %
Subscripts	
bf	Base fluid
nf	Nanofluid
np	Particles in nanosize
w	Wall of pipe
i	Inlet
o	Output
b	Bulk fluid

References

1. Sadri, R.; Mallah, A.R.; Hosseini, M.; Ahmadi, G.; Kazi, S.N.; Dabbagh, A.; Yeong, C.H.; Ahmad, R.; Yaakup, N.A. CFD Modeling of Turbulent Convection Heat Transfer of Nanofluids Containing Green Functionalized Graphene Nanoplatelets Flowing in a Horizontal Tube: Comparison with Experimental Data. *J. Mol. Liq.* **2018**, *269*, 152–159. [CrossRef]
2. Mohammad, R.S.; Aldlemy, M.S.; Al Hassan, S.; Abdulla, A.I.; Scholz, M.; Yaseen, Z.M. Frictional Pressure Drop and Cost Savings for Graphene Nanoplatelets Nanofluids in Turbulent Flow Environments. *Nanomaterials* **2021**, *11*, 3094. [CrossRef] [PubMed]
3. Gonçalves, H.M.R.; Pereira, R.F.P.; Lepleux, E.; Pacheco, L.; Valente, A.J.M.; Duarte, A.J.; de Zea Bermudez, V. Non-Newtonian Thermosensitive Nanofluid Based on Carbon Dots Functionalized with Ionic Liquids. *Small* **2020**, *16*, 1907661. [CrossRef]

4. MR Gonçalves, H.; Neves, S.A.F.; Duarte, A.; de Zea Bermudez, V. Nanofluid Based on Carbon Dots Functionalized with Ionic Liquids for Energy Applications. *Energies* **2020**, *13*, 649. [CrossRef]
5. Sofiah, A.G.N.; Samykano, M.; Pandey, A.K.; Kadirgama, K.; Sharma, K.; Saidur, R. Immense Impact from Small Particles: Review on Stability and Thermophysical Properties of Nanofluids. *Sustain. Energy Technol. Assess.* **2021**, *48*, 101635. [CrossRef]
6. Esfe, M.H.; Kamyab, M.H.; Toghraie, D. Statistical Review of Studies on the Estimation of Thermophysical Properties of Nanofluids Using Artificial Neural Network (ANN). *Powder Technol.* **2022**, *400*, 117210. [CrossRef]
7. Adun, H.; Kavaz, D.; Dagbasi, M. Review of Ternary Hybrid Nanofluid: Synthesis, Stability, Thermophysical Properties, Heat Transfer Applications, and Environmental Effects. *J. Clean. Prod.* **2021**, *328*, 129525. [CrossRef]
8. Bumataria, R.K.; Chavda, N.K.; Panchal, H. Current Research Aspects in Mono and Hybrid Nanofluid Based Heat Pipe Technologies. *Heliyon* **2019**, *5*, e01627. [CrossRef]
9. Ambreen, T.; Kim, M.-H. Heat Transfer and Pressure Drop Correlations of Nanofluids: A State of Art Review. *Renew. Sustain. Energy Rev.* **2018**, *91*, 564–583. [CrossRef]
10. Alawi, O.A.; Kamar, H.M.; Hussein, O.A.; Mallah, A.R.; Mohammed, H.A.; Khiadani, M.; Roomi, A.B.; Kazi, S.N.; Yaseen, Z.M. Effects of Binary Hybrid Nanofluid on Heat Transfer and Fluid Flow in a Triangular-Corrugated Channel: An Experimental and Numerical Study. *Powder Technol.* **2022**, *395*, 267–279. [CrossRef]
11. Abdelrazek, A.H.; Alawi, O.A.; Kazi, S.N.; Yusoff, N.; Chowdhury, Z.; Sarhan, A.A.D. A New Approach to Evaluate the Impact of Thermophysical Properties of Nanofluids on Heat Transfer and Pressure Drop. *Int. Commun. Heat Mass Transf.* **2018**, *95*, 161–170. [CrossRef]
12. Murshed, S.M.S.; Estellé, P. A State of the Art Review on Viscosity of Nanofluids. *Renew. Sustain. Energy Rev.* **2017**, *76*, 1134–1152. [CrossRef]
13. Azmi, W.H.; Sharma, K.V.; Mamat, R.; Najafi, G.; Mohamad, M.S. The Enhancement of Effective Thermal Conductivity and Effective Dynamic Viscosity of Nanofluids—A Review. *Renew. Sustain. Energy Rev.* **2016**, *53*, 1046–1058. [CrossRef]
14. Khanafer, K.; Vafai, K. A Critical Synthesis of Thermophysical Characteristics of Nanofluids. *Int. J. Heat Mass Transf.* **2011**, *54*, 4410–4428. [CrossRef]
15. Alawi, O.A.; Kamar, H.M.; Mallah, A.R.; Mohammed, H.A.; Sabrudin, M.A.S.; Newaz, K.M.S.; Najafi, G.; Yaseen, Z.M. Experimental and Theoretical Analysis of Energy Efficiency in a Flat Plate Solar Collector Using Monolayer Graphene Nanofluids. *Sustainability* **2021**, *13*, 5416. [CrossRef]
16. Sadeghinezhad, E.; Togun, H.; Mehrali, M.; Sadeghi Nejad, P.; Tahan Latibari, S.; Abdulrazzaq, T.; Kazi, S.N.; Metselaar, H.S.C. An Experimental and Numerical Investigation of Heat Transfer Enhancement for Graphene Nanoplatelets Nanofluids in Turbulent Flow Conditions. *Int. J. Heat Mass Transf.* **2015**, *81*, 41–51. [CrossRef]
17. Zubir, M.N.M.; Badarudin, A.; Kazi, S.N.; Huang, N.M.; Misran, M.; Sadeghinezhad, E.; Mehrali, M.; Syuhada, N.I.; Gharehkhani, S. Experimental Investigation on the Use of Reduced Graphene Oxide and Its Hybrid Complexes in Improving Closed Conduit Turbulent Forced Convective Heat Transfer. *Exp. Therm. Fluid Sci.* **2015**, *66*, 290–303. [CrossRef]
18. Solangi, K.H.; Amiri, A.; Luhur, M.R.; Ghavimi, S.A.A.; Zubir, M.N.M.; Kazi, S.N.; Badarudin, A. Experimental Investigation of the Propylene Glycol-Treated Graphene Nanoplatelets for the Enhancement of Closed Conduit Turbulent Convective Heat Transfer. *Int. Commun. Heat Mass Transf.* **2016**, *73*, 43–53. [CrossRef]
19. Yarmand, H.; Gharehkhani, S.; Shirazi, S.F.S.; Amiri, A.; Alehashem, M.S.; Dahari, M.; Kazi, S.N. Experimental Investigation of Thermo-Physical Properties, Convective Heat Transfer and Pressure Drop of Functionalized Graphene Nanoplatelets Aqueous Nanofluid in a Square Heated Pipe. *Energy Convers. Manag.* **2016**, *114*, 38–49. [CrossRef]
20. Sadri, R.; Hosseini, M.; Kazi, S.N.; Bagheri, S.; Abdelrazek, A.H.; Ahmadi, G.; Zubir, N.; Ahmad, R.; Abidin, N.I.Z. A Facile, Bio-Based, Novel Approach for Synthesis of Covalently Functionalized Graphene Nanoplatelet Nano-Coolants toward Improved Thermo-Physical and Heat Transfer Properties. *J. Colloid Interface Sci.* **2018**, *509*, 140–152. [CrossRef]
21. Qi, C.; Wang, G.; Yan, Y.; Mei, S.; Luo, T. Effect of Rotating Twisted Tape on Thermo-Hydraulic Performances of Nanofluids in Heat-Exchanger Systems. *Energy Convers. Manag.* **2018**, *166*, 744–757. [CrossRef]
22. Qi, C.; Fan, F.; Pan, Y.; Liu, M.; Yan, Y. Effects of Turbulator with Round Hole on the Thermo-Hydraulic Performance of Nanofluids in a Triangle Tube. *Int. J. Heat Mass Transf.* **2020**, *146*, 118897. [CrossRef]
23. Kaood, A.; Hassan, M.A. Thermo-Hydraulic Performance of Nanofluids Flow in Various Internally Corrugated Tubes. *Chem. Eng. Process. Process Intensif.* **2020**, *154*, 108043. [CrossRef]
24. Alawi, O.A.; Sidik, N.A.C.; Kazi, S.N.; Najafi, G. Graphene Nanoplatelets and Few-Layer Graphene Studies in Thermo-Physical Properties and Particle Characterization. *J. Therm. Anal. Calorim.* **2019**, *135*, 1081–1093. [CrossRef]
25. Alawi, O.A.; Mallah, A.R.; Kazi, S.N.; Sidik, N.A.C.; Najafi, G. Thermophysical Properties and Stability of Carbon Nanostructures and Metallic Oxides Nanofluids: Experimental Approach. *J. Therm. Anal. Calorim.* **2018**, *135*, 1545–1562. [CrossRef]
26. Buongiorno, J.; Venerus, D.C.; Prabhat, N.; McKrell, T.; Townsend, J.; Christianson, R.; Tolmachev, Y.V.; Keblinski, P.; Hu, L.; Alvarado, J.L. A Benchmark Study on the Thermal Conductivity of Nanofluids. *J. Appl. Phys.* **2009**, *106*, 94312. [CrossRef]
27. Fernandez-Seara, J.; Uhía, F.J.; Sieres, J.; Campo, A. A General Review of the Wilson Plot Method and Its Modifications to Determine Convection Coefficients in Heat Exchange Devices. *Appl. Therm. Eng.* **2007**, *27*, 2745–2757. [CrossRef]
28. Gnielinski, V. New Equations for Heat and Mass Transfer in the Turbulent Flow in Pipes and Channels. *NASA STI/Recon Tech. Rep. A* **1975**, *75*, 8–16.

29. Colebrook, C.F. Turbulent Flow in Pipes, with Particular Reference to the Transition Region between the Smooth and Rough Pipe Laws. *J. Inst. Civ. Eng.* **1939**, *11*, 133–156. [CrossRef]
30. Petukhov, B.S. Heat Transfer and Friction in Turbulent Pipe Flow with Variable Physical Properties. *Adv. Heat Transf.* **1970**, *6*, 503–564. [CrossRef]
31. Sadeghinezhad, E.; Mehrali, M.; Tahan Latibari, S.; Mehrali, M.; Kazi, S.N.; Oon, C.S.; Metselaar, H.S.C. Experimental Investigation of Convective Heat Transfer Using Graphene Nanoplatelet Based Nanofluids under Turbulent Flow Conditions. *Ind. Eng. Chem. Res.* **2014**, *53*, 12455–12465. [CrossRef]
32. Blasius, H. Grenzschichten in Flussigkeiten Mit Kleiner Reibung. *Z. Math. Phys.* **1908**, *56*, 1–37.
33. Das, S.K.; Putra, N.; Thiesen, P.; Roetzel, W. Temperature Dependence of Thermal Conductivity Enhancement for Nanofluids. *J. Heat Transfer* **2003**, *125*, 567–574. [CrossRef]
34. Amiri, A.; Ahmadi, G.; Shanbedi, M.; Etemadi, M.; Mohd Zubir, M.N.; Chew, B.T.; Kazi, S.N. Heat Transfer Enhancement of Water-Based Highly Crumpled Few-Layer Graphene Nanofluids. *RSC Adv.* **2016**, *6*, 105508–105527. [CrossRef]
35. Taylor, J. *Introduction to Error Analysis, the Study of Uncertainties in Physical Measurements*, 2nd ed.; University Science Books: New York, NY, USA, 1997; ISBN 9780935702750.
36. Ramires, M.L.V.; Nieto de Castro, C.A.; Nagasaka, Y.; Nagashima, A.; Assael, M.J.; Wakeham, W.A. Standard Reference Data for the Thermal Conductivity of Water. *J. Phys. Chem. Ref. Data* **1995**, *24*, 1377–1381. [CrossRef]
37. Aravind, S.S.J.; Baskar, P.; Baby, T.T.; Sabareesh, R.K.; Das, S.; Ramaprabhu, S. Investigation of Structural Stability, Dispersion, Viscosity, and Conductive Heat Transfer Properties of Functionalized Carbon Nanotube Based Nanofluids. *J. Phys. Chem. C* **2011**, *115*, 16737–16744. [CrossRef]
38. Pak, B.C.; Cho, Y.I. Hydrodynamic and Heat Transfer Study of Dispersed Fluids with Submicron Metallic Oxide Particles. *Exp. Heat Transf.* **1998**, *11*, 151–170. [CrossRef]
39. Zhang, L.; Chen, L.; Liu, J.; Fang, X.; Zhang, Z. Effect of Morphology of Carbon Nanomaterials on Thermo-Physical Characteristics, Optical Properties and Photo-Thermal Conversion Performance of Nanofluids. *Renew. Energy* **2016**, *99*, 888–897. [CrossRef]
40. Elias, M.M.; Mahbubul, I.M.; Saidur, R.; Sohel, M.R.; Shahrul, I.M.; Khaleduzzaman, S.S.; Sadeghipour, S. Experimental Investigation on the Thermo-Physical Properties of Al2O3 Nanoparticles Suspended in Car Radiator Coolant. *Int. Commun. Heat Mass Transf.* **2014**, *54*, 48–53. [CrossRef]
41. Khajeh Arzani, H.; Amiri, A.; Arzani, H.K.; Rozali, S.B.; Kazi, S.N.; Badarudin, A. Toward Improved Heat Transfer Performance of Annular Heat Exchangers with Water/Ethylene Glycol-Based Nanofluids Containing Graphene Nanoplatelets. *J. Therm. Anal. Calorim.* **2016**, *126*, 1427–1436. [CrossRef]
42. Selvam, C.; Mohan Lal, D.; Harish, S. Thermal Conductivity and Specific Heat Capacity of Water–Ethylene Glycol Mixture-Based Nanofluids with Graphene Nanoplatelets. *J. Therm. Anal. Calorim.* **2017**, *129*, 947–955. [CrossRef]
43. Cengel, Y.A.; Klein, S.; Beckman, W. *Heat Transfer: A Practical Approach*; McGraw-Hill: New York, NY, USA, 2003. [CrossRef]
44. Kazi, S.N.; Duffy, G.G.; Chen, X.D. Validation of Heat Transfer and Friction Loss Data for Fibre Suspensions in a Circular and a Coaxial Pipe Heat Exchanger. *Int. J. Therm. Sci.* **2014**, *79*, 146–160. [CrossRef]
45. Mehrali, M.; Sadeghinezhad, E.; Rosen, M.A.; Tahan Latibari, S.; Mehrali, M.; Metselaar, H.S.C.; Kazi, S.N. Effect of Specific Surface Area on Convective Heat Transfer of Graphene Nanoplatelet Aqueous Nanofluids. *Exp. Therm. Fluid Sci.* **2015**, *68*, 100–108. [CrossRef]
46. Wensel, J.; Wright, B.; Thomas, D.; Douglas, W.; Mannhalter, B.; Cross, W.; Hong, H.; Kellar, J.; Smith, P.; Roy, W. Enhanced Thermal Conductivity by Aggregation in Heat Transfer Nanofluids Containing Metal Oxide Nanoparticles and Carbon Nanotubes. *Appl. Phys. Lett.* **2008**, *92*, 023110. [CrossRef]
47. Michaelides, E.E. Brownian Movement and Thermophoresis of Nanoparticles in Liquids. *Int. J. Heat Mass Transf.* **2015**, *81*, 179–187. [CrossRef]
48. Shanbedi, M.; Amiri, A.; Rashidi, S.; Heris, S.Z.; Baniadam, M. Thermal Performance Prediction of Two-Phase Closed Thermosyphon Using Adaptive Neuro-Fuzzy Inference System. *Heat Transf. Eng.* **2015**, *36*, 315–324. [CrossRef]

nanomaterials

MDPI

Article

Experimental Investigation on Stability, Viscosity, and Electrical Conductivity of Water-Based Hybrid Nanofluid of MWCNT-Fe_2O_3

Solomon O. Giwa [1], Mohsen Sharifpur [2,3,*], Mohammad H. Ahmadi [4], S. M. Sohel Murshed [5,*] and Josua P. Meyer [2]

1 Department of Mechanical Engineering, Olabisi Onabanjo University, Ibogun 112104, Nigeria; sologiwa2002@yahoo.com
2 Department of Mechanical and Aeronautical Engineering, University of Pretoria, Pretoria 0002, South Africa; josua.meyer@up.ac.za
3 Department of Mechanical Engineering, University of Science and Culture, Tehran 1461968151, Iran
4 Faculty of Mechanical Engineering, Shahrood University of Technology, Shahrood 3619995161, Iran; mohammadhosein.ahmadi@gmail.com
5 Center for Innovation, Technology and Policy Research (IN+), Department of Mechanical Engineering, Instituto Superior Técnico, University of Lisbon, 1049-001 Lisbon, Portugal
* Correspondence: mohsen.sharifpur@up.ac.za (M.S.); smurshed@tecnico.ulisboa.pt (S.M.S.M.)

Abstract: The superiority of nanofluid over conventional working fluid has been well researched and proven. Newest on the horizon is the hybrid nanofluid currently being examined due to its improved thermal properties. This paper examined the viscosity and electrical conductivity of deionized water (DIW)-based multiwalled carbon nanotube (MWCNT)-Fe_2O_3 (20:80) nanofluids at temperatures and volume concentrations ranging from 15 °C to 55 °C and 0.1–1.5%, respectively. The morphology of the suspended hybrid nanofluids was characterized using a transmission electron microscope, and the stability was monitored using visual inspection, UV–visible, and viscosity-checking techniques. With the aid of a viscometer and electrical conductivity meter, the viscosity and electrical conductivity of the hybrid nanofluids were determined, respectively. The MWCNT-Fe_2O_3/DIW nanofluids were found to be stable and well suspended. Both the electrical conductivity and viscosity of the hybrid nanofluids were augmented with respect to increasing volume concentration. In contrast, the temperature rise was noticed to diminish the viscosity of the nanofluids, but it enhanced electrical conductivity. Maximum increments of 35.7% and 1676.4% were obtained for the viscosity and electrical conductivity of the hybrid nanofluids, respectively, when compared with the base fluid. The obtained results were observed to agree with previous studies in the literature. After fitting the obtained experimental data, high accuracy was achieved with the formulated correlations for estimating the electrical conductivity and viscosity. The examined hybrid nanofluid was noticed to possess a lesser viscosity in comparison with the mono-particle nanofluid of Fe_2O_3/water, which was good for engineering applications as the pumping power would be reduced.

Keywords: nanofluids; Fe_2O_3 nanoparticle; multiwalled carbon nanotubes; viscosity; electrical conductivity; hybrid nanofluids

Citation: Giwa, S.O.; Sharifpur, M.; Ahmadi, M.H.; Sohel Murshed, S.M.; Meyer, J.P. Experimental Investigation on Stability, Viscosity, and Electrical Conductivity of Water-Based Hybrid Nanofluid of MWCNT-Fe_2O_3. *Nanomaterials* **2021**, *11*, 136. https://doi.org/10.3390/nano11010136

Received: 1 December 2020
Accepted: 1 January 2021
Published: 8 January 2021

1. Introduction

The exceptional thermal and flow properties exhibited by nanofluid compared with those of conventional thermal transporting media have projected this special fluid as a subject of intense global research. Early studies in this context have measured the viscosity and thermal conductivity of nanofluids with diverse base fluids (ethylene glycol (EG), water, propylene glycol, glycerol, etc.) and found that these properties of the nanofluids were enhanced in relation to the base fluids [1–10]. However, underlying mechanisms for such enhancements of these properties of nanofluids, particularly for thermal conductivity, are not yet explored, and there are also controversies and inconsistencies in literature results [10–13]. Outside the viscosity and thermal conductivity, the electrical conductivity (EC), specific heat capacity, dielectric, and density of various nanofluids prepared from different nanoparticles (Cu, MgO, CuO, CNT, SiO_2, ZnO, TiO_2, Al_2O_3, Fe_2O_3, Fe_3O_4, spinels etc.) and dispersed in several base fluids (water, propylene glycol, ethylene glycol, bio-glycol, palm oil, glycerol, ionic fluid, coconut oil, engine oil, etc.) have been subsequently examined at various mass/volume concentrations or fractions for different temperature ranges [14–20].

Hybridization of different types of nanoparticles to prepare hybrid nanofluids was first investigated by Jana and co-workers [21]. The idea behind this was to augment the thermal conductivity of nanofluid over that of conventional fluids. Jana and co-workers [21] synthesized aqueous carbon nanotubes (CNTs), Cu and Au nanofluids, and Cu-CNT and Au-CNT nanofluids and measured their thermal conductivity. They noticed that the hybrid nanofluids yielded a lower thermal conductivity relative to the mono-particle nanofluids. Suresh et al. [22] examined the thermal conductivity and viscosity of water-based Al_2O_3-Cu (90:10) nanofluids with varying volume concentrations (0.1–2%) at ambient temperature. The enhancement of the thermal conductivity by 1.47–12.11% and viscosity by 8–115%, in comparison with water, was reported. Both properties were noticed to improve as the volume concentration increased. Conflicting with the result of Jana et al. [21], Suresh et al. [22] showed that hybrid nanofluids have higher thermal conductivity than mono-particle nanofluids (Al_2O_3/water). Similarly, Chen et al. [23], who used water-based Ag-MWCNT nanofluid, reported lower thermal conductivity for water-based MWCNT nanofluid when compared with the hybrid nanofluid. The thermal conductivity of the water-based multiwalled CNT and γ-Al_2O_3 (at an equal weight ratio (1:1)) nanofluids for different volume concentrations (0–1%) at the room temperature was examined [24]. A maximum enhancement of 20.68% was achieved with 1 vol%.

Esfe et al. [25] used an equal volume concentration of water-based mono-particle nanofluids (Al_2O_3 and MWCNTs) to formulate water-based Al_2O_3-MWCNT nanofluids in an effort to measure the thermal conductivity at temperatures of 303–323 K. They noticed that the thermal conductivity was augmented when temperature and volume concentration increased in comparison with water. The rheological study of engine oil (EO)-based hybrid nanofluids (Al_2O_3-MWCNTs (75:25%)) was studied under varying shear rates (1333–13,333 s^{-1}), temperatures (25–50 °C), and volume concentrations (0–1%) by Dardan et al. [26]. The hybrid nanofluids were noticed to exhibit Newtonian behaviors. The viscosity improved with volume concentration increase and reduced with temperature rise. The highest enhancement of 46% was observed with 1 vol%. Afrand and co-workers [27] measured the viscosity of EO-based SiO_2-MWCNT hybrid nanofluids (at equal volumes of nanoparticles) under varying temperatures (25–60 °C) and volume concentrations (0.0625–1%). The hybrid nanofluids showed an enhancement of viscosity as the volume concentrations increased. The hybrid nanofluids were observed to be higher than those of SiO_2/EO and MWCNT/EO nanofluids with a maximum enhancement of 37.4% for 1 vol% at 60 °C.

In the work of Megatif et al. [28], equal weights of CNT-TiO_2 nanoparticles (0.1, 0.15, and 2.0 wt%) were suspended in water to synthesize the hybrid nanofluids, and the thermal conductivity, density, viscosity, and specific heat capacity were determined at 25–40 °C. These properties were found to be improved for the hybrid nanofluids compared to the mono-particle CNT nanofluids. The viscosity, specific heat capacity, and density of the hybrid nanofluids diminished with rising temperature, whereas the thermal conductivity followed the reverse trend. The rheological behavior of MWCNT-SiO_2 (50:50 vol%)/EG-water (50:50 vol%) nanofluids was examined under varying shear rates (0.612–122.3 s^{-1}), volume concentrations (0.0625–2%), and temperatures of 27.5–50 °C [29]. The nanofluids examined were found to demonstrate shear-thinning flow as the power-law index was below unity. Recently, Kakavandi and Akbari [30] studied the thermal conductivity of EG-water-based MWCNT-SiC/EG using a similar base fluid and ratio of nanoparticles as the work of Eshgarf and Afrand [29] with concentrations of 0–0.75 vol% and temperatures of 25–50 °C. The thermal conductivity was improved by 33% at 0.75 vol% and 50 °C when compared with water-EG.

Esfe et al. [31] experimentally determined the thermal conductivity of DWCNT-ZnO/EG (10:90) with concentration and temperature ranges of 0.045–1.9 vol% and 30–50 °C, respectively. At 50 °C and 1.9 vol%, the thermal conductivity was augmented by 24.9%. They revealed that the addition of 10% DWCNT nanoparticles to 90% ZnO nanoparticles to formulate the hybrid nanofluids caused the thermal conductivity of EG-based ZnO nanofluids to be enhanced. Additionally, their cost analysis showed that it was more economical to use hybrid nanofluids than mono-particle nanofluids. Moldoveanu et al. [32] studied the thermal conductivity of aqueous mono-particle (TiO_2 and Al_2O_3 for 1–3 vol%) and hybrid (Al_2O_3 (0.05 vol%)-TiO_2 (0.05–2.5 vol%) nanofluids at temperatures of 20 to 50 °C. The measured property was enhanced by 10.7–14.1%, 8.5–17.7%, and 15.3–19.2% for aqueous TiO_2, Al_2O_3, and Al_2O_3-TiO_2 nanofluids, respectively. An investigation into the impact of variation in the volume concentration (0–2.3%) and temperature (25–50 °C) on the thermal conductivity of EG-based hybrid nanofluids of functionalized MWCNT-Fe_3O_4 (at equal volumes) was carried out by Harandi et al. [33]. The thermal conductivity of the hybrid nanofluid was augmented by 30% at 50 °C and 2.3 vol%. The rheological behavior of an identical hybrid nanofluid (at equal amounts) and temperature range like that of Harandi et al. [33] was performed using shear rates of 12.24–73.44 s^{-1} and volume concentration of 0.1–1.8% [34]. The result showed that the nanofluids exhibited Newtonian behavior for volume concentrations of 0.1–0.8% and non-Newtonian flow for the nanofluids beyond 0.8 vol%.

Shi et al. [35] investigated the thermophysical properties (viscosity, thermal conductivity, and specific thermal capacity) of Fe_3O_4 and Fe_3O_4-MWCNT nanofluids with 0.25 vol%. They reported higher thermal conductivity and viscosity and lower specific heat capacity for Fe_3O_4-MWCNT nanofluid compared to Fe_3O_4. The introduction of MWCNT particles to formulate the hybrid nanofluid was observed to enhance the thermal conductivity and viscosity but attenuated the specific heat capacity as the Fe_3O_4 nanoparticles possess a better specific heat capacity than the MWCNT particles. Recently, a study on the thermal properties (viscosity, density, surface tension, specific heat capacity, and thermal conductivity) of three-nanocomponent water-based hybrid nanofluids was conducted by Mousavi et al. [36]. Hybrid nanofluids of CuO-MgO-TiO_2/deionized water (DIW) were formulated in five different mixture ratios with volume concentrations of 0.1–0.5 vol%, and the thermal properties were measured at temperatures of 15–60 °C. The results showed that the hybrid nanofluid with a mixture ratio of 60:30:10 (CuO-MgO-TiO_2) was the best as it had the lowest viscosity (36.4% for 0.5 vol% at 60 °C), highest thermal conductivity (78.6% for 0.1 vol% at 15 °C), and lowest surface tension when compared with DIW. Goodarzi et al. [37] investigated the behavior and viscosity of EO-based ZnO-MWCNT (25:75) nanofluids under changing shear rates (666.5–13,300 s^{-1}), temperatures (5–55 °C), and volume concentrations (0.05–0.8%). They reported a Newtonian flow for all the samples and at the studied temperatures. A temperature rise was observed to reduce the viscosity, whereas increasing the concentration improved the viscosity of the hybrid nanofluids, with an enhancement of 5–20%.

The above literature survey supports the fact that the hybridization of nanoparticles engaged in formulating hybrid nanofluids yielded an improvement in their thermal properties. However, there are limited studies in the open literature regarding the stability and the thermal properties of hybrid ferrofluids. In addition, there is a scarcity of documentation on the thermophysical properties of MWCNT nanoparticle-based ferrofluids (F_3O_4, Fe_2O_3, Co_3O_4, etc.) at different mixing ratios of bi-nanoparticles. This present study involved an experimental measurement of the electrical conductivity and viscosity of MWCNT-Fe_2O_3/DIW nanofluids with a bi-nanoparticle mixing ratio of 80:20 (weight % basis) for volume concentrations and temperatures ranging from 0.05% to 1.5% and 15 °C to 55 °C, respectively. Furthermore, research progress in this context revealed that there is a notable dearth of knowledge on the electrical conductivity of hybrid nanofluids in the public domain.

2. Methodology

2.1. Materials

Nanoparticles of γ-Fe_2O_3 and functionalised MWCNTs were used in this work. The γ-Fe_2O_3 nanoparticles (20–30 nm diameter as specified by the manufacturer) were sourced from Nanostructured and Amorphous Materials Inc., TX, USA, while the f-MWCNT nanoparticles (length: 10–30 µm; outer diameter: 10–20 nm and inner diameter: 3–5 nm) were purchased from MKnano Company, ON, Canada. Sodium dodecyl sulphate (SDS) with the purity of ≥98.5% bought from Sigma-Aldrich, Berlin, Germany was used as a surfactant to affect the stability of the studied hybrid nanofluids. The thermal properties of the base fluid and nanoparticles used in this study are provided in Table 1. It is noted that some properties and parameters of nanomaterials are provided in the company's product datasheets.

Table 1. Thermophysical properties of studied materials at room temperature.

Properties	Deionized Water [38]	Multi-Walled Carbon Nanotube [38]	γ-Fe_2O_3 [39,40]
Density (kg/m^3)	997	2100	5242
Thermal conductivity (W/m·K)	0.613	2000	20
Heat capacity (J/kg·K)	4179	710	681

2.2. Equipment

When suspending the bi-nanoparticles in the deionized (DIW) water, a sonicator (Hielscher, Germany) was employed to homogenize the mixture. A programmable water bath (LAUDA ECO RE1225, Berlin, Germany) was deployed to keep the mixture at a low temperature during sonication and used to keep the nanofluids at the specific temperatures while measuring the thermophysical properties. Other pieces of equipment, such as a digital weighing balance (Radwag AS 220.R2 (Radom, Poland) with an accuracy of ±0.01 g), pH meter (Jenway 3510, Staffordshire, UK; –2 to 19.999 range with ±0.003 accuracy), vibro-viscometer (SV-10; A&D, Tokyo, Japan; with ±3% accuracy), UV–visible spectrophotometer (Jenway, Staffordshire, UK), transmission electron microscope (JEOL JEM-2100F, Tokyo, Japan)—dry sample type, and electrical conductivity meter (EUTECH Instrument (Singapore) with ±1% accuracy), were used for various purposes in this work.

2.3. Hybrid Nanofluid Preparation and Stability

In the formulation of the hybrid nanofluids (Fe_2O_3 (80%) and MWCNTs (20%)), a two-step method was used. To ensure proper stability, the pH and electrical conductivity of the formulated MWCNT-Fe_2O_3/DIW nanofluids were monitored while SDS amounts and sonication parameters (amplitude, frequency, and sonication time) were optimized at a 0.1% volume concentration and room temperature (20 °C). SDS to bi-nanoparticle weight ratios of 0.4–1.0 were examined. After obtaining the optimum values of amplitude, frequency, sonicating time, and the ratio of SDS to hybrid nanoparticle weight (dispersion fraction), hybrid nanofluids of various volume concentrations (0.1–1.5%) were formulated according to Equation (1).

$$\varphi = \left(\frac{X_{Fe_2O_3}\left(\frac{M}{\rho}\right)_{Fe_2O_3} + X_{MWCNT}\left(\frac{M}{\rho}\right)_{MWCNT}}{X_{Fe_2O_3}\left(\frac{M}{\rho}\right)_{Fe_2O_3} + X_{MWCNT}\left(\frac{M}{\rho}\right)_{MWCNT} + \left(\frac{M}{\rho}\right)_{DIW}} \right) \quad (1)$$

where X = ratio of each nanoparticle type; M = weight of material; ρ = density of material.

The morphology and dispersion of the bi-particles in the hybrid nanofluids were monitored using TEM. Viscosity, UV–visible spectrophotometry, and visual inspection methods were employed to monitor the nanofluids' stability. The nanofluid can be considered stable if one of the properties has not changed during the period of stability check. However, for our case, both the UV–visible spectrophotometry and viscosity techniques were carried out for 43 h while the visual inspection was done at weekly intervals for a month.

2.4. Electrical Conductivity and pH

The electrical conductivity meter was first calibrated with the use of standard calibration fluid supplied by the manufacturer. On calibration, the standard fluid was measured at 25 °C (in triplicates) and an average of 1413.6 $\mu S \cdot cm^{-1}$ was recorded. Thereafter, the electrical conductivity of DIW and hybrid nanofluids was determined at temperatures of 15–55 °C (at 10 °C intervals). Buffer solutions were used to calibrate the pH meter at a pH values of 4, 7, and 10. Thereafter, the measurement of pH values of the hybrid nanofluids was conducted. The uncertainty of the electrical conductivity and pH measurements was 6.2% and 0.04%, respectively. In comparison with DIW, the relative electrical conductivity and the improvement recorded for MWCNT-Fe_2O_3/DIW nanofluids were evaluated as expressed in Equations (2) and (3), respectively.

$$\sigma_{rel} = \frac{\sigma_{hnf}}{\sigma_{bf}} \quad (2)$$

$$\sigma_{enhan}(\%) = \left(\frac{\sigma_{hnf} - \sigma_{bf}}{\sigma_{bf}}\right) \times 100 \quad (3)$$

where

σ_{hnf} = measured electrical conductivity (hybrid nanofluids) and
σ_{bf} = measured electrical conductivity (DIW).

2.5. Viscosity

Calibration of the vibro-viscometer was carried out prior to viscosity measurement of the DIW and MWCNT-Fe_2O_3/DIW nanofluids at temperatures of 15 °C to 55 °C. A percent error of 1.6% was estimated when the measured viscosity of DIW was compared to the standard viscosity of water. The uncertainty of the viscosity measurement was 7.02%. Relative viscosity and improvement of the viscosity of MWCNT-Fe_2O_3/DIW nanofluids compared with DIW were evaluated using Equations (4) and (5), respectively.

$$\mu_{rel} = \frac{\mu_{hnf}}{\mu_{bf}} \quad (4)$$

$$\mu_{enhan}(\%) = \left(\frac{\mu_{hnf} - \mu_{bf}}{\mu_{bf}}\right) \times 100 \quad (5)$$

where

μ_{hnf} = measured viscosity of hybrid nanofluids and
μ_{bf} = measured viscosity of DIW.

The expression in Equation (6) was used to evaluate the margin of deviation (MOD) of the thermal properties of MWCNT-Fe_2O_3/DIW nanofluids.

$$MOD(\%) = \left(\frac{V_{Exp.} - V_{Pred.}}{V_{Exp.}} \times 100\right) \quad (6)$$

where

$V_{Exp.}$ = experimental data and
$V_{Pred.}$ = predicted data.

It is worth stating that the measured variable (Z), temperature (*T*), weight of hybrid nanoparticles (*m*), and volume of DIW were sources of error associated with the measurement of pH, μ, and σ. The errors were propagated using Equation (7) to estimate the uncertainty related to pH, μ, and σ.

$$U(\%) = \pm\sqrt{\left(\frac{\Delta Z}{Z}\right)^2 + \left(\frac{\Delta T}{T}\right)^2 + \left(\frac{\Delta m}{m}\right)^2 + \left(\frac{\Delta V}{V}\right)^2} \tag{7}$$

3. Results and Discussion

3.1. Preparation of Hybrid Nanofluids

The preparation of stable hybrid nanofluids is a very vital step prior to their characterization and thermophysical property measurement, hence the need to optimize several parameters toward achieving this. The electrical conductivity has been reported as a viable property that can be employed to achieve the critical micelle concentration (CMC) of the surfactant utilized for nanofluid preparation [41,42]. The point of inflection of this property is known to be the CMC. In this study, the optimum ratio of SDS to hybrid nanoparticle weight (dispersion fraction), sonicator amplitude, and sonicating time were determined through the monitoring of the pH and electrical conductivity of MWCNT-Fe_2O_3/DIW nanofluids (0.1 vol%) at room temperature, as presented in Figures 1 and 2. A turning point (inflection) was observed for the pH and electrical conductivity at an optimum dispersion fraction of 0.5 (Figure 1). Again, the point of inflection was noticed at 120 min for the electrical conductivity in determining the optimum sonication time (Figure 2). The optimum sonication time also indicated a turning point for pH, leading to the optimum (lowest) pH for the MWCNT-Fe_2O_3/DIW nanofluid. In this present study, the hybrid nanofluids (at all volume concentrations) were prepared by sonicating at an amplitude of 70%, frequency of 70%, and sonication time of 120 min using a dispersion fraction of 0.5.

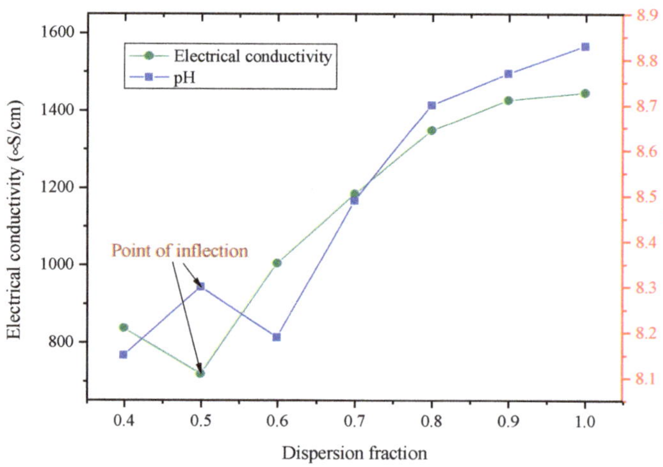

Figure 1. Optimum dispersion fraction of sodium dodecyl sulphate (SDS) in hybrid nanofluids via electrical conductivity and pH monitoring.

3.2. Stability of Hybrid Nanofluid

The morphology and dispersion of the hybrid nanoparticles are presented in Figure 3. A good suspension of the hybrid nanoparticles was observed. The UV–visible, viscosity, and visual methods were used to check the stability of the prepared hybrid nanofluids. As it is widely used for studying the stability of nanofluids [7,8], the absorbance of our hybrid nanofluids was measured using UV–visible spectrophotometry in order to assess their stability status. Figure 4 displays the viscosity and absorbance of the hybrid nanofluid (0.5 vol%) for 2580 min (43 h). The absorbance was around 3.2 with a peak wavelength of 261 nm, while the viscosity (at 15 °C) was around 2.0 mPas for the monitored period. These parameters (absorbance and viscosity) depicted the stability of the nanofluid, as a straight line relatively parallel to the horizontal was noticed for each parameter. An absorbance range of 3.0–3.8 with wavelengths of 287–264 nm was measured for the hybrid nanofluids at varying volume concentrations (0.1–1.5%). It can be noticed that the absorbance increased with a rise in the volume concentration, which was in agreement with previous studies [43,44]. It was observed that the increased suspension of the bi-nanoparticles into DIW altered the values of both the absorbance and the wavelength. A careful visual inspection of the samples (even titling the sample vial) was also done and no sedimentation was noticed after a month upon inspection (Figure 5).

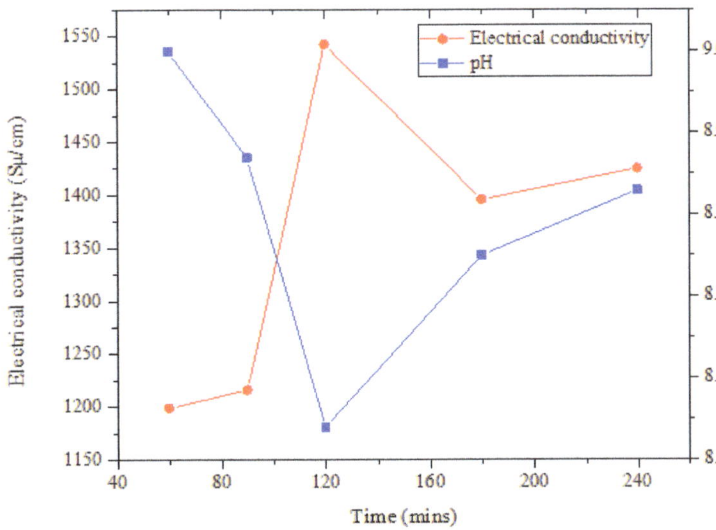

Figure 2. Optimum sonication time of hybrid nanofluids via electrical conductivity and pH monitoring.

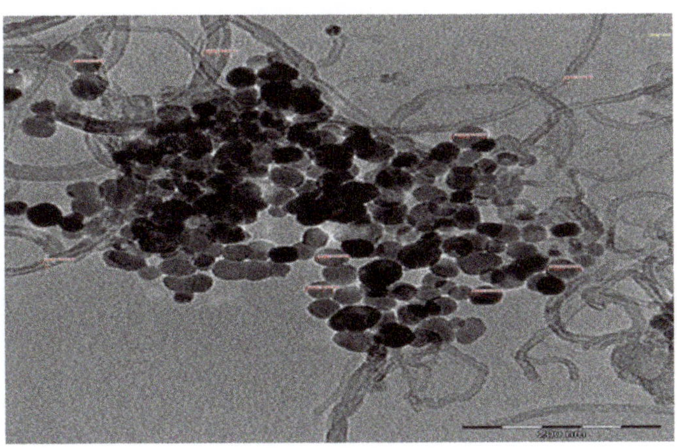

Figure 3. Morphology of the hybrid nanofluid using TEM.

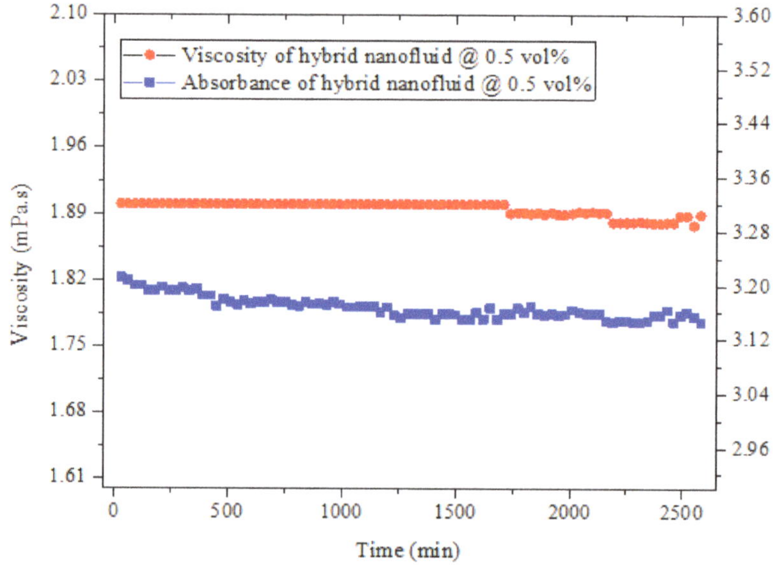

Figure 4. Stability monitoring of 0.5 vol% hybrid nanofluid using viscosity and absorbance.

Figure 5. Visual stability of the hybrid nanofluids.

3.3. *Electrical Conductivity of Hybrid Nanofluid*

The capability of an aqueous solution to allow the passage of an electric current with the application of a potential difference has been termed "electrical conductivity". The dispersion of nanoparticles into base fluids (having known electrical conductivity values) to form nanofluids resulted in improved electrical conductivity in the base fluids due to the increased presence and mobility of electric charges. The hybrid nanofluids of MWCNT-Fe_2O_3/DIW, as electrically conducting fluids, have been investigated for their electrical conductivity at varying temperatures and volume concentrations, as illustrated in Figures 6 and 7, respectively. The electrical conductivity of MWCNT-Fe_2O_3/DIW nanofluids was enhanced significantly as the volume concentration increased (Figure 6). This can be linked to an increase in the electric charges in the MWCNT-Fe_2O_3/DIW nanofluids as the volume concentration rose. The electrical conductivity was noticed to be linearly dependent on the volume concentration of the nanofluids. Subject to a temperature increase, a slight enhancement of electrical conductivity was noticed (Figure 6). This observation is well illustrated in Figure 7, in which the electrical conductivity was improved linearly as the temperature increased. Consequently, the electrical conductivity of the nanofluids was directly proportional to the volume concentration and temperature, which was consistent with earlier studies reported in the literature for mono-particle and hybrid nanofluids [14,38,45–48]. At 55 °C and 1.5 vol%, maximum electrical conductivity (4139 μS/cm) was obtained which was considerably higher than the values of 19.0 μS/cm (for Fe_3O_4/water nanofluid at 0.6 vol% and 60 °C) and 1127–1265 μS/cm (Al_2O_3-MWCNT (80:20)/DIW nanofluid at 0.1 vol% and 50 °C) reported by Bagheli et al. [49] and Giwa et al. [38], respectively. However, Giwa et al. [48] published a range of 640–4570 μS/cm for Al_2O_3-Fe_2O_3 (75:25)/DIW nanofluid (at 0.75 vol% and 50 °C), which had a maximum value slightly higher than that reported for MWCNT-Fe_2O_3 (20:80)/DIW nanofluid in the present work. The electrical conductivity values were found to be strongly connected to the types of nanoparticles and the mixture ratios used in formulating the hybrid nanofluids.

In Figure 8, the relative electrical conductivity of MWCNT-Fe_2O_3/DIW nanofluids at varying volume concentrations is presented. The continued suspension of Fe_2O_3 and MWCNT nanoparticles into DIW demonstrated a considerable increase in the relative electrical conductivity, while an increment in temperature slightly enhanced this property. The obtained result was in line with the work of Adio et al. [14], in which the relative electrical conductivity of MgO-EG nanofluids was enhanced with an increase in the nanofluid volume concentration. However, for their study, the temperature only increased relative electrical conductivity up to 30 °C, after which it declined. The obtained relative effective conductivity of MWCNT-Fe_2O_3/DIW nanofluids ranged from 3.95 to 17.76 for all the concentrations (vol.) at 55 °C.

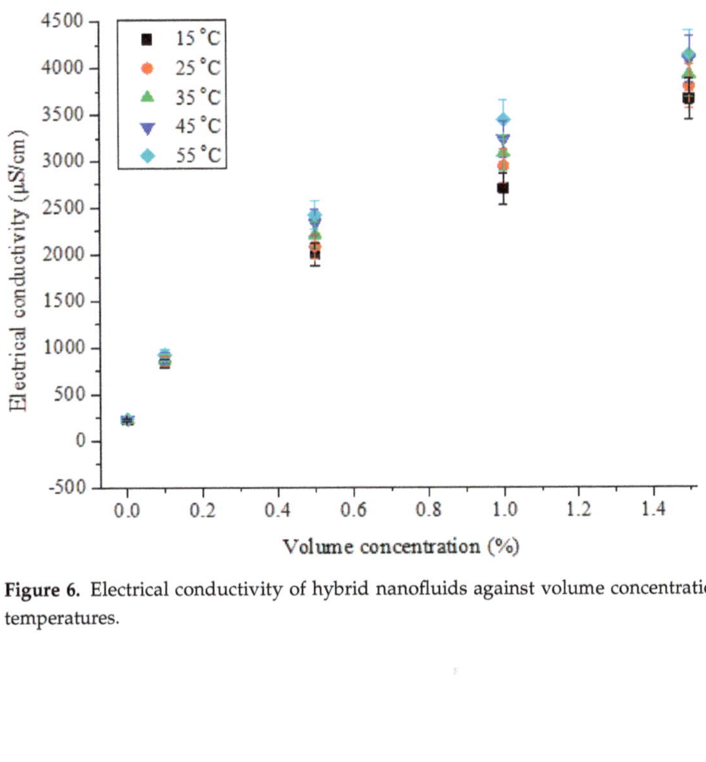

Figure 6. Electrical conductivity of hybrid nanofluids against volume concentration at different temperatures.

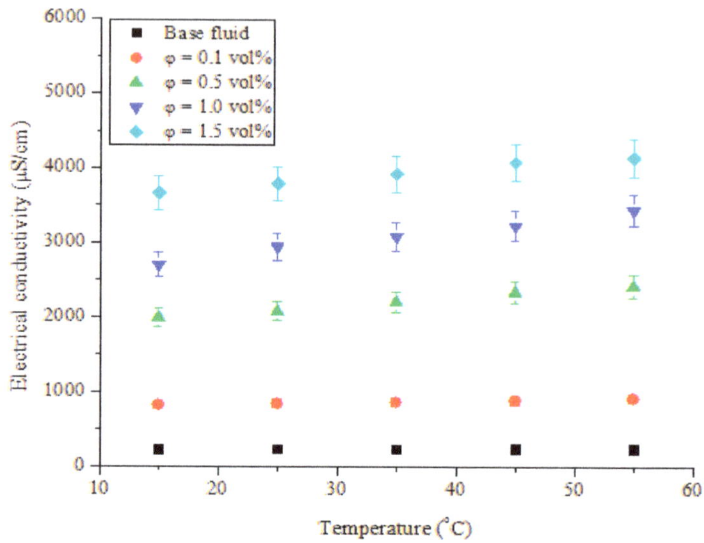

Figure 7. Electrical conductivity of hybrid nanofluids against temperature at different volume concentrations.

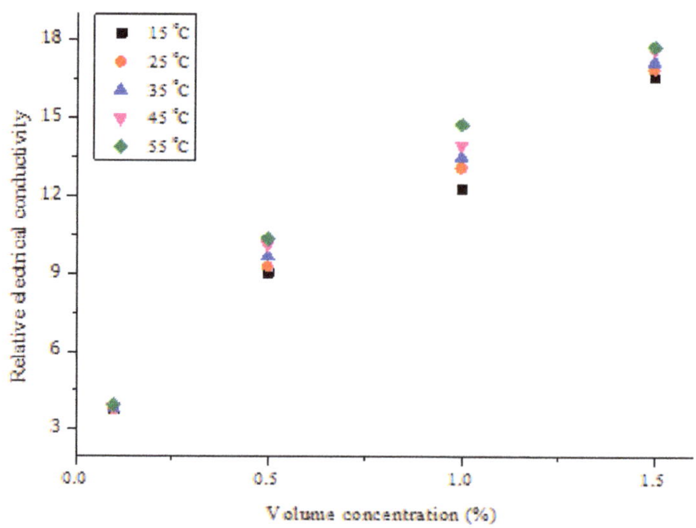

Figure 8. Relative electrical conductivity of hybrid nanofluids against volume concentration at different temperatures.

The electrical conductivity enhancement recorded by the addition of the hybrid nanoparticles to DIW at varying temperatures and volume concentrations is provided in Figure 9. In relation to DIW, the electrical conductivity of MWCNT-Fe_2O_3/DIW nanofluid was augmented significantly as the volume concentration rose but with a small enhancement as the temperature surged. It can be observed in Figure 9 that a relatively linear correlation existed between the enhancement of electrical conductivity and nanofluid volume concentration. A similar trend was noticed for the electrical conductivity enhancement with temperature, though with slight improvement. This result was found to agree with previous works on the electrical conductivity enhancement of nanofluids [44,47,50]. However, some studies reported either independence from temperature in the enhancement or a reduction in electrical conductivity with a temperature increase [43,47,51]. Maximum enhancement was recorded at 55 °C for all samples of the hybrid nanofluids. Therefore, at 55 °C and 1.5 vol%, an enhancement of 1676.4% was attained.

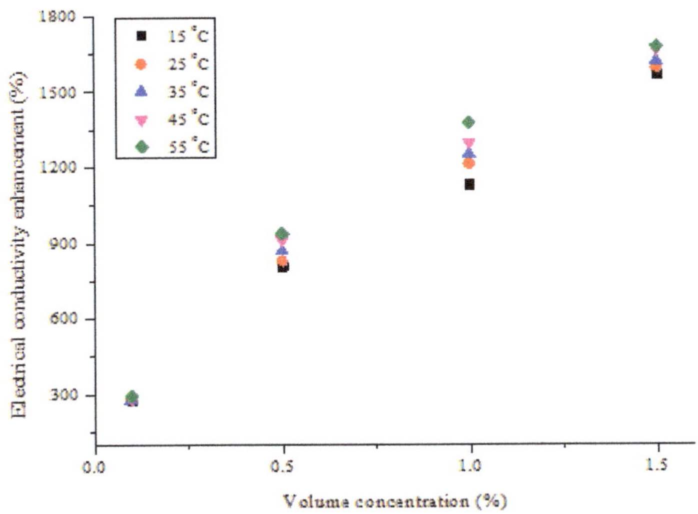

Figure 9. Electrical conductivity enhancement of hybrid nanofluids against volume concentration at different temperatures.

In comparison to previous studies, Bagheli et al. [49] (Fe_3O_4/water nanofluid; 60 °C and 0.5 vol%), Sundar et al. [47] (nanodiamond-nickel/water nanofluid; 0.1 vol% and 65 °C), Kumar et al. [44] (MWCNT/water nanofluid; 0.6 vol% and 50 °C), Mehrali et al. [43] (nitrogen-doped graphene/water nanofluid; 60 °C and 0.06 wt%), Adio et al. [14] (EG-based MgO nanofluid; 0.5 vol% and 25 °C), Giwa et al. [38] (Al_2O_3-MWCNT (80:20)/DIW nanofluid; 0.1 vol% and 50 °C), and Giwa et al. [48] (Al_2O_3-Fe_2O_3 (75:25)/DIW nanofluid; 0.75 vol% and 50 °C) measured enhancements of 360%, 853.15%, 1814.96%, 190.57%, 6000%, 134.12–255.34%, and 163.37–1692.16%, respectively, for the electrical conductivity of different mono-particles and hybrid nanofluids. The nanofluid types, size of nanoparticles, volume/weight concentration or fraction, mixing ratio (for hybrid nanofluid), and temperature used in the various studies may be responsible for the variation in the results. However, a comparison of the works of Kumar et al. [44] and Bagheli et al. [49] with those of Giwa et al. [38] and Giwa et al. [48] supported the finding in the present study for MWCNT-Fe_2O_3/DIW nanofluid with regard to the augmentation of electrical conductivity because of the hybridization of bi-nanoparticles.

3.4. *Viscosity of Hybrid Nanofluid*

An examination of the influence of the studied temperatures and volume concentrations on the viscosity of MWCNT-Fe_2O_3/DIW nanofluids was carried out. The viscosity of the hybrid nanofluids under varying volume concentrations and temperatures is shown in Figures 10 and 11, respectively. An appreciation of the volume concentration of MWCNT-Fe_2O_3 (20:80)/DIW nanofluid was observed to enhance the viscosity in a linear pattern (Figure 10). This was because of the higher density of the hybrid nanoparticles in relation to DIW. An increment in the volume concentration due to the amount of hybrid nanoparticles suspended in DIW was noticed to enhance the viscosity of the nanofluid. Nanofluid viscosity was also observed to be dependent on the temperature, as depicted in Figure 11. The increasing change in the temperature was found to lessen the viscosity of the nanofluids. From Figures 10 and 11, it can be observed that the influence of temperature on the viscosity of the hybrid nanofluids was more than that of volume concentration. Thus, the viscosity was dependent on both variables. The obtained results agreed with the works of Nadooshan et al. [34], Mehrali et al. [43], Adio et al. [45,52], Giwa et al. [38], Sharifpur et al. [53], and Giwa et al. [54], for the viscosity–temperature and viscosity–volume concentration relationships. In the present study, the viscosity of MWCNT-Fe_2O_3/DIW nanofluids ranged from 0.65 to 1.36 mPas for the ranges of volume concentration and temperature investigated. This was slightly higher than the range of 0.51 to 1.11 mPas and 0.57 to 1.13 mPas published by Gangadevi and Vinayagam [55] and Giwa et al. [48] for Al_2O_3-CuO/water nanofluid (0.2 vol% and at 20–60 °C) and Al_2O_3-Fe_2O_3/DIW nanofluid (0.75 vol% and at 20–50 °C).

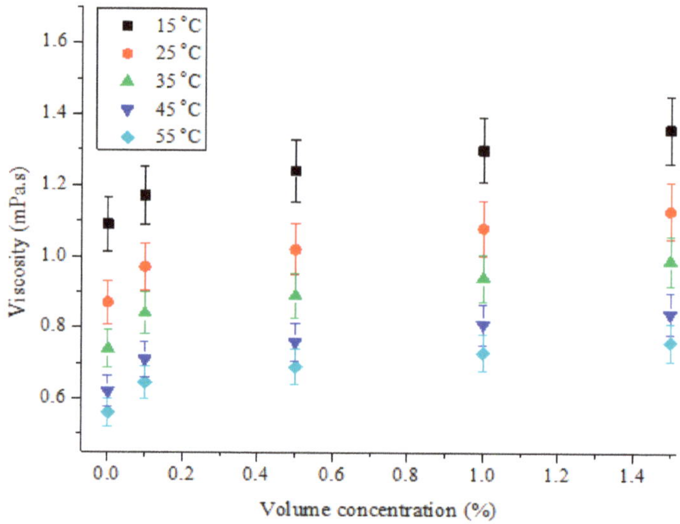

Figure 10. Viscosity of hybrid nanofluids against volume concentration at different temperatures.

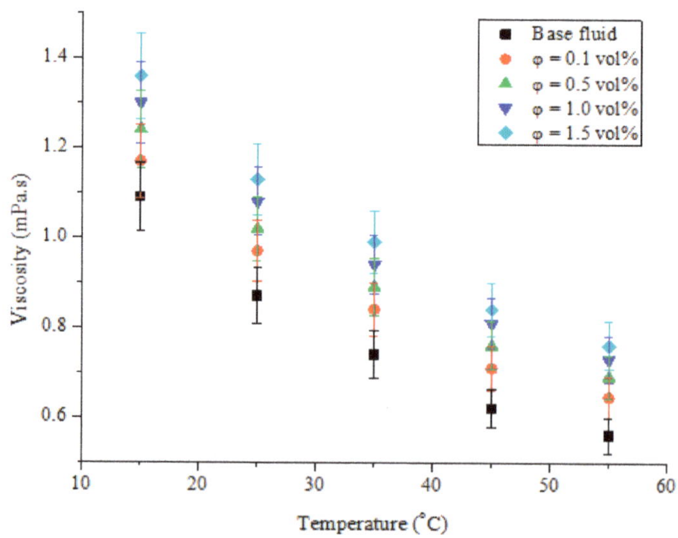

Figure 11. Viscosity of hybrid nanofluids against temperature at different volume concentrations.

As shown in Figure 12, the relative viscosity of MWCNT-Fe_2O_3/DIW nanofluids was compared with DIW for the studied volume concentrations. An increment in the concentration of the bi-nanoparticles was found to enhance the relative viscosity of the nanofluid. A seemingly linear correlation was observed between volume concentration and relative viscosity. In addition, the relative viscosity of MWCNT-Fe_2O_3/DIW nanofluids was noticed to increase as the temperature increased. The observed trend agreed with the data published by Nadooshan et al. [34] (Fe_3O_4-MWCNT/EG nanofluid), Adio et al. [45] (Al_2O_3-glycerol nanofluid), and Zawawi et al. [56] (hybrid nanofluids). However, Dardan et al. [26] reported the reverse of the trend noticed in the present work, as the relative viscosity of Al_2O_3-MWCNT/EO nanofluid was reduced with temperatures at 35–50 °C, for measurements spanning 25–50 °C. At 55 °C, the relative viscosity of the nanofluids examined in this work increased from 1.152 (0.1 vol%) to 1.357 (1.5 vol%).

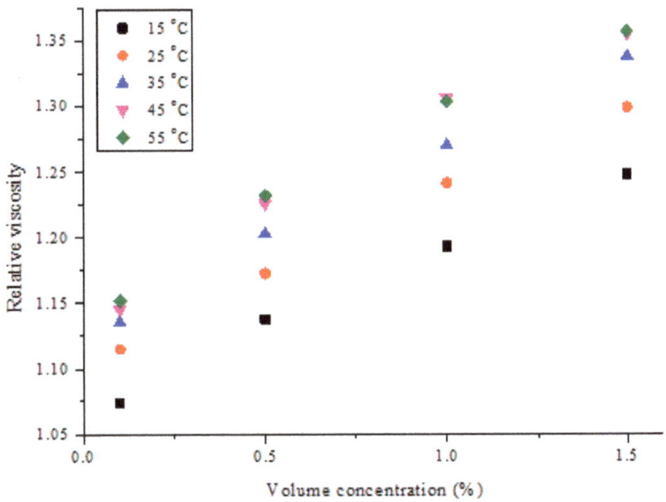

Figure 12. Relative viscosity of hybrid nanofluids against volume concentration at different temperatures.

The percentage enhancement of the hybrid nanofluid viscosity compared with DIW under varying temperatures and volume concentrations is illustrated in Figure 13. It can be noticed that a rise in the volume concentration of MWCNT-Fe_2O_3 (20:80)/DIW nanofluids resulted in substantial viscosity enhancement when compared with DIW. Relatively linear enhancement of the viscosity was noticed for MWCNT-Fe_2O_3 (20:80)/DIW nanofluids, as the concentration and temperature rose when compared with DIW. For this work, the highest viscosity enhancement of 35.7% was estimated for the hybrid nanofluids compared to DIW. Previous studies have recorded viscosity enhancements of 58% (MWCNT; 1 vol%) [57], 20.5% (Al_2O_3-TiO_2/PAG; 0.1 vol%) [56], 43.52% (MWCNT-CuO/EO; 1 vol%) [58], 46% (Al_2O_3-MWCNT/EO; 1.0 vol%) [26], 24.56% (Al_2O_3-MWCNT/DIW; 0.1 vol%) [38], 20% (ZnO-MWCNT/EO; 0.8 vol%) [37], 36.4% (CuO-MgO-TiO_2/DIW; 0.5 vol%) [36], and 43.64% (Al_2O_3-Fe_2O_3/DIW; 0.75 vol%) [48] for mono-particle and hybrid nanofluids, when compared with the respective base fluids. The results from the present work showed a relatively lower viscosity enhancement in relation to earlier studies, which can be attributed to the types of hybrid nanoparticles and base fluids utilized in preparing MWCNT-Fe_2O_3 (20:80)/DIW nanofluids. MWCNT nanoparticles are known to have significantly lower density in comparison to metal oxide-based nanoparticles.

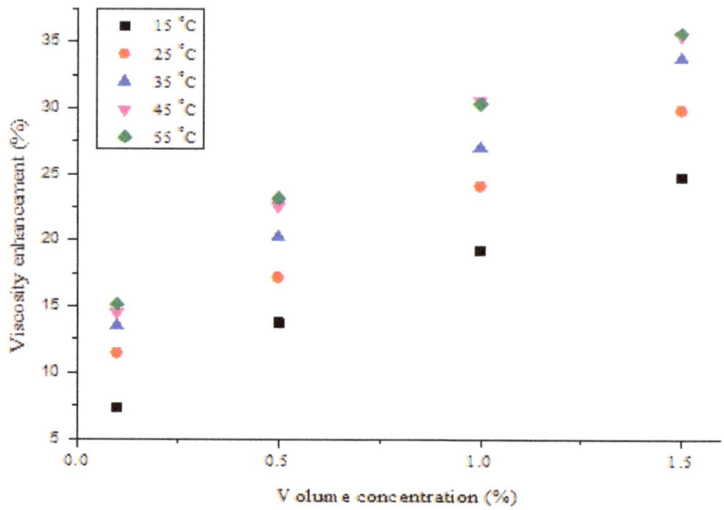

Figure 13. Viscosity enhancement of hybrid nanofluids against volume concentration at different temperatures.

3.5. Correlation Development

Formulated classical models and theoretical correlations for estimating various thermal properties of different nanofluids have been demonstrated by numerous studies to be inadequate for their estimation [20,36,51,53,59]. They are found to either underestimate or overestimate the measured data of the thermal properties of nanofluids. Research progress showed that there is a need to formulate correlations to effectively estimate the thermal properties of MWCNT-Fe_2O_3 (20:80)/DIW nanofluid. The uniqueness of the physicochemical properties of the diverse nanoparticles and base fluids used to prepare hybrid nanofluids and the fitting of the experimental data to estimate the thermal properties are now becoming more important and relevant.

Curve fittings of the experimental data of the relative electrical conductivity and relative viscosity garnered for MWCNT-Fe_2O_3 (20:80)/DIW nanofluids were performed. The formulas developed from the data of relative electrical conductivity and viscosity as dependencies of temperature and volume concentration are expressed in Equations (8) and (9), respectively. For the electrical conductivity correlation, R^2 = 0.968, coefficient of corelation (R) = 0.984, root mean square error (RMSE) = 0.16, and mean absolute percentage error (MAPE) = 11.085, while for the viscosity correlation (linear regression), R^2 = 0.966, R = 0.983, RMSE = 0.166, and MAPE = 8.721. These variables revealed relatively high coefficients of determination and correlation coefficients with significantly low errors for both correlations.

$$\frac{\sigma_{hnf}}{\sigma_{bf}} = 2.757 + 0.032T + 9.287\varphi \quad (8)$$

$$\frac{\mu_{hnf}}{\mu_{bf}} = 1.031 + 0.0025T + 0.1386\varphi \quad (9)$$

The correlation from the work of Ganguly et al. [60] and that derived using Equation (8) for the relative electrical conductivity of MWCNT-Fe_2O_3 (20:80)/DIW at 35 °C are plotted in Figure 14. It is obvious that the existing correlation, as published in the literature, overestimated the experimental data of the electrical conductivity of MWCNT-Fe_2O_3 (20:80)/DIW nanofluid. This was because the existing correlation was formulated using data from a different nanofluid (Al_2O_3/water) to that utilized (MWCNT-Fe_2O_3/water) in this study, hence, it could not adequately predict the property. The correlation relating the experimental and predicted data of the relative electrical conductivity of MWCNT-Fe_2O_3

(20:80)/DIW nanofluid is presented in Figure 15. A straight line was noticed to relate the predicted and experimental values, showing a good correlation between both data sets. The MOD for the correlation was ±3.48%.

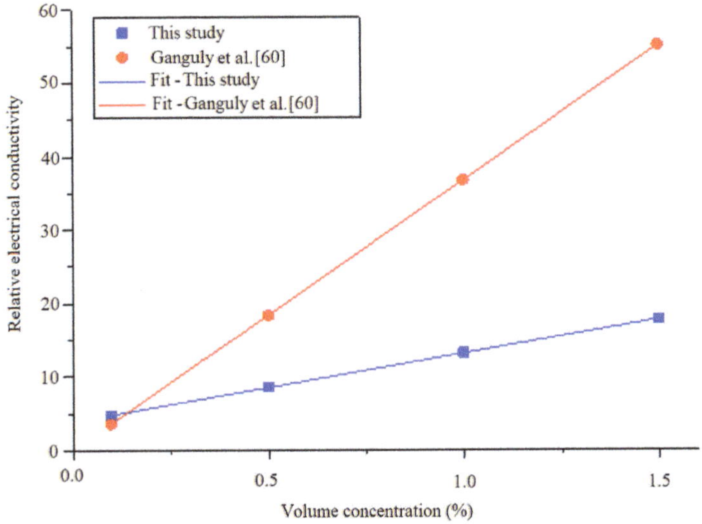

Figure 14. Developed correlations for electrical conductivity compared to an existing correlation at different temperatures.

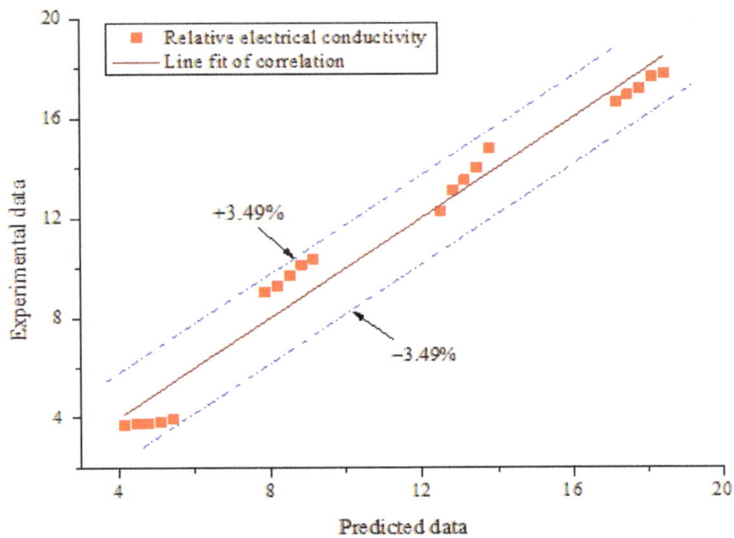

Figure 15. Correlation of experimental and predicted data (electrical conductivity).

A plot of the fitted relative viscosity correlation for this work and those formulated in previous studies at 35 °C is presented in Figure 16 [26,61,62]. It shows that none of the existing correlations for estimating the viscosity of mono-particle and hybrid nanofluids could fit the obtained experimental data. They all overestimated the relative viscosity of the investigated MWCNT-Fe_2O_3/DIW nanofluid. The obtained relative viscosity of MWCNT-Fe_2O_3/DIW nanofluid was overestimated using the experiment-derived correlation for Fe_2O_3/DIW nanofluid [62]. This revealed a reduction (49.8%) in the viscosity of MWCNT-Fe_2O_3 (20:80)/DIW nanofluid because of the hybridization of Fe_2O_3 nanoparticles with MWCNT nanoparticles. With the evident reduction in the viscosity of MWCNT-Fe_2O_3/DIW nanofluid, the use of this hybrid nanofluid is favorable for engineering applications in terms of lower pumping power. A linear relationship was found to occur between the experimental and predicted data, as presented in Figure 17. An MOD of ±0.01% was found.

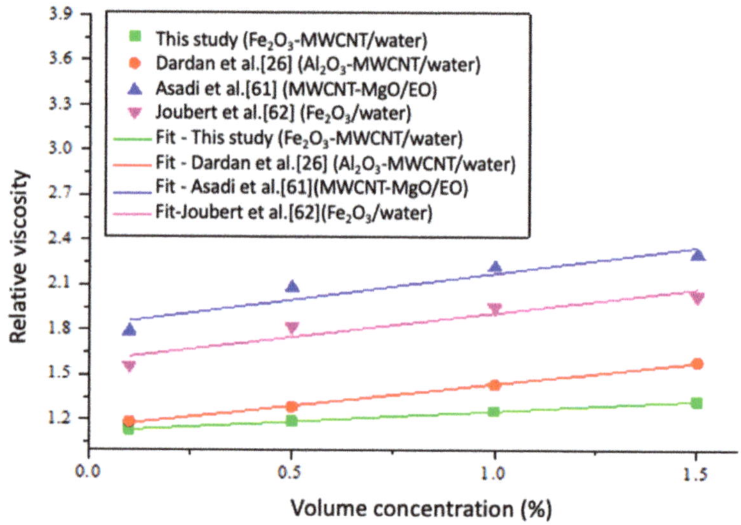

Figure 16. Developed correlations for viscosity compared to existing correlations at different temperatures.

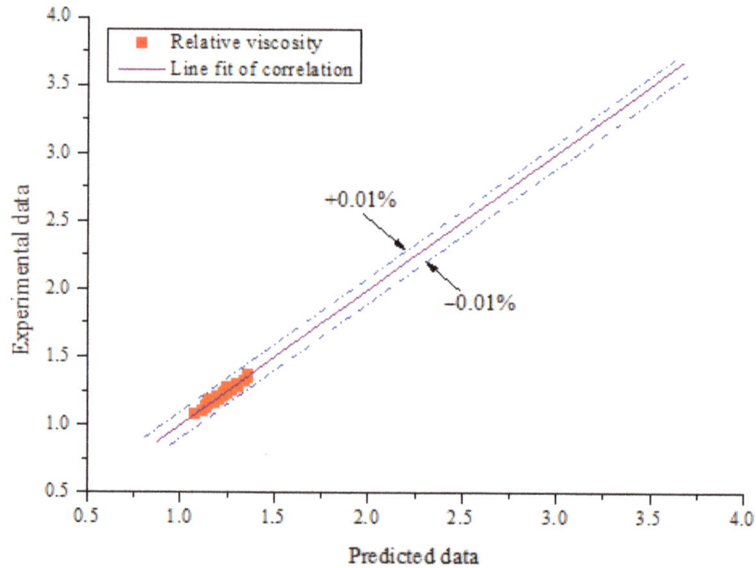

Figure 17. Correlation of experimental and predicted data (viscosity).

4. Conclusions

Stable MWCNT-Fe_2O_3/DIW nanofluids have been prepared and investigated for their electrical conductivity and viscosity at varying temperatures and volume concentrations. Increasing the bi-nanoparticles' concentration enhanced both the viscosity and electrical conductivity of MWCNT-Fe_2O_3/DIW nanofluids. Additionally, increasing the nanofluid temperature augmented the electrical conductivity, whereas the viscosity was reduced. Relative to the base fluid, maximum enhancements of 1676.4% and 35.7% were achieved for the electrical conductivity and viscosity of MWCNT-Fe_2O_3/DIW nanofluids, respectively. The obtained results were noticed to be consistent with previous studies in the literature. The hybridization of MWCNT and Fe_2O_3 nanoparticles to prepare MWCNT-Fe_2O_3 (20:80)/DIW nanofluid was proven to cause a reduction of viscosity, which is advantageous in engineering applications of the fluid. To predict the viscosity and electrical conductivity of the studied MWCNT-Fe_2O_3/DIW nanofluids, correlations have been developed.

Author Contributions: Conceptualization, S.O.G. and M.S.; methodology, S.O.G. and M.S.; investigation, S.O.G.; data analysis, S.O.G. and M.H.A.; writing—original draft preparation, S.O.G. and M.H.A.; writing—review and editing, J.P.M. and S.M.S.M.; supervision, M.S., J.P.M. and S.M.S.M.; funding acquisition, J.P.M. and S.M.S.M. All authors have read and agreed to the published version of the manuscript.

Funding: This work was also partially supported by the Fundação para a Ciência e Tecnologia (FCT), Portugal through grant: PTDC/NAN-MAT/29989/2017 (NanCOM).

Acknowledgments: The supports of the National Research Foundation (South Africa) under the Renewable and Sus-tainable Energy Doctoral Scholarships and of the FCT, Portugal through project-NanoCOM are hereby acknowledged.

Conflicts of Interest: The authors declare no conflict of interest.

References

1. Masuda, H.; Ebata, A.; Teramae, K.; Hishinuma, N. Alteration of thermal conductivity and viscosity of liquid by dispersing ultra-fine particles. *Netsu. Bussei.* **1993**, *7*, 227–233. [CrossRef]
2. Eastman, J.A.; Choi, S.U.S.; Li, S.; Yu, W.; Thompson, L.J. Anomalously increased effective thermal conductivities of ethylene glycol-based nanofluids containing copper nanoparticles. *Appl. Phys. Lett.* **2001**, *78*, 718–720. [CrossRef]
3. Murshed, S.M.S.; Leong, K.C.; Yang, C. Investigations of Thermal Conductivity and Viscosity of Nanofluids. *Int. J. Therm. Sci.* **2008**, *47*, 560–568. [CrossRef]
4. Murshed, S.M.S.; Leong, K.C.; Yang, C. Thermophysical and Electrokinetic Properties of Nanofluids—A Critical Review. *Appl. Therm. Eng.* **2008**, *28*, 2109–2125. [CrossRef]
5. Prasher, R.; Song, D.; Wang, J.; Phelan, P. Measurements of nanofluid viscosity and its implications for thermal applications. *Appl. Phys. Lett.* **2006**, *89*, 133108. [CrossRef]
6. Abareshi, M.; Sajjadi, S.H.; Zebarjad, S.M.; Goharshadi, E.K. Fabrication, characterization, and measurement of viscosity of α-Fe_2O_3-glycerol nanofluids. *J. Mol. Liq.* **2011**, *163*, 27–32. [CrossRef]
7. Murshed, S.M.S.; Nieto de Castro, C.A. *Nanofluids: Synthesis, Properties and Applications*; Nova Science Publishers Inc.: New York, NY, USA, 2014.
8. Cacua, K.; Murshed, S.M.S.; Pabón, E.; Buitrago, R. Dispersion and thermal conductivity of TiO_2/water nanofluid. *J. Therm. Ana. Calor.* **2020**, *140*, 109–114. [CrossRef]
9. Namburu, P.K.; Kulkarni, D.P.; Misra, D.; Das, D.K. Viscosity of copper oxide nanoparticles dispersed in ethylene glycol and water mixture. *Exp. Therm. Fluid Sci.* **2007**, *32*, 397–402. [CrossRef]
10. Murshed, S.M.S.; Estell, P. A State of the Art Review on viscosity of nanofluids. *Renew. Sustain. Energy Rev.* **2017**, *76*, 1134–1152. [CrossRef]
11. Murshed, S.M.S.; Nieto de Castro, C.A. Contribution of Brownian motion in thermal conductivity of nanofluids. *Proc. World Cong. Eng.* **2011**, *3*, 6–8.
12. Murshed, S.M.S. Correction and comment on "thermal conductance of nanofluids: Is the controversy over?". *J. Nanopart. Res.* **2009**, *11*, 511–512. [CrossRef]
13. Murshed, S.M.S.; Nieto de Castro, C.A. Predicting the thermal conductivity of nanofluids-Effect of Brownian motion of nanoparticles. *J. Nanofluid.* **2012**, *1*, 180–185. [CrossRef]
14. Adio, S.A.; Sharifpur, M.; Meyer, J.P. Factors affecting the pH and electrical conductivity of MgO-ethylene glycol nanofluids. *Bull. Mater. Sci.* **2015**, *38*, 1345–1357. [CrossRef]

15. Sharifpur, M.; Yousefi, S.; Meyer, J.P. A new model for density of nanofluids including nanolayer. *Int. Commun. Heat Mass Transf.* **2016**, *78*, 168–174. [CrossRef]
16. Nabati Shoghl, S.; Jamali, J.; Keshavarz Moraveji, M. Electrical conductivity, viscosity, and density of different nanofluids: An experimental study. *Exp. Therm. Fluid Sci.* **2016**, *74*, 339–346. [CrossRef]
17. Nieto de Castro, C.A.; Murshed, S.M.S.; Lourenço, M.J.V.; Santos, F.J.V.; Lopes, M.L.M.; França, J.M.P. Enhanced thermal conductivity and specific heat capacity of carbon nanotubes ionanofluids. *Int. J. Therm. Sci.* **2012**, *62*, 34–39. [CrossRef]
18. Fal, J.; Barylyak, A.; Besaha, K.; Bobitski, Y.V.; Cholewa, M.; Zawlik, I.; Szmuc, K.; Cebulski, J. Experimental investigation of electrical conductivity and permittivity of SC-TiO_2-EG nanofluids. *Nanoscale Res. Lett.* **2016**, *11*, 375. [CrossRef]
19. Nor, S.; Azis, N.; Jasni, J.; Kadir, M.; Yunus, R.; Yaakub, Z. Investigation on the electrical properties of palm oil and coconut oil based TiO_2 nanofluids. *IEEE Trans. Dielectr. Electr. Insul.* **2017**, *24*, 3432–3442. [CrossRef]
20. Abdolbaqi, M.K.; Azmi, W.H.; Mamat, R.; Sharma, K.V.; Najafi, G. Experimental investigation of thermal conductivity and electrical conductivity of bioglycol—Water mixture based Al_2O_3 nanofluid. *Appl. Therm. Eng.* **2016**, *102*, 932–941. [CrossRef]
21. Jana, S.; Salehi-Khojin, A.; Zhong, W.H. Enhancement of fluid thermal conductivity by the addition of single and hybrid nano-additives. *Thermochim. Acta.* **2007**, *462*, 45–55. [CrossRef]
22. Suresh, S.; Venkitaraj, K.P.; Selvakumar, P.; Chandrasekar, M. Synthesis of Al_2O_3-Cu/water hybrid nanofluids using two step method and its thermo physical properties. *Colloids Surf. A Physicochem. Eng. Asp.* **2011**, *388*, 41–48. [CrossRef]
23. Chen, L.; Yu, W.; Xie, H. Enhanced thermal conductivity of nanofluids containing Ag/MWNT composites. *Powder Technol.* **2012**, *231*, 18–20. [CrossRef]
24. Abbasi, S.M.; Rashidi, A.; Nemati, A.; Arzani, K. The effect of functionalisation method on the stability and the thermal conductivity of nanofluid hybrids of carbon nanotubes/gamma alumina. *Ceram Int.* **2013**, *39*, 3885–3891. [CrossRef]
25. Hemmat Esfe, M.; Saedodin, S.; Yan, W.M.; Afrand, M.; Sina, N. Erratumto: Study on thermal conductivity of water-based nanofluids with hybrid suspensions of CNTs/Al_2O_3 nanoparticles. *J. Therm. Anal. Calorim.* **2016**, *125*, 565. [CrossRef]
26. Dardan, E.; Afrand, M.; Meghdadi Isfahani, A.H. Effect of suspending hybrid nano-additives on rheological behavior of engine oil and pumping power. *Appl. Therm. Eng.* **2016**, *109*, 524–534. [CrossRef]
27. Afrand, M.; Nazari Najafabadi, K.; Akbari, M. Effects of temperature and solid volume fraction on viscosity of SiO_2-MWCNTs/SAE40 hybrid nanofluid as a coolant and lubricant in heat engines. *Appl. Therm. Eng.* **2016**, *102*, 45–54. [CrossRef]
28. Megatif, L.; Ghozatloo, A.; Arimi, A.; Shariati-Niasar, M. Investigation of laminar convective heat transfer of a novel TiO_2-carbon nanotube hybrid water-based nanofluid. *Exp. Heat Transf.* **2016**, *29*, 124–138. [CrossRef]
29. Eshgarf, H.; Afrand, M. An experimental study on rheological behavior of non-Newtonian hybrid nano-coolant for application in cooling and heating systems. *Exp. Therm. Fluid Sci.* **2016**, *76*, 221–227. [CrossRef]
30. Kakavandi, A.; Akbari, M. Experimental investigation of thermal conductivity of nanofluids containing of hybrid nanoparticles suspended in binary base fluids and propose a new correlation. *Int. J. Heat Mass Transf.* **2018**, *124*, 742–751. [CrossRef]
31. Esfe, M.H.; Esfandeh, S.; Afrand, M.; Rejvani, M.; Rostamian, S.H. Experimental evaluation, new correlation proposing and ANN modeling of thermal properties of EG based hybrid nanofluid containing ZnO-DWCNT nanoparticles for internal combustion engines applications. *Appl. Therm. Eng.* **2018**, *133*, 452–463. [CrossRef]
32. Moldoveanu, G.M.; Minea, A.A.; Huminic, G.; Huminic, A. Al_2O_3/TiO_2 hybrid nanofluids thermal conductivity: An experimental approach. *J. Therm. Anal. Calorim.* **2019**, *137*, 583–592. [CrossRef]
33. Harandi, S.S.; Karimipour, A.; Afrand, M.; Akbari, M.; D'Orazio, A. An experimental study on thermal conductivity of F-MWCNTs-Fe_3O_4/EG hybrid nanofluid: Effects of temperature and concentration. *Int. Commun. Heat Mass Transf.* **2016**, *76*, 171–177. [CrossRef]
34. Nadooshan, A.A.; Eshgarf, H.; Afrand, M. Measuring the viscosity of Fe_3O_4-MWCNTs/EG hybrid nanofluid for evaluation of thermal efficiency: Newtonian and non-Newtonian behavior. *J. Mol. Liq.* **2018**, *253*, 169–177. [CrossRef]
35. Shi, L.; He, Y.; Hu, Y.; Wang, X. Thermophysical properties of Fe_3O_4@CNT nanofluid and controllable heat transfer performance under magnetic field. *Energy Convers. Manag.* **2018**, *177*, 249–257. [CrossRef]
36. Mousavi, S.M.; Esmaeilzadeh, F.; Wang, X.P. Effects of temperature and particles volume concentration on the thermophysical properties and the rheological behavior of CuO/MgO/TiO_2 aqueous ternary hybrid nanofluid Experimental investigation. *J. Therm. Anal. Calorim.* **2019**. [CrossRef]
37. Goodarzi, M.; Toghraie, D.; Reiszadeh, M.; Afrand, M. Experimental evaluation of dynamic viscosity of ZnO–MWCNTs/engine oil hybrid nanolubricant based on changes in temperature and concentration. *J. Therm. Anal. Calorim.* **2019**, *136*, 513–525. [CrossRef]
38. Giwa, S.O.; Sharifpur, M.; Meyer, J.P. Experimental study of thermo-convection performance of hybrid nanofluids of Al_2O_3-MWCNT/water in a differentially heated square cavity. *Int. J. Heat Mass Transf.* **2020**, *148*, 119072. [CrossRef]
39. Mussatti EMerlini, C.; de Oliveira Barra, G.M.; Güths, S.; de Oliveira, A.P.N.; Siligardi, C. Evaluation of the properties of Iron oxide-filled castor oil polyurethane. *Mater. Res.* **2013**, *16*, 65–70. [CrossRef]
40. Korolev, V.V.; Arefyev, I.M.; Blinov, A.V. Heat capacity of superfine oxides of iron under applied magnetic fields. *J. Therm. Anal. Calorim.* **2008**, *92*, 697–700. [CrossRef]
41. Sharker, K.K.; Islam, M.N.; Das, S. Counterion effect on Krafft temperature and related properties of octadecyltrimethylammonium bromide. *J. Surfactants Deterg.* **2017**, *20*, 923–932. [CrossRef]
42. Topallar, H.; Karadag, B. Mechanism of micelle formation in sodium dodecyl sulfate and cetyltrimethylammonium bromide. *J. Surfactants Deterg.* **1998**, *1*, 49–51. [CrossRef]

43. Mehrali, M.; Sadeghinezhad, E.; Latibari, S.T.; Mehrali, M.; Togun, H.; Zubir, M.N.M.; Kazi, S.N.; Metselaar, H.S.C. Preparation, characterization, viscosity, and thermal conductivity of nitrogen-doped graphene aqueous nanofluids. *J. Mater. Sci.* **2014**, *49*, 7156–7171. [CrossRef]
44. Kumar, P.G.; Kumaresan, V.; Velraj, R. Stability, viscosity, thermal conductivity, and electrical conductivity enhancement of multi-walled carbon nanotube nanofluid using gum arabic. *Fuller. Nanotub. Carbon Nanostruct.* **2017**, *25*, 230–240. [CrossRef]
45. Adio, S.A.; Sharifpur, M.; Meyer, J.P. Investigation into effective viscosity, electrical conductivity, and pH of γ-Al_2O_3-glycerol nanofluids in Einstein concentration regime. *Heat Transf. Eng.* **2015**, *36*, 1241–1251. [CrossRef]
46. Said, Z.; Allagui, A.; Abdelkareem, M.A.; Alawadhi, H.; Elsaid, K. Acid-functionalized carbon nanofibers for high stability, thermoelectrical and electrochemical properties of nanofluids. *J. Colloid Interface Sci.* **2018**, *520*, 50–57. [CrossRef] [PubMed]
47. Sundar, L.S.; Shusmitha, K.; Singh, M.K.; Sousa, A.C.M. Electrical conductivity enhancement of nanodiamond-nickel (ND-Ni) nanocomposite based magnetic nanofluids. *Int. Commun. Heat Mass Transf.* **2014**, *57*, 1–7. [CrossRef]
48. Giwa, S.O.; Sharifpur, M.; Goodarzi, M.; Alsulami, H.; Meyer, J.P. Influence of base fluid, temperature, and concentration on the thermophysical properties of hybrid nanofluids of alumina—Ferrofluid: Experimental data, modeling through enhanced ANN, ANFIS, and curve fitting. *J. Therm. Anal. Calorim.* **2020**. [CrossRef]
49. Bagheli, S.; Fadafan, H.K.; Orimi, R.L.; Ghaemi, M. Synthesis and experimental investigation of the electrical conductivity of water based magnetite nanofluids. *Powder Technol.* **2015**, *274*, 426–430. [CrossRef]
50. Minea, A.A.; Luciu, R.S. Investigations on electrical conductivity of stabilized water based Al_2O_3 nanofluids. *Microfluid. Nanofluid.* **2012**, *13*, 977–985. [CrossRef]
51. Khdher, A.M.; Sidik, N.A.C.; Hamzah, W.A.W.; Mamat, R. An experimental determination of thermal conductivity and electrical conductivity of bio glycol based Al_2O_3 nanofluids and development of new correlation. *Int. Commun. Heat Mass Transf.* **2016**, *73*, 75–83. [CrossRef]
52. Adio, S.A.; Sharifpur, M.; Meyer, J.P. Influence of ultrasonication energy on the dispersion consistency of Al_2O_3–glycerol nanofluid based on viscosity data, and model development for the required ultrasonication energy density. *J. Exp. Nanosci.* **2016**, *11*, 630–649. [CrossRef]
53. Sharifpur, M.; Adio, S.A.; Meyer, J.P. Experimental investigation and model development for effective viscosity of Al_2O_3-glycerol nanofluids by using dimensional analysis and GMDH-NN methods. *Int. Commun. Heat Mass Transf.* **2015**, *68*, 208–219. [CrossRef]
54. Giwa, S.O.; Sharifpur, M.; Meyer, J.P. Effects of uniform magnetic induction on heat transfer performance of aqueous hybrid ferrofluid in a rectangular cavity. *Appl. Therm. Eng.* **2020**, *170*, 115004. [CrossRef]
55. Gangadevi, R.; Vinayagam, B.K. Experimental determination of thermal conductivity and viscosity of different nanofluids and its effect on a hybrid solar collector. *J. Therm. Anal. Calorim.* **2019**, *136*, 199–209. [CrossRef]
56. Zawawi, N.N.M.; Azmi, W.H.; Redhwan, A.A.M.; Sharif, M.Z.; Samykano, M. Experimental investigation on thermo-physical properties of metal oxide composite nanolubricants. *Int. J. Refrig.* **2018**, *89*, 11–21. [CrossRef]
57. Garbadeen, I.D.; Sharifpur, M.; Slabber, J.M.; Meyer, J.P. Experimental study on natural convection of MWCNT-water nanofluids in a square enclosure. *Int. Commun. Heat Mass Transf.* **2017**, *88*, 1–8. [CrossRef]
58. Hemmat Esfe, M.; Sarlak, M.R. Experimental investigation of switchable behavior of CuO-MWCNT (85–15%)/10W-40 hybrid nano-lubricants for applications in internal combustion engines. *J. Mol. Liq.* **2017**, *242*, 326–335. [CrossRef]
59. Saeedi, A.H.; Akbari, M.; Toghraie, D. An experimental study on rheological behavior of a nanofluid containing oxide nanoparticle and proposing a new correlation. *Phys. E Low Dimens. Syst. Nanostruct.* **2018**, *99*, 285–293. [CrossRef]
60. Ganguly, S.; Sikdar, S.; Basu, S. Experimental investigation of the effective electrical conductivity of aluminum oxide nanofluids. *Powder Technol.* **2009**, *196*, 326–330. [CrossRef]
61. Asadi, A.; Asadi, M.; Rezaei, M.; Siahmargoi, M.; Asadi, F. The effect of temperature and solid concentration on dynamic viscosity of MWCNT/MgO (20–80)–SAE50 hybrid nano-lubricant and proposing a new correlation: An experimental study. *Int. Commun. Heat Mass Transf.* **2016**, *78*, 48–53. [CrossRef]
62. Joubert, J.C.; Sharifpur, M.; Solomon, A.B.; Meyer, J.P. Enhancement in heat transfer of a ferrofluid in a differentially heated square cavity through the use of permanent magnets. *J. Magn. Magn. Mater.* **2017**, *443*, 149–158. [CrossRef]

nanomaterials

Article

Modelling Thermal Conduction in Polydispersed and Sintered Nanoparticle Aggregates

Nikolaos P. Karagiannakis [1], Eugene D. Skouras [1,2] and Vasilis N. Burganos [1,*]

[1] Institute of Chemical Engineering Sciences (ICE-HT), Foundation for Research and Technology, Hellas (FORTH), GR-26504 Patras, Greece; nick_karag@iceht.forth.gr (N.P.K.); Eugene.Skouras@iceht.forth.gr (E.D.S.)

[2] Department of Mechanical Engineering, University of the Peloponnese, GR-26334 Patras, Greece

* Correspondence: vbur@iceht.forth.gr; Tel.: +30-261-096-5215

Abstract: Nanoparticle aggregation has been found to be crucial for the thermal properties of nanofluids and their performance as heating or cooling agents. Most relevant studies in the literature consider particles of uniform size with point contact only. A number of forces and mechanisms are expected to lead to deviation from this ideal description. In fact, size uniformity is difficult to achieve in practice; also, overlapping of particles within aggregates may occur. In the present study, the effects of polydispersity and sintering on the effective thermal conductivity of particle aggregates are investigated. A simulation method has been developed that is capable of producing aggregates made up of polydispersed particles with tailored morphological properties. Modelling of the sintering process is implemented in a fashion that is dictated by mass conservation and the desired degree of overlapping. A noticeable decrease in the thermal conductivity is observed for elevated polydispersity levels compared to that of aggregates of monodisperse particles with the same morphological properties. Sintered nanoaggregates offer wider conduction paths through the coalescence of neighbouring particles. It was found that there exists a certain sintering degree of monomers that offers the largest improvement in heat performance.

Keywords: nanofluid; heat conduction; effective thermal conductivity; particle aggregates; polydispersity; sintering

Citation: Karagiannakis, N.P.; Skouras, E.D.; Burganos, V.N. Modelling Thermal Conduction in Polydispersed and Sintered Nanoparticle Aggregates. *Nanomaterials* **2022**, *12*, 25. https://doi.org/10.3390/nano12010025

Academic Editor: S. M. Sohel Murshed

Received: 17 November 2021
Accepted: 18 December 2021
Published: 22 December 2021

1. Introduction

Numerous contemporary applications are based on the incorporation of nanoparticles into convectional fluids [1,2]. The resulting nanofluids may have drastically increased thermal properties and reduced sedimentation. Nanofluids and nanoparticles are increasingly used in a variety of fields. Heating and cooling systems using nanofluids show significant improvement in energy consumption, while new medical techniques are developing using nanoparticles in bioliquids. [3,4]. A large number of recent publications study the potential use of nanofluids in multidisciplinary fields [5,6]. Although much effort has been placed on the correlation of the heat transfer properties of nanofluids with the underlying phenomena, it appears that there is no widely accepted explanation of their behaviour or a reliable way to predict their heat conduction properties [7,8].

Many models have been developed to predict the effective conductivity of nanoparticles. The effect of Brownian motion, interfacial resistance, the existence of nanolayers, and the aggregation mechanism have been discussed in detail in the literature [9–13]. A large increase in effective thermal conductivity has been detected experimentally when nanoparticles are organized in small aggregates [14,15]. The increased contact of the particles within the aggregate was found to facilitate heat transfer compared to fully dispersed particles. On the contrary, larger mass aggregates have a negative effect on the stability of the nanofluid and, therefore, on heat transport properties [16]. Lotfizadeh et al. [11], Prasher et al. [17], Evans et al. [18], and Liao et al. [19], among others, developed models to predict the thermal conductivity of nanofluids based on the morphology of the aggregates. These works showed that the configuration of the nanoparticles and the morphological parameters of the aggregates can alter the effective conductivity of the nanofluids noticeably. A typical assumption made in several studies is that the nanoparticles are of the same size and the neighbouring monomers are mainly at single point contact.

In real conditions, samples always come with a certain distribution in particle size [20–22], which is held responsible for altering the thermal properties of nanofluids [21]. It has been noticed that polydispersity may occur during dispersion of the nanoparticles in the base fluid [23,24]. Chon et al. [21] used commercial, uniform nanoparticles for the preparation of nanofluids. They measured the size distribution after dispersion and found a significant deviation. In fact, it is technically challenging to synthesize and disperse a large quantity of highly monodispersed nanoparticles [25]. Zhiting et al. [26] studied the effect of polydispersity, among other parameters, on the heat transfer coefficient of nanocomposites with molecular dynamics simulations. They concluded that polydispersity negatively affects the effective conductivity. However, the nature of the used method does not allow the simulation of large systems, such as that of aggregated nanoparticles.

Strong electrostatic forces between particles, collision of particles during the formation of the aggregates, and high-temperature environments are some of the factors that contribute to a certain degree of overlapping between particles [27,28]. Two methods have been widely used for the preparation of nanofluids. The one-step nanoparticle production and blending method produces nanofluids with increased stability and offers elevated thermal conductivity, but has a relatively high production cost and is not yet suitable for large-scale production [29]. The two-step method is one that proceeds in two sequential steps, namely, the step of separate production of nanoparticles followed by suspension in a base fluid [29]. Commercial nanoparticles are usually found in powder form. The production of nanoparticles commences with the creation of a nanoparticle suspension from their precursors. This suspension is dried using various methods and, eventually, a powdered form of nanoparticles is obtained [23,30]. Thermal decomposition of organic precursors is a well-established process for the fabrication of solid nanoparticles [23]. It is temperature dependent and has been reported to result in sintered and/or polydispersed particles [23,24,30]. Other methods include nanoparticle production in the gas phase [31,32]. During the creation of these nanoparticles, formation processes, including surface reactions, condensation, coagulation and sintering, are some of the key mechanisms that take place [32–35]. The kinetics of each process determines the final structure morphology, which can vary among spherical particles, agglomerates, or compact aggregates [34]. The resulting system usually includes partially coalesced particles with sintering necks [32,35].

Attempts to model the aforementioned process usually start with the formation of an aggregate using a stochastic method. Eggersdorfer et al. [36] modelled the sintering process in aggregates, formed by Diffusion Limited Aggregation (DLA), Diffusion Limited Cluster–Cluster Aggregation (DLCCA), and Ballistic Aggregation (BA). The driving force for sintering was the minimization of free energy. They noted that, during sintering, primary particles approach each other. Sander et al. [31] presented an analytical description of the underlying phenomena during the production of nanoparticles, such as coagulation and sintering. The primary particles were modelled having a spherical shape and a polydispersed size, with each particle described by its sintering level and radius. The final structure has been compared with experimental data and transmission electron microscopy (TEM) images. An overlapping algorithm, developed by Brasil et al. [27], has studied the effect of sintering on the morphological properties of the aggregates, such as the fractal dimension and the radius of gyration. Schmid et al. [33,37] have developed a model for aggregates subjected to coagulation and sintering. The sintered aggregates were presented as the result of successive overlapping of spherical, primary particles.

The previous discussion underlines two major open issues in the study of the thermal conductivity of aggregated nanoparticles. Even though the effect of aggregation has been extensively studied, there is a dearth of research dealing with the effect of polydispersity of the nanoparticles within the aggregate. Moreover, sintering and the concomitant partial coalescence are most likely to occur in nanoparticle systems, yet the study of their effects on heat conduction remains a challenging, open field. These configurations complicate the determination of heat transfer properties and, therefore, reliable simulations of transport phenomena are required.

The present work examines the effects of polydispersity and sintering of particles on the effective thermal conductivity of nanofluids that contain particle aggregates. To this end, the method developed by the authors [38] for reconstructing particle aggregates is extended to include particle overlapping due to sintering as well as non-uniform particle size as a realistic outcome of nanofluid preparation. Among the merits of this method is the algorithmically rapid reconstruction of agglomerated systems with predetermined properties, namely, the fractal dimension and the average number of particles in the aggregates. As a case study, the particle size here follows the normal distribution and the standard deviation is expressed as a fraction of the mean size. Moreover, a technique has been developed to simulate sintered aggregates. The sintering process is expected to change the particle position and size while, naturally, preserving the mass of the working sample. An overlap parameter and the morphology of the primary aggregate determine the final morphology. The effective thermal conductivity is calculated through the temperature distribution obtained from the solution of the heat transfer equation. The Meshless Local Petrov–Galerkin (MLPG) method [39–41] is used here as it was shown to provide stable and fast solutions to particulate systems even with point contact. The Discretisation-Corrected Particle Strength Exchange (DC PSE) method [42,43] is used to approach the field function and its derivatives, while the meshless nature of the method allows local increase of the domain discretisation at the interface between the base fluid and solid particles.

The effect of the overlap parameter and the polydispersity level of particle aggregates on the thermal conductivity is studied by changing the number of particles in the aggregate, the fractal dimension of the aggregates, and the volume fraction of the particles. The effective conductivity of the polydispersed nanoaggregates, as predicted by the present method, is compared to the effective conductivity of the corresponding systems, as these result from the Diffusion Limited Aggregation (DLA) method. Moreover, the effective conductivity of aggregates consisting of polydispersed particles is compared with that of aggregates of monodispersed particles, keeping all other morphological parameters constant. Notable deviations between monodispersed and polydispersed cases are observed and discussed. In addition, the effect of sintering is examined by varying the overlap parameter. The results are compared with predictions of analytic expressions from the literature.

2. Modelling Polydispersed and Sintered Aggregates

Modelling of fractal aggregates is the first stage in the process of correlating numerically their physical properties with their morphological parameters. In reality, aggregation is a very complex process, being sensitive to parameters such as the temperature, the physical properties of the particles and the solvent, the polydispersity extent, the primary particle shapes, etc. [11,12]. A well-established approach to generate fractal structures numerically is the usage of various stochastic methods, such as Diffusion Limited Aggregation (DLA), Diffusion Limited Cluster–Cluster Aggregation (DLCCA), and Reaction Limited Aggregation (RLA) [43,44]. The validity of these methods has been verified through comparison of the resulting structures with experimental data [45,46]. Different physical processes are engaged in different aggregation models. A very common description of such clustering relies on the fractal dimension, d_f, using the relation between the number of particles in the aggregate, N, and basic cluster-size characteristics [45,46]: N.

$$N = k_g (R_g / r_p)^{d_f}, \tag{1}$$

where r_p is the mean radius of the primary particles, k_g is the structure factor, and R_g is the radius of gyration of the aggregate [12,32,47]:

$$R_g = \sqrt{\frac{\sum_i^N m_i(\boldsymbol{r}_i - \boldsymbol{r}_c)^2}{\sum_i m_i}}, \tag{2}$$

where $\boldsymbol{r}_i$ is the position vector of the centre of mass of particle i, m_i is its mass, and $\boldsymbol{r}_c$ is the position vector of the centre of mass of the aggregate.

In order to study the effect of polydispersity on thermal conductivity, the work developed in [38] has been extended to include polydispersed particles. This technique offered fast convergence of the algorithm for the representation of agglomerated systems with predetermined properties. For the sake of completeness, the major steps of the algorithm are mentioned below. The primary input of the algorithm consists of the volume fraction of the particles, the fractal dimension or a range of values around it, and the average number of particles per aggregate or a range of values around it. A random deposition of a particle initiates the process. A new particle stochastically appears on the surface of the particle, and the process is repeated, with the restriction of no overlapping between any pair of particles. During the process, certain restrictions are imposed as described in [38], aiming at the convergence of the fractal dimension to the target value or range of values. It has been shown that the desired fractal dimension can be achieved with only a few particles. The process ends when the aggregate acquires a predefined number of particles. Then, another particle appears at a random place in the computational domain, and the aforementioned process is repeated for the formation of a second cluster. The whole algorithm is repeated as many times as needed to satisfy the desired number of aggregates in the working domain. Further analytical descriptions of the method are presented in [38].

In the present study, all simulations take place in a cubic box of length l. All spatial parameters and variables are normalized with this quantity. Thus, the particle radius is related with the volume fraction (f_p), the number of particles in the aggregate (N), and the number of aggregates (N_c), as follows:

$$r_p = \sqrt[3]{\frac{3f_p}{4\pi\sum_i^{N_c} N_i}}, \tag{3}$$

For monodispersed particles, the definition of r_p is straightforward. For polydispersed particles, in the present extended approach, this value is set as the mean radius of the particles, r_0. The exact radius of each particle is randomly sampled from a prescribed normal distribution. The deviation of the particle size distribution is expressed as a fraction of the particle radius. The probability density function (p) for the particle radius is shown in Figure 1 for two different standard deviations. In order to avoid negative values, a threshold was imposed at zero radius. For symmetry reasons, another threshold is set at the value $r_{p,max} = 2r_0$. The maximum deviation of the particle radius, in this study, is $\sigma = 0.5r_0$. For higher polydispersity levels a lognormal distribution can be used instead, in order to avoid a large number of negative values and adhere to more realistic particle size distributions.

The resulting system may have a volume fraction different from the desired one. This is an important issue in polydispersed particle systems since large sizes may be sampled as the tail additions to a cluster. Depending on whether the desired volume fraction is smaller or greater than expected, particles will be added or removed. To remove a particle, an aggregate is randomly selected and the last-added particle is removed. To add a particle, an aggregate is randomly selected and a new particle with a size that is sampled from the prescribed distribution is added at a random location on the surface of a randomly selected particle of the aggregate, with restrictions in order to satisfy the fractal dimension and the non-overlapping condition, as described in [38] for uniformly sized particles. The process is repeated until the predetermined volume fraction is achieved.

With this methodology, critical quantities such as the volume fraction, the mean radius of the particles, and the fractal dimension remain within their prescribed bounds while a small deviation is maintained in the number of particles per aggregate.

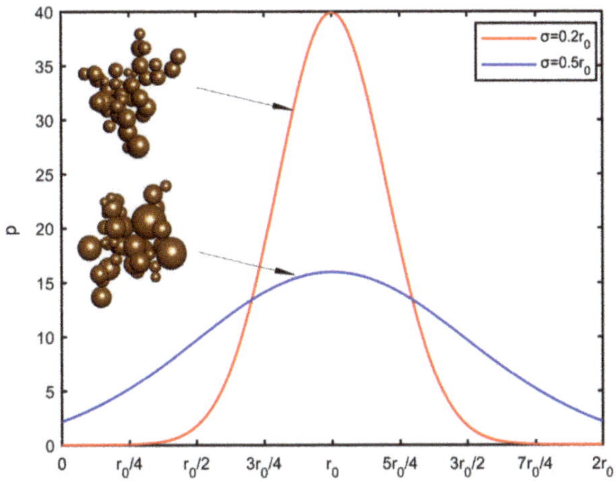

Figure 1. The probability density function, p, of the particle radius distribution for different standard deviation values (red $\sigma = 0.2r_0$, blue $\sigma = 0.5r_0$) and a representative visualization of the resulting aggregates.

The structure factor, k_g, is strongly dependent on the polydispersity levels and the degree of overlapping of the particles [48,49]. Eggersdorfer et al. [49] and Tomchuk et al. [48] studied the effect of polydispersity on the fractal dimension and the structure factor, considering a wide range in the number of particles in each aggregate, the prescribed deviation of the particle size, and the aggregation model that is used. They noted a reduction in the structure factor for increased polydispersity for aggregates formed by the DLCCA algorithm and the Ballistic Aggregation model. Independently of the agglomeration mechanism, in the limiting case of infinitely polydispersed particles, the structure factor tends to unity [49]. For monodispersed particles with fractal dimension ranging between 1.7 and 2.5, the structure factor can be considered constant, $k_g = 1.5$ [38,50]. In the present case, for polydispersed particles, the structure factor changes linearly with the deviation of the particle radius, taking values between 1.2 and 1.5 [49].

During sintering, particles are expected to increase their radius and come closer to each other [37]. Typical simulations of coagulation and sintering of nanoaggregates include an overlapping step, where neighbouring particles penetrate each other, and a growth step, during which particle size increases to maintain mass and volume. This process captures the redistribution of mass in the free surface of the aggregate and offers a realistic representation of the final morphology [33]. Assuming aggregates consisting of monodispersed particles and following this methodology, an overlapping coefficient is defined as [27]:

$$\delta = 1 - \frac{d}{2R_p}, \tag{4}$$

where d is the final distance of the centres of the neighbour particles, and R_p is the final radius of the particles. The initial radius of the particles can be calculated from Equation (3).

At initial stages of the sintering algorithm, aggregates are forced to collapse to their centre of mass by the penetration coefficient, δ_p, while the sizes of the particles remain constant. The penetration coefficient relates the final distance of the neighbouring particles to the initial radius ($\delta_p = 1 - \frac{d}{2r_p}$). Obviously, this process causes a mass loss. In a second step, the particle sizes are increased, in order to reproduce the volume fraction at its prescribed value. The growth coefficient relates the initial and the final particle radius to the final distance of the neighbouring particles ($\delta_R = \left(\frac{R_p - r_p}{R_p r_p}\right) d/2$). Combining these definitions, the final radius of the particles can be related to the initial radius, the overlapping coefficient, and the penetration coefficient, as follows:

$$R_p = \frac{r_p}{1-\delta}(1-\delta_p), \tag{5}$$

The overlapping coefficient, δ, is the sum of the penetration (δ_p) and the growth (δ_R) coefficients: $\delta = \delta_p + \delta_R$. If $\delta = 1$, the aggregates are totally sintered (i.e., every aggregate merged into a single particle), whereas $\delta = 0$ indicates that particles are in point contact. The permissible values for the penetration (δ_p) and growth (δ_R) coefficients range from zero to δ.

The generation of sintered aggregates initiate with the determination of the volume fraction, the number of particles in the aggregate, the fractal dimension, and the overlapping coefficient. After the formation of each aggregate, a series of trial simulation scenarios are used to evaluate the values of δ_p and R_p. Evidently, each combination results in a different volume fraction ($f_{p,\delta}$). In order to determine the appropriate combination, for each aggregate the value of the penetration coefficient varies from zero to δ. For each δ_p value, the final radius (Equation (5)) is calculated. Finally, the volume fraction of the resulting system is compared with the initial volume fraction (f_p), ($\beta = 100 \cdot |f_{p,\delta} - f_p| / f_{p,\delta}$) and its dependence on δ_p, as shown in Figure 2. The δ_p value with the smallest acceptable error is selected. Following the aforementioned technique, mass conservation is secured for each aggregate of the system, for the entire range of the overlapping coefficient. In Figure 2, the percentage error in volume fraction (β) is represented as a function of the penetration coefficient (δ_p), for different values of the overlapping coefficient and the morphological characteristics of the initial aggregate. It is shown that a unique combination of δ_p and R_p results in a system with the same volume fraction. This methodology, in addition to achieving the desired overlapping coefficient and volume fraction, has the advantage of being straightforward and fast during calculations. Needless to say, in real conditions the final structure during sintering may be different from that of the overlapping spheres, due to the appearance of neck effects and redistribution of mass that will eventually differentiate the structure from the one that is simulated here.

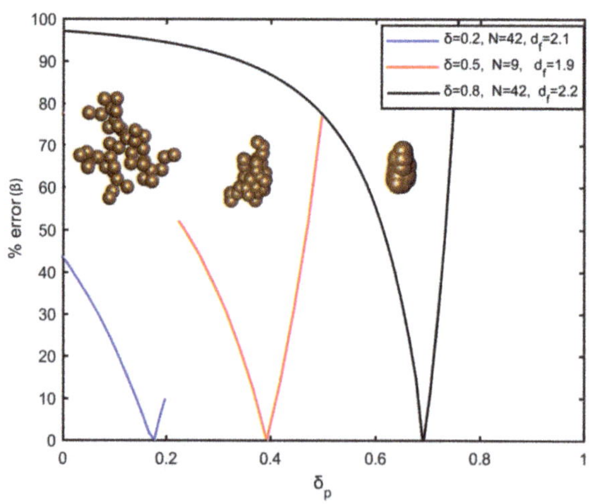

Figure 2. The percentage error in volume fraction (β) varying with the penetration coefficient for different values of the fractal dimension, the number of particles in the aggregate, and the overlapping coefficient. A representative visualization of each aggregate structure is also shown.

3. Effective Conductivity Calculation

The reconstructed aggregates that are obtained following the algorithm of the previous section are used as input to heat transport modelling. A constant temperature difference is imposed along the vertical axis of the nanofluid, whereas the rest of the boundaries are considered periodic. For the numerical solution of the heat transport equation, the Meshless Local Petrov–Galerkin (MLPG) method is used [41]. It has been shown to offer concrete advantages to more conventional methods in particulate systems with several contact areas, as is the case here. Differential equations are integrated into local subdomains, a fact that facilitates the increase of discretisation in regions of the geometries where steep gradients are developed. In the nanofluid case, the effective conductivity changes drastically at the interface of nanoparticles and the base fluid. The shape of the subdomains alters the performance of the method, with cubic sectors having been proved to increase the stability of the method [40]. The DC PSE approach is chosen as the trial function, while a step function is used as test function for the integration [42,43]. A set of cubic grids digitizes the domain and the integrals are calculated with the Gauss quadrature method [43]. In each Ω_x subdomain, the dimensionless weak form of the energy equation is given by the relation [43]:

$$\left(k_{rp}-1\right)\int_{\partial\Omega_x}\Phi\nabla T_{\hat{n}}d(\partial\Omega_x)+\int_{\partial\Omega_x}\nabla T_{\hat{n}}d(\partial\Omega_x)=0, \tag{6}$$

where Φ is a spatial step function defined as unit in the particle phase and zero elsewhere, and $k_{rp}=\frac{k_p}{k_f}$ is the ratio of the conductivity of the particles to that of the base fluid. The solution of the heat transfer equation determines the temperature throughout the computational domain. Then, the calculation of the dimensionless effective conductivity is straightforward from $k_{eff}=\overline{\int\limits_S k\frac{\partial T}{\partial n}dS}$, where the surface S is vertical to the heat flow. A detailed description of the approach, the respective variables and integrals, the mesh construction, and the conductivity calculation can be found in a previous work by the authors [43].

This method is capable of calculating the effective conductivity of large particle systems. The aggregates are considered stationary and the heat conduction equation is solved within the computational domain. A typical simulation contains about 500–1000 particles organized into aggregates. Modelling of aggregates in heat transfer processes is performed with in-house meshless CFD methods implemented in Matlab kernels, as described in [38]. The computational time for the reconstruction of the aggregates is also provided in [38], along with comparison with other aggregation models. The effective conductivity calculations used herein have been shown in [43] to reduce the computational cost, compared with other numerical models and commercial software. A typical run for 1000 particles requires ~5 mil. nodes and ~1.5 ks on an Intel(R) Xeon(R) Silver 4116 CPU at 2.10 GHz using 12 cores.

Moreover, the present method can be extended to include calculations of the effective conductivity of different nanoparticle shapes. If the equation of the external surface of the particles is simple, the application is straightforward; otherwise the aggregation algorithm should be modified rather drastically, especially for non-convex surfaces.

The corresponding predictions of analytical models for the effective thermal conductivity are presented next. A well-known model for the conductivity of dispersed particles is Maxwell's effective medium theory. Maxwell developed an expression for the effective conductivity, k_{eff}, of a suspension of solid spheres in liquid [51]:

$$k_{eff}=\frac{k_p+2k_f+2\left(k_p-k_f\right)f_p}{k_p+2k_f-2\left(k_p-k_f\right)f_p}k_f, \tag{7}$$

where k_f is the conductivity of the base fluid, f_p is the volume fraction of the particles, and k_p is the conductivity of the particles. According to this equation, the process is apparently not sensitive to the size or the arrangement of the dispersed phase. However, this relation is not symmetric. By replacing k_p with k_f and f_p with $(1 - f_p)$, the effective conductivity calculation changes drastically [11]:

$$k_{eff} = \frac{k_f + 2k_p + 2\left(k_f - k_p\right)(1 - f_p)}{k_f + 2k_p - 2\left(k_f - k_p\right)(1 - f_p)} k_p, \tag{8}$$

Equation (7) accurately predicts well-dispersed particles at low volume fraction, while Equation (8) has been used to describe the conductivity of aggregate structures. In this perspective, the particles are considered as a solid network, with the base fluid enclosed in some regions. In any case, Equations (7) and (8) estimate the lower and upper bounds of the conductivity of an inhomogeneous medium, respectively [52,53].

The majority of the attempts to develop a model for the conductivity of colloidal clusters include a two-step approximation. The clusters are considered as spheres, with an effective conductivity (k_e) and an effective volume fraction (f_e). For the calculation of k_e, many relations from the effective medium theory have been applied [11,52–54]. The size of these effective particles is usually set equal to the radius of gyration of the aggregates [11,55]. In this case, the effective volume fraction (f_e), can be expressed as:

$$f_e = \frac{4\pi}{3} \sum_{i}^{N_c} {R_{gi}}^3, \tag{9}$$

where N_c is the number of aggregates in the solution and R_{g_i} is the (dimensionless) radius of gyration of each aggregate. The volume fraction of the solid phase inside the aggregates is $f_i = f_p/f_e$ [17]. In this work, k_e is calculated from the upper limit (Equation (8)) by swapping f_p with f_i, while the Maxwell relation is used (Equation (7)) in a second step, by replacing k_p with k_e and f_p with f_e.

4. Results and Discussion

4.1. Comparison with Other Aggregation Models

Figure 3 shows the comparison between the effective conductivity of aggregates containing polydispersed particles, as extracted with the use of the method developed here, and the results of the DLA method. The volume fraction of particles is $f_p = 0.1$ and the standard deviation of the particle size is $\sigma = 0.5r_0$, where r_0 is the mean radius of the particles. The aggregates consist of $N = 42$ particles and the thermal conductivity of the particles is considered $k_{rp} = \frac{k_p}{k_f}$ larger than that that of the base fluid by a factor of 100. The simulation points are the averages of 10 realizations with the same morphological characteristics. In an earlier work by the authors, it was indicated that the mechanism of aggregation does not affect the effective conductivity for monodispersed particles. The same conclusion is drawn here for the case of polydispersed particles. However, the morphological characteristics of the aggregates appear to be significant for the effective thermal conductivity. Similar to the behaviour of aggregates of monodispersed particles, the thermal conductivity decreases with an increase in the fractal dimension.

4.2. Dependence of Conductivity on the Fractal Dimension, the Number of Particles in the Aggregate, and the Polydispersity of the Particles

In an earlier work by the authors [38], the effect of aggregation on the thermal conductivity of monodispersed particles was studied by varying the fractal dimension and the number of particles in the aggregate. A considerable increase in the thermal conductivity has been shown, even for aggregates consisting of a small number of particles. However, in the present work, a different behaviour is noticed upon the introduction of polydispersity in the particle size. The results are presented next.

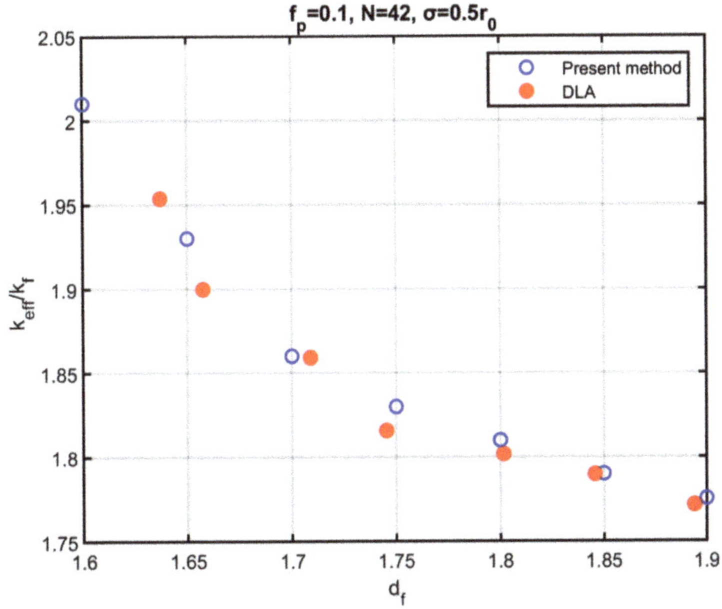

Figure 3. Effective thermal conductivity as a function of the fractal dimension of aggregates consisting of polydispersed particles by the present method (open, blue symbols) and by simulations from the DLA method (filled, red symbols).

Figure 4 shows the dimensionless effective thermal conductivity of a nanofluid as a function of the fractal dimension, for two volume fractions, namely $f_p = 0.03$ (Figure 4a,b) and for $f_p = 0.1$ (Figure 4c). Two different values of the number of particles per aggregate are also investigated, namely $N = 42$ (Figure 4a,c) and $N = 9$ (Figure 4b,c). To enable a comparison, the mean radius of the particles for the polydispersed cases was set equal to the ones in corresponding cases of monodispersed particles. The standard deviation of the particle size varies from $\sigma = 0.1r_0$ to $\sigma = 0.5r_0$ in Figure 4a,b, whereas in Figure 4c it is kept at $\sigma = 0.5r_0$. Figure 4 also shows the thermal conductivity predictions of two analytical models, namely, the two-step Maxwell (Equations (7) and (8)) and the single-step Maxwell model (Equation (7)). Every simulation point is the average of 10 realizations with the same parameters. The conductivity ratio of the nanoparticles and the base fluid is chosen to be 130 ($k_{rp} = 130$), which is a representative value for several practical nanofluids, such as water–Fe, engine oil–Al_2O_3, and water–CuO.

The effective conductivity decreased as the fractal dimension increased in all cases studied here. The level of reduction was affected by the polydispersity degree, the volume fraction, and the number of particles per aggregate. For cohesive aggregates, corresponding to relatively high fractal dimension ($d_f = 2.5$), polydispersity did not affect the effective conductivity, whereas for smaller values of the fractal dimension and high polydispersity level, a notable variation was observed. More specifically, the thermal conductivity decreased with the increase of the deviation of the particle radius compared with monodisperse particle cases. However, a small deviation ($\sigma = 0.1r_0$, $\sigma = 0.2r_0$) affected the thermal conductivity only slightly (Figure 4a,b). A 10% reduction was observed for polydispersity degree $\sigma = 0.5r_0$ and volume fraction $f_p = 0.1$ (Figure 4c), whereas the reduction was only about 5% for volume fraction $f_p = 0.03$ (Figure 4a,b).

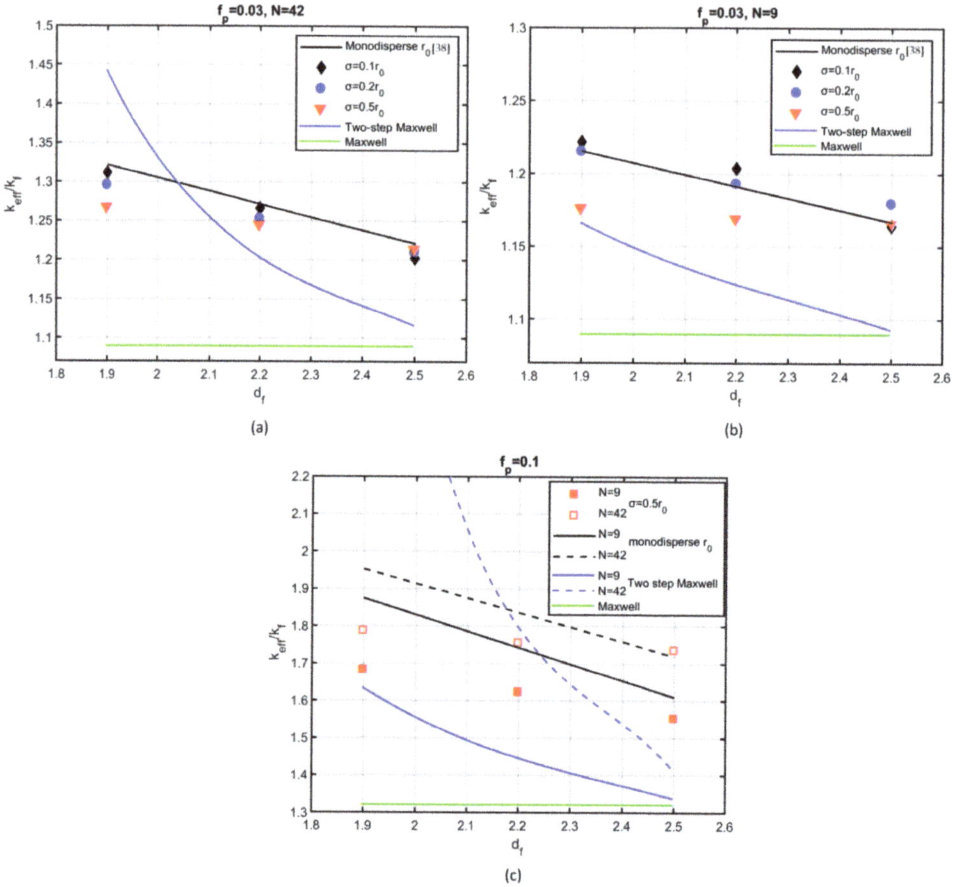

Figure 4. Effective thermal conductivity as a function of fractal dimension for: (**a**) volume fraction $f_P = 0.03$ and number of particles per aggregate $N = 42$; (**b**) $f_P = 0.03$, $N = 9$; (**c**) $f_P = 0.1$, $N = 9$ (solid lines, solid symbols), $N = 42$ (dashed lines, open symbols). Black lines: monodispersed cases. Blue lines: two-step Maxwell model. Green line: Maxwell model. Symbols: polydispersed cases.

The previous results are in contrast to the increase of the projected area of an aggregate as the polydispersity level is increased [32]. Particle size distribution affects heat transfer in two distinct ways. Large particles accelerate heat transfer, while small particles hinder it. According to the present results, the smaller particles act as regulators of heat transport; therefore, the effective conductivity is reduced. Experimental works have observed that polydispersity of the nanoparticles has a significant impact on the thermal properties of the nanofluid [56]. Specifically, the largest enhancement has been found for highly monodisperse particles [57].

The single-step Maxwell relation (Equation (7)) remains insensitive to the fractal dimension. However, the two-step Maxwell model, although affected by the fractal dimension and the number of particles in the aggregate, cannot predict the effective conductivity of such systems. It is worth noting that aggregation increased the effective conductivity of the nanofluid significantly in all cases studied. For highly polydispersed particles, when organized into aggregates consisting of, say, $N = 9$ particles per aggregate, a 10% volume fraction (shown in Figure 4c) resulted in a 70% increase of the thermal conductivity. The higher the particles per aggregate, the higher the conductivity increase.

4.3. *Effect of Sintering*

Figure 5a,b portray the dependence of the thermal conductivity on the overlapping coefficient (δ) for nanoparticles that are organised into aggregates containing $N = 9$ and $N = 50$ particles, with volume fraction $f_p = 0.03$ (Figure 5a) and $f_p = 0.1$ (Figure 5b). The overlapping coefficient ranges from $\delta = 0$, which indicates particles at single-point contact, to $\delta = 1$, which corresponds to degeneration of the aggregate to a single particle. The corresponding thermal conductivity predictions of analytical models are also presented.

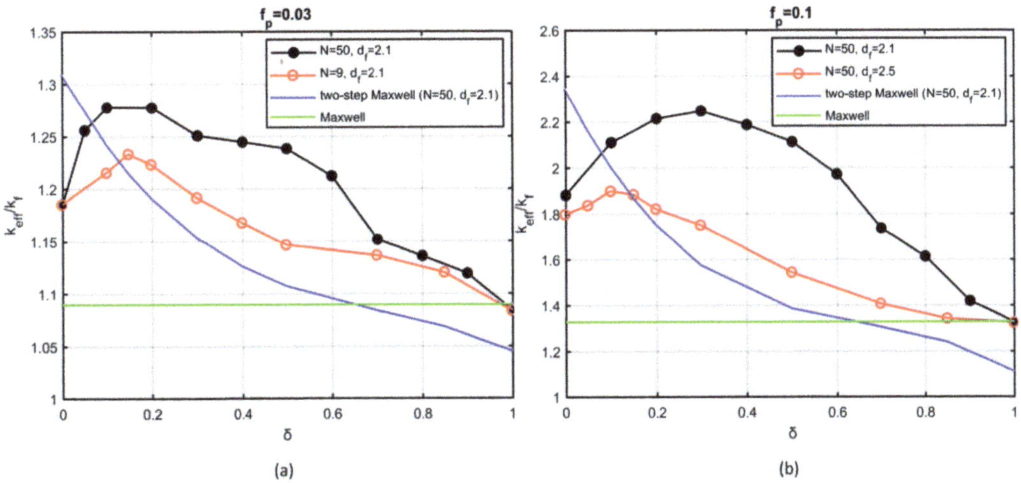

Figure 5. Effective thermal conductivity as a function of overlapping coefficient: (**a**) volume fraction $f_p = 0.03$; (**b**) volume fraction $f_p = 0.1$. Solid, black circles: $N = 50$, $d_f = 2.1$. Open, red circles: (**a**) $N = 9$, $d_f = 2.1$; (**b**) $N = 50, d_f = 2.5$. Blue lines: two-step Maxwell model. Green lines: Maxwell model.

Upon the introduction of sintering, the effective conductivity is found to increase with the sintering level, up to a maximum value. Further sintering beyond that point has a negative effect on conduction until the value of Maxwell's model is obtained, for $\delta = 1$. The value of the overlapping coefficient that offers the highest conductivity increase changes with the volume fraction and the number of particles per aggregate. A 15% maximum increase is shown for $f_p = 0.1$, $N = 42$, and $\delta = 0.25$ (Figure 5b). On the other hand, the conductivity calculated by the two-step Maxwell model decreases monotonically upon increase of the overlapping coefficient, due to the monotonic decrease of the radius of gyration (R_g).

The results showed that controlled aggregation and sintering can offer significantly improved thermal properties to nanofluids. From a physical point of view, higher values of the overlapping coefficient create an increased number of conduction pathways, which are also wider and longer, thus facilitating conduction along the macroscopic direction of heat transport. At the same time, the size of the aggregates decreases. For low sintering levels, the former factor prevails, whereas for high values of the overlapping coefficient, the aggregate tends to degenerate to an isolated body, thus reducing the thermal conductivity drastically.

5. Conclusions

The effect of particle size polydispersity and the sintering level on the thermal conductivity of aggregated nanoparticles was studied in the present paper. It was shown that both parameters examined here have the potential to change the heat performance of nanofluids drastically.

A method for reconstructing aggregates with the desired polydispersity degree was developed, satisfying simultaneously the requirements for certain morphological characteristics of the aggregate, namely, the fractal dimension and the number of particles in the aggregate. Particle sintering in aggregates was simulated for monodispersed cases and encoded as an overlapping mechanism in two steps: a penetration step and a growth step. In order to ensure mass conservation, the progression of each step was controlled through the minimisation of the error in the volume fraction of the sintered aggregate compared with the volume fraction of the initial aggregate. A meshless method with local refinement was used for the solution of the heat transfer equation and was found to be stable for the complex systems that were studied here. This is of key importance in the present problem as it allows using relatively large working domains that contain overlapping particles or particles at point contact with others and being able to extract statistically meaningful conclusions.

The effective thermal conductivity was calculated for aggregates that resulted from the present method of aggregation of polydispersed particles, then compared with the thermal conductivity of aggregates that were constructed with the Diffusion Limited Aggregation (DLA) method. The dependence of the effective thermal conductivity on the fractal dimension was found to be in good agreement with that in DLA method aggregates, which indicates that the proposed method produces aggregates that are thermally equivalent to those resulting from methods that describe the physical process of particle aggregation. Consequently, one can employ the present method for the investigation of the behaviour of nanofluids in heat transport problems, taking advantage of the increased simplicity of the aggregation algorithm and its rapid convergence to the final configuration.

The variation of the effective thermal conductivity was investigated over a wide range of fractal dimension values, number of particles per aggregate, and standard deviation of the particle size. Compared to fully dispersed particles, aggregation was shown to increase the thermal conductivity in all cases studied here. Small radius deviation does not substantially change the thermal conductance compared to monodispersed cases; however, a further increase of polydispersity leads to a clear reduction of the effective conductivity. On the contrary, strong polydispersity leads to an increase in the projected area, which implies an increase of heat transfer. A possible explanation for our result could be the existence of small particles within the aggregate that hinder heat transfer. This result is qualitatively confirmed by experimental measurements [56,57] according to which nanofluids consisting of particles with low polydispersity levels have higher heat performance compared to particles with high polydispersity.

The two-step Maxwell model predicts a monotonic decrease of the effective conductivity with increasing fractal dimension; however, large deviations from the numerical results were found for most of the cases examined here.

The effect of sintering of the aggregates was investigated and quantified as a function of the overlapping coefficient. Sintered aggregates have a lower effective size than the original aggregates, so a reduction in the effective conductivity should be expected. At the same time, sintering increases the heat conduction by forming larger heat pathways. This interplay yields a maximum in the thermal conductivity as a function of the degree of coalescence. The precise value of the overlapping coefficient that provides the highest conductivity increase depends on the morphological properties and the volume fraction of the initial aggregates. The present study indicates that the conditions of the production and dispersion of nanoparticles have a major impact on the thermal properties of the nanofluids. This is a possible explanation for the large deviations that have been observed between experimental works. Nanofluids with monodispersed particles, which are organized into aggregates with small overlapping, offer the highest heat transfer coefficient over the range of parameter values that were examined here. The results and conclusions of this work are also relevant to nanocomposite materials that contain polydispersed particle inclusions, which are organized in aggregates either at simple contact or in sintered form.

Author Contributions: Conceptualization, N.P.K. and V.N.B.; funding acquisition, V.N.B.; investigation, N.P.K., E.D.S. and V.N.B.; methodology, N.P.K.; project administration, V.N.B.; software, N.P.K.; supervision, V.N.B.; validation, N.P.K.; visualization, N.P.K.; writing—original draft, N.P.K.; writing—review and editing, E.D.S. and V.N.B. All authors have read and agreed to the published version of the manuscript.

Funding: This research was funded by the project "Materials and Processes for Energy and Environment Applications-AENAO" (MIS 5002556) which is implemented under the "Action for the Strategic Development on the Research and Technological Sector", funded by the Operational Programme "Competitiveness, Entrepreneurship and Innovation" (NSRF 2014-2020) and co-financed by Greece and the European Union (European Regional Development Fund).

Institutional Review Board Statement: Not applicable.

Informed Consent Statement: Not applicable.

Data Availability Statement: Not applicable.

Acknowledgments: Computational infrastructure was provided by the Institute of Chemical Engineering Sciences, Foundation for Research and Technology, Hellas (FORTH/ICE-HT).

Conflicts of Interest: The authors declare no conflict of interest. The funders had no role in the design of the study; in the collection, analyses, or interpretation of data; in the writing of the manuscript, or in the decision to publish the results.

References

1. Choi, S.U.; Eastman, J.A. *Enhancing Thermal Conductivity of Fluids with Nanoparticles;* Argonne National Lab.: Argonne, IL, USA, 1995.
2. Wen, D.; Lin, G.; Vafaei, S.; Zhang, K. Review of nanofluids for heat transfer applications. *Particuology* **2009**, *7*, 141–150. [CrossRef]
3. Mahdi, R.A.; Mohammed, H.A.; Munisamy, K.M.; Saeid, N.H. Review of convection heat transfer and fluid flow in porous media with nanofluid. *Renew. Sustain. Energy Rev.* **2015**, *41*, 715–734. [CrossRef]
4. Nagaraja, S.; Yang, V.; Adamovich, I. Multi-scale modelling of pulsed nanosecond dielectric barrier plasma discharges in plane-to-plane geometry. *J. Phys. D Appl. Phys.* **2013**, *46*, 155205. [CrossRef]
5. Sarafraz, M.M.; Tlili, I.; Tian, Z.; Bakouri, M.; Safaei, M.R.; Goodarzi, M. Thermal Evaluation of Graphene Nanoplatelets Nanofluid in a Fast-Responding HP with the Potential Use in Solar Systems in Smart Cities. *Appl. Sci.* **2019**, *9*, 2101. [CrossRef]
6. Agista, M.N.; Guo, K.; Yu, Z. A State-of-the-Art Review of Nanoparticles Application in Petroleum with a Focus on Enhanced Oil Recovery. *Appl. Sci.* **2018**, *8*, 871. [CrossRef]
7. Estellé, P.; Halelfadl, S.; Thierry, M. Thermal conductivity of CNT water based nanofluids: Experimental trends and models overview. *J. Therm. Eng.* **2015**, *1*, 381–390. [CrossRef]
8. Kleinstreuer, C.; Feng, Y. Experimental and theoretical studies of nanofluid thermal conductivity enhancement: A review. *Nanoscale Res. Lett.* **2011**, *6*, 229. [CrossRef]
9. Jang, S.P.; Choi, S.U. Role of Brownian motion in the enhanced thermal conductivity of nanofluids. *Appl. Phys. Lett.* **2004**, *84*, 4316–4318. [CrossRef]
10. Koo, J.; Kleinstreuer, C. A new thermal conductivity model for nanofluids. *J. Nanopart. Res.* **2004**, *6*, 577–588. [CrossRef]
11. Lotfizadeh, S.; Matsoukas, T. A continuum Maxwell theory for the thermal conductivity of clustered nanocolloids. *J. Nanopart. Res.* **2015**, *17*, 262. [CrossRef]
12. Xuan, Y.; Li, Q.; Hu, W. Aggregation structure and thermal conductivity of nanofluids. *AIChE J.* **2003**, *49*, 1038–1043. [CrossRef]
13. Yu, W.; Choi, S. The role of interfacial layers in the enhanced thermal conductivity of nanofluids: A renovated Maxwell model. *J. Nanopart. Res.* **2003**, *5*, 167–171. [CrossRef]
14. Putnam, S.A.; Cahill, D.G.; Braun, P.V.; Ge, Z.; Shimmin, R.G. Thermal conductivity of nanoparticle suspensions. *J. Appl. Phys.* **2006**, *99*, 084308. [CrossRef]
15. Zhang, X.; Gu, H.; Fujii, M. Experimental study on the effective thermal conductivity and thermal diffusivity of nanofluids. *Int. J. Thermophys.* **2006**, *27*, 569–580. [CrossRef]
16. Ghadimi, A.; Saidur, R.; Metselaar, H.S.C. A review of nanofluid stability properties and characterization in stationary conditions. *Int. J. Heat Mass Transf.* **2011**, *54*, 4051–4068. [CrossRef]
17. Prasher, R.; Phelan, P.E.; Bhattacharya, P. Effect of Aggregation Kinetics on the Thermal Conductivity of Nanoscale Colloidal Solutions (Nanofluid). *Nano Lett.* **2006**, *6*, 1529–1534. [CrossRef] [PubMed]
18. Evans, W.; Prasher, R.; Fish, J.; Meakin, P.; Phelan, P.; Keblinski, P. Effect of aggregation and interfacial thermal resistance on thermal conductivity of nanocomposites and colloidal nanofluids. *Int. J. Heat Mass Transf.* **2008**, *51*, 1431–1438. [CrossRef]
19. Liao, J.; Zhang, A.; Qing, S.; Zhang, X.; Luo, Z. Investigation on the aggregation structure of nanoparticle on the thermal conductivity of nanofluids by molecular dynamic simulations. *Powder Technol.* **2022**, *395*, 584–591. [CrossRef]

20. Pastoriza-Gallego, M.J.; Casanova, C.; Legido, J.L.; Piñeiro, M.M. CuO in water nanofluid: Influence of particle size and polydispersity on volumetric behaviour and viscosity. *Fluid Phase Equilib.* **2011**, *300*, 188–196. [CrossRef]
21. Chon, C.H.; Kihm, K.D.; Lee, S.P.; Choi, S.U. Empirical correlation finding the role of temperature and particle size for nanofluid (Al_2O_3) thermal conductivity enhancement. *Appl. Phys. Lett.* **2005**, *87*, 153107. [CrossRef]
22. Beck, M.P.; Yuan, Y.; Warrier, P.; Teja, A.S. The effect of particle size on the thermal conductivity of alumina nanofluids. *J. Nanopart. Res.* **2009**, *11*, 1129–1136. [CrossRef]
23. Odularu, A.T. Metal Nanoparticles: Thermal Decomposition, Biomedicinal Applications to Cancer Treatment, and Future Perspectives. *Bioinorg. Chem. Appl.* **2018**, *2018*, 9354708. [CrossRef] [PubMed]
24. Yeap, S.P. Permanent agglomerates in powdered nanoparticles: Formation and future prospects. *Powder Technol.* **2018**, *323*, 51–59. [CrossRef]
25. Cui, H.; Feng, Y.; Ren, W.; Zeng, T.; Lv, H.; Pan, Y. Strategies of large scale synthesis of monodisperse nanoparticles. *Recent Pat. Nanotechnol.* **2009**, *3*, 32–41. [CrossRef]
26. Tian, Z.; Hu, H.; Sun, Y. A molecular dynamics study of effective thermal conductivity in nanocomposites. *Int. J. Heat Mass Transf.* **2013**, *61*, 577–582. [CrossRef]
27. Brasil, A.M.; Farias, T.L.; Carvalho, M.G.; Koylu, U.O. Numerical characterization of the morphology of aggregated particles. *J. Aerosol Sci.* **2001**, *32*, 489–508. [CrossRef]
28. Brasil, A.M.; Farias, T.L.; Carvalho, M.G. A recipe for image characterization of fractal-like aggregates. *J. Aerosol Sci.* **1999**, *30*, 1379–1389. [CrossRef]
29. Yu, W.; Xie, H. A review on nanofluids: Preparation, stability mechanisms, and applications. *J. Nanomater.* **2012**, *2012*, 1–17. [CrossRef]
30. Pawłowski, R.; Kiełbasiński, K.; Sobik, P.; Pawłowski, B.; Wita, H.; Konefał, R.; Auguścik, M.; Pajor-Świerzy, A.; Szałapak, J.; Krzemiński, J. Obtaining of silver nanopowders by the thermal decomposition of fatty silver salts with various chain length. *Mater. Res. Express* **2019**, *6*, 065046. [CrossRef]
31. Sander, M.; West, R.H.; Celnik, M.S.; Kraft, M. A Detailed Model for the Sintering of Polydispersed Nanoparticle Agglomerates. *Aerosol Sci. Technol.* **2009**, *43*, 978–989. [CrossRef]
32. Eggersdorfer, M.L.; Kadau, D.; Herrmann, H.J.; Pratsinis, S.E. Aggregate morphology evolution by sintering: Number and diameter of primary particles. *J. Aerosol Sci.* **2012**, *46*, 7–19. [CrossRef] [PubMed]
33. Schmid, H.-J.; Al-Zaitone, B.; Artelt, C.; Peukert, W. Evolution of the fractal dimension for simultaneous coagulation and sintering. *Chem. Eng. Sci.* **2006**, *61*, 293–305. [CrossRef]
34. Artelt, C.; Schmid, H.J.; Peukert, W. On the impact of accessible surface and surface energy on particle formation and growth from the vapour phase. *J. Aerosol Sci.* **2005**, *36*, 147–172. [CrossRef]
35. Koch, W.; Friedlander, S.K. The effect of particle coalescence on the surface area of a coagulating aerosol. *J. Colloid Interface Sci.* **1990**, *140*, 419–427. [CrossRef]
36. Eggersdorfer, M.L.; Kadau, D.; Herrmann, H.J.; Pratsinis, S.E. Multiparticle Sintering Dynamics: From Fractal-Like Aggregates to Compact Structures. *Langmuir* **2011**, *27*, 6358–6367. [CrossRef] [PubMed]
37. Schmid, H.-J.; Tejwani, S.; Artelt, C.; Peukert, W. Monte Carlo simulation of aggregate morphology for simultaneous coagulation and sintering. *J. Nanopart. Res.* **2004**, *6*, 613–626. [CrossRef]
38. Karagiannakis, N.P.; Skouras, E.D.; Burganos, V.N. Modelling Thermal Conduction in Nanoparticle Aggregates in the Presence of Surfactants. *Nanomaterials* **2020**, *10*, 2288. [CrossRef]
39. Atluri, S.; Han, Z.; Rajendran, A. A new implementation of the meshless finite volume method, through the MLPG "mixed" approach. *CMES Comput. Model. Eng. Sci.* **2004**, *6*, 491–514.
40. Karagiannakis, N.P.; Bourantas, G.C.; Kalarakis, A.N.; Skouras, E.D.; Burganos, V.N. Transient thermal conduction with variable conductivity using the Meshless Local Petrov–Galerkin method. *Appl. Math. Comput.* **2016**, *272*, 676–686. [CrossRef]
41. Wu, X.-H.; Tao, W.-Q. Meshless method based on the local weak-forms for steady-state heat conduction problems. *Int. J. Heat Mass Transf.* **2008**, *51*, 3103–3112. [CrossRef]
42. Schrader, B. *Discretization-Corrected PSE Operators for Adaptive Multiresolution Particle Methods*; ETH: Zurich, Switzerland, 2011.
43. Karagiannakis, N.P.; Bali, N.; Skouras, E.D.; Burganos, V.N. An Efficient Meshless Numerical Method for Heat Conduction Studies in Particle Aggregates. *Appl. Sci.* **2020**, *10*, 739. [CrossRef]
44. Meakin, P.; Jullien, R. The effects of restructuring on the geometry of clusters formed by diffusion-limited, ballistic, and reaction-limited cluster–cluster aggregation. *J. Chem. Phys.* **1988**, *89*, 246–250. [CrossRef]
45. Tence, M.; Chevalier, J.; Jullien, R. On the measurement of the fractal dimension of aggregated particles by electron microscopy: Experimental method, corrections and comparison with numerical models. *J. Phys.* **1986**, *47*, 1989–1998. [CrossRef]
46. Koeylue, U.; Xing, Y.; Rosner, D.E. Fractal morphology analysis of combustion-generated aggregates using angular light scattering and electron microscope images. *Langmuir* **1995**, *11*, 4848–4854. [CrossRef]
47. Xiong, C.; Friedlander, S. Morphological properties of atmospheric aerosol aggregates. *Proc. Natl. Acad. Sci. USA* **2001**, *98*, 11851–11856. [CrossRef]
48. Tomchuk, O.V.; Avdeev, M.V.; Bulavin, L.A. Modeling fractal aggregates of polydisperse particles with tunable dimension. *Colloids Surf. A* **2020**, *605*, 125331. [CrossRef]

49. Eggersdorfer, M.L.; Pratsinis, S.E. The structure of agglomerates consisting of polydisperse particles. *Aerosol Sci. Technol.* **2012**, *46*, 347–353. [CrossRef]
50. Meakin, P.; Jullien, R. Structural readjustment effects in cluster-cluster aggregation. *J. Phys.* **1985**, *46*, 1543–1552. [CrossRef]
51. Torquato, S.; Haslach, H.W., Jr. Random Heterogeneous Materials: Microstructure and Macroscopic Properties. *Appl. Mech. Rev.* **2002**, *55*, B62–B63. [CrossRef]
52. Keblinski, P.; Prasher, R.; Eapen, J. Thermal conductance of nanofluids: Is the controversy over? *J. Nanopart. Res.* **2008**, *10*, 1089–1097. [CrossRef]
53. Eapen, J.; Rusconi, R.; Piazza, R.; Yip, S. The Classical Nature of Thermal Conduction in Nanofluids. *J. Heat Transf.* **2010**, *132*, 102402. [CrossRef]
54. Bourantas, G.C.; Skouras, E.D.; Loukopoulos, V.C.; Burganos, V.N. Heat transfer and natural convection of nanofluids in porous media. *Eur. J. Mech. B Fluids* **2014**, *43*, 45–56. [CrossRef]
55. Prasher, R.; Evans, W.; Meakin, P.; Fish, J.; Phelan, P.; Keblinski, P. Effect of aggregation on thermal conduction in colloidal nanofluids. *Appl. Phys. Lett.* **2006**, *89*, 143119. [CrossRef]
56. Karthikeyan, N.R.; Philip, J.; Raj, B. Effect of clustering on the thermal conductivity of nanofluids. *Mater. Chem. Phys.* **2008**, *109*, 50–55. [CrossRef]
57. Vinod, S.; Philip, J. Role of field-induced nanostructures, zippering and size polydispersity on effective thermal transport in magnetic fluids without significant viscosity enhancement. *J. Magn. Magn. Mater.* **2017**, *444*, 29–42. [CrossRef]

nanomaterials

MDPI

Article

Comparisons of Numerical and Experimental Investigations of the Thermal Performance of Al_2O_3 and TiO_2 Nanofluids in a Compact Plate Heat Exchanger

Wagd Ajeeb and S M Sohel Murshed *

IDMEC, Department of Mechanical Engineering, Instituto Superior Técnico, University of Lisbon, 1049-001 Lisbon, Portugal
* Correspondence: smurshed@tecnico.ulisboa.pt

Abstract: This study reports the thermal performance of Al_2O_3 and TiO_2 nanofluids (NFs) flowing inside a compact plate heat exchanger (CPHE) by comparing the experimental and numerical investigations. The NF samples were prepared for five concentrations each of Al_2O_3 and TiO_2 nanoparticles dispersed in distilled water (DW) as a base fluid (BF). The stability of NF samples was ensured, and their viscosity and thermal conductivity were measured. Firstly, the experimental measurements were performed for the heat transfer and fluid flow of the NFs in the plate heat exchanger (PHE) system and then the numerical investigation method was developed for the same PHE dimensions and operation conditions of the experimental investigation. A finite volume method (FVM) and single-phase fluid were used for numerical modelling. The obtained experimental and numerical results show that the thermal performance of the CPHE enhances by adding nanoparticles to the BFs. Furthermore, numerical predictions present lower values of convection heat transfer coefficients than the experimental measurements with a maximum deviation of 12% at the highest flow rate. Nevertheless, the numerical model is suitable with acceptable accuracy for the prediction of NFs through PHE and it becomes better for relatively small particles' concentrations and low flow rates.

Keywords: plate heat exchanger; heat transfer; Al_2O_3 and TiO_2 nanofluids; numerical model

Citation: Ajeeb, W.; Murshed, S.M.S. Comparisons of Numerical and Experimental Investigations of the Thermal Performance of Al_2O_3 and TiO_2 Nanofluids in a Compact Plate Heat Exchanger. *Nanomaterials* **2022**, *12*, 3634. https://doi.org/10.3390/nano12203634

Academic Editors: Henrich Frielinghaus and Rajinder Pal

Received: 18 September 2022
Accepted: 13 October 2022
Published: 17 October 2022

1. Introduction

In the last decades, a strong trend in the industry was shown toward miniaturization and natural resource management, mainly the energy and materials sources (enhancing the energy efficiency of the systems and reducing the equipment sizes are considered important roads to natural resource management reduce the demand for energy sources and minerals). In this, heat transfer systems are widely spread in industry applications, and the development of heat exchangers continues. The latter gave rise to advanced heat exchangers called "compact heat exchangers" involving compact plate heat exchangers (CPHEs) that contain channels with relatively small mean hydraulic diameters, presenting miniature dimensions but with higher heat transfer effectiveness [1,2]. The unique design of plate heat exchangers (PHEs) consists of several plates separating two different cold and hot fluids providing large heat transfer surfaces between them [3]. So far, several configuration types of PHEs were developed for enhancing the heat transfer effectiveness according to the requirements of industrial applications such as aircraft, electronics, chemicals, and other applications that contain cooling and heating equipment. The wavy shape of the plates in the chevron PHEs causes turbulent fluid flow inside the channels even for low Reynolds numbers (*Re*) [4], thus offering better heat transfer effectiveness compared to other normal heat exchangers [5]. However, micro/mini channels/passages that are used in the compact heat exchanger models provide bigger heat transfer surfaces but also higher pressure drops that need higher pumping power in the system [6]. Moreover, high heat loads are presented in those compact heat exchangers which require an innovative method for intensively absorbing the heat from the surfaces. For this purpose, recent research in the field of thermal management systems, namely in compact heat exchangers, indicates to NFs as superior thermal fluids for enhancing the intensification of the heat transfer method [7]. Various types of nanoparticles have been used to produce NFs by mixing them with conventional heat transfer fluids, presenting various characteristics [8,9]. Besides the developed thermal conduction property of the NFs [10–12], there is an increase in the viscosity levels too, which is not preferable for heat exchangers due to the higher pumping power required [10–12]. Nevertheless, NFs were recommended to be used for compact heat exchangers to improve their thermal characteristics [13]. Moreover, a considerable recommendation was given to Al_2O_3 nanoparticles for the preparation of NFs due to achieving good dispersion and improvements in the heat transfer effectiveness [14–16]. An empirical and numerical research by Awais et al. [17] for Al_2O_3 NFs flow in a heat sink heat exchanger reported a good improvement of 17% to the thermal performance of the heat sink. Moreover, Choi et al. [18] reported a 6.9% improvement in the thermal performance of the radiator by using Al_2O_3 NFs as a coolant instead of the BF for a high-level power system. Another study by Huang et al. [19] studied the performance of PHE with Al_2O_3 and MWCNT NFs and an inconsiderable heat transfer improvement for both NFs was reported in comparison with BF, i.e., water. Furthermore, an empirical study by Mare et al. [20] presented a better cooling performance for the used CNTs NF than Al_2O_3 NF for the fluid flow through PHE.

So far, numerical examination methods such as computational fluid dynamic (CFD) tools showed good flexibility and a big advantage to be used for studying the heat transfer characteristics of heat exchangers by different numerical modelling methods [21,22]. Yet, numerical modelling has been used to simulate the performance of NFs for heat transfer of NFs flow through uniform mini-channels [23] and micro-channels [24], and it showed respectable agreement with experimental measurements for several conditions such as non-Newtonian rheology behavior for the NFs [25]. In addition, different computational methods were presented in the literature to investigate the behavior of NF flows for different applications such as solar energy systems, electronics, and automotive [26,27]. Ahmed et al. [28] numerically tested the thermal performance of Cu NFs in the isothermally corrugated channel and a considerable heat transfer upgrading was reported by using NF instead of water for *Re* between 100 and 1000. Other researchers have numerically tested the performance of other types of NFs. For example, Shirzad et al. [29] investigated Al_2O_3, CuO and TiO_2 NFs in PHE for *Re* between 1000 and 8000. In their investigation, while Al_2O_3 NF had the best heat transfer values for low *Re*, TiO_2 NF had the better performance in heat transfer for high *Re*. Bahiraei et al. [30] numerically tested the flow of Al_2O_3 NFs in micro PHE and different shapes of particles were used at 1.0 vol.%. concentration and *Re* of 500. Platelet-shaped Al_2O_3 particles presented the best heat transfer rates. There is a numerical research study conducted by Tiwari et al. [31] on PHE works with CeO_2 and Al_2O_3 NFs as homogeneous fluids using the CFD tools (ANSYS-FLUENT). The numerical predictions were well matched with the experimental measurements and better performance was found for CeO_2 NFs as coolants. Generally, preparation and stability influence the thermal performance in any heat transfer systems, such as PHE [7] and heat pipes [32].

In addition, the literature shows a single-phase numerical technique as a common method to simulate the NF's behavior in heat exchangers and a good agreement is usually presented when numerical predictions are compared with experimental measurements [7]. Nevertheless, reported numerical investigations on NFs for compact heat exchangers were not supported by enough validation and comparison with experimental measurements [33,34], and the thermophysical properties of the used NFs are mostly obtained theoretically by using mixture laws and without studying the stability of NFs. Furthermore, the real mechanism behind the deviation between the numerical and experimental measurements was not identified, and the thermal behavior of the NFs in the flows was not explained. Therefore, the current study intended to perform firstly a thorough experimental determination of the thermophysical properties of Al_3O_2 and TiO_2 NFs with low particle concentrations

(0.01–0.2 vol.%) to be suitably tailored in the numerical simulations through the corrugated channel of compact chevron PHE. Furthermore, a careful experimental investigation was carried out for the heat transfer of the NFs flows in the hot loop of a compact PHE system. The details of the numerical methodology (geometry dimensions, boundary conditions, flow rates, etc.) were determined based on the experimental investigation and the equivalent working conditions were applied for an accurate comparison between the numerical and experimental measurements.

2. Nanofluids Preparation and Properties Characterization

In this study, a two-step preparation method was followed to produce the Al_2O_3 and TiO_2 NFs samples. The primary sizes (diameter) of both Al_2O_3 and TiO_2 nanoparticle were <50 nm and 20 nm, respectively, and their purity was 99.5% (as provided by IoLiTec, Heilbronn, Germany). The nanoparticles were dispersed into the BF (DW) for the volumetric concentration of 0.01, 0.05, 0.1, 0.15, and 0.2 vol.% (equivalent to 0.04, 0.2, 0,4, 0.6, and 0.8% of the mass fractions). First, nanoparticles were precisely weighed using a KERN ABS 80-4N scale and then dispersed into the base fluid. The good dispersion of nanoparticles into the base fluid was obtained after employing a magnetic stirring process for 15 min in a first moment followed by an ultrasonication process for 25 min using a probe-type ultrasonicator (Hielscher UP200Ht) at an amplitude of 60%, power of 110 W and 40 kHz frequency, to improve the dispersion and stability of the nanoparticles into the fluids. Moreover, the viscosity and thermal conductivity (k) of the NFs were directly measured after the preparation to ensure very good stability during the measurements. The higher thermophysical properties of NFs, mainly the thermal conductivity and viscosity, are considered key factors that define the heat transfer characteristics when NFs are employed to flow through the heat exchangers channels/passages. Therefore, these thermophysical properties of the NF samples were carefully measured and evaluated in the following sections. On other hand, specific heat (C_p) and density of NFs do not change significantly at low particle concentrations, and mixture rules are widely used for the determination of these properties based on the volume fraction of particles (φ) into the BF. Thus, the mixture rules (e.g., [35,36]) which were applied to NFs are given by Equation (1) for density and Equation (2) for C_p, respectively,

$$\rho_{nf} = \varphi\rho_p + (1-\varphi)\rho_{bf} \tag{1}$$

$$c_{p,nf} = \frac{\varphi\left(\rho C_p\right)_p + (1-\varphi)\left(\rho C_p\right)_{bf}}{\rho_{nf}} \tag{2}$$

where the subscripts nf represent NF, p the particle, and bf the BF.

2.1. Thermal Conductivity

In this study, the thermal conductivity of the Al_2O_3 and TiO_2 NFs was tested at room temperature (20 °C) by the transient hot-wire technique which showed good reliability for NFs measurements [37]. Several tests were performed for each NF with an interval time of 20 min. The resulted thermal conductivity' values of the Al_2O_3 and TiO_2 NFs are given in Figure 1 for several particles' concentrations, and they show good enhancements up to 4.25% and 7.34% for TiO_2 NF and Al_2O_3 NF at 0.2% particles' concentration, respectively, compared to the BF. Moreover, Al_2O_3 NFs reported higher enhancement in the thermal conductivity values compared to TiO_2 NFs at several values of φ. This is understandable as the thermal conductivity of Al_2O_3 has several times (~5 times) larger than that of the TiO_2.

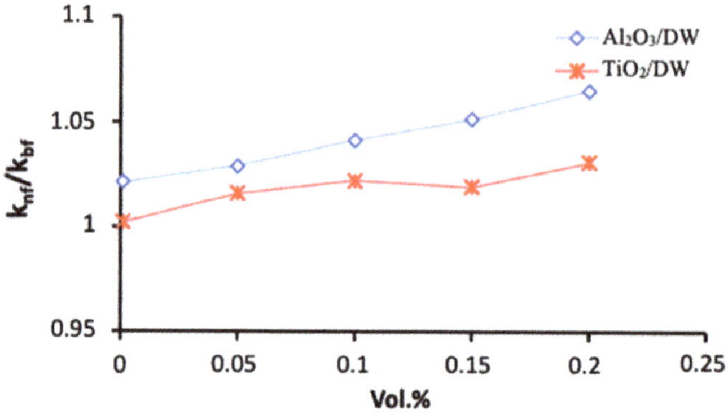

Figure 1. The enhancements of thermal conductivity of NFs as a function of volumetric concentrations of Al_2O_3 and TiO_2 nanoparticles.

2.2. Viscosity

The rheology of NFs is considered a unique parameter for understanding their hydraulic and thermal behavior in the flows through heat exchanges. Therefore, the viscosity and rheology of the NFs and BFs are determined using a rheometer (from Brookfield) with a thermostatic bath for several temperatures and shear rates. However, the viscosity values did not change with shear rate values for all the types of particles and their amounts into the BF, which indicates a Newtonian behavior for those NFs. The latter (Newtonian behavior) was also reported in the literature for similar NF samples [38,39].

On the other hand, the viscosity' values of Al_2O_3 and TiO_2 NFs are tested for several temperatures' values and φ. The resulted values are presented in Figure 2 show that the level of viscosity decreased by increasing the temperature value and raised by the increase in the particles' volume fraction for both types of NFs (Al_2O_3 and TiO_2 NFs), in similar behavior to most results in the literature [11,38].

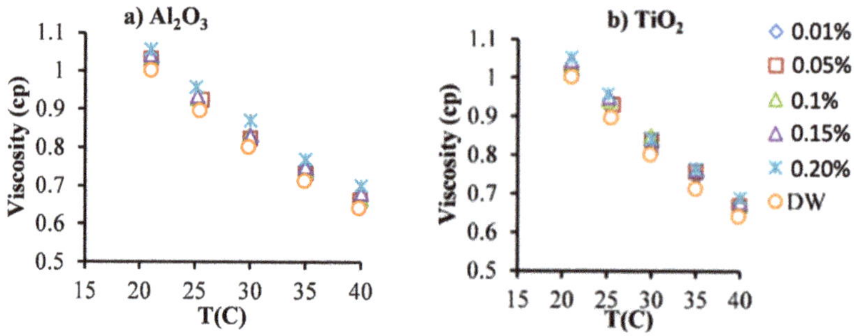

Figure 2. Viscosity of NFs as a function of temperature for two types of nanoparticles: (**a**) Al_2O_3, and (**b**) TiO_2.

It can be noticed a close values of viscosity for TiO_2 and Al_2O_3 at the same particles' concentrations and an increase of about 2.15% for 0.01 vol.% and of about 6.54% for 0.2 vol.% in comparison with the BF. Moreover, the decrease in the viscosity' value due to the rise in the temperature is significant: up to around 33.6% for 0.2 vol.% when the temperature was increased from the lowest value of 21 °C to the highest value of 40 °C. The latter findings (viscosity results) were anticipated and agreed with the data in the literature for similar NFs [11,38].

3. Experimental Heat Exchanger System and Methods

The experimental rig of the CPHE was established for the fluid flow and heat transfer of nanofluids as presented in Figure 3. The CPHE system contains an open loop for the cold fluid (using only water that goes to the drain after passing the CPHE) and a hot fluid loop for the NF flow, which includes a tank with heater, pump, 2 flow meters, differential pressure sensor, 4 thermocouples for temperature measurements, and DAQ linked to PC for collecting data. Details about this investigational setup and working principle can be found in an earlier study [40]. The components' accuracy of the investigational setup was firstly checked with DW as a well-known fluid and calibration procedure for the CPHE system was conducted [40]. A temperature of 40 °C was set for the NF at the inlet of the CPHE. The stabilization of the set flows and temperature were insured before recording the data by the data acquisition system considering 2 s interval time. In short, the fluids and NFs samples are heated in a tank then they flow into the hot loop passing the flowmeter and the CPHE to return to the tank. The temperatures of the fluid at the inlets and outlets of the CPHE are determined through four thermocouples and used to calculate the convection heat transfer coefficient (CHTC) for each sample at each flow rate.

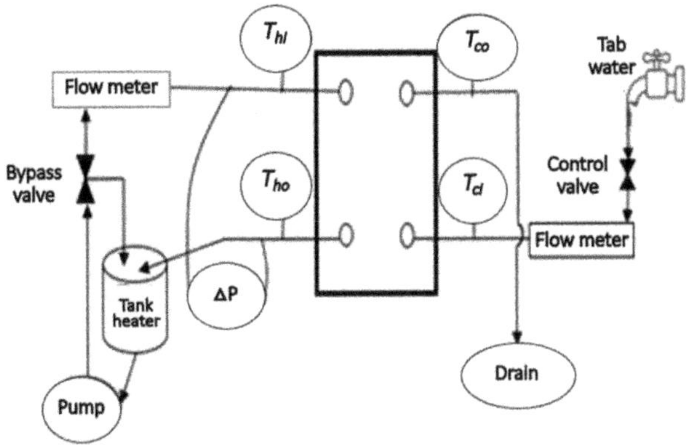

Figure 3. A flow diagram of the setup used for the experimental investigation.

The heat absorbed from the hot loop (Q_h) and moved to the cold loop (Q_c) is determined by Equations (3) and (4), and the average heat Q is assessed by Equation (5).

$$Q_h = \dot{m}_h C_{p,h}(T_{hi} - T_{ho}) \tag{3}$$

$$Q_c = \dot{m}_c C_{p,c}(T_{ci} - T_{co}) \tag{4}$$

$$Q = (Q_h + Q_c)/2 \tag{5}$$

T_{hi} and T_{ho} represent the temperatures values at the inlet and outlet of the hot loop of the PHE, respectively. In addition, T_{ci} and T_{co} represent the temperatures values at the inlet and outlet of the cold loop of the PHE, respectively. Moreover, the overall convection heat transfer coefficient (U) is found by Equations (6) and (7).

$$U = \frac{Q}{A \cdot LMTD} \tag{6}$$

A is the convection heat transfer area and $LMTD$ is the log mean temperature difference.

$$LMTD = \frac{(T_{ho} - T_{ci}) - (T_{hi} - T_{co})}{ln\frac{(T_{ho} - T_{ci})}{(T_{hi} - T_{co})}} \tag{7}$$

Then, the convection heat transfer coefficient (CHTC) for NFs in the hot loop (h_h) is defined by Equation (8):

$$U = \frac{1}{h_h} + \frac{\delta}{k_{pl}} + \frac{1}{h_c} \quad (8)$$

where δ is the thickness of the plate of the heat exchanger, k_{pl} represents the thermal conductivity of the plate's material. Furthermore, h_c is the CHTC for the water in the cold loop and it is theoretically predicted based on the heat exchanger design by Equation (9) [41] that has been used and validated in previous studies [42,43] for similar conditions.

$$Nu = 0.348Re^{0.663}pr^{0.33} \quad (9)$$

Then h_c is determined from the definition of Nu as given by Equation (10),

$$h_c = k \times Nu/D_h \quad (10)$$

D_h is the hydraulic diameter of the channel in the CPHE ($D_h = 2b$).

4. Numerical Modelling

The numerical investigation methodology was developed using the CFD tools of the ANSYS-FLUENT software package [44]. The numerical modelling was created based on the physics principles and the investigational procedure of the problem. The NFs samples were deemed single-phase fluids and the FVM method was employed. The previously determined (reported in the previous section) thermophysical properties of the NF samples were inserted into the numerical model corresponding to each NF sample. Moreover, a second-order upwind method is chosen for the convection and diffusing matters. The pressure and velocity of each fluid flow was linked by the SIMPLE method [45]. The simulations of the numerical approach are accomplished with achieving residual errors below 10^{-6} for the calculations of the governing equations.

4.1. Geometric Configuration and Boundary Conditions

The dimensions of the numerical domain were chosen to approach and focus as much as possible on the physics of the problem of flows and heat transfer of the NFs in the PHE. Therefore, only the loop that contains the NFs' flow was considered, while heat conditions of the cold loop (where the water is flowing at a constant flow rate during all the measurements) was obtained experimentally and applied in the numerical modelling into the boundary conditions of the NFs flows. The latter helps to avoid the possible numerical errors caused by the complex design of the PHE and reduces the time costs of the simulations by using a smaller number of nodes in the mesh of the numerical domain. Moreover, only one channel was chosen from the 5 hot channels in the hot loop for the simulations to increase the accuracy of the numerical investigations, mainly for NFs based on low particles concentrations. Therefore, the channel was designed to have similar dimensions to the channel of the PHE used in the investigational setup and the numerical domain was established as a two-dimensional (2D) corrugated channel with 0.278 m length (L_{ch}) and 2.4 mm distance between channel's plates (b) as it is presented in Figure 4. Moreover, the heat flux boundary conditions on the wall were found based on the measured data obtained for the cold loop, where the absorbed heat by the cold loop (Q_c) was found by Equation (4). Thus, the heat absorbance value for one channel (Q_{c1}) can be calculated ($Q_{c1} = Q_c/5$), considering the assumption of the current study of having equal flows and heat rates in the five channels of the cold loop, then the heat flux per unit of area on the wall of the channel can be found in Equation (11).

$$q = Q_c/A' \quad (11)$$

where A' is the convection heat transfer area for one plate channel.

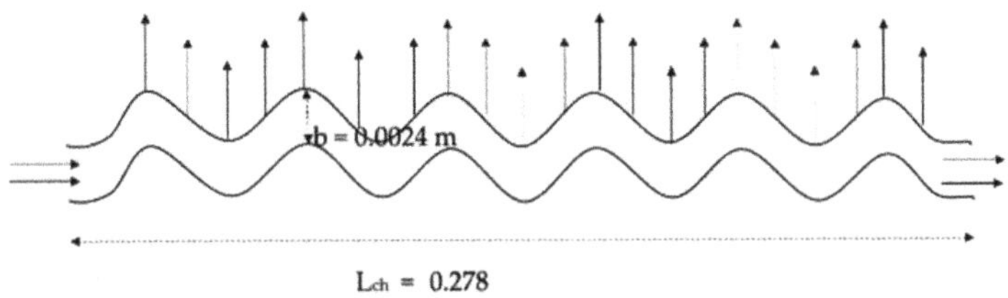

Figure 4. Schematic diagram of the numerical domain.

However, the numerical investigation methodology was established considering incompressible turbulent fluid flow. The thermophysical properties of the NFs and the BF were defined and used in the model. The inlet temperature (T_{in}) was considered at 40 °C similar to the experimental measurement conditions for the hot loop. Moreover, velocity (u) at the channel inlet was determined based on the measurement procedures for each flow rate (0.03–0.93 L/s).

4.2. Governing Equations and Calculation

In this study, the equations of Navier–Stokes and energy are used in the numerical investigation methodology as governing equations for running the simulations for the NFs flowing through the corrugated channel of PHE. The turbulent model called as Realizable K-ε found in the FLUENT-ANSYS package is adapted, where it is considered a developed form of K-ε turbulence model proposed by Shih et at. [46] applying a recent formulation for eddy-viscosity.

The dimensional governing equations for the current study conditions are as follows in Equations (12)–(14),

Continuity equation:

$$\nabla\cdot(\rho V) = 0 \tag{12}$$

Momentum equation:

$$\nabla\cdot(\rho \boldsymbol{V}\boldsymbol{V}) = -\nabla P + \nabla\cdot(\mu \nabla \boldsymbol{V}) \tag{13}$$

Energy equation:

$$\nabla\cdot(\rho C_p V T) = \nabla\cdot(k \nabla T) \tag{14}$$

where $\boldsymbol{V}$ represents the vector of velocity, P is the pressure, ρ is the density of the fluid and μ is the viscosity of the fluid. Moreover, the kinetic energy of turbulence (K) and the dissipation rate (ε) are attained by the following transport equations (Equations (15) and (16)):

$$\nabla\cdot(\rho V K) = \nabla\cdot\left[\left(\mu + \frac{\mu_t}{\sigma_K}\right)\nabla K\right] + G_K - \rho\varepsilon \tag{15}$$

$$\nabla\cdot(\rho V \varepsilon) = \nabla\cdot\left[\left(\mu + \frac{\mu_t}{\sigma_\varepsilon}\right)\nabla \varepsilon\right] + \rho C_{1\varepsilon} S\varepsilon - \rho C_{2\varepsilon}\frac{\varepsilon^2}{K + \sqrt{v\varepsilon}} \tag{16}$$

In these equations, G_k symbolizes the generation of turbulence kinetic energy term for the gradients in velocities. $C_{2\varepsilon}$ and $C_{1\varepsilon}$ are equation' constants. σ_K and σ_ε represent the turbulent Prandtl numbers for K and ε, respectively. S is the average strain rate [44]. Moreover, the value of the eddy viscosity of the used Realizable model is not constant, and it is calculated from the following Equation (17):

$$\mu_t = \rho\cdot C_\mu\cdot\frac{K^2}{\varepsilon} \tag{17}$$

where C_μ is a factor associated with the eddy viscosity. However, the model constants ($C_{2\varepsilon}$, σ_K, σ_ε, A_0 and A_s) have been formed to ensure the good performance of the numerical for the turbulent flow conditions. The mentioned constants of the numerical model are provided as:

$$C_{2\varepsilon} = 1.90,\ A_0 = 4.040,\ \sigma_K = 1.00,\ \sigma_\varepsilon = 1.20 \text{ and } A_s = \sqrt{6}\cos\varphi.$$

However, the simulations were carried out for the BFs and the NFs for the various particle concentrations (0.01, 0.05, 0.1 and 0.15 and 0.2 vol.%) flowing through the corrugated channel of the hot loop in the PHE in the similar working conditions (flow rates, temperature value, and thermophysical properties) of the experimental measurements. The temperature outcomes of the different samples at the outlet of the corrugated channel (T_{ho}) were collected and used to calculate the corresponding CHTC at each flow. The heat removed from the hot loop (Q_h) is found by Equation (3). Meanwhile, the values of the temperatures of the channel in a cold loop and the heat transferred (T_{ci}, T_{co}, and Q_c) were already known from the experimental measurements. Then the mean heat transfer (Q) was calculated as in Equation (5). Moreover, the overall heat transfer coefficient (U) was found by Equations (6) and (7). Then, the CHTC for the BFs and NFs in the hot loop (h_h) can be determined by Equation (8). Where the h_c is the CHTC for the water in the cold loop that was defined in the experimental investigation.

4.3. Mesh Optimisation and Validation

The independency of the mesh was ensured by testing several meshes by increasing the number of the nodes and comparing it with experimental measurements for water as it has a well-known hydraulic-thermal behavior. The tested meshes were 112621 (Mesh 1), 168441 (Mesh 2), 219282 (Mesh 3), and 303101 (Mesh 4): nodes were tested for several flow rates at the equivalent working conditions of the experimental investigation. The obtained numerical predictions are presented in Figure 5 for the CHTC and they show good accuracy for the mesh of 219282 nodes in comparison with the experimental measurements (Experimental data) with a deviation of around 6%. Moreover, it was found that the numerical predictions of CHTC were not changing for node numbers higher than 219282 nodes. Therefore, the mesh of 219282 nodes was selected for the intended simulations of this study to ensure the independency of the results from the mesh. A part of the mesh is presented in Figure 6, which shows the uniform and smooth structure of the mesh for accurate simulations of the turbulent flows.

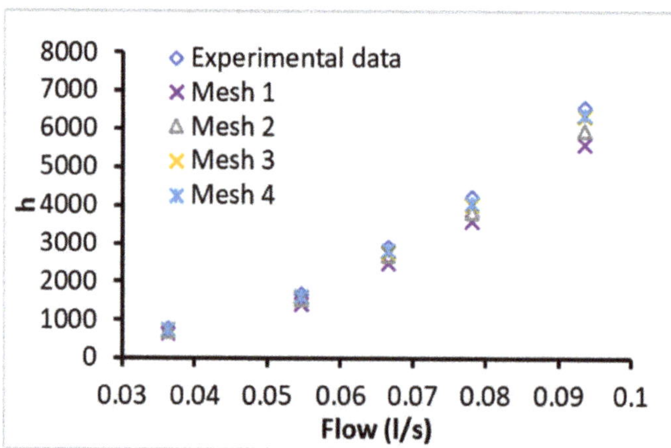

Figure 5. Mesh independency and validation for CHTC (h) of BF as a function of flow rate.

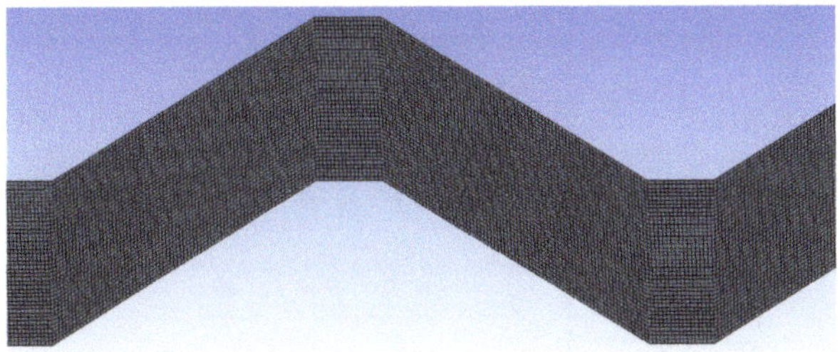

Figure 6. A view of a section of the mesh of the numerical domain.

Furthermore, the resulted profiles of the velocity and temperature fields, at the entrance and end of the channel of the PHE, are presented in Figure 7 which shows good consistency with the fluid flow characteristics under the PHE problem conditions. It can be noticed from Figure 7 that the values of the temperature at the entrance of the channel starts high and it becomes lower in the outlet area of the channel with lower values of temperature near the upper wall. Moreover, the velocity profile shows the boundary layer of the flow where there are lower values of velocity near the walls which make closely a parabolic profile of the velocity in the channel.

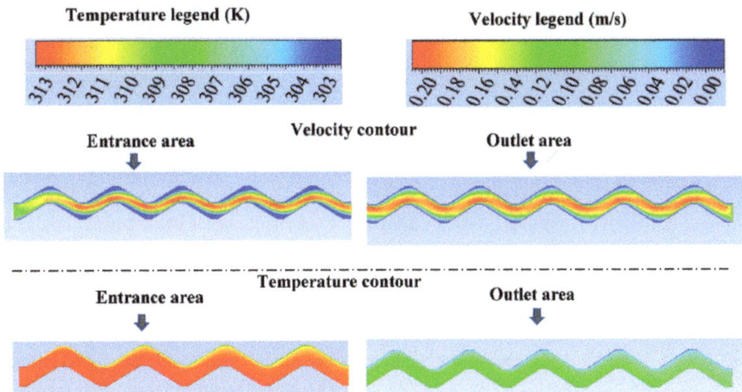

Figure 7. The numerical section of study (test) with the temperature and velocity profiles at the entrance and outlet of the channel.

5. Heat Transfer Results and Discussion

The experimental and numerical results of the heat transfer performance of two types of NFs (Al_2O_3 and TiO_2 NFs) flowing through the hot loop of the CPHE are presented in Figure 8. The heat transfer enhancements of the NFs are determined in comparison with the corresponding BFs. The outcomes show good improvement with the increase in loading of nanoparticles for all flow rates (Figure 8), and better thermal performance for Al_2O_3 compared to TiO_2 NFs for a deviation of 9% at 2.0 vol.% particles that decreases with reducing the concentration of the particles. The maximum average enhancement of about 24.6% is observed for the highest particle concentration (0.2 vol.%) through the experimental measurements (Figure 8) of Al_2O_3 NF. It was anticipated as the enhanced thermal conductivity of nanofluids was found to increase with increasing the amount of nanoparticles into the BF. Moreover, it is due to the fact that Al_2O_3 NFs exhibit higher thermal conductivity compared to TiO_2 NFs (as presented in Figure 1).

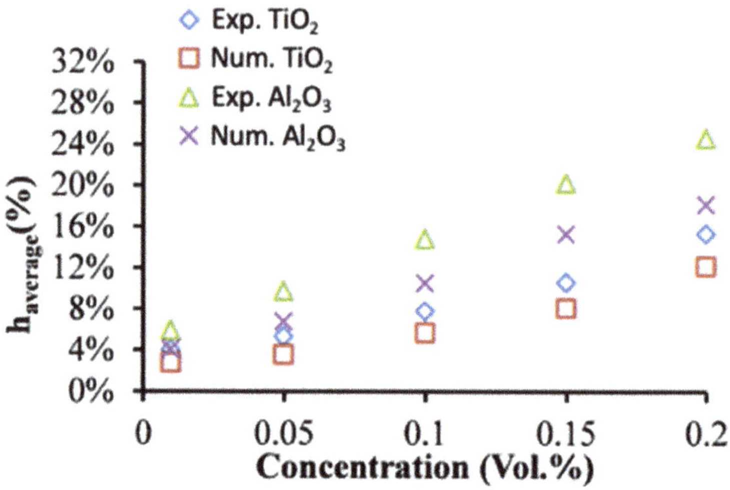

Figure 8. Enhancement of average heat transfer coefficient of DW-based Al_2O_3 and TiO_2 NFs obtained experimentally and numerically as a function of nanoparticle concentration.

Therefore, the numerical findings (Num. in Figure 8) and experimental measurements (Exp. in Figure 8) confirm that the dispersion of nanoparticles characterized by superior thermal conduction such as Al_2O_3 and TiO_2 particles with a conventional heat transfer fluid (e.g., DW in this study) enhances the heat transfer performance, due to the greater thermal conductivity of the NFs. As presented in Figure 8, an increase in nanoparticles' concentrations led to higher enhancement value in CHTC. It should be also noted that the temperature and velocity profiles and their boundary layer developments into the flow regime are influenced by the advanced properties of the NFs performing better heat transfer between the two loops of the PHE, especially in the conditions of the current study where the NFs are operating in the hot loop under high temperature that leads to having higher thermal conduction characteristic and lower viscosity values of the NFs. The latter explains the excellent heat transfer improvement that reached around 24.6% for 0.2 vol.% of Al_2O_3 and 15.3% for 0.2 vol.% of TiO_2. On other hand, the random movement of the nanoparticle in the turbulent flow regime and the possible migration of the nanoparticles into the flow inside the PHE (as appeared in the experimental investigations) can cause further development in the heat transfer rates [47]. These mentioned factors can also be the reasons for the higher enhancements of the CHTCs for Al_2O_3 and TiO_2 NFs than the enhancements of their thermal conductivity (e.g., Figure 1). Several relevant studies have also highlighted that the improved convection heat transfer for the laminar flow of Al_2O_3 NFs through a horizontal tube mainly due to the migration phenomena of nanoparticles into the flow [48]. The latter is proved by the current study when the experimental results were compared with the numerical results where the impact of the nanoparticles' movements and the migration phenomena are not considered for the numerical approach (nanofluids are simulated as single-phase fluids) but it exists in the experimental investigation. However, the advanced heat transfer performance of Al_2O_3 NFs for different types of heat exchangers was widely reported in the literature [12,49,50] as well as for TiO_2 NFs [51,52] in agreement with the finding of the current study. Furthermore, Tiwari et al. [42] found in their experimental investigation of CeO_2, Al_2O_3, TiO_2, and SiO_2 NFs for gasketed PHE that there was heat transfer boost for all the particles' types with maximum values for CeO_2 NFs, whereas the Al_2O_3 NFs showed better heat transfer enhancement than TiO_2 NFs. Nevertheless, the findings and discussion can vary among different researchers even for the same NF type because of many parameters related to the concentration and morphology of the particles,

BF, flow rate, the operation temperature value, and the method of preparing the NFs and the type of heat exchanger and its dimensions.

On the other hand, the differences between the numerical and the experimental results are shown in Figure 9. The numerical results show relatively lower heat transfer enhancements in comparison with experimental data with a deviation between 1.0% and 3.3% for TiO_2 NFs and a deviation between 1.6% and 7.2% for Al_2O_3 NFs.

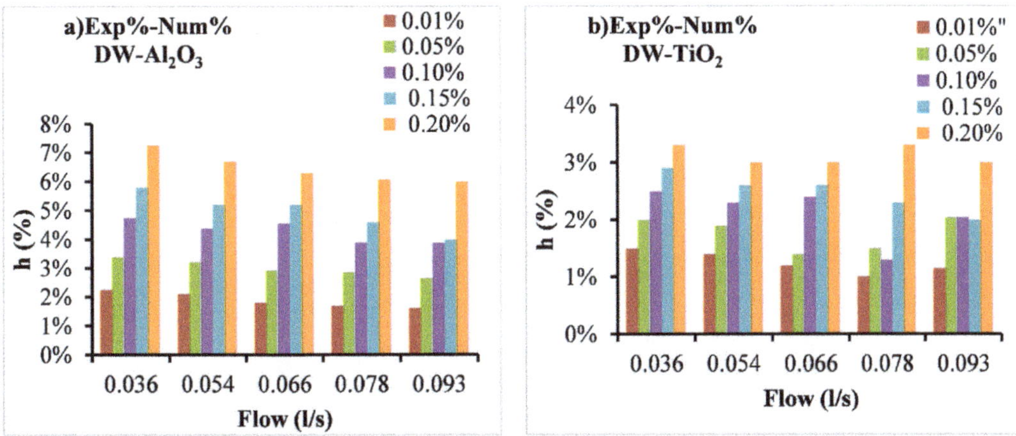

Figure 9. The deviation between experimental (Exp) and numerical (Num) findings of the enhancement of CHTC (h%) as a function of flow rate of (**a**) Al_2O_3 and (**b**) TiO_2 NFs.

Moreover, the deviation, in most cases, increases with the decrease in flow rates and the rise in particles concentrations (Figure 9). The latter refers to the existence of some factors in experimental investigations responsible for extra heat transfer enhancement than the ones in numerical investigations. Those factors are mainly related to the nanoparticle's movements in the flow which is not considered in the numerical investigation methodology. However, at higher flow rates the hydraulic impact of the flow on convection heat transfer performance becomes higher than the impact of the thermal conductivity increases in the NFs which led to a slightly lower impact of the nanoparticle's movements on heat transfer. On other hand, the increase in the deviation between the numerical and experimental data refers to the increase in the impact of nanoparticles' movements with increasing the concentration due to their influence on hydraulic and thermal boundary layer development leading to higher heat transfer levels. However, some previous studies have conducted both numerical and experimental examinations on the NFs through PHE, such as a study by Pantzali et al. [53] for CuO NFs in a miniature PHE and they mentioned better heat transfer levels at low flow rates reaching the overall heat transfer enhancement of about 29.41%. Their numerical results were in good agreement with the experimental results, demonstrating CFD as a reliable tool for investigating NFs in PHE. Moreover, Bhattad et al. [54] have numerically and empirically examined the behavior of hybrid NF (Al_2O_3 + MWCNT/water) in PHE and reported an increase in CHTC of 39.16%. Their numerical predictions were in good agreement with the experimental results with a smaller deviation. The latter conclusion is also agreed with the findings of a numerical study by Tiwari et al. [31] using CeO_2 and Al_2O_3 NFs in PHE. Therefore, the numerical results of the current study showed significant advantages to define the parameters that cannot be determined when only the experimental methods are used. The parameters such as the nanoparticles movements and their impact on the fluid flow and heat transfer through the channels of PHE can't be defined by the traditional experimental investigation methods. The numerical results of the current study allowed to isolate (comparing with the experimental data of heat transfer) those important parameters that are considered responsible for the extra enhancement of the heat transfer.

6. Conclusions

In this study, numerical simulations and experimental measurements on the flow and heat transfer performance of Al_2O_3 and TiO_2 NFs in a compact PHE are carried out. The PHE system was established and verified for the operation of the flows of two fluids (cold and hot) in separated loops. The Al_2O_3 and TiO_2 NFs are prepared and their thermophysical properties such as thermal conductivity and viscosity are measured. The improvements in thermal conductivity caused by adding different concentrations of nanoparticles are presented, and maximum enhancements of about 7.30% for Al_2O_3 NF and 4.20% for TiO_2 NF at 0.2 vol.% concentration were found. Additionally, the viscosity was found to increase by increasing concentration of nanoparticles for both NFs, and there was a decrease with increasing the temperature. The numerical methodology is developed using the CFD tools of the ANSYS-FLUENT software for the similar physical conditions of the experimental method.

The heat transfer investigations on the Al_2O_3 and TiO_2 NFs through the hot loop of the compact plate heat exchanger were conducted for several flow rates at an inlet temperature of 40 °C, and the heat transfer characteristics were empirically determined. The experimental and numerical data obtained for both NFs regarding CHTC enhancements were compared for several flows. Good enhancements of the heat transfer were found for both NFs and it was found to increase with the concentration of particles for all the flow rates. However, Al_2O_3 NFs showed better enhancement compared to TiO_2 NFs. The maximum enhancement of heat transfer (24.6%) was observed for the Al_2O_3 NF at the highest particle concentration (0.2 vol.%) through the experimental measurements.

Moreover, the numerical results show lower heat transfer enhancements in comparison with the experimental measurements with a deviation between 1.0% and 3.3% for TiO_2 NFs and a deviation between 1.6% and 7.2% for Al_2O_3 NFs. The deviation was changed based on the particles' concentration and the flow rate. The latter is presumed due to the nanoparticles' movements in the flow which is not considered in the numerical investigation and led to extra heat transfer enhancement.

Finally, the experimental and numerical findings of flows of Al_2O_3 and TiO_2 NFs showed good heat transfer enhancements in the compact PHE. This study helps for a better understanding of the heat transfer performance and mechanisms of the NFs' behavior through such PHE as well as highlights the benefits of using CFD tools for modelling NFs with clarification of their thermal characteristics.

Author Contributions: Conceptualization, W.A. and S.M.S.M.; formal analysis, W.A. and S.M.S.M.; investigation, W.A.; writing—original draft preparation, W.A.; funding acquisition, S.M.S.M.; supervision, S.M.S.M.; writing—review and editing, S.M.S.M. All authors have read and agreed to the published version of the manuscript.

Funding: This research was funded by Fundação para a Ciência e a Tecnologia (FCT) through project PTDC/NAN- MAT/29989/2017.

Institutional Review Board Statement: Not applicable.

Informed Consent Statement: Not applicable.

Acknowledgments: This work has also been supported by Fundação para a Ciência e a Tecnologia (FCT) through IDMEC under LAETA, project UIDB/50022/2020.

Conflicts of Interest: The authors declare no conflict of interest.

Nomenclature

Al_2O_3	Alumina
b	Distance between the channels' plates (mm)
BF	Base Fluid
C_p	Specific heat of the fluid (J/kg·K)
CFD	Computational Fluid Dynamic
CPHE	Compact plate heat exchanger
D_h	Hydraulic diameter of the channel (m)
DW	Distilled Water
h	Convection Heat transfer coefficient ($W/m^2{\cdot}K$)
CHTC	Convection Heat transfer coefficient ($W/m^2{\cdot}K$)
k	Thermal conductivity of the fluid (W/m·K)
LMTD	Log mean temperature difference
$\dot{m}$	Mass flow (kg/s)
NF	Nanofluid
PHE	plate heat exchanger
Q	Heat rate (W)
T	Temperature (K)
U	Overall convection heat transfer coefficient ($W/m^2{\cdot}K$)
Greek symbols	
μ	The viscosity (mPa·s)
ρ	The density (kg/m^3)
δ	Plate thickness (mm)
φ	Volume fraction of particles
Subscripts	
bf	Base fluid
c	Cold
f	Fluid
h	Hot
i	Inlet
nf	Nanofluid
o	Outlet
p	Particle
pl	Plate

References

1. da Silva, R.P.P.; Mortean, M.V.V.; de Paiva, K.V.; Beckedorff, L.E.; Oliveira, J.L.G.; Brandão, F.G.; Monteiro, A.S.; Carvalho, C.S.; Oliveira, H.R.; Borges, D.G.; et al. Thermal and hydrodynamic analysis of a compact heat exchanger produced by additive manufacturing. *Appl. Therm. Eng.* **2021**, *193*, 116973. [CrossRef]
2. Abou Elmaaty, T.M.; Kabeel, A.E.; Mahgoub, M. Corrugated plate heat exchanger review. *Renew. Sustain. Energy Rev.* **2017**, *70*, 852–860. [CrossRef]
3. Mota, F.A.S.; Carvalho, E.P.; Ravagnani, M.A.S.S. Modeling and Design of Plate Heat Exchanger. In *Heat Transfer Studies and Applications*, 1st ed.; Open: Rijeka, Croatia, 2015. [CrossRef]
4. Jafari, A.; Sadeghianjahromi, A.; Wang, C.C. Experimental and numerical investigation of brazed plate heat exchangers—A new approach. *Appl. Therm. Eng.* **2022**, *200*, 117694. [CrossRef]
5. Zhang, J.; Zhu, X.; Mondejar, M.E.; Haglind, F. A review of heat transfer enhancement techniques in plate heat exchangers. *Renew. Sustain. Energy Rev.* **2019**, *101*, 305–328. [CrossRef]
6. Liu, H.; Guo, H.; Xie, Z.; Sang, L. Numerical investigations for optimizing a novel micro-channel sink with perforated baffles and perforated walls. *Int. Commun. Heat Mass Transf.* **2021**, *126*, 105342. [CrossRef]
7. Ajeeb, W.; Murshed, S.M.S. Nanofluids in Compact Heat exchangers for Thermal Applications: A State-of-the-Art Review. *Therm. Sci. Eng. Prog.* **2022**, *30*, 101276. [CrossRef]
8. Murshed, S.M.S.; Leong, K.C.; Yang, C. Thermophysical and electrokinetic properties of nanofluids—A critical review. *Appl. Therm. Eng.* **2008**, *28*, 2109–2125. [CrossRef]
9. Roque, D.; Ajeeb, W.; Murshed, S.M.S.; Pereira, J.M.C. Forced Convection Heat Transfer Characteristics of Al_2O_3 Nanofluids in a Minichannel- An Experimental Study. *J. Phys. Conf. Ser. 8th Eur. Therm. Sci. Conf.* **2021**, *2116*, 012056. [CrossRef]
10. Ajeeb, W.; Silva, R.R.S.T.; Murshed, S.M.S. Rheology and Thermal Conductivity of Three Metallic Oxides Nanofluids. In Proceedings of the 7th World Congress on Momentum, Heat and Mass Transfer (MHMT'22), Lisbon, Portugal, 7–9 April 2022; pp. 7–10.

11. Murshed, S.M.S.; Estellé, P. A state of the art review on viscosity of nanofluids. *Renew. Sustain. Energy Rev.* **2017**, *76*, 1134–1152. [CrossRef]
12. Ajeeb, W.; Roque, D.; Murshed, S.M.S. Convective Heat Transfer Characteristics of Al_2O_3 Nanofluid in Minitube. In Proceedings of the 5th–6th Thermal and Fluids Engineering Conference (TFEC), New Orleans, LA, USA, 26–28 May 2021; ASTFE American Society of Thermal and Fluids Engineers: New Orleans, LA, USA, 2021; pp. 1181–1190.
13. Ajeeb, W.; Murshed, S.M.S. An Overview of Performance and Application of Nanofluids in Compact Heat Exchangers. In Proceedings of the 5th–6th Thermal and Fluids Engineering Conference (TFEC), New Orleans, LA, USA, 26–28 May 2021; ASTFE American Society of Thermal and Fluids Engineers: New Orleans, LA, USA, 2020; pp. 1171–1180.
14. Nikulin, A.; Moita, A.S.; Moreira, A.L.N.; Murshed, S.M.S.; Huminic, A.; Grosu, Y.; Faik, A.; Nieto-Maestre, J.; Khliyeva, O. Effect of Al_2O_3 nanoparticles on laminar and transient flow of isopropyl alcohol. *Int. J. Heat Mass Transf.* **2019**, *130*, 1032–1044. [CrossRef]
15. Elbadawy, I.; Fayed, M. Reliability of Al_2O_3 nanofluid concentration on the heat transfer augmentation and resizing for single and double stack microchannels. *Alexandria Eng. J.* **2020**, *59*, 1771–1785. [CrossRef]
16. Ajeeb, W.; Roque, D.; Murshed, S.M.S. Thermal-Hydraulic Performance of Al_2O_3 Nanofluids under Laminar Flow in a Minitube. *Int. J. Energy a Clean Environ.* **2022**, *23*, 39–59. [CrossRef]
17. Awais, A.A.; Kim, M.H. Experimental and numerical study on the performance of a minichannel heat sink with different header geometries using nanofluids. *Appl. Therm. Eng.* **2020**, *171*, 115125. [CrossRef]
18. Choi, T.J.; Kim, S.H.; Jang, S.P.; Yang, D.J.; Byeon, Y.M. Heat transfer enhancement of a radiator with mass-producing nanofluids (EG/water-based Al_2O_3 nanofluids) for cooling a 100 kW high power system. *Appl. Therm. Eng.* **2020**, *180*, 115780. [CrossRef]
19. Huang, D.; Wu, Z.; Sunden, B. Pressure drop and convective heat transfer of Al_2O_3/water and MWCNT/water nanofluids in a chevron plate heat exchanger. *Int. J. Heat Mass Transf.* **2015**, *89*, 620–626. [CrossRef]
20. Maré, T.; Halelfadl, S.; Sow, O.; Estellé, P.; Duret, S.; Bazantay, F. Comparison of the thermal performances of two nanofluids at low temperature in a plate heat exchanger. *Exp. Therm. Fluid Sci.* **2011**, *35*, 1535–1543. [CrossRef]
21. Gao, J.; Hu, Z.; Yang, Q.; Liang, X.; Wu, H. Fluid flow and heat transfer in microchannel heat sinks: Modelling review and recent progress. *Therm. Sci. Eng. Prog.* **2022**, *29*, 101203. [CrossRef]
22. Ajeeb, W.; Oliveira, M.S.A.; Martins, N.; Murshed, S.M.S. Numerical Approach for Fluids flow and Thermal Convection in Microchannels. *J. Phys. Conf. Ser.* **2021**, *2116*, 012049. [CrossRef]
23. Ajeeb, W.; Oliveira, M.S.A.; Martins, N.; Abreu, B. Experimental and Numerical Study of Convective Heat Transfer and Laminar Flow of a MWCNTs Nanofluid in a Horizontal Tube. *J. Nanofluid* **2019**, *8*, 132–142. [CrossRef]
24. Ajeeb, W.; Oliveira, M.S.A.; Martins, N.; Murshed, S.M.S. Forced convection heat transfer of non-Newtonian MWCNTs nanofluids in microchannels under laminar flow. *Int. Commun. Heat Mass Transf.* **2021**, *127*, 105495. [CrossRef]
25. Ajeeb, W.; Oliveira, M.S.A.; Martins, N.; Murshed, S.M.S. Performance evaluation of convective heat transfer and laminar flow of non-Newtonian MWCNTs in a circular tube. *Therm. Sci. Eng. Prog.* **2021**, *25*, 101029. [CrossRef]
26. Mahian, O.; Kolsi, L.; Amani, M.; Estellé, P.; Ahmadi, G.; Kleinstreuer, C.; Marshall, J.S.; Taylor, R.A.; Abu-Nada, E.; Rashidi, S.; et al. Recent advances in modeling and simulation of nanofluid flows—Part II: Applications. *Phys. Rep.* **2019**, *791*, 1–59. [CrossRef]
27. Ajeeb, W.; Murshed, S.M.S. Numerical Study of Convective Heat Transfer Performance, entropy generation and energy efficiency of Al and Al_2O_3 Nanofluids in Minichannel. *J. Nanofluids* **2022**, *in press*. [CrossRef]
28. Ahmed, M.A.; Shuaib, N.H.; Yusoff, M.Z.; Al-Falahi, A.H. Numerical investigations of flow and heat transfer enhancement in a corrugated channel using nanofluid. *Int. Commun. Heat Mass Transf.* **2011**, *38*, 1368–1375. [CrossRef]
29. Shirzad, M.; Soheil, S.; Ajarostaghi, M.; Aghajani, M.; Sedighi, K. Improve the thermal performance of the pillow plate heat exchanger by using nanofluid: Numerical simulation. *Adv. Powder Technol.* **2019**, *30*, 1356–1365. [CrossRef]
30. Bahiraei, M.; Monavari, A. Thermohydraulic characteristics of a micro plate heat exchanger operated with nano fluid considering different nanoparticle shapes. *Appl. Therm. Eng.* **2020**, *179*, 115621. [CrossRef]
31. Tiwari, A.K.; Ghosh, P.; Sarkar, J.; Dahiya, H.; Parekh, J. Numerical investigation of heat transfer and fluid flow in plate heat exchanger using nano fluids. *Int. J. Therm. Sci.* **2014**, *85*, 93–103. [CrossRef]
32. Cacua, K.; Buitrago-Sierra, R.; Herrera, B.; Pabón, E.; Murshed, S.M. Nanofluids' stability effects on the thermal performance of heat pipes: A critical review. *J. Therm. Anal. Calorim.* **2019**, *136*, 1597–1614. [CrossRef]
33. Halelfadl, S.; Adham, A.M.; Mohd-Ghazali, N.; Maré, T.; Estellé, P.; Ahmad, R. Optimization of thermal performances and pressure drop of rectangular microchannel heat sink using aqueous carbon nanotubes based nanofluid. *Appl. Therm. Eng.* **2014**, *62*, 492–499. [CrossRef]
34. Awais, M.; Ullah, N.; Ahmad, J.; Sikandar, F.; Ehsan, M.M.; Salehin, S.; Bhuiyan, A.A. Heat transfer and pressure drop performance of Nanofluid: A state-of- the-art review. *Int. J. Thermofluids* **2021**, *9*, 100065. [CrossRef]
35. Pak, B.C.; Cho, Y.I. Hydrodynamic and heat transfer study of dispersed fluids with submicron metallic oxide particles. *Exp. Heat Transf.* **1998**, *11*, 151–170. [CrossRef]
36. Xuan, Y.; Roetzel, W. Conceptions for heat transfer correlation of nanofluids. *Int. J. Heat Mass Transf.* **2000**, *43*, 3701–3707. [CrossRef]
37. Murshed, S.M.S.; Leong, K.C.; Yang, C. Enhanced thermal conductivity of TiO_2-Water based nanofluids. *Int. J. Therm. Sci.* **2005**, *44*, 367–373. [CrossRef]

38. Murshed, S.M.S.; Santos, F.J.V.; Nieto de Castro, C.A. Investigations of viscosity of silicone oil-based semiconductor nanofluids. *J. Nanofluids* **2013**, *2*, 261–266. [CrossRef]
39. Abdullah, A.M.; Chowdhury, A.R.; Yang, Y.; Vasquez, H.; Moore, H.J.; Parsons, J.G.; Lozano, K.; Gutierrez, J.J.; Martirosyan, K.S.; Uddin, M.J. Tailoring the viscosity of water and ethylene glycol based TiO_2 nanofluids. *J. Mol. Liq.* **2020**, *297*, 111982. [CrossRef]
40. Ajeeb, W.; da Silva, R.R.T.; Murshed, S.S. Experimental investigation of heat transfer performance of Al_2O_3 nanofluids in a compact plate heat exchanger. *Appl. Therm. Eng.* **2022**, *218*, 119321. [CrossRef]
41. Kakaç, S.; Liu, H. *Heat Exchangers: Selection, Rating and Thermal Design*, 2nd ed.; CRC Press LLC: Boca Raton, FL, USA, 2002.
42. Tiwari, A.K.; Ghosh, P.; Sarkar, J. Performance comparison of the plate heat exchanger using different nanofluids. *Exp. Therm. Fluid Sci.* **2013**, *49*, 141–151. [CrossRef]
43. Behrangzade, A.; Mahdi, M. The effect of using nano-silver dispersed water based nanofluid as a passive method for energy efficiency enhancement in a plate heat exchanger. *Appl. Therm. Eng.* **2016**, *102*, 311–317. [CrossRef]
44. ANSYS. *FLUENT 18.2 User's Guide*; ANSYS: Canonsburg, PA, USA, 2017.
45. Patankar, S. *Numerical Heat Transfer and Fluid Flow*; CRC press: NewYork, NY, USA, 1980; ISBN 978-08911652240.
46. Shih, T.-H.; Liou, W.W.; Shabbir, A.; Yang, Z.; Zhu, J. A New—Eddy-Viscosity Model for High Reynolds Number Turbulent Flows—Model Development and Validation. *Comput. Fluids* **1995**, *24*, 227–238. [CrossRef]
47. Hussien, A.A.; Abdullah, M.Z.; Yusop, N.M.; Al-Kouz, W.; Mahmoudi, E.; Mehrali, M. Heat transfer and entropy generation abilities of MWCNTs/GNPs hybrid nanofluids in microtubes. *Entropy* **2019**, *21*, 480. [CrossRef]
48. Wen, D.; Ding, Y. Experimental investigation into convective heat transfer of nanofluids at the entrance region under laminar flow conditions. *Int. J. Heat Mass Transf.* **2004**, *47*, 5181–5188. [CrossRef]
49. Murshed, S.M.S.; Nieto de Castro, C.A. Conduction and convection heat transfer characteristics of ethylene glycol based nanofluids—A review. *Appl. Energy* **2016**, *184*, 681–695. [CrossRef]
50. Kong, M.; Lee, S. Performance evaluation of Al_2O_3 nanofluid as an enhanced heat transfer fluid. *Adv. Mech. Eng.* **2020**, *12*, 1–13. [CrossRef]
51. Murshed, S.M.S.; Leong, K.C.; Yang, C.; Nguyen, N.T. Convective heat transfer characteristics of aqueous TiO_2 nanofluid under laminar flow conditions. *Int. J. Nanosci.* **2008**, *7*, 325–331. [CrossRef]
52. Shahrul, I.M.; Mahbubul, I.M.; Saidur, R.; Khaleduzzaman, S.S.; Sabri, M.F.M.; Rahman, M.M. Effectiveness study of a shell and tube heat exchanger operated with nanofluids at different mass flow rates. *Numer. Heat Transf. Part A Appl.* **2014**, *65*, 699–713. [CrossRef]
53. Pantzali, M.N.; Kanaris, A.G.; Antoniadis, K.D.; Mouza, A.A.; Paras, S.V. Effect of nanofluids on the performance of a miniature plate heat exchanger with modulated surface. *Int. J. Heat Fluid Flow* **2009**, *30*, 691–699. [CrossRef]
54. Bhattad, A.; Sarkar, J.; Ghosh, P. Discrete phase numerical model and experimental study of hybrid nano fluid heat transfer and pressure drop in plate heat exchanger. *Int. Commun. Heat Mass Transf.* **2018**, *91*, 262–273. [CrossRef]

nanomaterials

MDPI

Article

Numerical Study of Natural Convection Heat Transfer in a Porous Annulus Filled with a Cu-Nanofluid

Lingyun Zhang [1,2], Yupeng Hu [2,*] and Minghai Li [2,*]

1 School of Aerospace Engineering, Beijing Institute of Technology, Beijing 100081, China; zhanglingyun17@gscaep.ac.cn
2 Institute of Systems Engineering, China Academy of Engineering Physics, Mianyang 621999, China
* Correspondence: hjhuyp@caep.cn (Y.H.); 3120170782@bit.edu.cn (M.L.); Tel.: +86-0816-2494908 (Y.H.); +86-0816-2484335 (M.L.)

Abstract: Natural convection heat transfer in a porous annulus filled with a Cu nanofluid has been investigated numerically. The Darcy–Brinkman and the energy transport equations are employed to describe the nanofluid motion and the heat transfer in the porous medium. Numerical results including the isotherms, streamlines, and heat transfer rate are obtained under the following parameters: Brownian motion, Rayleigh number (10^3–10^5), Darcy number (10^{-4}–10^{-2}), nanoparticle volume fraction (0.01–0.09), nanoparticle diameter (10–90 nm), porosity (0.1–0.9), and radius ratio (1.1–10). Results show that Brownian motion should be considered. The nanoparticle volume fraction has a positive effect on the heat transfer rate, especially with high Rayleigh number and Darcy number, while the nanoparticle diameter has an inverse influence. The heat transfer rate is enhanced with the increase of porosity. The radius ratio has a significant influence on the isotherms, streamlines, and heat transfer rate, and the rate is greatly enhanced with the increase of radius ratio.

Keywords: natural convection; heat transfer; water-based nanofluid; Brownian motion; porous medium; Darcy–Brinkman equation; numerical simulation

Citation: Zhang, L.; Hu, Y.; Li, M. Numerical Study of Natural Convection Heat Transfer in a Porous Annulus Filled with a Cu-Nanofluid. *Nanomaterials* **2021**, *11*, 990. https://doi.org/10.3390/nano11040990

Academic Editor: S M Sohel Murshed

Received: 26 March 2021
Accepted: 8 April 2021
Published: 12 April 2021

1. Introduction

Fluid flow and heat transfer due to natural convection in porous annulus is one of the most considerable research issues due to its wide applications in science and engineering [1–3], such as thermal insulators, chemical catalytic convectors, thermal storage systems, geothermal energy utilization, electronic cooling, and nuclear reactor systems. Therefore, natural convection in porous annulus has been extensively investigated during the past decades. Caltagirone [4] was the first to study natural convection in a saturated porous medium bounded by two concentric, horizontal, isothermal cylinders experimentally and numerically. The author used the Christiansen effect in order to visualize a fluctuating three-dimensional thermal field for Rayleigh number exceeding some critical value. Rao et al. [5,6] performed steady and transient investigations on natural convection in a horizontal porous annulus heated from the inner face using Galerkin method. The effects of Rayleigh number and Darcy number on heat transfer characteristics were studied. In addition, the bifurcation point was obtained numerically, which compared very well with that from experimental observation.

Himasekhar [7] examined the two-dimensional bifurcation phenomena in thermal convection in horizontal, concentric annuli containing saturated porous media. The fluid motion is described by the Darcy–Oberbeck–Boussinesq equations, which were solved using regular perturbation expansion. The flow structure was obtained under different parameters, such as the radius ratio and Rayleigh–Darcy number. A parametric study was performed by Leong et al. [8] to investigate the effects of Rayleigh number, Darcy number, porous sleeve thickness, and relative thermal conductivity on heat transfer characteristics. Braga et al. [9] presented numerical computations for laminar and turbulent natural convection within a horizontal cylindrical annulus filled with a fluid saturated porous medium. Computations covered the range $25 < Ra_m < 500$ and $3.2 \times 10^{-4} > Da > 3.2 \times 10^{-6}$ and made use of the finite volume method. Khanafer et al. [10] carried out a numerical simulation in order to examine the parametric effects of Rayleigh number and radius ratio on the role played by natural convection heat transfer in the porous annuli. The model was governed by Darcy–Oberbeck–Boussinesq equations and solved using the Galerkin method. In order to investigate the buoyancy-induced flow as affected by the presence of the porous layer, Alloui and Vasseur [11] studied natural convection in a horizontal annular porous layer filled with a binary fluid under the influence of the Soret effect using the Darcy model with the Boussinesq approximation. Numerical solutions of the full governing equations are obtained for a wide range of the governing parameters, such as Rayleigh number, Lewis number, buoyancy ratio, radius ratio of the cavity, and normalized porosity.

Belabid and Cheddadi [12] solved natural convection heat transfer within a two-dimensional horizontal annulus filled with a saturated porous medium using ADI (Alternating Direction Implicit) finite difference method. This work placed emphasis on the mesh effect on the determination of the bifurcation point between monocellular and bicellular flows for different values of the aspect ratio. In a recent work, Belabid and Allali [13,14] studied the effects of a periodic gravitational and temperature modulation on the convective instability in a horizontal porous annulus. Results showed that the convective instability is influenced by the amplitude and the frequency of the modulation. Rostami et al. [15] provided a review of recent natural convection studies, including experimental and numerical studies. The effects of the parameters, such as nanoparticle addition, magnetic fields, and porous medium on the natural convection were examined.

The traditional fluids in engineering, such as water and mineral oils, have a primary limitation in the enhancement of heat transfer due to a rather low thermal conductivity. The term nanofluids, which was first put forward by Choi [16], is used to describe the mixture of nanoparticles and base fluid. Due to the relatively higher thermal conductivities, nanofluids are considered as an effective approach to meet some challenges associated with the traditional fluids [17]. In the last few decades, studies on the natural convection of nanofluids in enclosure were conducted by a number of researchers [18]. For instance, Jou and Tzeng [19] investigated the numerically natural convection heat transfer enhancement utilizing nanofluids in a two-dimensional enclosure. Results showed that increasing the buoyancy parameter and volume fraction of nanofluids caused an increase in the average heat transfer coefficient. Ghasemi et al. [20] simulated the natural convection heat transfer in an inclined enclosure filled with a CuO–water nanofluid. The effects of pertinent parameters such as Rayleigh number, inclination angle, and solid volume fraction on the heat transfer characteristics were studied. The results indicated that the heat transfer rate is maximized at a specific inclination angle depending on Rayleigh number and solid volume fraction.

In a related work, Abu-Nada and Oztop [21] found that the effect of nanoparticles concentration on Nusselt number was more pronounced at low volume fraction than at high volume fraction and the inclination angle could be a control parameter for nanofluid filled enclosure. Soleimani et al. [22] studied the natural convection heat transfer in a semi-annulus enclosure filled with a Cu–water nanofluid using the Control Volume based Finite Element Method. The numerical investigation was carried out for different governing parameters, such as Rayleigh number, nanoparticle volume fraction, and angle of turn for the enclosure. The results revealed that there was an optimal angle of turn in which the average Nusselt number was maximum for each Rayleigh number. Seyyedi et al. [23] simulated the natural convection heat transfer of Cu–water nanofluid in an annulus enclosure using the Control Volume-based Finite Element Method. The Maxwell–Garnetts and Brinkman models were employed to estimate the effect of thermal conductivity and viscosity of nanofluid. The results showed the effects of the governing parameters on the local Nusselt number, average Nusselt number, streamlines, and isotherms. Boualit et al. [24] and Liao [25] studied respectively natural convection heat transfer of Cu- and

Al_2O_3-water nanofluids in a square enclosure under the horizontal temperature gradient. Both the flow structure and the corresponding heat transfer characteristics at different Rayleigh numbers and nanoparticle volume fractions were obtained. Wang et al. [26] investigated numerically the natural convection in a partially heated enclosure filled with Al_2O_3 nanofluids. The results indicated that at low Rayleigh numbers, the heat transfer performance increased with nanoparticle volume fraction, while at high Rayleigh numbers, there existed an optimal volume fraction at which the heat transfer performance had a peak. In a recent work, Mi et al. [27] examined the effects of graphene nano-sheets (GNs) nanoparticles by comparing the thermal conductivity of graphene nano-sheets (GNs)/ethylene glycol (EG) nanofluid with EG thermal conductivity. Results showed that the presence of nanoparticles improved the thermal conductivity, and with increasing temperature, the effect of adding GNs was strengthened.

Although several numerical and experimental studies on the natural convection heat transfer were published, most of them concentrated on traditional fluids in cavities, and only a few of them consider a nanofluid in a porous annulus [28–30]. In the present work, steady natural convection heat transfer in a porous annulus filled with a Cu-nanofluid has been investigated, and the governing equations, including the Darcy–Brinkman equation, were solved using the Galerkin method. This paper presented a systematical examination on the effects of Brownian motion, solid volume fraction, nanoparticle diameter, Rayleigh number, Darcy number, porosity on the flow pattern, temperature distribution, and heat transfer characteristics. To the best of our knowledge, no study on this problem has been considered before, and accordingly, the current paper will address this topic.

2. Problem Formulation

2.1. Physical Description

We consider porous annulus filled with a Cu–water nanofluid between a horizontal inner and outer cylinder of radius r_i and r_o, respectively, as shown in Figure 1. The inner and outer cylinders are kept at uniform high temperature T_i and low temperature T_o, respectively. It is taken into consideration that the flow is two-dimensional, steady, and laminar due to the low velocity. The porous medium is considered as isotropic, homogeneous, and filled with a nanofluid, which is thermal equilibrium with the solid matrix, and the Darcy–Brinkman equation without inertia item is adopted. For the nanofluid, the effect of Brownian motion [31–33] is considered. The thermophysical properties of the base water, copper nanoparticles, and solid structure of the porous medium used in this study are presented in Tables 1 and 2 [34,35].

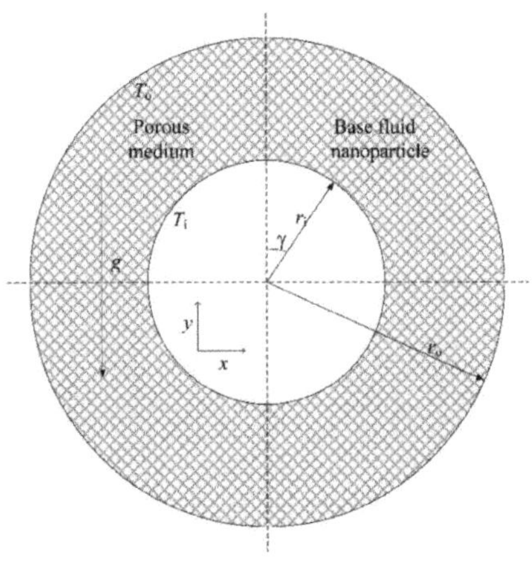

Figure 1. Sketch of problem geometry.

Table 1. Thermal physical properties of the base fluid (water), nanoparticle (Cu), and solid of the porous medium (glass balls).

Physical Properties	Base Fluid (Water)	Nanoparticle (Cu)	Porous (Glass Balls)
ρ [kg/m^3]	997.1	8933	2700
c_p [J/(kg·K)]	4179	385	840
k [W/(m·K)]	0.613	76.5	1.05
μ [kg/(m·s)]	0.001003	-	-
$\beta \times 10^5$ [1/K]	21	1.67	0.9

Table 2. Applied models for thermophysical properties of the nanofluid with or without Brownian motion.

Physical Properties	Applied Model	
	Without Brownian Motion	With Brownian Motion
k [W/(m·K)]	$\frac{k_{nf}}{k_{bf}} = \frac{k_{sp}+2k_{bf}-2\varphi(k_{bf}-k_{sp})}{k_{sp}+2k_{bf}+\varphi(k_{bf}-k_{sp})}$	$k_{nf} = k_{bf}\frac{k_{sp}+2k_{bf}-2\varphi(k_{bf}-k_{sp})}{k_{sp}+2k_{bf}+\varphi(k_{bf}-k_{sp})} + C\left(\rho c_p\right)_{nf}\frac{2k_b T_o}{\pi\mu_{nf}{d_{sp}}^2}\varphi d_{sp}$
μ [kg/(m·s)]	$\mu_{nf} = \mu_{bf}/(1-\phi)^{2.5}$	$\mu_{nf} = \mu_{bf}/[1 - 34.87(d_{sp}/d_{bf})^{-0.3}\phi^{1.03}]$;
ρ [kg/m^3]	$(\rho c_p)_{nf} = (1-\phi)(\rho c_p)_{bf} + \phi((\rho\ c_p)_{sp}$	
c_p [J/(kg·K)]	$\rho_{nf}(T) = (1-\phi)\rho_{bf}(T) + \phi\rho_{sp}$;	

$C = 3.6 \times 10^4$; $k_b = 1.38065 \times 10^{-23}$ J K^{-1}, $d_{bf} = 0.1[6M/(N\pi\rho_m)]^{1/3}$: $M = 0.018$ kg mol^{-1}, $N = 6.022 \times 10^{23}$ mol^{-1}, ρ, (ρc_p) and μ denote the density, heat capacitance, and dynamic viscosity, respectively, k is the thermal conductivity, ϕ is the nanoparticle volume fraction, and the subscripts bf, sp, and nf designate the base fluid, nanoparticle, and nanofluid.

2.2. Governing Equations and Boundary Conditions

By introducing the Boussinesq approximation, the governing equations for the heat transfer and fluid flow can be written as follows:

$$\frac{\partial u}{\partial x} + \frac{\partial v}{\partial y} = 0 \tag{1}$$

$$\left(u\frac{\partial u}{\partial x} + v\frac{\partial u}{\partial y}\right) = -\varepsilon\frac{1}{\rho_{nf}}\frac{\partial p}{\partial x} + \varepsilon\nu_{nf}\left(\frac{\partial^2 u}{\partial x^2} + \frac{\partial^2 u}{\partial y^2}\right) - \varepsilon^2\frac{\nu_{nf}}{K}u \tag{2}$$

$$\left(u\frac{\partial v}{\partial x} + v\frac{\partial v}{\partial y}\right) = -\varepsilon\frac{1}{\rho_{nf}}\frac{\partial p}{\partial y} + \varepsilon\nu_{nf}\left(\frac{\partial^2 v}{\partial x^2} + \frac{\partial^2 v}{\partial y^2}\right) - \varepsilon^2\frac{\nu_{nf}}{K}v + \varepsilon^2 g\beta(T - T_o) \tag{3}$$

$$(\rho c_p)_{nf}\left(u\frac{\partial T}{\partial x} + v\frac{\partial T}{\partial y}\right) = k_{mnf}\left(\frac{\partial^2 T}{\partial x^2} + \frac{\partial^2 T}{\partial y^2}\right) \tag{4}$$

where (x, y) are the Cartesian coordinates of the geometry, (u, v) are the velocity components, T and p are the temperature and pressure, respectively, ρ, (ρc_p), ν, β and α denote the density, heat capacitance, kinematic viscosity, thermal expansion coefficient, and thermal diffusion, respectively, k is the thermal conductivity, ε is the porosity, and K is the medium permeability. The subscripts nf and mnf designate nanofluid and porous medium filled with nanofluid.

The effective thermal conductivity of the porous medium filled with nanofluid can be as modeled as:

$$k_{mnf} = \varepsilon k_{nf} + (1-\varepsilon)k_s. \tag{5}$$

Using the following dimensional variables,
$(X,Y) = \frac{(x,y)}{(r_o - r_i)}$, $(U,V) = \frac{(u,v)(r_o-r_i)}{\alpha_{mnf}}$, $P = \frac{p(r_o-r_i)^2}{\rho_{nf}{\alpha_{mnf}}^2}$, $\theta = \frac{T-T_o}{T_i-T_o}$, $\alpha_{mnf} = \frac{k_{mnf}}{(\rho c_p)_{nf}}$, $Ra = \frac{g\beta(T_i-T_o)(r_o-r_i)^3}{v_{nf}\alpha_{mnf}}$, $Pr = \frac{v_{nf}}{\alpha_{mnf}}$, $Da = \frac{k}{L^2}$.

The governing Equations (1)–(4) reduce to a dimensionless form:

$$\frac{\partial U}{\partial X} + \frac{\partial U}{\partial Y} = 0 \tag{6}$$

$$U\frac{\partial U}{\partial X} + V\frac{\partial U}{\partial Y} = -\varepsilon\frac{\partial P}{\partial X} + \varepsilon Pr\left(\frac{\partial^2 U}{\partial X^2} + \frac{\partial^2 U}{\partial Y^2}\right) - \varepsilon^2\frac{Pr}{Da}U \tag{7}$$

$$U\frac{\partial V}{\partial X} + V\frac{\partial V}{\partial Y} = -\varepsilon\frac{\partial P}{\partial X} + \varepsilon Pr\left(\frac{\partial^2 V}{\partial X^2} + \frac{\partial^2 V}{\partial Y^2}\right) - \varepsilon^2\frac{Pr}{Da}U + \varepsilon^2 RaPr\theta \tag{8}$$

$$U\frac{\partial \theta}{\partial X} + V\frac{\partial \theta}{\partial Y} = \frac{1}{Pr}\left(\frac{\partial^2 \theta}{\partial X^2} + \frac{\partial^2 \theta}{\partial Y^2}\right) \tag{9}$$

where (X, Y) are the dimensionless Cartesian coordinates of the geometry, (U, V) are the dimensionless velocity components, and θ and P are the temperature and pressure, respectively. Ra, Pr, and Da denote respectively the Raleigh number, Prandtl number, and Darcy number.

The boundary conditions for this problem are as follows:

On the outer cylinder surface:

$$\theta = 0. \tag{10}$$

On the inner cylinder surface:

$$\theta = 1. \tag{11}$$

On the outer and inner cylinder surfaces:

$$U = V = 0. \tag{12}$$

The local and overall heat transfer rate along the inner cylinder surface are estimated using the local Nusselt number and average Nusselt number, respectively:

$$Nu_{\text{loc}} = \left.\frac{\partial \theta}{\partial N}\right|_S = \sqrt{\left(\frac{\partial \theta}{\partial X}\right)^2 + \left(\frac{\partial \theta}{\partial Y}\right)^2} \tag{13}$$

$$Nu_{\text{avg}} = \frac{1}{S}\int_0^S Nu_{\text{loc}} \tag{14}$$

where S is the non-dimensional length along the inner cylinder surface.

3. Numerical Procedure

The dimensionless governing Equations (6)–(9) together with the boundary conditions (10) and (11) have been solved numerically using the commercial software tool, which is known as COMSOL Multiphysics (Version 5.5, COMSOL Inc., Stockholm, Sweden). The software employs the Galerkin finite element method, which enforces the orthogonality of residuals to all basis functions in a basis. In Galerkin formulation, weighting functions are chosen to become identical to basis functions [36]. In this paper, we have employed a segregated and parallel direct (Pardiso) solver to solve those equations. As convergence criteria, 10^{-6} has been chosen for all dependent variables.

3.1. Grid Generation and Independence Test

In the finite element method, grid generation is the technique to discretize the computational domain into subdomains. In this study, we have adopted unstructured triangular elements in the interior and structured quadrilateral elements on the boundary. Figure 2 is the grid generation of the structure with a legend of quality measure. A quality of 1 represents the best possible grid quality, and a quality of 0 represents the worst possible grid quality. For purpose of ensuring the grid independence of the numerical solution, different grid levels in the Comsol Multiphysics are examined. As shown in Table 3, the average Nusselt number of on the inner cylinder at different grids is presented for Cu nanofluid when the nanoparticle volume fraction (ϕ) is 0.5, nanoparticle diameter (d_{sp}) is 50 nm, Rayleigh number (Ra) is 10^5, porosity (ε) is 0.5, and Darcy number (Da) is 10^{-2}. The difference between normal and extremely fine is within 1.35%. By comprehensive considering the calculation accuracy and cost, the extra fine level is chosen in this study.

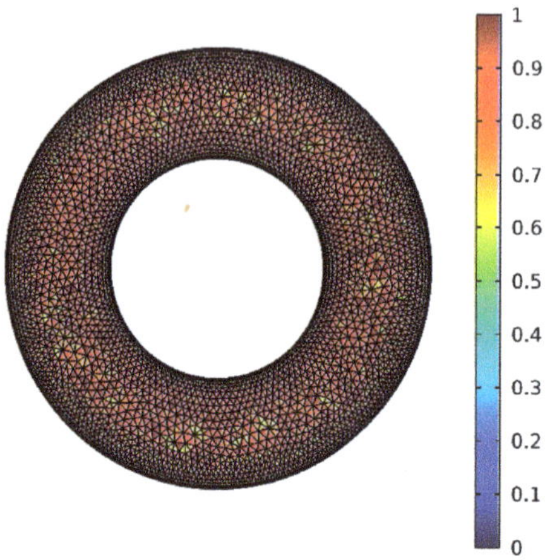

Figure 2. Grid generation of the structure with a legend of quality measure.

Table 3. Comparison of the average Nusselt number for Cu nanofluid at different levels when $\phi = 0.5$, $d_{sp} = 50$ nm, $Ra = 10^5$, $\varepsilon = 0.5$, and $Da = 10^{-2}$.

Level	Number of Elements	Minimum Quality	Average Quality	Nu_{avg}
normal	894	0.4632	0.7779	2.2466
fine	1344	0.4938	0.8284	2.2632
finer	1882	0.5084	0.8305	2.2668
extra fine	6394	0.5276	0.8424	2.2766
extremely fine	17858	0.5053	0.8466	2.2769

3.2. Code Validation

Due to the lack of experimental data for conjugate heat transfer of nanofluid in an annulus filled with porous medium, we have compared our results with the numerical results of Abhishek Kumar Singh and Tanmay Basak et al. [37] for a square cavity filled with base fluid ($\phi = 0$). The comparisons are presented in Table 4, when $\varepsilon = 0.4$, and the deviations are 0%, 0%, and 0.84% with $Ra = 10^3$, 10^4, and 10^5, respectively. When $\varepsilon = 0.9$, the deviations are 0.5%, 1.2%, and 1.3% with $Ra = 10^3$, 10^4, and 10^5, respectively. It is clear that the current results are in good agreement with the earlier work, and the maximum deviation is 1.3%. In addition, it can be seen that the deviation is increased with the increase of Rayleigh number and porosity. The validation work has enhanced the confidence in the numerical solution of the current study.

Table 4. Comparison of the average Nusselt number with those of Singh et al. [37] for different Rayleigh number when $\varphi = 0$, $\varepsilon = 0.4, 0.9$, $Da = 10^{-2}$, and $Pr = 1$.

Darcy Number (*Da*)	Reyleigh Number (*Ra*)	Average Nusselt Number (Nu_{avg})					
		$\varepsilon = 0.4$			$\varepsilon = 0.9$		
		Ref [37]	Present Work	Diff(%)	Ref [37]	Present Work	Diff(%)
10^{-2}	10^5	2.983	3.008	0.84	3.91	3.95	1.3
	10^4	1.408	1.408	0	1.64	1.66	1.2
	10^3	1.01	1.01	0	1.023	1.028	0.5

4. Results and Discussion

In this section, numerical simulations are carried out to investigate the flow and heat transfer characteristics of the nanofluid filled in the porous annulus. The results present the effects of several parameters, such as Brownian motion, nanoparticle diameter d_{sp} (10–90 nm), nanoparticle volume fraction ϕ (0.01–0.09), Rayleigh number Ra (10^3–10^5),

Darcy number Da (10^{-4}–10^{-2}), porosity ε (0.1–0.9), and radius ratio RR (1.1–10) on the isotherms and streamlines, the local Nusselt number (Nu_{loc}), and the average Nusselt number (Nu_{avg}).

4.1. Effects of Brownian Motion

In this section, we have investigated the effect of Brownian motion on the heat transfer characteristics using two applied models in Table 2. One ignores the Brownian motion of the nanoparticles. The other takes Brownian motion into consideration and the modified effective thermal conductivity and effective dynamic viscosity are adopted. Figure 3 presents the effect of Brownian motion on the average Nusselt number along the inner wall under different parameters. From Figure 3a, the average Nusselt number increases generally as Brownian motion is considered. With the increase of nanoparticle volume fraction, the influence of Brownian motion becomes more noticeable. In addition, with the decrease of nanoparticle diameter, the influence of Brownian motion becomes more remarkable. For instance, when $Ra = 5 \times 10^3$, $Da = 10^{-2}$, $d_{sp} = 90$ nm, Nu_{avg} with Brownian motion increased by 0.21% compared with that without Brownian motion at $\phi = 0.01$, while the growth rate is 2.7% at $\phi = 0.09$. When $Ra = 5 \times 10^3$, $Da = 10^{-2}$, $d_{sp} = 10$ nm, Nu_{avg} with Brownian motion increased by 1.98% compared with that without Brownian motion at $\phi = 0.01$, while the growth rate is 23.76% at $\phi = 0.09$. Comparing Figure 3a,b and Figure 3b with Figure 3c, the effect of Brownian motion becomes more remarkable with the increase of Rayleigh number and the decrease of Darcy number. For example, when $Da = 10^{-2}$, $\phi = 0.09$, $d_{sp} = 10$ nm, Nu_{avg} with Brownian motion increased by 23.76% compared with that without Brownian motion at $Ra = 5 \times 10^3$, while the growth rate is 35.92% at $Ra = 5 \times 10^4$. When $Ra = 5 \times 10^4$, $\phi = 0.09$, $d_{sp} = 10$ nm, Nu_{avg} with Brownian motion increased by 35.92% compared with that without Brownian motion at $Da = 10^{-2}$, while the growth rate is 46.29% at $Da = 10^{-3}$. Therefore, the effect of Brownian motion on the natural convection heat transfer of the nanofluid should be considered. Furthermore, the positive effect of Brownian motion on the overall heat transfer rate is different at different parameters.

4.2. Effects of Nanoparticle Volume Fraction

Figure 4 shows the isotherms and streamlines for different nanoparticle volume fraction and Rayleigh number at $Da = 10^{-2}$, $d_{sp} = 50$ nm, $\varepsilon = 0.5$, and $RR = 2$. The color scales on the left represent the dimensionless temperature and those on the right represent the dimensionless velocity, as in other sections of the article. From Figure 4, for $Ra = 10^3$, the isotherms have a uniform distribution; this is because the buoyancy force is weak compared with the viscous force, and it indicates that the heat transfer in the annulus is dominated by thermal conduction. The effect of volume fraction on the isotherms is weak. For $Ra = 10^4$, a slight thermal disturbance appeared, which indicates that the flow is enhanced and the transition from conduction to natural convection takes place. In this case, the effect of volume fraction becomes more important. For $Ra = 10^5$, the isotherms are almost horizontally distributed, especially when $\phi = 0.9$, which means that the natural convection heat transfer turns out to be more significant and the effect of volume fraction is more pronounced. With the increase of Rayleigh number, the streamlines become denser near the walls and the cell becomes bigger and has a tendency to move upward due to the enhanced buoyance force. In addition, for $Ra = 10^3$ and $Ra = 10^4$, the effect of volume fraction on the streamlines is very weak, while for $Ra = 10^5$, the effect is more pronounced. Figure 5 displays the effect of volume fraction on the overall heat transfer rate along the inner wall at different Rayleigh numbers. The figure shows that an increase in volume fraction leads to heat transfer enhancement for all considered Rayleigh numbers, and the effect of volume fraction is more pronounced when the Rayleigh number is high.

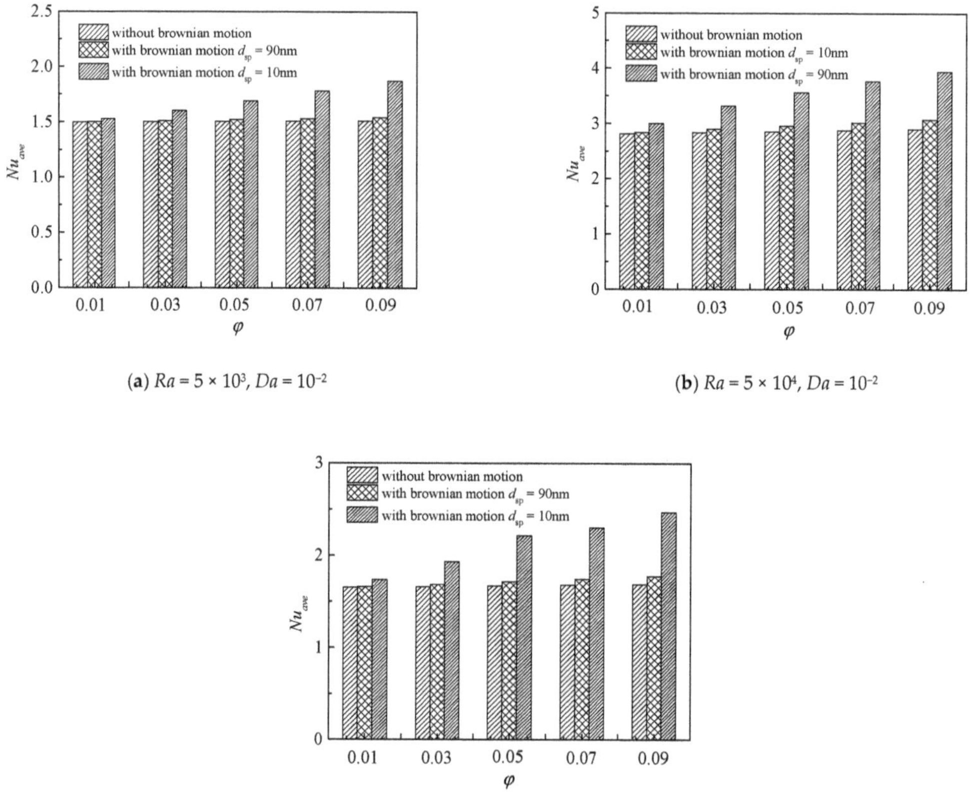

(**a**) $Ra = 5 \times 10^3$, $Da = 10^{-2}$ (**b**) $Ra = 5 \times 10^4$, $Da = 10^{-2}$

(**c**) $Ra = 5 \times 10^4$, $Da = 10^{-3}$

Figure 3. Effects of Brownian motion on the average Nusselt number (Nu_{avg}) along the inner wall for different nanoparticle volume fractions (φ) and nanoparticle diameters (d_{sp}) at $\varepsilon = 0.5$, $RR = 2$, and (**a**) $Ra = 5 \times 10^3$, $Da = 10^{-2}$, (**b**) $Ra = 5 \times 10^4$, $Da = 10^{-2}$, (**c**) $Ra = 5 \times 10^4$, $Da = 10^{-3}$.

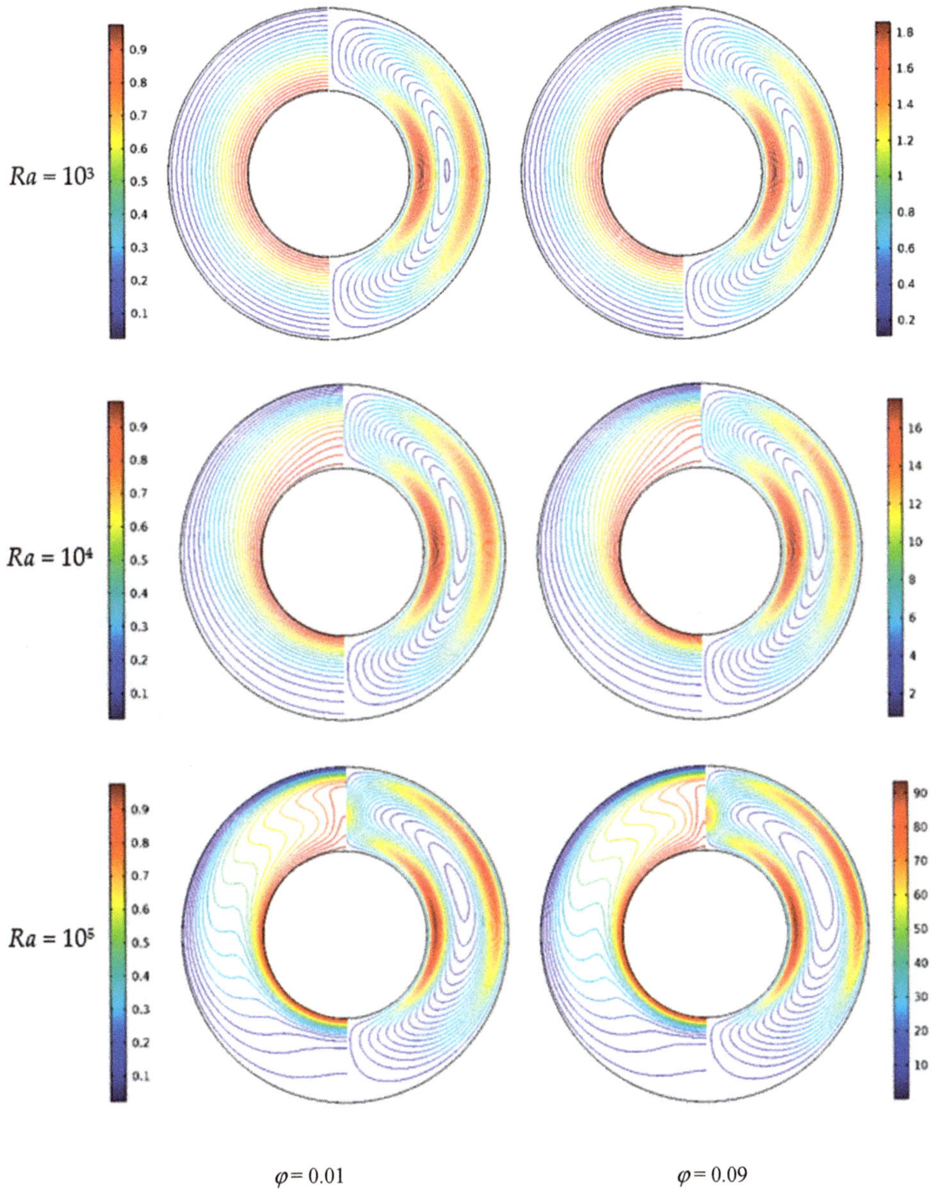

Figure 4. Isotherms (left) and streamlines (right) for Cu–water nanofluid for different nanoparticle volume fractions (ϕ) and Rayleigh numbers (Ra) at d_{sp} = 50 nm, Da = 0.01, ε = 0.5, and RR = 2.

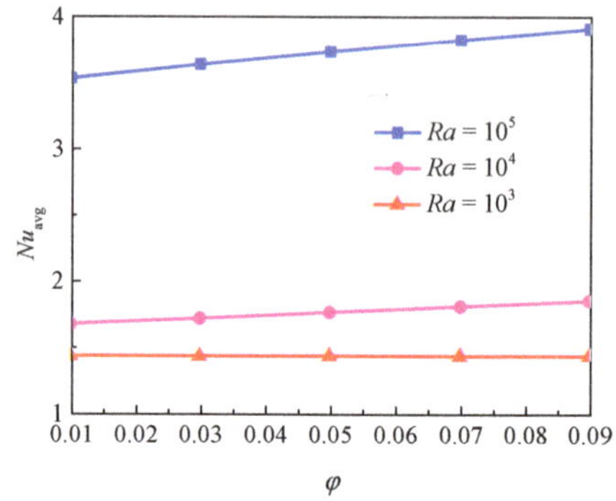

Figure 5. Average Nusselt number (Nu_{avg}) of the inner wall for different nanoparticle volume fractions (ϕ) and Rayleigh numbers (Ra) at $Da = 10^{-2}$, d_{sp} = 50 nm, $\varepsilon = 0.5$, and $RR = 2$.

Figure 6 illustrates the isotherms and streamlines for different nanoparticle volume fraction and Darcy number at d_{sp} = 50 nm, $Ra = 10^5$, $\varepsilon = 0.5$, and $RR = 2$. From Figure 6, for $Da = 10^{-4}$, the isotherms have a uniform distribution due to the low permeability, and it indicates that the heat transfer in the annulus is dominated by thermal conduction. The effect of volume fraction on the isotherms is slight. For $Da = 10^{-3}$, due to the enhanced permeability, the flow in the annulus is strengthened, and the transition from conduction to natural convection takes place. In this case, the effect of volume fraction becomes more important. For $Da = 10^{-2}$, the isotherms are almost horizontally distributed, especially when $\phi = 0.9$, which means the natural convection heat transfer plays a significant role and the effect of volume fraction is more pronounced. With the increase of Darcy number, the streamlines become denser near the walls and the cells become bigger and have a tendency to move upward due to the enhanced flow. In addition, for $Da = 10^{-4}$ and $Da = 10^{-3}$, the effect of volume fraction on the streamlines is very weak, while for $Da = 10^{-2}$, the effect is more pronounced. Figure 7 displays the effect of volume fraction on the overall heat transfer rate along the inner wall at different Darcy numbers. The figure shows that an increase in volume fraction leads to heat transfer enhancement for all considered Darcy numbers and the effect of volume fraction is increased with the increase of Darcy number.

Figure 8 presents the evolution of the local Nusselt number along the inner wall for different nanoparticle volume fractions at $Ra = 10^5$, $Da = 10^{-2}$, d_{sp} = 50 nm, $\varepsilon = 0.5$, and $RR = 2$. The increase of Nu_{loc} in the whole region means that the local heat transfer rate is enhanced. It can be found that the maximum Nu_{loc} occurred at $\gamma = 180°$, which means the natural convection heat transfer is more intense in the bottom half of the inner wall. In addition, the heat transfer is enhanced with the increase of volume fraction.

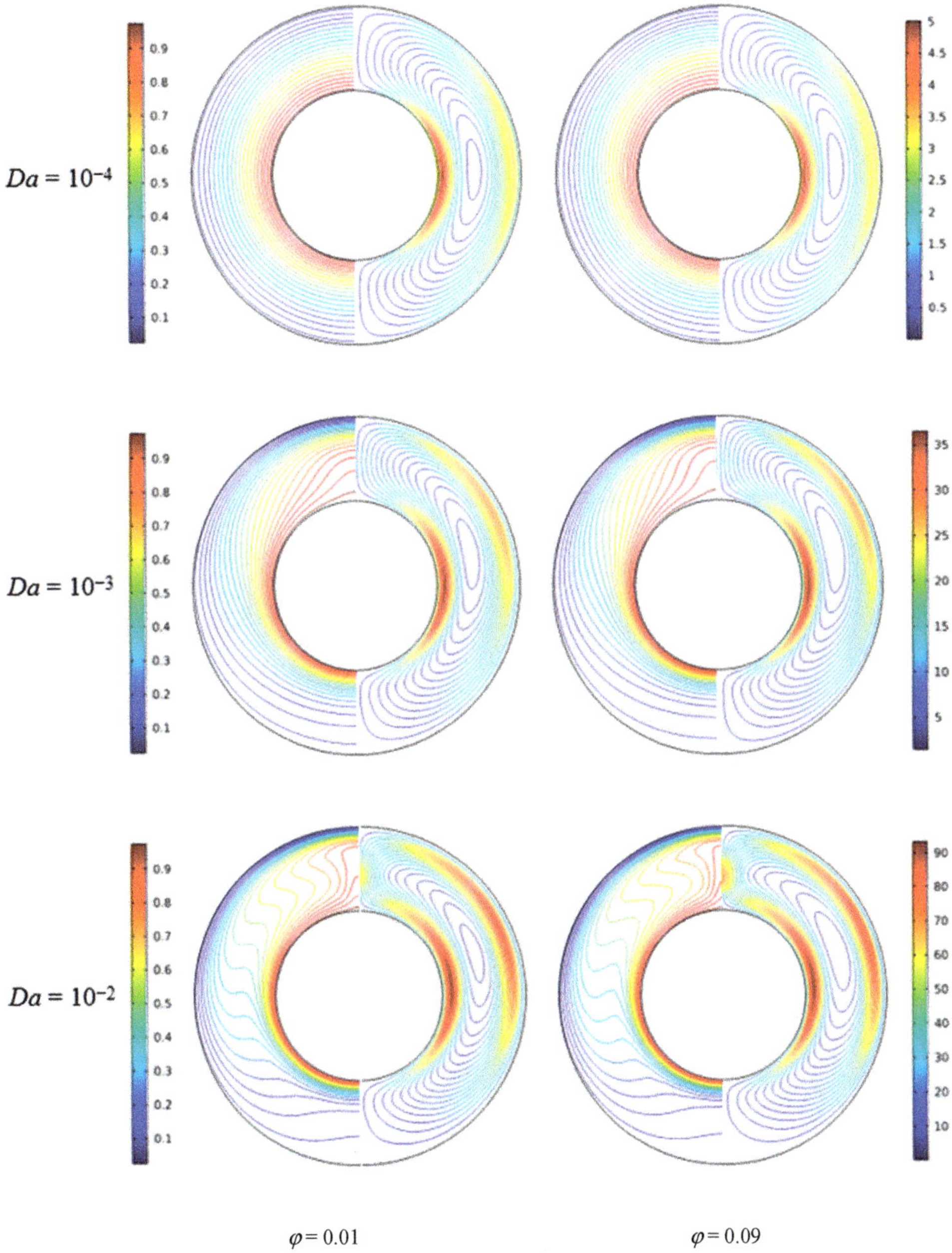

Figure 6. Isotherms (left) and streamlines (right) for Cu–water nanofluid for different nanoparticle volume fractions (ϕ) and Darcy numbers (Da) at $Ra = 10^5$, $d_{sp} = 50$ nm, $\varepsilon = 0.5$, and $RR = 2$.

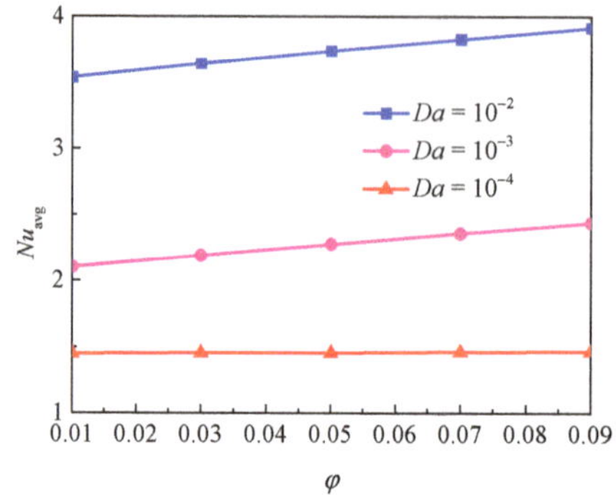

Figure 7. Average Nusselt number (Nu_{avg}) of the inner wall for different nanoparticle volume fractions (ϕ) and Darcy numbers (Da) at Ra= 10^5, d_{sp} = 50 nm, ε = 0.5, and RR = 2.

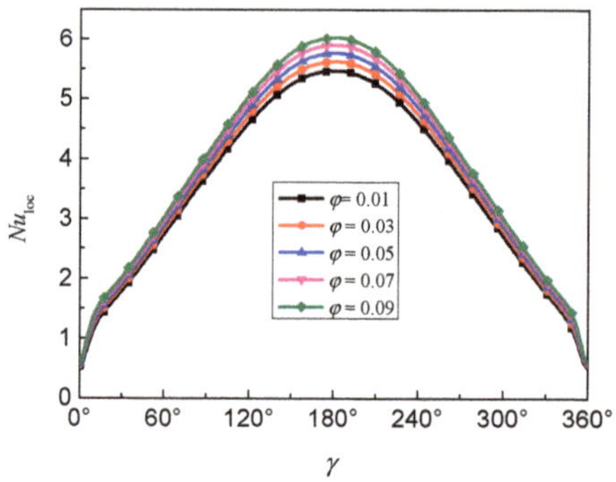

Figure 8. Local Nusselt number (Nu_{loc}) along the inner wall for different nanoparticle volume fractions (ϕ) at $Ra = 10^5$, $Da = 10^{-2}$, d_{sp} = 50 nm, ε = 0.5, and RR = 2.

4.3. *Effects of Nanoparticle Diameter*

Figure 9 shows the isotherms and streamlines for different nanoparticle diameters and Rayleigh numbers at $Da = 10^{-2}$, $\phi = 0.05$, $\varepsilon = 0.5$ and $RR = 2$. From Figure 9, for $Ra = 10^3$, the isotherms have a uniform distribution due to the weak buoyancy force, and it indicates that the heat transfer in the annulus is dominated by thermal conduction. The nanoparticle diameter has a weak effect on the isotherms. For $Ra = 10^4$, the natural convection heat transfer is strengthened due to the enhanced buoyancy force. The isotherms near the top half of the inner wall are disturbed, and the change of the isotherms distribution is more remarkable at $d_{sp} = 10$ compared with $d_{sp} = 90$. For $Ra = 5 \times 10^4$, the isotherms are almost horizontally distributed, especially when $d_{sp} = 10$, which means that the natural convection dominates the heat transfer and the effect of the nanoparticle diameter is more pronounced. In addition, for $Ra = 10^3$ and $Ra = 10^4$, the effect of nanoparticle diameter on the streamlines is very weak, while for $Ra = 10^5$, the effect is more pronounced, and the cell becomes bigger and has a tendency to move upward. Figure 10 displays the effect of nanoparticle diameter on the overall heat transfer rate along the inner wall at different Rayleigh numbers. The figure shows that an increase in the nanoparticle diameter leads to reduced heat transfer for all considered Rayleigh numbers. For $Ra = 10^3$, the effect of nanoparticle diameter is less pronounced. For $Ra = 5 \times 10^4$, the effect of nanoparticle diameter is remarkable, especially when it is at a low value. For instance, Nu_{avg} decreased by 11.79% from $d_{sp} = 10$ to $d_{sp} = 30$ and decreased by 5.81% from $d_{sp} = 30$ to $d_{sp} = 90$ at $Ra = 5 \times 10^4$, while Nu_{avg} decreased by 0.48% from $d_{sp} = 10$ to $d_{sp} = 30$ and decreased by 0.12% from $d_{sp} = 30$ to $d_{sp} = 90$ at $Ra = 10^4$.

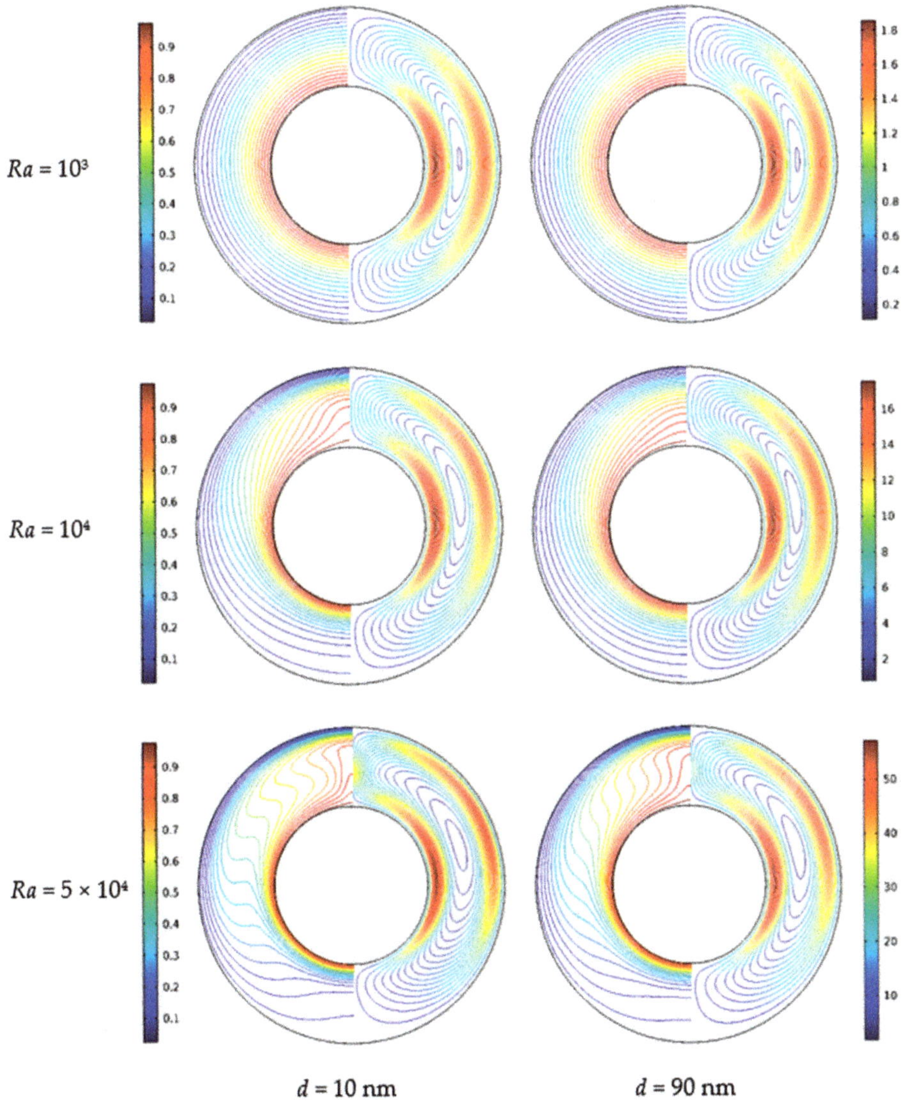

Figure 9. Isotherms (left) and streamlines (right) for Cu–water nanofluid for different nanoparticle diameters (d_{sp}) and Rayleigh numbers (*Ra*) at $Da = 10^{-2}$, $\phi = 0.05$, $\varepsilon = 0.5$, and $RR = 2$.

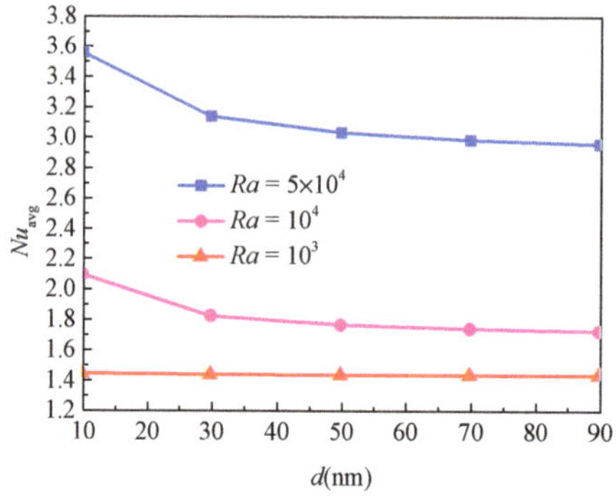

Figure 10. Average Nusselt number (Nu_{avg}) of the inner wall for different nanoparticle diameters (d_{sp}) and Rayleigh numbers (Ra) at $Da = 10^{-2}$, $\phi = 0.05$, $\varepsilon = 0.5$, and $RR = 2$.

Figure 11 illustrates the isotherms and streamlines for different nanoparticle diameter and Darcy numbers at $Ra = 5 \times 10^4$, $\phi = 0.05$, $\varepsilon = 0.5$, and $RR = 2$. From Figure 11, for $Da = 10^{-4}$, the isotherms have a uniform distribution due to the low permeability, which indicates that the flow is weak and thermal conduction plays a leading role. The effect of nanoparticle diameter on the heat transfer is weak. For $Da = 10^{-3}$, due to the enhanced permeability, the natural convection heat transfer is strengthened. For $Da = 10^{-2}$, the isotherms are almost horizontally distributed, especially when $d_{sp} = 10$, which means that the natural convection heat transfer plays a significant role and the effect of nanoparticle diameter is more pronounced. In addition, for $Da = 10^{-4}$ and $Da = 10^{-3}$, the effect of nanoparticle diameter on the streamlines is very weak, while for $Da = 10^{-2}$, the effect is more pronounced; the cell becomes bigger and has a tendency to move upward. Figure 12 displays the effect of the nanoparticle diameter on the overall heat transfer rate along the inner wall at different Darcy numbers. The figure shows that an increase in the nanoparticle diameter leads to heat transfer enhancement for all considered Darcy numbers, and it can be found that the effect of nanoparticle diameter is increased with the increase of the Darcy number.

Figure 13 presents the evolution of the local Nusselt number along the inner wall for different nanoparticle diameters at $Ra = 10^5$, $Da = 10^{-2}$, $\phi = 0.05$, $\varepsilon = 0.5$, and $RR = 2$. It can be found that the local Nusselt number has a symmetrical distribution due to the symmetrical geometry and boundary conditions, and the maximum Nu_{loc} occurred at $\gamma = 180^\circ$, which means that the natural convection heat transfer is more intense in the bottom half of the inner wall. Furthermore, the local Nusselt number is increased with the decrease of nanoparticle diameter, which indicates that the nanoparticle diameter at high value will weaken the natural convection heat transfer.

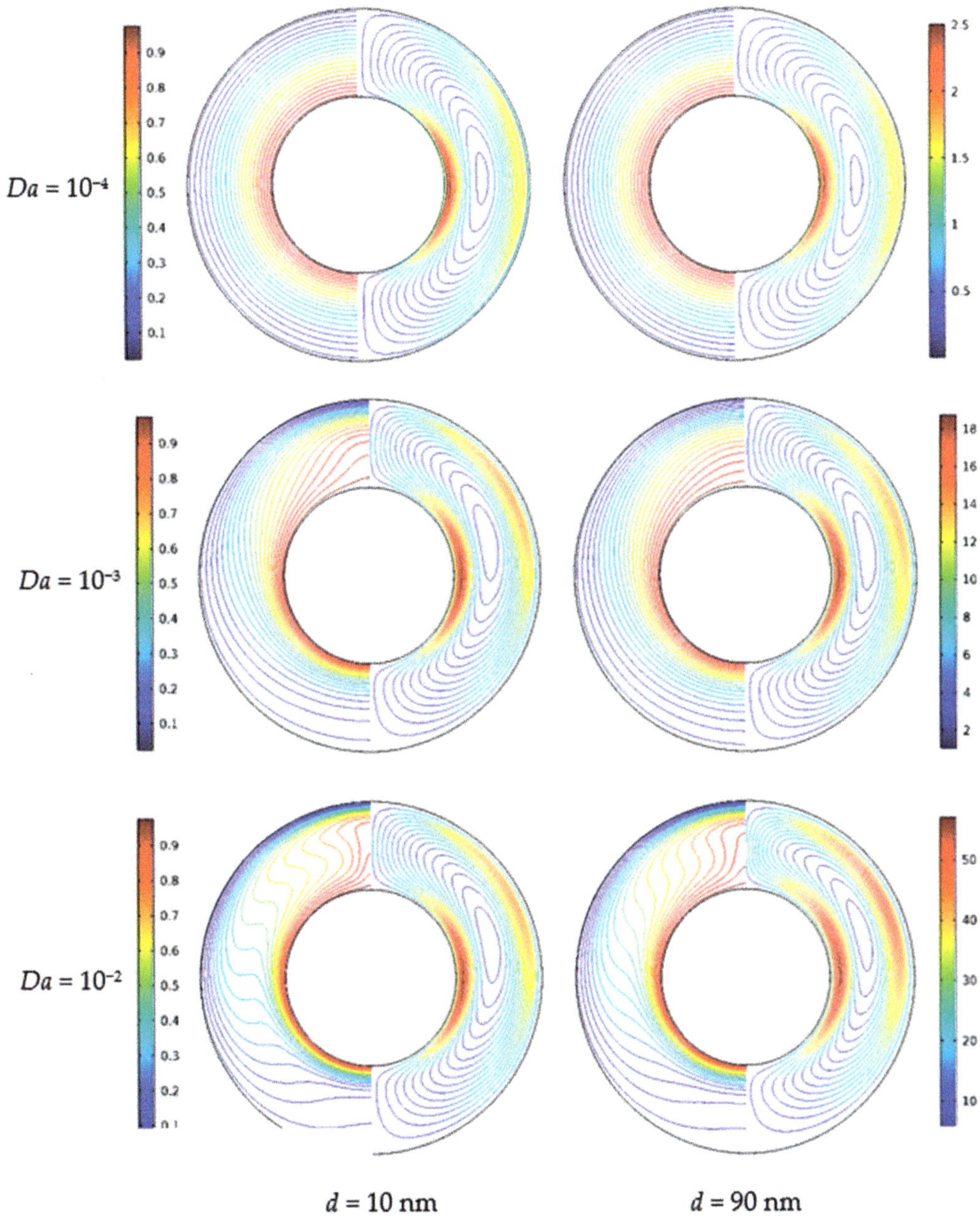

Figure 11. Isotherms (left) and streamlines (right) for Cu–water nanofluid for different nanoparticle diameters (d_{sp}) and Darcy numbers (Da) at $Ra = 5 \times 10^4$, $\phi = 0.05$, $\varepsilon = 0.5$, and $RR = 2$.

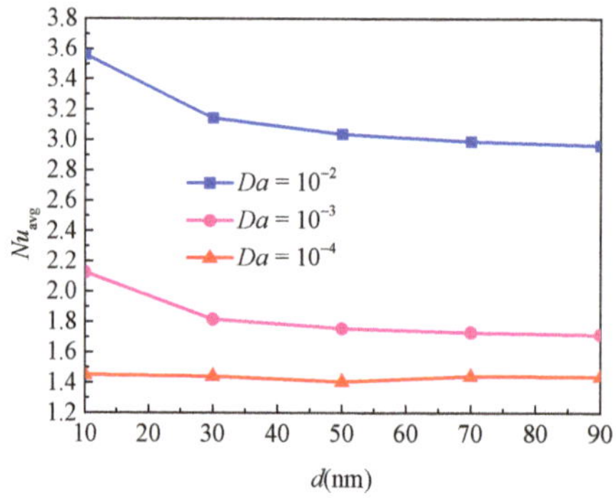

Figure 12. Average Nusselt number (Nu_{loc}) of the inner wall for different nanoparticle diameters (d_{sp}) and Darcy numbers (Da) at $Ra = 5 \times 10^4$, $\phi = 0.05$, $\varepsilon = 0.5$, and $RR = 2$.

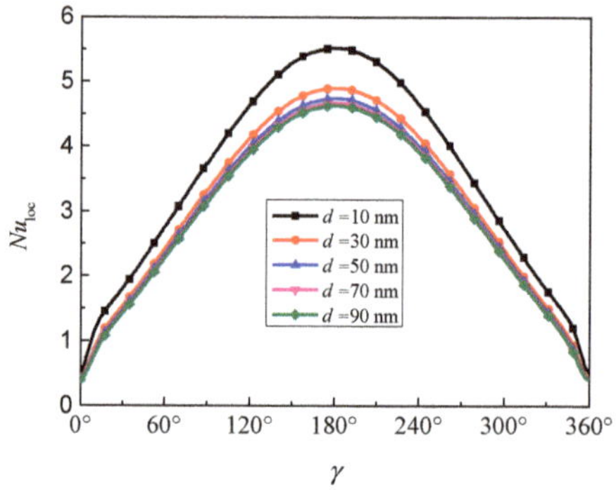

Figure 13. Local Nusselt number (Nu_{loc}) along the inner wall for different nanoparticle diameters (d_{sp}) at $Ra = 10^5$, $Da = 10^{-2}$, $\phi = 0.05$, $\varepsilon = 0.5$, and $RR = 2$.

4.4. Effects of Porosity

Figure 14 shows the isotherms and streamlines for different porosity and Rayleigh numbers at $d_{sp} = 50$ nm, $Da = 10^{-2}$, and $RR = 2$. From Figure 14, for $Ra = 10^3$, the isotherms have a uniform distribution, and the isotherms are almost unchanged when the porosity increased from $\varepsilon = 0.1$ to $\varepsilon = 0.9$. This is because at low Rayleigh numbers, the buoyancy force is very weak, and the heat transfer in the annulus is dominated by thermal conduction. The porosity has a slight effect on the isotherms. For $Ra = 10^4$, the isotherms have a similar trend with that for $Ra = 10^3$ at $\varepsilon = 0.1$, while the isotherms have an obvious change at $\varepsilon = 0.9$. For $Ra = 5 \times 10^4$, the natural convection heat transfer plays a leading role due to the enhanced flow, and the heat transfer is more intense at $\varepsilon = 0.9$ compared with $\varepsilon = 0.1$. The streamlines have a similar distribution, but some details are different. For $Ra = 10^3$, the streamlines have little change at $\varepsilon = 0.1$ and $\varepsilon = 0.9$. For $Ra = 10^4$ and $Ra = 5 \times 10^4$, it can be found the cells become bigger and have a tendency to move upward when the porosity increases. Figure 15 displays the effect of porosity on the overall heat transfer rate along the inner wall at different Rayleigh numbers. It can be found that a continuous increase of the overall heat transfer rate occurs with the increase of porosity for all considered Rayleigh numbers and the effect of porosity is more pronounced at high Rayleigh numbers.

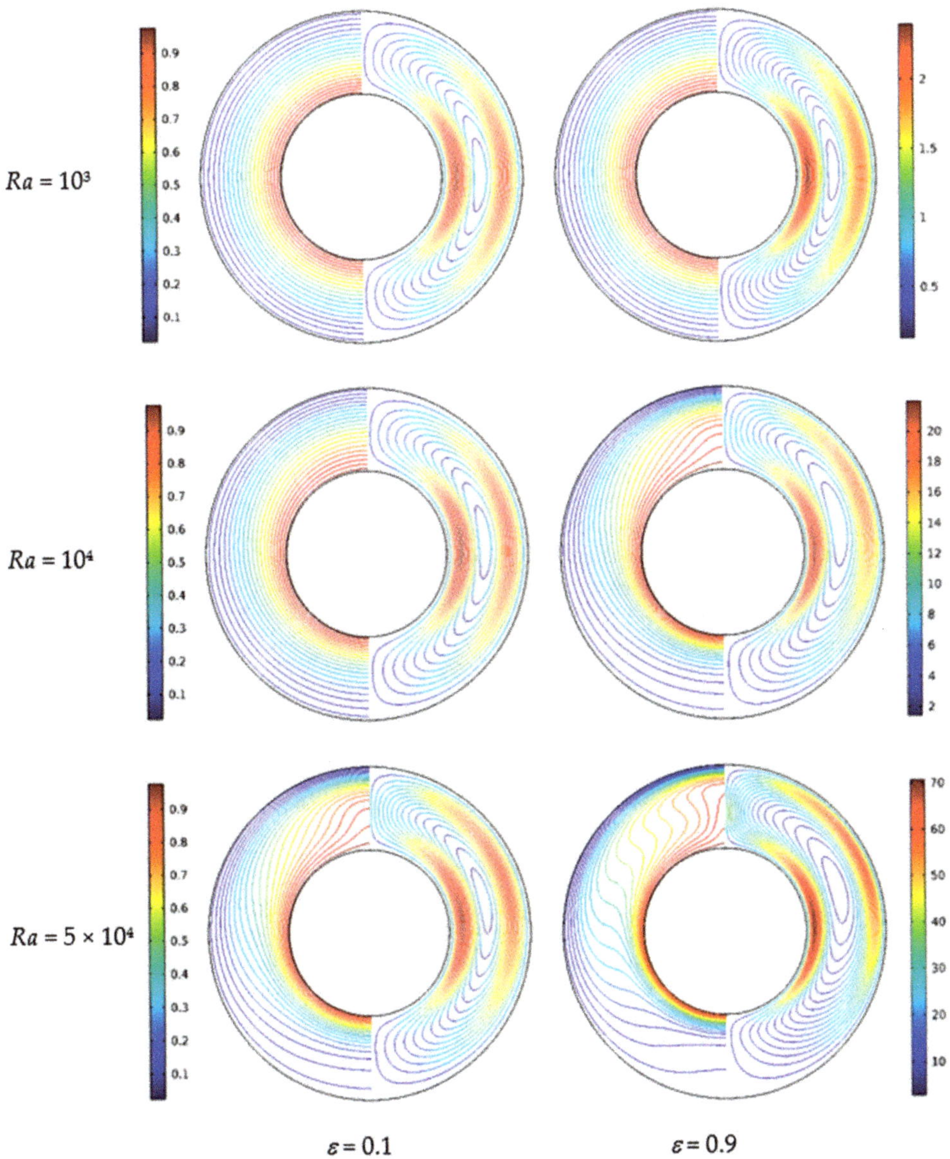

Figure 14. Isotherms (left) and streamlines (right) for Cu–water nanofluid for different porosity (ε) and Rayleigh numbers (Ra) at $Da = 10^{-2}$, $\phi = 0.05$, $d_{sp} = 50$ nm, and $RR = 2$.

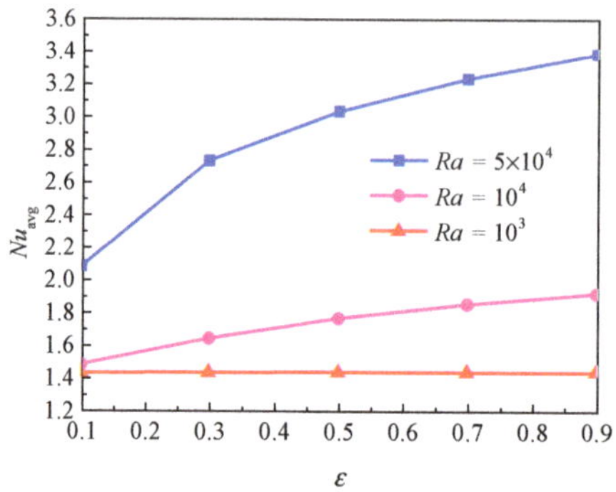

Figure 15. Average Nusselt number (Nu_{avg}) of the inner wall for different porosity (ε) and Rayleigh numbers (Ra) at $Da = 10^{-2}$, $\phi = 0.05$, $d_{sp} = 50$ nm, and $RR = 2$.

Figure 16 displays the isotherms and streamlines for different porosity and Darcy numbers at $Ra = 5 \times 10^4$, $\phi = 0.05$, $d_{sp} = 50$ nm, and $RR = 2$. From Figure 16, for $Da = 10^{-4}$, the isotherms have a uniform distribution at $\varepsilon = 0.1$ and $\varepsilon = 0.9$; the reason is that the flow is constrained by the low permeability. In this case, porosity has little effect on the isotherms. For $Da = 10^{-3}$, due to the enhanced permeability, the flow in the annulus is strengthened, and a disturbance occurs in the isotherms. For $Da = 10^{-2}$, the isotherms at $\varepsilon = 0.1$ and $\varepsilon = 0.9$ have a remarkable difference. The isotherms are almost horizontally distributed at $\varepsilon = 0.9$, which means that the overall heat transfer is dominated by natural convection. For $Da = 10^{-4}$, the streamlines are almost the same at $\varepsilon = 0.1$ and $\varepsilon = 0.9$. For $Da = 10^{-3}$ and $Da = 10^{-2}$, the cell becomes bigger and has a tendency to move upward when porosity increases from $\varepsilon = 0.1$ to $\varepsilon = 0.9$. Figure 17 displays the effect of porosity on the overall heat transfer rate along the inner wall at different Darcy number. The figure shows that an increase in porosity leads to heat transfer enhancement for all considered Darcy numbers, and the effect of porosity is more pronounced at high Darcy numbers.

Figure 18 presents the evolution of the local Nusselt number along the inner wall for different porosity at $Ra = 5 \times 10^4$, $Da = 10^{-2}$, $\phi = 0.05$, $\varepsilon = 0.5$, $d_{sp} = 50$ nm, and $RR = 2$. The local Nusselt number has a symmetrical distribution due to the symmetrical geometry and boundary conditions, and the maximum Nu_{loc} occurred at $\gamma = 180°$. Furthermore, the local Nusselt number is increased with the increase of porosity, so it can be concluded that the porosity has a positive effect on the overall heat transfer rate.

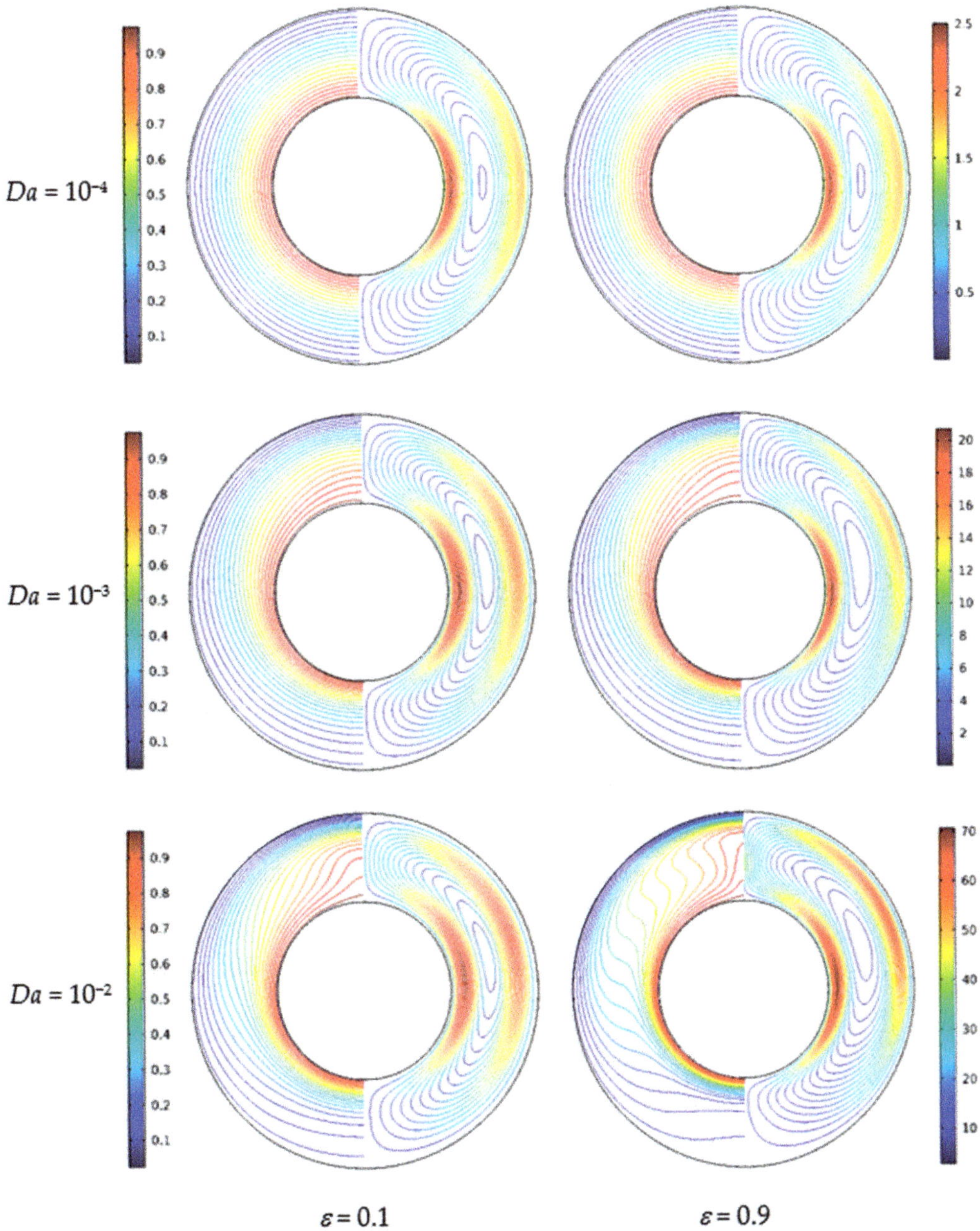

Figure 16. Isotherms (left) and streamlines (right) for Cu–water nanofluid for different porosity (ε) and Darcy numbers (Da) at $Ra = 5 \times 10^4$, $\phi = 0.05$, $d_{sp} = 50$ nm, and $RR = 2$.

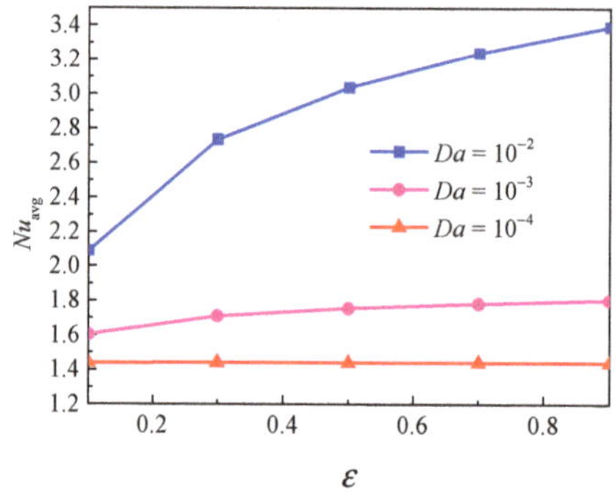

Figure 17. Average Nusselt number (Nu_{avg}) of the inner wall for different porosity (ε) and Darcy numbers (Da) at $Ra = 5 \times 10^4$, $\phi = 0.05$, $d_{sp} = 50$ nm, and $RR = 2$.

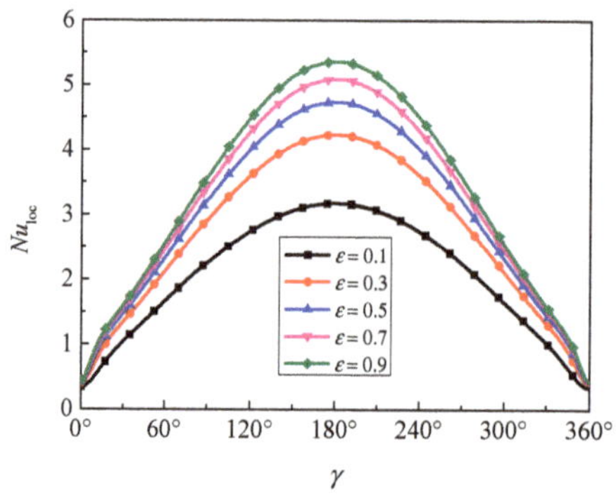

Figure 18. Local Nusselt number (Nu_{loc}) along the inner wall for different porosity (ε) at $Ra = 5 \times 10^4$, $Da = 10^{-2}$, $\phi = 0.05$, $d_{sp} = 50$ nm, and $RR = 2$.

4.5. Effects of Radius Ratio

Figure 19 illustrates the isotherms and streamlines for different radius ratio at $Ra = 5 \times 10^4$, $Da = 10^{-2}$, $\phi = 0.05$, $d_{sp} = 50$ nm, and $\varepsilon = 0.5$. The radius ratio is an important parameter that affects the fluid flows and natural convection heat transfer according the previous research [10–12]. From Figure 19, when the radius ratio increased from $RR = 1.2$ to $RR = 6$, the flow is strengthened, a main cell is formed in the middle of the annulus, and the cell becomes bigger and has a tendency to move upward. It is worth noting that the color scales on the right have a downward trend with the increase of RR, which is inverse to the above conclusion. In fact, the reason is that the change of RR leads to a change of the characteristic length (r_o–r_i). The isotherms distribution is changed from vertical to horizontal, which means that the heat transfer is dominated by natural convection. When $RR = 1.2$, multicellular flow structures are formed. Figure 20 displays the effect of radius ratio on the overall heat transfer rate along the inner wall. It can be found that heat transfer in the annulus is enhanced with the increase of radius ratio from $RR = 1.3$ to $RR = 6$, but a fall occurred in the range of $RR = 1.2$ to $RR = 1.3$. A conclusion can be drawn that a bifurcation point exists when the radius ratios is in the range of $RR = 1.2$ to $RR = 1.3$ for the considered parameters.

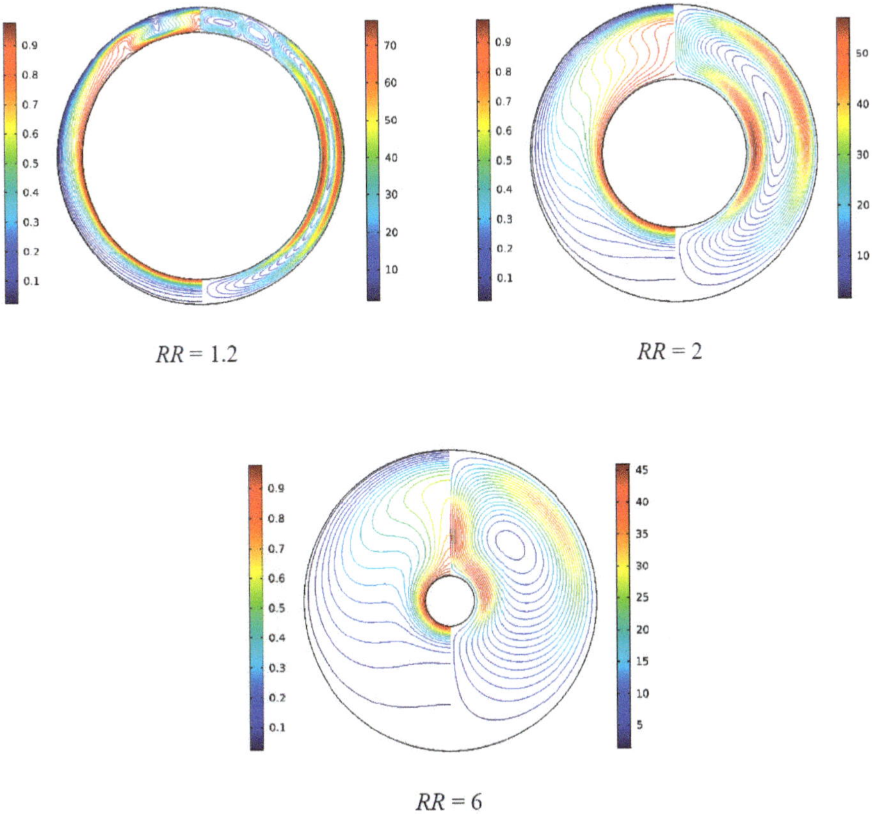

Figure 19. Isotherms (left) and streamlines (right) for Cu–water nanofluid for different radius ratios (*RR*) at $Ra = 5 \times 10^4$, $Da = 10^{-2}$, $\phi = 0.05$, $d_{sp} = 50$ nm, and $\varepsilon = 0.5$.

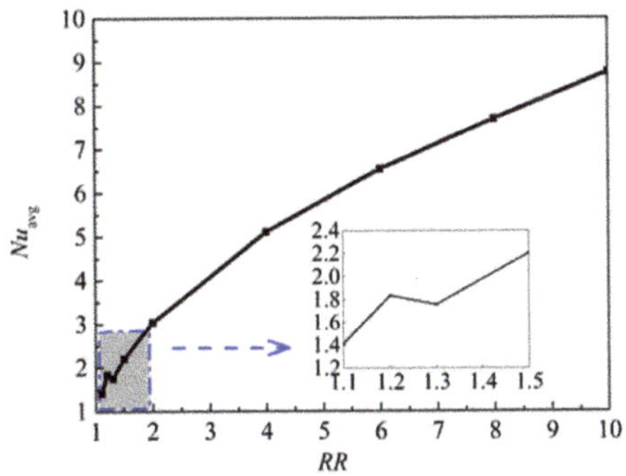

Figure 20. Average Nusselt number (Nu_{avg}) of the inner wall for different radius ratios (*RR*) at $Ra = 5 \times 10^4$, $Da = 10^{-2}$, $\phi = 0.05$, $d_{sp} = 50$ nm, and $\varepsilon = 0.5$.

5. Conclusions

Natural convection heat transfer in a porous annulus filled with a Cu-Nanofluid has been investigated numerically. The effects of Brownian motion, solid volume fraction, nanoparticle diameter, Rayleigh number, Darcy number, porosity on the flow pattern, temperature distribution, and heat transfer characteristics are discussed in detail. The following conclusions could be drawn:

(1) Brownian motion should be considered in the natural convection heat transfer of a nanofluid.

The effect of Brownian motion becomes more remarkable with the increase of Rayleigh number and nanoparticle volume fraction, while it is less pronounced with the increase of Darcy number and nanoparticle diameter.

(2) The increase of nanoparticle volume fraction results in an improvement of the overall heat transfer rate, and the effect is more remarkable when the Rayleigh and Darcy numbers are at high value.

(3) Increasing the nanoparticle diameter has a negative effect on the overall heat transfer rate and the effect has a limit when the nanoparticle diameter reaches a high value.

(4) The porosity affects the flow pattern, temperature distribution, and heat transfer rate. The flow motion is limited when the porosity is too low. The heat transfer rate is strengthened with the increase of porosity, especially with high Rayleigh and Darcy numbers.

(5) The radius ratio has a significant influence on the isotherms, streamlines, and heat transfer rate. The rate is greatly enhanced with the increase of radius ratio. Additionally, when the radius ratio is too low, multicellular flow structures are formed, and a bifurcation point exists.

Author Contributions: Conceptualization, Y.H. and M.L.; Methodology, L.Z.; Software, L.Z.; Validation, Y.H., L.Z. and M.L.; Formal Analysis, L.Z.; Investigation, Y.H.; Resources, Y.H.; Data Curation, L.Z.; Writing—Original Draft Preparation, L.Z.; Writing—Review & Editing, L.Z.; Visualization, L.Z.; Supervision, Y.H.; Project Administration, M.L.; Funding acquisition Administration, M.L. All authors have read and agreed to the published version of the manuscript.

Funding: This research was funded by National Nature Science Foundation of China, grant number [51706213].

Data Availability Statement: Data available in a publicly accessible repository.

Conflicts of Interest: The authors declare no conflict of interest.

Nomenclature

RR	radius ratio
c_p	thermal capacity, J/(kg × K)
d	nanoparticle diameter, m
Da	Darcy number
g	acceleration due to gravity, m/s^2
k	thermal conductivity, W/(m × K)
K	medium permeability, m^2
Nu	Nusselt number
p	pressure, Pa
P	dimensionless pressure
Pr	Prandtl number
r	radius, m
Ra	Rayleigh number
T	temperature, K
(u, v)	velocity, m/s
(U, V)	dimensionless velocity
(x, y)	cartesian coodinates
(X, Y)	dimensionless cartesian coodinates

Greek symbols	
α	thermal diffusivity, m^2/s
β	thermal expansion coefficient, K^{-1}
γ	angle, deg
ε	porosity
θ	dimensionless temperature
ρ	density, Kg/m^3
μ	dynamic viscosity, $kg/(m \times s)$
ν	kinematic viscosity, m^2/s
ϕ	nanoparticle volume fraction
Subscripts	
avg	average
bf	base fluid
i, o	inner, outer
loc	local
mnf	porous medium filled with nanofluid
nf	nanofluid
s	porous medium
sp	nanoparticle

References

1. Esfe, M.H.; Bahiraei, M.; Hajbarati, H.; Valadkhani, M. A comprehensive review on convective heat transfer of nanofluids in porous medium: Energy-related and thermalhydraulic characteristics. *Appl. Them. Eng.* **2020**, *178*, 115487. [CrossRef]
2. Das, D.; Roy, M.; Basak, T. Studies on natural convection within enclosures of various (non-square) shapes—A review. *Int. J. Heat Mass Transf.* **2017**, *106*, 356–406. [CrossRef]
3. Dawood, H.K.; Mohammed, H.A.; Sidik, N.A.C.; Munisamy, K.M.; Wahid, M.A. Forced, Natural and Mixed-convection Heat Transfer and Fluid Flow in Annulus: A Review. Int. Commun. *Int. Commun. Heat Mass Transf.* **2015**, *62*, 45–57. [CrossRef]
4. Caltagirone, J.P. Thermoconvective instabilities in a porous medium bounded by two concentric horizontal cylinders. *J. Fluid Mech.* **1976**, *76*, 337–362. [CrossRef]
5. Rao, Y.F.; Fukuda, K.; Hasegawa, S. Steady and transient analyses of natural convection in a horizontal porous annulus with the Galerkin method. In Proceedings of the 4th AIAA/ASME Thermophysics Heat Transfer Conference, Boston, MA, USA, 2–4 June 1986.
6. Rao, Y.F.; Fukuda, K.; Hasegawa, S. A numerical study of three-dimensional natural convection in a horizontal porous annulus with Galerkin method. *Int. J. Heat Mass Transf.* **1988**, *31*, 695–707. [CrossRef]
7. Himasekhar, K. Two-dimensional bifurcation phenomena in thermal convection in horizontal concentric annuli containing saturated porous media. *J. Fluid Mech.* **1988**, *187*, 267–300. [CrossRef]
8. Leong, J.C.; Lai, F.C. Natural convection in a concentric annulus with a porous sleeve. *Int. J. Heat Mass Transf.* **2006**, *49*, 3016–3027. [CrossRef]
9. Braga, E.J.; Lemos, M. Simulation of turbulent natural convection in a porous cylindrical annulus using a macroscopic two equation Model. *Int. J. Heat Mass Transf.* **2006**, *49*, 4340–4351. [CrossRef]
10. Khanafer, K.; Al-Amiri, A.; Pop, I. Numerical analysis of natural convection heat transfer in a horizontal annulus partially filled with a fluid-saturated porous substrate. *Int. J. Heat Mass Transf.* **2008**, *51*, 1613–1627. [CrossRef]
11. Alloui, Z.; Vasseur, P. Natural convection in a horizontal annular porous cavity saturated by a binary mixture. *Comput. Therm. Sci.* **2011**, *3*, 407–417. [CrossRef]
12. Belabid, J.; Cheddadi, A. Multicellular flows induced by natural convection in a porous horizontal cylindrical annulus. *Phys. Chem. News* **2013**, *70*, 67–71.
13. Belabid, J.; Allali, K. Influence of gravitational modulation on natural convection in a horizontal porous annulus. *J. Heat Transf.* **2017**, *139*, 022502.1–022502.6. [CrossRef]
14. Belabid, J.; Allali, K. Effect of temperature modulation on natural convection in a horizontal porous annulus. *Int. J. Therm. Sci.* **2020**, *151*, 106273. [CrossRef]
15. Rostami, S.; Aghakhani, S.; Pordanjani, A.H.; Afrand, M.; Shadloo, M.S. A review on the control parameters of natural convection in different shaped cavities with and without nanofluid. *Processes* **2020**, *8*, 1011. [CrossRef]
16. Choi, S.U.S. Enhancing thermal conductivity of fluids with nanoparticles. In Proceedings of the 1995 ASME International Mechanical Engineering Congress and Exposition, San Francisco, CA, USA, 12–17 November 1995; Volume 66, pp. 99–105.
17. Zhou, W.N.; Yan, Y.Y.; Xu, J.L. A lattice Boltzmann simulation of enhanced heat transfer of nanofluids. *Int. Commun. Heat Mass Transf.* **2014**, *55*, 113–120. [CrossRef]
18. Abouali, O.; Ahmadi, G. Computer simulations of natural convection of single phase nanofluids in simple enclosures: A critical review. *Appl. Therm. Eng.* **2012**, *36*, 1–13. [CrossRef]

19. Jou, R.Y.; Tzeng, S.C. Numerical research of nature convective heat transfer enhancement filled with nanofluids in a rectangular enclosure. *Int. Commun. Heat Mass Transf.* **2006**, *33*, 727–736. [CrossRef]
20. Ghasemi, B.; Aminossadati, S.M. Natural convection heat transfer in an inclined enclosure filled with a water–CuO nanofluid. *Numer. Heat Transf. Part A* **2009**, *55*, 807–823. [CrossRef]
21. Abu-Nada, E.; Oztop, H.F. Effects of inclination angle on natural convection in enclosures filled with Cu-Water nanofluid. *Int. J. Heat Fluid Flow* **2009**, *30*, 669–678. [CrossRef]
22. Soleimani, S.; Sheikholeslami, M.; Ganji, D.D.; Gorji-Bandpay, M. Natural convection heat transfer in a nanofluid filled semi-annulus enclosure. *Int. Commun. Heat Mass Transf.* **2012**, *39*, 565–574. [CrossRef]
23. Seyyedi, S.M.; Dayyan, M.; Soleimani, S.; Ghasemi, E. Natural convection heat transfer under constant heat flux wall in a nanofluid filled annulus enclosure. *Ain Shams Eng. J.* **2015**, *6*, 267–280. [CrossRef]
24. Boualit, A.; Zeraibi, N.; Chergui, T.; Lebbi, M.; Boutina, L.; Laouar, S. Natural convection investigation in square cavity filled with nanofluid using dispersion model. *Int. J. Hydrogen Energy* **2017**, *42*, 8611–8623. [CrossRef]
25. Liao, C. Heat transfer transitions of natural convection flows in a differentially heated square enclosure filled with nanofluids. *Int. J. Heat Mass Transf.* **2017**, *115*, 625–634. [CrossRef]
26. Wang, L.; Yang, X.G.; Huang, C.S.; Chai, Z.H.; Shi, B.C. Hybridlattice Boltzmann-TVD simulation of natural convection of nanofluids in a partially heated square cavity using Buongiorno's model. *Appl. Therm. Eng.* **2019**, *146*, 318–327. [CrossRef]
27. Mi, A.; Ts, B.; Ymcc, D.; Hma, E.; Gc, F.; Rk, G. Comprehensive study concerned graphene nano-sheets dispersed in ethylene glycol: Experimental study and theoretical prediction of thermal conductivity. *Powder Technol.* **2021**, *386*, 51–59.
28. Sheremet, M.A.; Pop, I. Free convection in a porous horizontal cylindrical annulus with a nanofluid using Buongiorno's model. *Comput. Fluids* **2015**, *118*, 182–190. [CrossRef]
29. Sheremet, M.A.; Pop, I. Natural convection in a horizontal cylindrical annulus filled with a porous medium saturated by a nanofluid using tiwari and das' nanofluid model. *Eur. Phys. J. Plus* **2015**, *130*, 107. [CrossRef]
30. Belabid, J.; Belhouideg, S.; Allali, K.; Mahian, O.; Abu-Nada, E. Numerical simulation for impact of copper/water nanofluid on thermo-convective instabilities in a horizontal porous annulus. *J. Therm. Anal. Calorim.* **2019**, *138*, 1515–1525. [CrossRef]
31. Abu-Nada, E. Effects of variable viscosity and thermal conductivity of Al_2O_3-water nanofluid on heat transfer enhancement in natural convection. *Int. J. Heat Fluid Flow* **2009**, *30*, 679–690. [CrossRef]
32. Cianfrini, M.; Corcione, M.; Quintino, A. Natural convection heat transfer of nanofluids in annular spaces between horizontal concentric cylinders. *Appl. Therm. Eng.* **2011**, *31*, 4055–4063. [CrossRef]
33. Goudari, S.; Shekaramiz, M.; Omidvar, A.; Golab, E.; Karimipour, A. Nanopaticles migration due to thermophoresis and Brownian motion and its impact on Ag-MgO/Water hybrid nanofluid natural convection. *Powder Technol.* **2020**, *375*, 493–503. [CrossRef]
34. Bourantas, G.C.; Skouras, E.D.; Loukopoulos, V.C.; Burganos, V.N. Heat transfer and natural convection of nanofluids in porous media. *Eur. J. Mech. B Fluid* **2014**, *43*, 45–56. [CrossRef]
35. Hu, Y.P.; Li, Y.R.; Lu, L.; Mao, Y.J.; Li, M.H. Natural convection of water-based nanofluids near the density maximum in an annulus. *Int. J. Thermal. Sci.* **2020**, *152*, 106309. [CrossRef]
36. Wang, B.; Shih, T.M.; Chen, X.; Chang, R.G.; Wu, C.X. Cascade-like and cyclic heat transfer characteristics affected by enclosure aspect rations for low Prandtl numbers. *Int. J. Heat Mass Transf.* **2018**, *124*, 131–140. [CrossRef]
37. Singh, A.K.; Basak, T.; Nag, A.; Roy, S. Heatlines and thermal management analysis for natural convection within inclined porous square cavities. *Int. J. Heat Mass Transf.* **2015**, *87*, 583–597. [CrossRef]

nanomaterials

Article

Enhanced Heat Transfer for NePCM-Melting-Based Thermal Energy of Finned Heat Pipe

Sameh E. Ahmed [1,2], Aissa Abderrahmane [3], Sorour Alotaibi [4,*], Obai Younis [5,6], Radwan A. Almasri [7] and Wisam K. Hussam [8]

1 Department of Mathematics, Faculty of Science, King Khalid University, Abha 62529, Saudi Arabia; sehassan@kku.edu.sa
2 Department of Mathematics, Faculty of Science, South Valley University, Qena 83523, Egypt
3 Laboratoire de Physique Quantique de la Matière et Modélisation Mathématique (LPQ3M), University of Mascara, Mascara 29000, Algeria; a.aissa@univ-mascara.dz
4 Mechanical Engineering Department, College of Engineering and Petroleum, Kuwait University, P.O. Box 5969, Safat 13060, Kuwait
5 Department of Mechanical Engineering, College of Engineering at Wadi Addwaser, Prince Sattam Bin Abdulaziz University, Wadi Addwaser 11991, Saudi Arabia; oubeytaha@hotmail.com
6 Department of Mechanical Engineering, Faculty of Engineering, University of Khartoum, Khartoum 11111, Sudan
7 Department of Mechanical Engineering, College of Engineering, Qassim University, Buraydah 51452, Saudi Arabia; masri.radwan@qec.edu.sa
8 School of Engineering, Australian College of Kuwait, Safat 12000, Kuwait; w.alsaadi@ack.edu.kw
* Correspondence: sr.alotaibi@ku.edu.kw

Citation: Ahmed, S.E.; Abderrahmane, A.; Alotaibi, S.; Younis, O.; Almasri, R.A.; Hussam, W.K. Enhanced Heat Transfer for NePCM-Melting-Based Thermal Energy of Finned Heat Pipe. *Nanomaterials* **2022**, *12*, 129. https://doi.org/10.3390/nano12010129

Academic Editor: S. M. Sohel Murshed

Received: 7 December 2021
Accepted: 28 December 2021
Published: 31 December 2021

Abstract: Using phase change materials (PCMs) in energy storage systems provides various advantages such as energy storage at a nearly constant temperature and higher energy density. In this study, we aimed to conduct a numerical simulation for augmenting a PCM's melting performance within multiple tubes, including branched fins. The suspension contained Al_2O_3/n-octadecane paraffin, and four cases were considered based on a number of heated fins. A numerical algorithm based on the finite element method (FEM) was applied to solve the dimensionless governing system. The average liquid fraction was computed over the considered flow area. The key parameters are the time parameter ($100 \leq t \leq 600$ s) and the nanoparticles' volume fraction ($0\% \leq \varphi \leq 8\%$). The major outcomes revealed that the flow structures, the irreversibility of the system, and the melting process can be controlled by increasing/decreasing number of the heated fins. Additionally, case four, in which eight heated fins were considered, produced the largest average liquid fraction values.

Keywords: melting process; PCM; FEM; shell designs; tubes; branched fins; latent heat

1. Introduction

In modern times, energy storage and conservation solutions are as important as ever due to depletion and climate change challenges. Phase change materials (PCMs) are at the forefront in this domain due to their low cost and availability [1–3].

Due to the continuous development of PCM research, their range of applications has widened: PCMs are now integrated into automotive applications [4], energy storage applications [5], and heat exchangers [6]. Allouhi et al. [7] employed PCMs to optimize an energy storage system for solar water heaters operated by households in rural regions to fulfil their hot water needs at night. Huang et al. [8] used a dual-phase-change material to design a heat sink to minimize its cost and weight. Carmona et al. [9] used a PCM to improve the energy efficiency of a water storage tank for domestic applications. Their numerical model was validated with an experimental study. They found that including 40% of PCM inside the water storage tank increased its efficiency by 16%. PCMs are also used to conserve energy in the building sector as they are integrated into walls [10] and roofs [11], and can be used in indoor applications such as house furniture or decorations [12] to maintain a favorable room temperature. They can also be used in building envelopes as PCM-to-air heat exchangers where air discharges and charges the PCM with energy [13,14]. Padala et al. [15] explored the use of a PCM in a mixture for masonry blocks to determine the ideal mix proportions to attain the best mechanical, durability-related, and thermal characteristics of these mixes. Chen et al. [16] investigated the thermal performance of a thermo-activated PCM mixture integrated into a wall for energy saving purposes during winter. Gholamibozanjani et al. [17] experimentally examined incorporating PCMs to conserve energy inside huts over the seasons of a year. The results proved that the PCM storage units were able to reduce the heating/cooling energy requirements: the accumulative energy-saving varied between 10% and 40% during the year. Tyagi et al. [18] evaluated the performance of a PCM-based energy storage unit comprising panels charged by sunlight during the daytime then used during the night as a heat source for a test room. Violidakis et al. [19] discussed the use of an ultra-high-temperature PCM such as silicon in residential buildings. They possess excellent thermal conductivity and high latent heat thermal energy, achieving greater energy density and capacity. Frazzica et al. [20] produced and characterized mortar PCMs, then they introduced a new experimental setup to evaluate its thermal performance. However, PCMs still face challenges such as long discharging and changing times and low heat transfer rate due to their low thermal conductivity [21]. Thus, new technologies must be developed to improve their thermal properties and heat transfer performance. The most intuitive solution to enhance a PCM's melting process is to extend the contact surface. Dmitruk et al. [22] studied a cylindrical PCM-based heat storage system equipped with a pin-fin structure to improve the heat transfer rate within this system. After multiple charging/discharging cycles, the experimental and numerical results showed the positive effect of the pin-fin structure on PCM performance. Sathe et al. [23] numerically analyzed the thermal-hydraulic performance a PCM melting inside a tilted finned container with a top heating mode. They observed that decreasing the inclination angles and extending the surface and fins increased the melting time for all the PCMs. Nie et al. [24] studied the thermal performance of a PCM as part of a composite containing fumed silica and graphene that was used to enhance the capability of a portable box to maintain cold temperature. According to their results, adding 1 wt% of graphene and 4 wt% of fumed silica enhanced the thermal conductivity of the PCM composite by 55.4% and helped to eliminate the PCM leakage problem. Izgi et al. [25] studied the solidification process of a PCM inside a three-dimensional cylinder. They focused on finding controlling parameters for this phenomenon. From the results, they observed that the diameter of the cylinder influenced the energy discharge and the solidification times. Sweidan et al. [26] performed multiple numerical computations to investigate the effectiveness of various techniques (multiple PCM layers, mingle PCM with highly conductive fins, and PCM-saturated metal foam) in improving PCM performance. Tarigond et al. [27] used iron scrap additives to boost the thermal performance of the PCM inside a thermal energy storage system for hot water. The results demonstrated the positive impact of these additives on the performance of the PCM, as the yield of hot water was improved by 25% compared to the control system. Ahmed et al. [28] proposed a novel design for a cascaded-layered PCM as a cost effective solution for medium-temperature industrial applications. From the results, they indicated the best volume fraction arrangement for thermal energy storage. Selimefendigil et al. [29] used FEM to evaluate the free convection of a copper oxide/water nano-liquid within a square cavity with PCM attached to its vertical wall. They found that by increasing the height of the PCM from 0.2 to 0.8 H, both the local and average Nusselt numbers decreased by 42.14%. Abu-Hamdeh et al. [30] performed a three-dimensional examination of the paraffin wax melting process in an ellipsoidal pipe with a hotter inner pipe. They discovered that the location of the inner pipe and the temperature differential influence the time required for the PCM to melt.

Nowadays, the most effective approaches for improving the thermal performance of PCMs are using a porous medium or metal foam as a support matrix for the PCM and adding high thermal conductivity nanoparticles to the PCM [31]. However, the simplest the most used method involves embedding fins in the PCM containers. Over the years, several studies have been conducted on the influence of fins on heat transfer in various media, including PCMs [32–35].

Jeong et al. [36] reported that the thermal properties of three different shape-stabilized PCM composites were enhanced by adding exfoliated graphite nanoplatelets (xGnP). Nóbrega et al. [37] investigated the influence of adding fins to a tube filled with a nano-PCM mixture. The findings suggested that the fins helped decrease the solidification time by 9.1%, and increasing the concentration of nanoparticles improved the solidification process regardless of the presence of fins. Raj et al. [38] packed a nano-enriched PCM inside a wall-less heat sink used in a thermal management application. The study results revealed that the addition of MWCNTs and GnP nanoparticles to the FS-PCM enhanced its thermal conductivity by 61.73% and 84.48%, respectively, thus improving the heat transfer performance. Jourabian et al. [39] investigated the melting behavior of ice as a PCM within a horizontal elliptical tube filled with nickel-steel porous media. Das et al. [40] addressed the form stability of a PCM subjected to numerous cycles of charging and discharging by producing a novel biocomposite-based PCM. The testing results indicated that using biochar from *Eichornia crassipes* as a supporting matrix mix with a small amount of aluminum metal powder enhanced the thermal conductivity of the PCM 17.27 times and improved its overall stability. Vennapusa et al. [41] tested six lightweight support materials to improve the form stability of a caprylic-acid-based PCM. From the results, they concluded that using expanded perlite as a support matrix for the PCM achieved the highest enthalpy and thermal buffering.

Rathore et al. [42] added expanded vermiculite and expanded graphite to improve the thermophysical properties of a low-cost PCM. The study results revealed that the thermal conductivity increased by 114.4% when the PCM was loaded with 7% of EG. Moreover, the enhanced PCM preserved its thermal properties even after 1000 charging and discharging cycles. Qureshi et al. [43] employed TPMS-based metal foams with a 3D-printed structures as a skeleton for MFPCMs to improve the thermal conductivity of traditional PCMs. According to their results, the thermal conductivity of MFPCMs was strongly affected by the cell type and its unique shape, in addition to the cell porosity. Combining the two approaches, Nada et al. [44] discussed enhancing paraffin wax as a PCM in an energy storage unit by using a carbon foam matrix and MWCNTs additives. Mehryan et al. [45] studied the non-Newtonian behavior in space between two coaxial pipes filled with metal foam. They reported that a 54% reduction in the melting time could be achieved by lowering the power law index from 1 to 0.6.

The literature review revealed that entropy formation during the charging process has been not well-investigated. As a result, in this study, we used GFEM to replicate NEPCM's second-law behavior during melting. The contours of entropy component, temperature, and melt fraction are shown.

2. Methodology and Problem Definition

Figure 1 depicts an illustration of the current computational models. Al_2O_3/n-octadecane paraffin was used as the working fluid. Fins were modeled in four different configurations. Figure 2 depicts the specifics of four examples. At the start, all fluids had a solidus temperature. The fins and inner cylinder are maintained at hot temperature (312 K), and the outer cylinder is adiabatic.

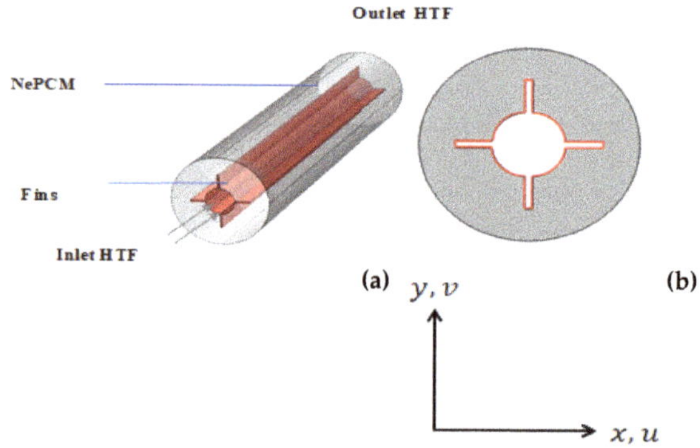

Figure 1. Computational setup of LTESS: (**a**) 3D model (**b**) 2D cross-section.

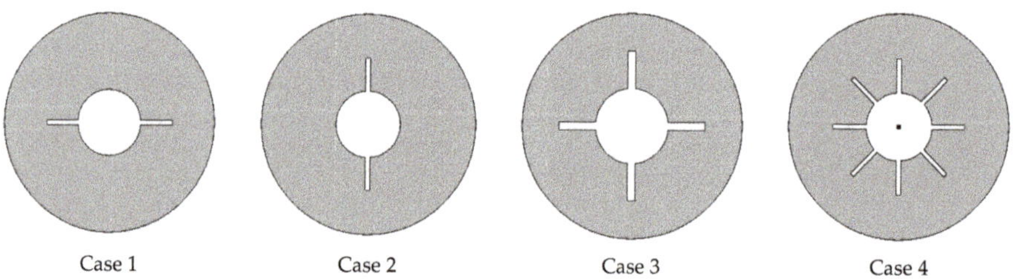

Figure 2. Different cases considered in this study.

2.1. Problem Formulation

To model this transitory process, we assumed that the flow was Newtonian and laminar. To account for the gravity force effect, Boussinesq estimation was used. The calculation formulas are as follows [46–49]:

$$\nabla \cdot \vec{V} = 0 \tag{1}$$

$$\left(\frac{\partial v}{\partial t} + \vec{V} \cdot \nabla v\right) = vC\frac{(\lambda - 1)^2}{\varepsilon + \lambda^3} + \frac{1}{\rho_{nf}}\left(-\nabla P + \mu_{nf}\nabla^2 v\right) + \frac{1}{\rho_{nf}}(\rho\beta)_{nf}g\left(T - T_{ref}\right) \tag{2}$$

$$\frac{\partial u}{\partial t} + \vec{V} \cdot \nabla u = uC\frac{(\lambda - 1)^2}{\varepsilon + \lambda^3} + \frac{1}{\rho_{nf}}\left(-\nabla P + \mu_{nf}\nabla^2 u\right) \tag{3}$$

$$(\rho C_p)_{nf}\frac{\partial(\rho L\lambda)_{nf}}{\partial t} + (\rho C_p)_{nf}\frac{\partial T}{\partial t} - k_{nf}\nabla^2 T = -(\rho C_p)_{nf}\vec{V} \cdot \nabla T \tag{4}$$

We considered $\varepsilon = 10^{-3}$ and $C = 10^5$.

To forecast the NEPCM attributes, a single-phase model was used:

$$(\rho C_p)_f^{-1}(\rho C_p)_{nf} = (1 - \phi) + \phi(\rho C_p)_s(\rho C_p)_f^{-1} \tag{5}$$

$$\rho_{nf} = \phi\rho_s + \rho_f(1 - \phi) \tag{6}$$

$$(\rho\beta)_{nf} = \phi(\rho\beta)_s + (1 - \phi)(\rho\beta)_f \tag{7}$$

$$(\rho L)_f = \frac{(\rho L)_{nf}}{(1 - \phi)} \tag{8}$$

$$k_{nf} = \frac{2k_f + 2\phi\left(k_s - k_f\right) + k_p}{k_p - \phi\left(k_s - k_f\right) + 2k_f} k_f \tag{9}$$

$$\mu_{nf} = \frac{\mu_f}{(1-\phi)^{2.5}} \tag{10}$$

Table 1 lists the characteristics of both nanoparticles and the PCM.

Table 1. Properties of the PCM and alumina [33].

Property	Al_2O_3	n-octadecane
ρ (kg/m^3)	3970	770
$\beta \times 10^5$ (K^{-1})	0.85	91
k (w/mK)	40	0.157
L (j/kg)	—	242.9
Fusion (C)	—	28×10^3
$\mu \times 10^3$ (Pa s)	—	3.79
C_p (j/kg K)	765	2189

The enthalpy is formulated as [50]:

$$h = h_{ref} + \int_{T_{ret}}^{T} (C_p)_{nf} dT \tag{11}$$

$$\lambda = \begin{cases} 1 & T < T_l \\ \frac{T-T_s}{T_l-T_s} & T_s < T < T_l, H_e = h + \lambda L \\ 0 & T < T_s \end{cases} \tag{12}$$

The formulas of $S_{\text{gen, total}}$, $S_{\text{gen, th}}$, and $S_{\text{gen, f}}$ are:

$$\begin{aligned} S_{\text{gen,total}} &= S_{\text{gen,th}} + S_{\text{gen,f}} \\ &= \frac{k_{nf}}{T^2}\left[\left(\frac{\partial T}{\partial x}\right)^2 + \left(\frac{\partial T}{\partial y}\right)^2\right] \\ &+ \frac{\mu_{nf}}{T}\left\{2\left[\left(\frac{\partial u_x}{\partial x}\right)^2 + \left(\frac{\partial u_y}{\partial y}\right)^2\right] + \left(\frac{\partial u_x}{\partial y} + \frac{\partial u_y}{\partial x}\right)^2\right\} \end{aligned} \tag{13}$$

The no-slip boundary conditions are subjected to the previous system ($u = v = T = 0$) on the outer boundaries, while on the inner fins $u = v = 0,\ T = T_h$.

2.2. GFEM Approach

The Galerkin finite element method handles the transformed coupled Equations (1)–(4), which include both the flow and heat transfer phenomena as well as the aforementioned boundary conditions. The governing equations' weak forms are given and discretized on a nonuniform structural grid. The results are then simulated using mathematical software. The process is described in full in [51,52]. The validation of the present code was obtained and is displayed in Figure 3 using the additional numerical results of Tan et al. [53]. We are confident in our results based on this figure.

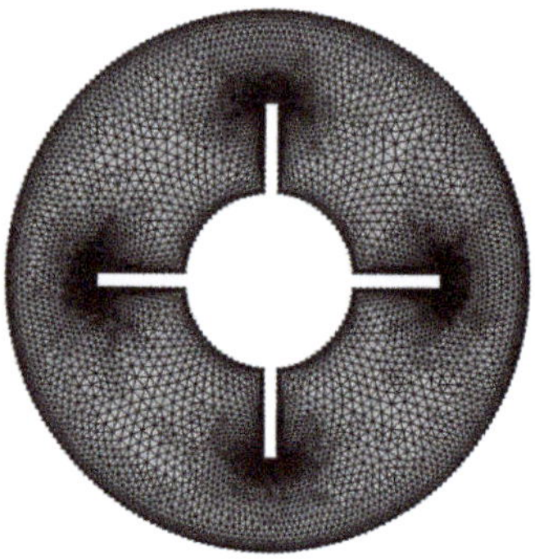

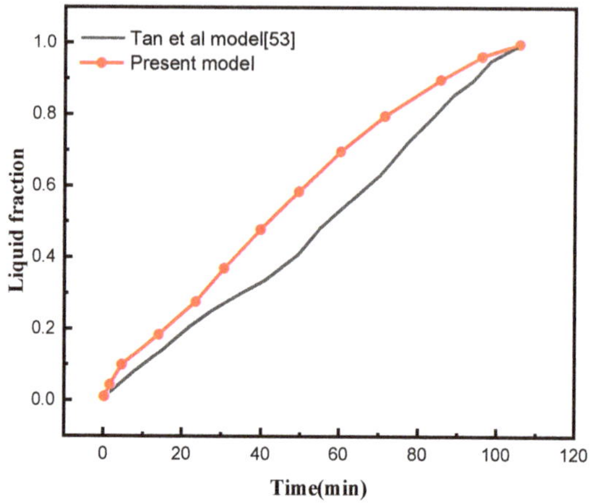

Figure 3. Mesh grid and of the present model in comparison to [53].

3. Results and Discussion

In this section, we illustrate and interpret the results obtained from examining the melting effects on the flow of a suspension that contains the phase change material (PCM). Here, the worked fluid was Al_2O_3/n-octadecane paraffin and the flow area was a cylinder tube including cross-section fins. Features of the temperature, velocity, Bejan number, and liquid fraction were examined for various heating cases, namely, a cylinder with two horizontal wings, a cylinder with two vertical wings, a cylinder with four wings, and a cylinder with eight heated fins. The variations in the time of required ranged from 100 and 600 s and the values of the nanoparticles volume fraction were considered between $\varphi = 0\%$ and 8%. To present a comprehensive investigation, the average values of the liquid fraction β, Bejan number Be_{avg}, and the rate of the heat transfer Nu_{avg} over time are presented graphically for a wide range of the considered parameters. Additionally, the condition of the totally melted NEPCM (liquid fraction = 1) may be used to terminate the computations.

Figure 4 displays the temperature, velocity, local Bejan number, and the liquid fraction for various cases of inner heating. Notably, the temperature features concentrated around the fins in all the cases, with a cold zone indicated near the bottom of the outer cylinder. These temperature distributions achieved their maximum values in case 4 (eight wings) pointing to a decrease in the aforementioned cold zone at the bottom. We also observed that the increase in number of wings reduced the temperature differences; hence, both the temperature gradients and heat transfer rate diminished. Additionally, a clear reduction in the velocity values was noted as the number of wings increased. Physically, the increase in the number of wings enhances the complexity in the flow area; hence, the flow resistance is augmented. In the same context, the features of the Bejan number showed that the increase in the number of heated fins reduced the gradients of the temperature; hence, the fluid friction irreversibility became dominant. Furthermore, the melting zone was seen in the upper half of the domain for all the considered cases as the increase in number of heated fins enhanced the melted zone.

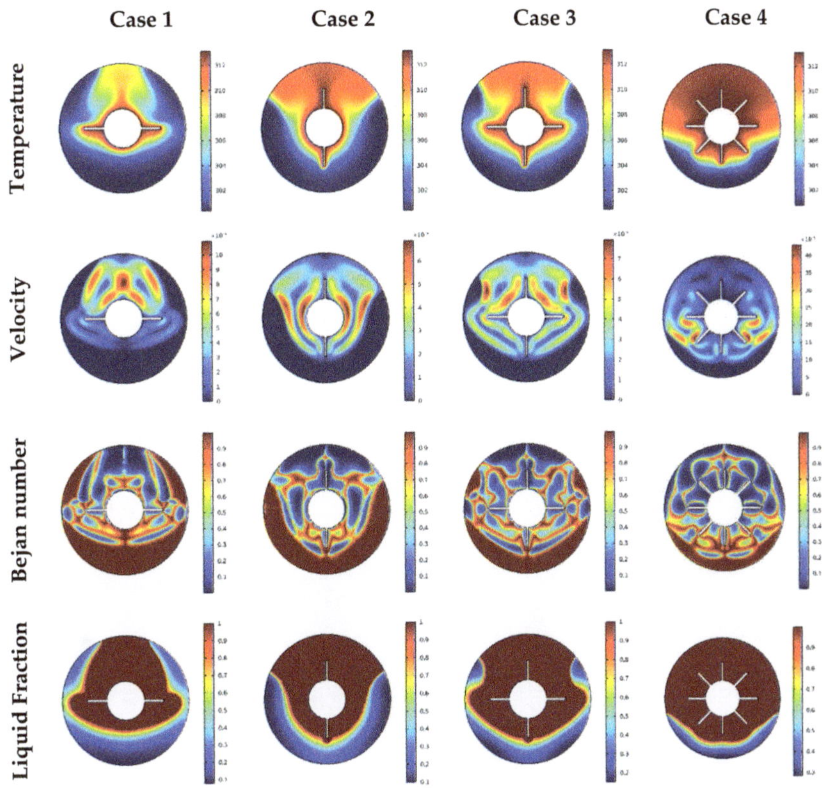

Figure 4. Temperature, velocity, Bejan number, and liquid fraction contour in different geometries.

The features of the temperature, velocity, local *Be* number, and local liquid fraction with variations in time are depicted in Figure 5. During these computations, case 3 used an inner cylinder with four wings. The results indicated that at the beginning of the calculations (small values of the time), the distributions of the temperature, velocity, and Bejan number occurred around the inner part heated, indicating a nonactive zone near the outer boundaries. Over time, the fluid started to carry and distribute the temperature throughout the whole domain. Therefore at $t = 600$ s, a good thermal domain was obtained with a higher velocity rate near the bottom of the outer boundaries. Additionally, for higher time values, the fluid friction irreversibility near the bottom dominated compared to the heat transfer irreversibility. From the physical viewpoint, this behavior is due to the gradients of the velocity that enhance with time, resulting in an augmentation in fluid friction irreversibility. Furthermore, a mushy zone was observed within the full flow domain with increasing time.

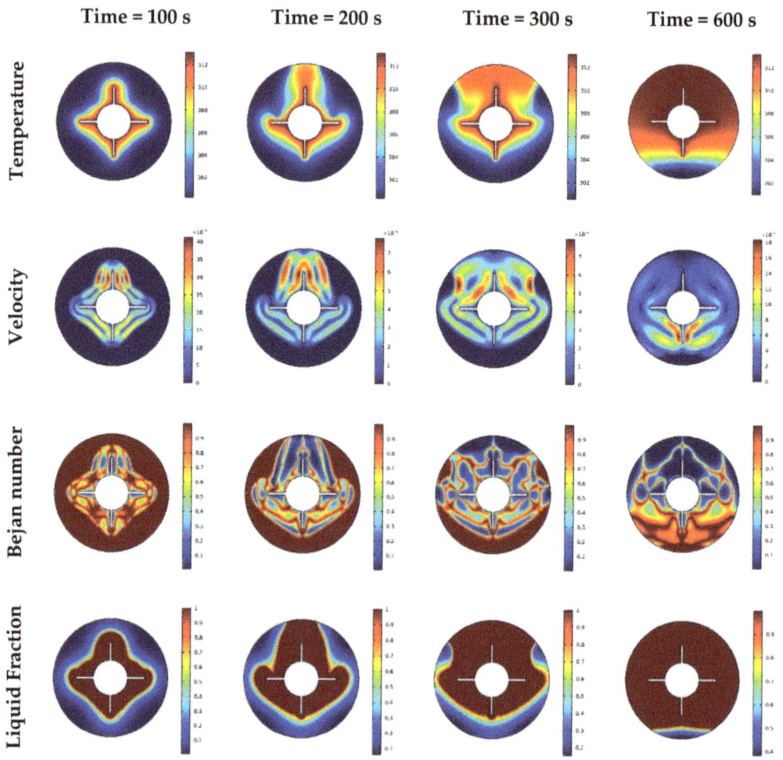

Figure 5. Temperature, velocity, Bejan number, and liquid fraction contour in different time steps.

Figure 6 shows the distributions of the temperature, velocity, local Bejan number, and liquid fraction under impacts of the volume fraction parameter φ. The inner heated cylinders with four wings were used in this case. We noted a low convective transport at higher values of φ due to the increase in the viscosity of the mixture. The results indicated diminishing velocity and temperature gradients with increasing φ. Additionally, the local Bejan number occurred around the wings instead of the bottom boundaries at low values of φ. Conversely, the increase in φ enhanced the mushy zone within the flow area until completely melted conditions were obtained at $\varphi \geq 0.04$.

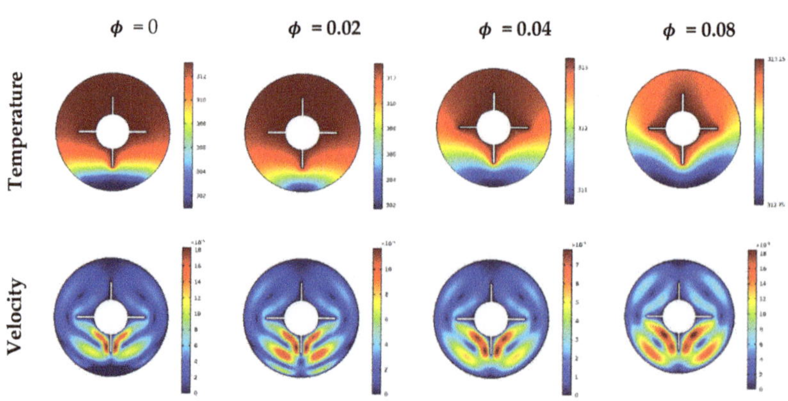

Figure 6. *Cont.*

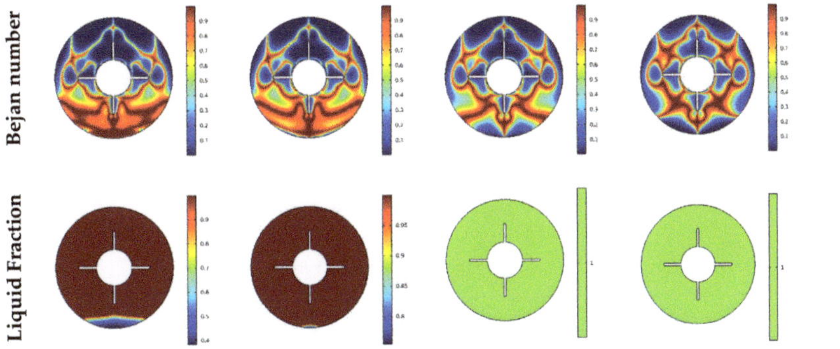

Figure 6. Temperature, velocity, Bejan number, and liquid fraction contour of nanoparticles' concentrations.

Figures 7 and 8 illustrate the profiles of the average liquid fraction β, average Bejan number Be_{avg}, and average Nusselt number Nu_{avg} under with different numbers of heated wings, time parameters, and volume fraction parameters. The results revealed that case 4, in which eight heated wings were assumed, produced the highest average liquid fraction values. However, the average Bejan and Nusselt numbers decreased as the number of heated wings increased. Additionally, the average rate of heat transfer diminished as φ increased due to the decrease in the temperature gradients. Furthermore, higher values of φ caused the irreversibility of the heat transfer to be dominant compared to the fluid friction irreversibility. Finally, we observed that the increase in the volume fraction parameter φ enhanced the mushy zone and, hence, the average liquid fraction rose.

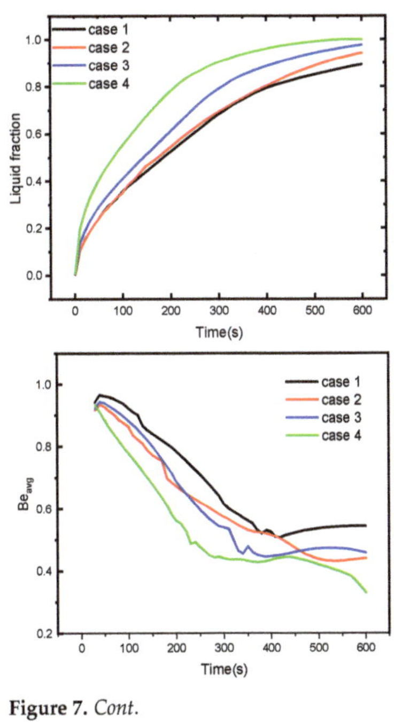

Figure 7. *Cont.*

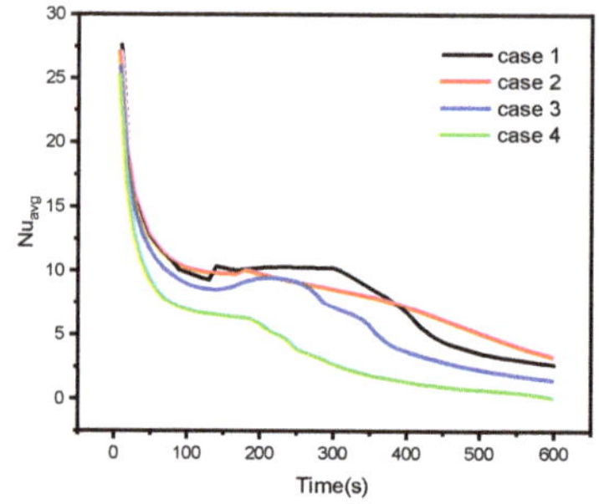

Figure 7. The influences of geometry on the liquid fraction, Nusselt number, and Bejan number.

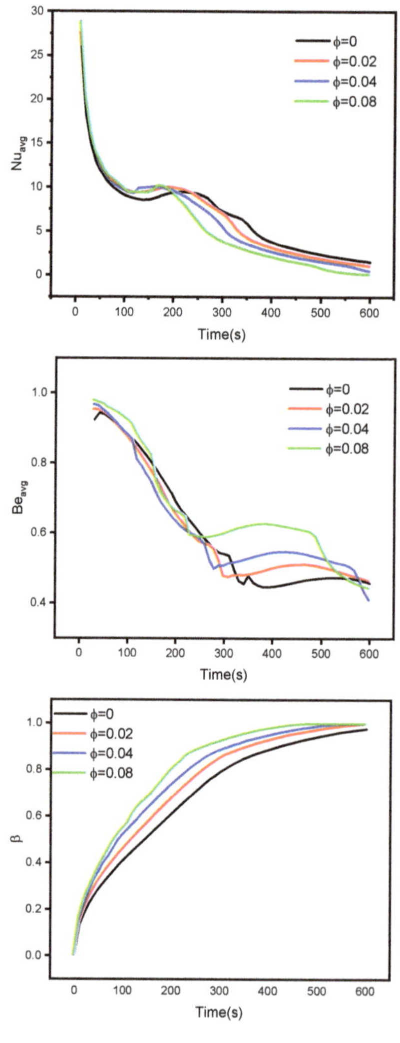

Figure 8. The influences of nanoparticles' concentration on liquid fraction, Nusselt. Number, and Bejan number.

4. Conclusions

This paper presented a numerical investigation into the impacts of melting on the convective flow of phase change materials within cylindrical tubes containing cross-shape heated sections. Four cases were considered based on the number of heated wings, namely, case 1 (two horizontal wings), case 2 (two vertical wings), case 3 (four wings), and case 4 (eight heated wings). The unsteady case was considered and completely melted conditions were assumed. The finite element method (FEM) with the Poisson pressure equation was applied to solve the governing system. The following are our major findings:

- Distributions of the temperature, velocity, and Bejan number increase as with an increasing number of heated wings due to the augmentation in the buoyancy convective case. Additionally, the melted area was controlled for the most of the flow domain in case 4.
- For small time values, the increases in temperature, velocity, and liquid fraction occur around the inner heated shapes, but over time, a good isothermal and melted flow domain is obtained.
- Increases in φ cause an enhancement in the dynamic viscosity of the mixture; hence, the velocity decreases as φ increases.
- With time, the irreversibility due to fluid friction becomes more dominant compare to heat transfer irreversibility.

Author Contributions: Conceptualization, S.E.A. and A.A.; methodology, A.A.; software, A.A.; validation, W.K.H., R.A.A. and A.A.; formal analysis, O.Y., and S.A; investigation, S.E.A.; resources, A.A.; data curation, S.A.; writing—original draft preparation, S.A.; writing—review and editing, W.K.H., R.A.A., and O.Y.; visualization, W.K.H.; supervision, A.A.; project administration, A.A.; funding acquisition, S.A. All authors have read and agreed to the published version of the manuscript.

Funding: This research work did not receive any funds.

Institutional Review Board Statement: Not applicable.

Informed Consent Statement: Not applicable.

Data Availability Statement: All data are available upon request from any of the authors.

Conflicts of Interest: The authors declare no conflict of interest.

Nomenclature

g	Gravity
T_m	Fusion temperature
C	Mushy zone constant
NEPCM	Nanoenhanced PCM
L_f	Latent heat coefficient
k	Thermal conductivity
T_s	Solidus temperature
FVM	Finite volume method
T_l	Liquidus temperature
Greek symbols	
α	Thermal diffusivity (m^2/s)
ρ	Fluid density
ϕ	Nanoparticle volume fraction
Subscripts	
nf	NEPCM
f	Pure fluid

References

1. Mi, X.; Liu, R.; Cui, H.; Memon, S.A.; Xing, F.; Lo, Y. Energy and economic analysis of building integrated with PCM in different cities of China. *Appl. Energy* **2016**, *175*, 324–336. [CrossRef]

2. Souayfane, F.; Biwole, P.H.; Fardoun, F.; Achard, P. Energy performance and economic analysis of a TIM-PCM wall under different climates. *Energy* **2019**, *169*, 1274–1291. [CrossRef]
3. Dehmous, M.; Franquet, E.; Lamrous, N. Mechanical and thermal characterizations of various thermal energy storage concretes including low-cost bio-sourced PCM. *Energy Build.* **2021**, *241*, 110878. [CrossRef]
4. Jaguemont, J.; Omar, N.; Van den Bossche, P.; Mierlo, J. Phase-change materials (PCM) for automotive applications: A review. *Appl. Therm. Eng.* **2018**, *132*, 308–320. [CrossRef]
5. Magendran, S.S.; Khan, F.S.A.; Mubarak, N.; Vaka, M.; Walvekar, R.; Khalid, M.; Abdullah, E.; Nizamuddin, S.; Karri, R.R. Synthesis of organic phase change materials (PCM) for energy storage applications: A review. *Nano-Struct. Nano-Objects* **2019**, *20*, 100399. [CrossRef]
6. Dardir, M.; Panchabikesan, K.; Haghighat, F.; El Mankibi, M.; Yuan, Y. Opportunities and challenges of PCM-to-air heat exchangers (PAHXs) for building free cooling applications—A comprehensive review. *J. Energy Storage* **2019**, *22*, 157–175. [CrossRef]
7. Allouhi, A.; Msaad, A.A.; Amine, M.B.; Saidur, R.; Mahdaoui, M.; Kousksou, T.; Pandey, A.K.; Jamil, A.; Moujibi, N.; Benbassou, A. Optimization of melting and solidification processes of PCM: Application to integrated collector storage solar water heaters (ICSSWH). *Sol. Energy* **2018**, *171*, 562–570. [CrossRef]
8. Huang, P.; Wei, G.; Cui, L.; Xu, C.; Du, X. Numerical investigation of a dual-PCM heat sink using low melting point alloy and paraffin. *Appl. Therm. Eng.* **2021**, *189*, 116702. [CrossRef]
9. Carmona, M.; Rincón, A.; Gulfo, L. Energy and exergy model with parametric study of a hot water storage tank with PCM for domestic applications and experimental validation for multiple operational scenarios. *Energy Convers. Manag.* **2020**, *222*, 113189. [CrossRef]
10. Li, M.; Cao, Q.; Pan, H.; Wang, X.; Lin, Z. Effect of melting point on thermodynamics of thin PCM reinforced residential frame walls in different climate zones. *Appl. Therm. Eng.* **2020**, *188*, 116615. [CrossRef]
11. Kharbouch, Y.; Ouhsaine, L.; Mimet, A.; El Ganaoui, M. Thermal performance investigation of a PCM-enhanced wall/roof in northern Morocco. *Build. Simul.* **2018**, *11*, 1083–1093. [CrossRef]
12. Yun, B.Y.; Park, J.H.; Yang, S.; Wi, S.; Kim, S. Integrated analysis of the energy and economic efficiency of PCM as an indoor decoration element: Application to an apartment building. *Sol. Energy* **2020**, *196*, 437–447. [CrossRef]
13. Dardir, M.; Roccamena, L.; El Mankibi, M.; Haghighat, F. Performance analysis of an improved PCM-to-air heat exchanger for building envelope applications—An experimental study. *Sol. Energy* **2019**, *199*, 704–720. [CrossRef]
14. Mukhamet, T.; Kobeyev, S.; Nadeem, A.; Memon, S.A. Ranking PCMs for building façade applications using multi-criteria decision-making tools combined with energy simulations. *Energy* **2021**, *215*, 119102. [CrossRef]
15. Padala, S.K.; Deshpande, S.J.; Bhattacharjee, B. Assessment of setting characteristics, water absorption, thermal performance and compressive strength of energy-efficient phase change material (PCM)–ashcrete blocks. *Sadhana-Acad. Proc. Eng. Sci.* **2021**, *46*, 1–19. [CrossRef]
16. Chen, S.; Yang, Y.; Olomi, C.; Zhu, L. Numerical study on the winter thermal performance and energy saving potential of thermo-activated PCM composite wall in existing buildings. *Build. Simul.* **2020**, *13*, 237–256. [CrossRef]
17. Gholamibozanjani, G.; Farid, M. Application of an active PCM storage system into a building for heating/cooling load reduction. *Energy* **2020**, *210*, 118572. [CrossRef]
18. Tyagi, V.V.; Pandey, A.K.; Kothari, R.; Tyagi, S.K. Thermodynamics and performance evaluation of encapsulated PCM-based energy storage systems for heating application in building. *J. Therm. Anal. Calorim.* **2014**, *115*, 915–924. [CrossRef]
19. Violidakis, I.; Zeneli, M.; Atsonios, K.; Strotos, G.; Nikolopoulos, N.; Karellas, S. Dynamic modelling of an ultra high temperature PCM with combined heat and electricity production for application at residential buildings. *Energy Build.* **2020**, *222*, 110067. [CrossRef]
20. Frazzica, A.; Brancato, V.; Palomba, V.; La Rosa, D.; Grungo, F.; Calabrese, L.; Proverbio, E. Thermal performance of hybrid cement mortar-PCMs for warm climates application. *Sol. Energy Mater. Sol. Cells* **2019**, *193*, 270–280. [CrossRef]
21. Atinafu, D.G.; Ok, Y.S.; Kua, H.W.; Kim, S. Thermal properties of composite organic phase change materials (PCMs): A critical review on their engineering chemistry. *Appl. Therm. Eng.* **2020**, *181*, 115960. [CrossRef]
22. Dmitruk, A.; Naplocha, K.; Kaczmar, J.W.; Smykowski, D. Pin-fin metal alloy structures enhancing heat transfer in PCM-based heat storage units. *Heat Mass Transf.* **2020**, *56*, 2265–2271. [CrossRef]
23. Sathe, T.; Dhoble, A. Thermal analysis of an inclined heat sink with finned PCM container for solar applications. *Int. J. Heat Mass Transf.* **2019**, *144*, 118679. [CrossRef]
24. Nie, B.; Chen, J.; Du, Z.; Li, Y.; Zhang, T.; Cong, L.; Zou, B.; Ding, Y. Thermal performance enhancement of a phase change material (PCM) based portable box for cold chain applications. *J. Energy Storage* **2021**, *40*, 102707. [CrossRef]
25. Izgi, B.; Arslan, M. Numerical analysis of solidification of PCM in a closed vertical cylinder for thermal energy storage applications. *Heat Mass Transf.* **2020**, *56*, 2909–2922. [CrossRef]
26. Sweidan, A.H.; Heider, Y.; Markert, B. Modeling of PCM-based enhanced latent heat storage systems using a phase-field-porous media approach. *Contin. Mech. Thermodyn.* **2020**, *32*, 861–882. [CrossRef]
27. Tarigond, H.; Reddy, R.M.; Maheswari, C.U.; Reddy, E.S. Effect of iron scrap additives in stearic acid as PCM for thermal energy storage system. *J. Therm. Anal. Calorim.* **2020**, *141*, 2497–2510. [CrossRef]
28. Ahmed, N.; Elfeky, K.; Lu, L.; Wang, Q. Thermal performance analysis of thermocline combined sensible-latent heat storage system using cascaded-layered PCM designs for medium temperature applications. *Renew. Energy* **2020**, *152*, 684–697. [CrossRef]

29. Selimefendigil, F.; Oztop, H.F.; Chamkha, A.J. Natural convection in a CuO–water nanofluid filled cavity under the effect of an inclined magnetic field and phase change material (PCM) attached to its vertical wall. *J. Therm. Anal. Calorim.* **2019**, *135*, 1577–1594. [CrossRef]
30. Abu-Hamdeh, N.H.; Akbal, Ö.; Öztop, H.F.; Abusorrah, A.M.; Bayoumi, M.M. A three-dimensional computational analysis of ellipsoidal radiator with phase change. *Int. J. Numer. Methods Heat Fluid Flow* **2020**, *31*, 2072–2087. [CrossRef]
31. Rostami, S.; Afrand, M.; Shahsavar, A.; Sheikholeslami, M.; Kalbasi, R.; Aghakhani, S.; Shadloo, M.S.; Oztop, H.F. A review of melting and freezing processes of PCM/nano-PCM and their application in energy storage. *Energy* **2020**, *211*, 118698. [CrossRef]
32. Jayaprakash, M.C.; Alzahrani, H.A.; Sowmya, G.; Kumar, R.V.; Malik, M.Y.; Alsaiari, A.; Prasannakumara, B.C. Thermal distribution through a moving longitudinal trapezoidal fin with variable temperature-dependent thermal properties using DTM-Pade approximant. *Case Stud. Therm. Eng.* **2021**, *28*, 101697. [CrossRef]
33. Baslem, A.; Sowmya, G.; Gireesha, B.; Prasannakumara, B.; Rahimi-Gorji, M.; Hoang, N.M. Analysis of thermal behavior of a porous fin fully wetted with nanofluids: Convection and radiation. *J. Mol. Liq.* **2020**, *307*, 112920. [CrossRef]
34. Sowmya, G.; Gireesha, B.J.; Sindhu, S.; Prasannakumara, B.C. Investigation of Ti_6Al_4V and AA7075 alloy embedded nanofluid flow over longitudinal porous fin in the presence of internal heat generation and convective condition. *Commun. Theor. Phys.* **2020**, *72*, 025004. [CrossRef]
35. Sowmya, G.; Gireesha, B.J.; Prasannakumara, B.C. Scrutinization of different shaped nanoparticle of molybdenum disulfide suspended nanofluid flow over a radial porous fin. *Int. J. Numer. Methods Heat Fluid Flow* **2019**, *30*, 3685–3699. [CrossRef]
36. Jeong, S.G.; Jeon, J.; Chung, O.; Kim, S.; Kim, S. Evaluation of PCM/diatomite composites using exfoliated graphite nanoplatelets (xGnP) to improve thermal properties. *J. Therm. Anal. Calorim.* **2013**, *114*, 689–698. [CrossRef]
37. Nóbrega, C.R.E.S.; Ismail, K.A.R.; Lino, F.A.M. Thermal performance of bare and finned tubes submersed in nano-PCM mixture. *J. Braz. Soc. Mech. Sci. Eng.* **2021**, *43*, 1–14. [CrossRef]
38. Raj, C.R.; Suresh, S.; Vasudevan, S.; Chandrasekar, M.; Singh, V.K.; Bhavsar, R. Thermal performance of nano-enriched form-stable PCM implanted in a pin finned wall-less heat sink for thermal management application. *Energy Convers. Manag.* **2020**, *226*, 113466. [CrossRef]
39. Jourabian, M.; Farhadi, M.; Darzi, A.A.R. Heat transfer enhancement of PCM melting in 2D horizontal elliptical tube using metallic porous matrix. *Theor. Comput. Fluid Dyn.* **2016**, *30*, 579–603. [CrossRef]
40. Das, D.; Bordoloi, U.; Muigai, H.H.; Kalita, P. A novel form stable PCM based bio composite material for solar thermal energy storage applications. *J. Energy Storage* **2020**, *30*, 101403. [CrossRef]
41. Vennapusa, J.R.; Konala, A.; Dixit, P.; Chattopadhyay, S. Caprylic acid based PCM composite with potential for thermal buffering and packaging applications. *Mater. Chem. Phys.* **2020**, *253*, 123453. [CrossRef]
42. Rathore, P.K.S.; Shukla, S.K. Improvement in thermal properties of PCM/Expanded vermiculite/expanded graphite shape stabilized composite PCM for building energy applications. *Renew. Energy* **2021**, *176*, 295–304. [CrossRef]
43. Qureshi, Z.A.; Al-Omari, S.A.B.; Elnajjar, E.; Al-Ketan, O.; Abu Al-Rub, R. Using triply periodic minimal surfaces (TPMS)-based metal foams structures as skeleton for metal-foam-PCM composites for thermal energy storage and energy management applications. *Int. Commun. Heat Mass Transf.* **2020**, *124*, 105265. [CrossRef]
44. Nada, S.A.; Alshaer, W.G. Experimental investigation of thermal conductivity enhancement of carbon foam saturated with PCM and PCM/MWCNTs composite for energy storage systems. *Heat Mass Transf.* **2019**, *55*, 2667–2677. [CrossRef]
45. Mehryan, S.; Heidarshenas, M.H.; Hajjar, A.; Ghalambaz, M. Numerical study of melting-process of a non-Newtonian fluid inside a metal foam. *Alex. Eng. J.* **2020**, *59*, 191–207. [CrossRef]
46. Al-Kouz, W.; Aissa, A.; Koulali, A.; Jamshed, W.; Moria, H.; Nisar, K.S.; Mourad, A.; Abdel-Aty, A.H.; Khashan, M.M.; Yahia, I.S. MHD darcy-forchheimer nanofluid flow and entropy optimization in an odd-shaped enclosure filled with a (MWCNT-Fe_3O_4/water) using galerkin finite element analysis. *Sci. Rep.* **2021**, *11*, 1–15.
47. Al-Kouz, W.; Bendrer, B.A.-I.; Aissa, A.; Almuhtady, A.; Jamshed, W.; Nisar, K.S.; Mourad, A.; Alshehri, N.A.; Zakarya, M. Galerkin finite element analysis of magneto two-phase nanofluid flowing in double wavy enclosure comprehending an adiabatic rotating cylinder. *Sci. Rep.* **2021**, *11*, 1–15. [CrossRef]
48. Shahzad., F.; Jamshed, W.; Sathyanarayanan, S.U.D.; Aissa, A.; Madheshwaran, P.; Mourad, A. Thermal analysis on Darcy-Forchheimer swirling Casson hybrid nanofluid flow inside parallel plates in parabolic trough solar collector: An application to solar aircraft. *Int. J. Energy Res.* **2021**, *45*, 20812–20834. [CrossRef]
49. Mourad, A.; Aissa, A.; Mebarek-Oudina, F.; Jamshed, W.; Ahmed, W.; Ali, H.M.; Rashad, A.M. Galerkin finite element analysis of thermal aspects of Fe_3O_4-MWCNT/water hybrid nanofluid filled in wavy enclosure with uniform magnetic field effect. *Int. Commun. Heat Mass Transf.* **2021**, *126*, 105461. [CrossRef]
50. Bouzennada, T.; Mechighel, F.; Filali, A.; Ghachem, K.; Kolsi, L. Numerical investigation of heat transfer and melting process in a PCM capsule: Effects of inner tube position and Stefan number. *Case Stud. Therm. Eng.* **2021**, *27*, 101306. [CrossRef]
51. Ghalambaz, M.; Doostani, A.; Chamkha, A.J.; Ismael, M. Melting of nanoparticles-enhanced phase-change materials in an enclosure: Effect of hybrid nanoparticles. *Int. J. Mech. Sci.* **2017**, *134*, 85–97. [CrossRef]
52. Sheikholeslami, M.; Jafaryar, M.; Shafee, A.; Li, Z. Hydrothermal and second law behavior for charging of NEPCM in a two dimensional thermal storage unit. *Chin. J. Phys.* **2019**, *58*, 244–252. [CrossRef]
53. Tan, F.; Hosseinizadeh, S.; Khodadadi, J.; Fan, L. Experimental and computational study of constrained melting of phase change materials (PCM) inside a spherical capsule. *Int. J. Heat Mass Transf.* **2009**, *52*, 3464–3472. [CrossRef]

nanomaterials

Article

Natural Convection within Inversed T-Shaped Enclosure Filled by Nano-Enhanced Phase Change Material: Numerical Investigation

Aissa Abderrahmane [1], Mohammad Al-Khaleel [2,3,*], Abed Mourad [1], Houssem Laidoudi [4], Zied Driss [5], Obai Younis [6], Kamel Guedri [7] and Riad Marzouki [8,9]

1 Laboratoire de Physique Quantique de la Matière et Modélisation Mathématique (LPQ3M), University Mustapha Stambouli of Mascara, Mascara 29000, Algeria
2 Department of Mathematics, Khalifa University, Abu Dhabi 127788, United Arab Emirates
3 Department of Mathematics, Yarmouk University, Irbid 21163, Jordan
4 Faculty of Mechanical Engineering, University of Sciences and the Technology of Oran, Oran 31000, Algeria
5 Laboratory of Electromechanical Systems (LASEM), National School of Engineers of Sfax (ENIS), University of Sfax (US), B.P. 1173, Road Soukra km 3.5, Sfax 3038, Tunisia
6 Department of Mechanical Engineering, College of Engineering in Wadi Addwasir, Prince Sattam Bin Abdulaziz University, Wadi Addwasir 11911, Saudi Arabia
7 Mechanical Engineering Department, College of Engineering and Islamic Architecture, Umm Al-Qura University, P.O. Box 5555, Makkah 21955, Saudi Arabia
8 Chemistry Department, College of Science, King Khalid University, Abha 61413, Saudi Arabia
9 Chemistry Department, Faculty of Sciences of Sfax, University of Sfax, Sfax 3038, Tunisia
* Correspondence: mohammad.alkhaleel@ku.ac.ae

Abstract: Energy saving has always been a topic of great interest. The usage of nano-enhanced phase change material NePCM is one of the energy-saving methods that has gained increasing interest. In the current report, we intend to simulate the natural convection flow of NePCM inside an inverse T-shaped enclosure. The complex nature of the flow results from the following factors: the enclosure contains a hot trapezoidal fin on the bottom wall, the enclosure is saturated with pours media, and it is exposed to a magnetic field. The governing equations of the studied system are numerically addressed by the higher order Galerkin finite element method (GFEM). The impacts of the Darcy number (Da = 10^{-2}–10^{-5}), Rayleigh number (Ra = 10^3–10^6), nanoparticle volume fraction (φ = 0–0.08), and Hartmann number (Ha = 0–100) are analyzed. The results indicate that both local and average Nusselt numbers were considerably affected by Ra and Da values, while the influence of other parameters was negligible. Increasing Ra (increasing buoyancy force) from 10^3 to 10^6 enhanced the maximum average Nusselt number by 740%, while increasing Da (increasing the permeability) from 10^{-5} to 10^{-2} enhanced both the maximum average Nusselt number and the maximum local Nusselt number by the same rate (360%).

Keywords: magnetohydrodynamics; inversed T-shaped enclosure; NEPCM; natural convection; nanofluid

Citation: Abderrahmane, A.; Al-Khaleel, M.; Mourad, A.; Laidoudi, H.; Driss, Z.; Younis, O.; Guedri, K.; Marzouki, R. Natural Convection within Inversed T-Shaped Enclosure Filled by Nano-Enhanced Phase Change Material: Numerical Investigation. *Nanomaterials* **2022**, *12*, 2917. https://doi.org/10.3390/nano12172917

Academic Editor: S M Sohel Murshed

Received: 3 August 2022
Accepted: 19 August 2022
Published: 24 August 2022

1. Introduction

In the last several decades, the heat transport of nanofluids in various shapes with varying outset and boundary conditions has been a popular research point. This is explained by the fact that these geometries have been widely used in real-world applications, such as building thermal management, electronic device cooling, biochemical and food processing, and renewable energy applications [1–7]. Raizah et al. [8] examined nanofluid natural convection (NC) flow inside a V-shaped cavity saturated with porous media. The findings showed that the best-case scenario for porous media is a horizontal heterogeneous porous medium. The buoyancy force is augmented with a Rayleigh number increase, which improves convective transport. In a triangular fin-shaped cavity, Khan et al. [9] presented a computational investigation of the convective heat transport of a hybrid nanofluid. Their findings demonstrate that both the nanoparticles' Rayleigh number and solid volume percentage raise the local and average Nusselt numbers. In a uniquely formed cavity, Ghalambaz et al. [10] analyzed the heat transmission and irreversibility of a hybrid nanosuspension. The findings demonstrate that raising the nanoparticle concentration accelerates the rate of entropy formation for all Rayleigh number values. Asmadi et al. [11] investigated how a hybrid nanofluid transfers heat naturally by convection within a U-shaped container with varying heating configurations. The findings indicate that the continuous heating setting delivers the optimum heat removal performance, whereas the oscillating heating setting performed the poorest. The magnetohydrodynamic free convection flow and heat transfer in an angled U-shaped enclosure loaded with Cu-water nanofluid were studied by Nabwey et al. [12]. The findings demonstrate that the mean Nu increases with dimensionless heat source position but decreases with heat source length and Ha number. A numerical study on the free convection flow of a hybrid nanofluid in a reservoir with a trapezoidal form under the impact of partial magnetic fields was conducted by Geridonmez et al. [13]. According to the results, heat transmission and fluid movement are inhibited by the partial magnetic field's vast effect zone.

Conventional heating and cooling systems rely on a working fluid with a limited thermal capacity. Several systems incorporate a greater flow or bigger volume to address this problem, which is an inadequate solution in some applications. In this context, NEPCM is one of the effective approaches for increasing the thermal efficiency of various systems. In this encapsulation technique, PCM is sealed inside nanoshells to prevent leakage. In this structure, the PCM-containing core layer may store and release enormous amounts of energy during melting and solidification at a constant fusion temperature. The use of NEPCM rather than only heat transfer fluid offers several benefits in many applications. NEPCM benefits from both heat transfer fluid and PCM characteristics [14–17]. For instance, it was recently shown that these could improve thermal storage characteristics and also for applications such as triple tube heat exchangers [18] or shell-and-tube heat exchangers [19]. To keep the temperature of lithium-ion batteries (LIBs) stable between 35 and 45 °C, Cao et al. [20] utilized water loaded with NEPCM particles. They found that raising the Reynolds number value from 70 to 100 increased the rate of heat transmission of LIBs by 12.1–17.2%. Raising the volume fraction from 0 to 3% also increased the heat transmission rate by 8.2–13.6%. Mohammadpour et al. [21] investigated NEPCM slurry's hydrodynamic and heat transfer characteristics inside a microchannel heat sink featuring two circular synthetic jets. The thermal performance improvement is maximized at 28.5% at 0.2 nanoparticle volume fraction and 180 out-of-phase actuation, according to simulation findings. On the other hand, the figure of merit falls when the concentration of NEPCM rises. In a conical diffuser, Iachachene et al. [22] investigated the turbulent flow of Al_2O_3, NEPCM, and a mixture of the two. The NEPCM nanofluid had the lowest pressure drop and the largest heat transfer improvements inside the diffuser, according to the findings of this investigation. The Nusselt numbers of NEPCM/Al_2O_3 hybrid and Al_2O_3 nanofluids were enhanced by 10% and 6%, respectively, whereas the Nusselt number of NEPCM nanofluids were raised by 15%. Analytical research on the laminar flow and heat transmission of water jet impingement augmented with NEPCM slurry was conducted by Mohaghegh et al. [23]. The results indicate that NEPCM slurry may greatly increase the system's cooling performance by increasing the liquid jet's ability to store latent heat. However, an ideal NEPCM concentration results in the system's maximal cooling performance (15%). To improve total heat transmission and lessen pressure drop, Doshi et al. [24] developed a water-based NEPCM nanofluid within a new microchannel with a wavy and irregular shape. The result demonstrates that the presence of NEPCM nanoparticles lowers the fluid domain temperature. NEPCM slurry and more conductive materials combined with heat sinks lessen the influence of thermal and frictional entropy creation as well.

Nano-encapsulated phase change material (NEPCM) is compact, has a high specific surface area, is thermally reliable, and has a wide range of potential applications. However, most of the process conditions in use today are rather complex, making it challenging to create NEPCM with good microscopic morphology and outstanding thermal characteristics [25–29]. Liu et al. [30] created a series of NEPCM using a simple sol-gel technique, using disodium hydrogen phosphate dodecahydrate for the core material and silicon dioxide for the shell material. The encapsulation ratio and melting enthalpy of the produced NEPCM reached maximum values of 70.1% and 165.6 J/g, respectively. According to the data, the adequate component ratio and suitable reaction conditions contribute to NEPCM's superior microscopic morphology and thermal characteristics. Stearic acid (SA)/Ag nanocapsules were synthesized by Huanmei Yuan et al. [31] utilizing a Pickering emulsifier and a chemical reduction process. The results demonstrated that the thermal dependability of the nanocapsules was assessed after 2000 thermal cycles, during which time their latent heat marginally decreased by 0.55%. The fabrication of N-Hexacosane-encapsulated Titania phase change composite using a sol-gel approach was achieved by Khanna et al. [32]. The findings showed that the NEPCM was solidified and liquefied at 52.08 °C and 54.02 °C, respectively, with latent heats of 127.37 J/g and 142.09 J/g. The thermogravimetric curves showed that the composite's overall thermal stability increased with the increasing titanium concentration. In another study [33], by using a chemical technique, they established how to manufacture silica NEPCM layered between exfoliated-graphite nanosheets. The results indicated that there were no chemical processes that occurring in the phase transition material, which had a diameter of 120–220 nm. Furthermore, at 57.9 °C and 48.1 °C, respectively, with latent heats of 126.7 J/g and 117.6 J/g, the solid–liquid phase transition of the NEPCM nanocomposite was observed. After 300 heat cycles, the NEPCM composites showed very high durability against thermal deterioration and 15.74 W/m K thermal conductivity.

Lately, the suspension of NEPCMs as a novel type of nanofluid was examined in various enclosures. Cao et al. [34] investigated the free convective of NEPCM nanofluid within an insulated chamber with two pipes acting as cooler and heater sources with a constant temperature boundary condition. They demonstrated how the NEPCM phase transition happens at low Rayleigh numbers but that it has no bearing on the heat transmission rate. Instead, it is heavily linked to the thermal conductivity of the nanofluid. The free convective heat transport in NEPCM nanofluid within a square enclosure that has been differentially heated and rotated with a constant, uniform counterclockwise rotational velocity was studied by Alhashash et al. [35]. The rotation parameter primarily influences the number of cell circulations, the number of inner vortexes, and the strength of those vortexes. Higher rotational speed results in less NEPCM phase transition and slower heat transfer. In an angled L-shaped chamber, Sadeghi et al. [36] studied the free convection and entropy formation of NEPCM. The findings show that the micro-rotation parameter, Stefan number, and nondimensional fusion temperature all negatively affected the NC heat transfer of NEPCMs and decrease the Nuavg by up to 42%, 25%, and 15%, respectively. On the other hand, the Nuavg was increased by up to 36% when more nanoparticles were present. Zidan et al. [37] investigated the NEPCM–water mixture NC flow in a reversed T-shaped porous cavity with two heated corrugated baffles. This research shows that increasing the Raleigh number causes escalation velocity fields and phase change zone structural changes, whereas decreasing the Darcy number has the opposite impact. Hussain et al. [38] looked into the free convection of NEPCM in a grooved enclosure with an oval form and saturated with a porous medium. According to the study's findings, raising the Darcy parameter reduces the porous flow's resistance, which in turn enhances the streamline strength and nanofluid movement. Furthermore, the enlarged radius of the inner oval form creates a barrier in the grooved cavity, which slows the passage of the nanofluid within.

In the literature, there are no studies on the natural convection of NEPCM in an inversed T-shaped cavity, including a trapezoidal fin subjected to a magnetic field. A better understanding of the impact of the studied parameters on heat transfer rates would be beneficial for design engineers, as cavities saturated with porous media are widely founded in engineering applications, including (but not limited to) heat exchangers and electrical components. In this work, the natural convection of NEPCM in an enclosure saturated with porous media is handled by the high-order GFEM. The run simulations look at how viral parameters affect the contours of temperature, heat capacity ratio, and nanofluid velocity within an inverse T-shaped cavity. These parameters include the Darcy number, the concentration of NEPCM nanoparticles in the base fluid, the Rayleigh number, and the Hartmann number representing the intensity of the magnetic field.

2. Problem Formulation

As shown in Figure 1, we consider an inverted T-shaped porous cavity with a trapezoidal fin at the bottom. The porous cavity is loaded with a nanofluid consisting of water as a base fluid and nano-encapsulated PCM (NEPCM) as nanoparticles. This novel substance's nanoparticles are made up of an outer shell and an inner core. The core is made of nonadecane, while the exterior is mostly made of polyurethane. Their thermal properties are shown in Table 1. The fusion temperature, which is restricted to this range $T_h < T_f < T_c$ is what characterizes the single-particle core. The latent heat of the core and the phase transition temperatures are estimated to be 211 (kJ/kg) and 305 (K), respectively. Overall, the PCM cores' capacity to absorb, store, and release heat energy distinguishes them. When combined with a base liquid such as water, they also effectively transmit heat energy.

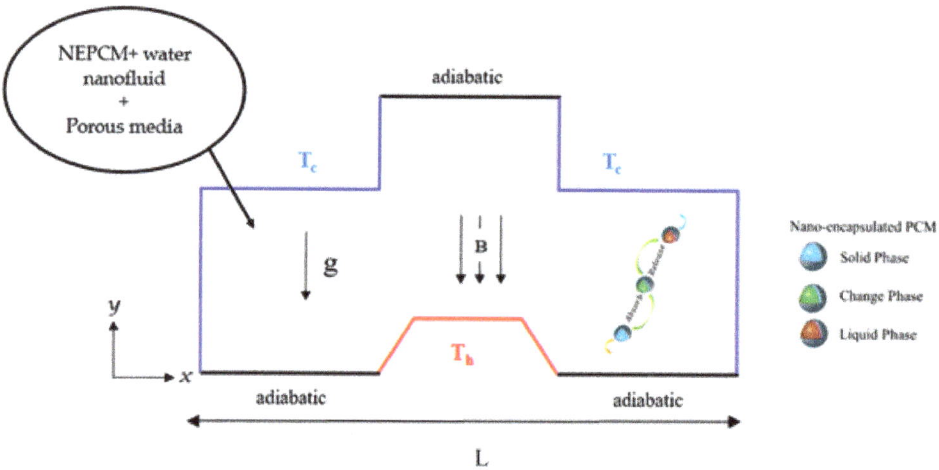

Figure 1. Physical problem.

Table 1. Thermophysical properties of the shell and core of the NEPCMs and the base fluid.

Material	β(K^{-1})	C(kJ/kg K)	k(W/m K)	ρ(kg/m^3)
Polyurethane: shell	17.28×10^{-5}	1.3177		786
Nonadecane: core		2.037		721
Water: base fluid	21×10^{-5}	4.179	0.613	997.1

The governing equations are as follows [39,40]:

The side walls are maintained at low temperature, T_c, while the trapezoidal fin walls are heated and kept at high temperature, Th with ($T_h > T_c$), and the other walls are thermally isolated. For this problem, the flow is assumed to be steady and laminar. The driving force in the geometry under study is the buoyancy force due to the temperature difference between the trapezoidal fin and the sidewalls. A uniform magnetic field is applied. It is believed that the effects of Joule heating, radiation, displacement currents, and viscous dissipation are insignificant. Natural convection is approximated using the Boussinesq approximation in the buoyancy element of the momentum equation. Pressure adjustments do not affect the density of nanoliquids. Temperature gradients, on the other hand, alter the density. The particles are distributed uniformly throughout the host fluid, and dynamic and thermal equilibrium between the nano-additives and the base fluid is established.

$$\nabla \cdot \mathrm{v} = 0 \tag{1}$$

$$\rho_b \cdot \nabla \mathrm{v} = -\nabla p + \nabla \cdot (\mu_b \nabla \mathrm{v}) + (\rho\beta)_b g(T - T_c) - \frac{\sigma_{nf} B_0^2 v}{\rho_{nf}} \tag{2}$$

$$(\rho C_p)_b \mathrm{v} \cdot \nabla T = \nabla \cdot (k_b \nabla T) \tag{3}$$

The current suspension is distinguished by its global density, which is expressed as [41]

$$\rho_b = (1 - \phi)\rho_f + \phi\rho_p \tag{4}$$

where the symbols f, and p denote the base fluid and the added nanoparticles, respectively.

The nanoparticle density of NePCM is provided below:

$$\rho_p = \frac{(1 + t)\rho_{\mathrm{co}}\rho_{\mathrm{sh}}}{\rho_{\mathrm{sh}} + i\rho_{\mathrm{co}}} \tag{5}$$

where the symbols ρ_{sh} , ρ_{co} and ι denote the shell's density, the core's density and the mass ratio of the core-shell ($\iota \sim 0.447$) [41], respectively.

Additionally, the core density of a PCM is the average of its solid and liquid phases. When using NEPCM, the water's specific heat capacitance value may be calculated as

$$C_{p,b} = \frac{(1 - \phi)(\rho C_p)_f + \phi(\rho C_p)_p}{\rho_p} \tag{6}$$

Heat capacitance is considered for a single-phase state, *ip*, *p*, is defined according to the following expression:

$$C_{p,p} = \frac{\left(C_{p,co} + \iota C_{p,sh}\right)\rho_{\mathrm{co}}\rho_{\mathrm{sh}}}{(\rho_{\mathrm{sh}} + \iota\rho_{\mathrm{co}})\rho_p} \tag{7}$$

The heat capacity of the inner substance (core) is considered the average of the heat capacities of both states, solid and fluid. This is because whether the nanoparticle's core is in a solid or fluid state, the latent heat is changed in the form of the heat capacity of the NEPCM. This new form of heat can be defined as follows [42]:

$$C_{P,p} = C_{P,co} + \frac{h_{sf}}{T_{Mr}} \tag{8}$$

$$C_{P,P} = C_{p,;00} + \left\{\frac{\pi}{2} \cdot \left(\frac{h_{sf}}{T_{Mr}} - C_{P,c0}\right) \cdot \sin\left(\pi \frac{T - T_{fu} + (T_{Mr}/2)}{T_{Mr}}\right)\right\} \tag{9}$$

$$C_{P,P} = C_{p,c0} + 2\left(\frac{h_{fs}}{T_{Mr}^2} - \frac{C_{p,C0}}{T_{Mr}}\right)\left(T - T_{fu} + \frac{T_{Mr}}{2}\right). \tag{10}$$

Most of the researchers employ linear interpolation to deal with the phase change due to it is simplicity. However, in this work, we chose to characterize phase change by employing the sine function to assure function continuity in the whole domain, where T_{Mr} is the range of the temperature. This interval circumvents the discontinuity in the stability of energy. The total heat capacity of the NEPCM core incorporating fusion temperature and the sensible is determined based on T_{Mr^*}

$$C_{p,p} = C_{p,co} + \left\{ \frac{\pi}{2} \cdot \left(\frac{h_{sf}}{T_{Mr}} - C_{p,c0} \right) \cdot \sin\left(\pi \frac{T - T_{fu} + (T_{Mr}/2)}{T_{Mr}} \right) \right\} \gamma \tag{11}$$

where

$$\gamma = \begin{cases} 0, T < T_{fu} - \frac{T_{Mr}}{2} \\ 1, T_{fu} - \frac{T_{Mr}}{2} < T < T_{fu} + \frac{T_{Mr}}{2} \\ 0, T > T_{fu} + \frac{T_{Mr}}{2} \end{cases} \tag{12}$$

The suspension's thermal volume expansion rate is calculated as

$$\beta_b = (1 - \phi)\beta_f + \phi\beta_{p}- \tag{13}$$

To estimate the thermal conductivity of a combination including nano-encapsulated particles, the following definitions are stated [43]:

$$\frac{k_b}{k_f} = 1 + Nc\phi. \tag{14}$$

The dynamic conductivity of suspension is

$$\frac{\mu_b}{\mu_f} = 1 + Nv\phi \tag{15}$$

where Nc and Nv in the above expressions define the numbers of thermal conductivity and viscosity, respectively.

The higher the thermal conductivity and viscosity values, the higher the increment in the mixture's thermal conductivity and dynamic viscosity (water and PCM particles). These constant values were established by Ghalambaz et al. [44] for several hybrid nanofluids and nanofluids. It is determined that these expressions are acceptable only for nanofluids if $\phi < 5\%$. The used quantities were considered in dimensionless form as follows:

$$X = \frac{x}{L'},$$

$$Y = \frac{y}{L'}$$

$$\delta = \frac{\delta^*}{L}$$

$$U = \frac{uL}{\alpha_f}$$

$$V = \frac{vL}{a_f}$$

$$P = \frac{p\ell^2}{\rho_f \alpha_f^2}$$

$$\Theta = \frac{T - T_c}{T_h - T_c}$$

The nondimensional mathematical formulations become

$$\frac{\partial U}{\partial X} + \frac{\partial V}{\partial Y} = 0 \tag{16}$$

$$\left(\frac{p_b}{\rho_f} \right) \left(U \frac{\partial U}{\partial X} + V \frac{\partial U}{\partial Y} \right) = \frac{\partial P}{\partial X} + \mathrm{Pr} \left(\frac{\mu_b}{\mu_f} \right) \left(\frac{\partial^2 U}{\partial X^2} + \frac{\partial^2 U}{\partial Y^2} \right) \tag{17}$$

$$\left(\frac{p_t}{\rho_f} \right) \left(U \frac{\partial V}{\partial X} + V \frac{\partial V}{\partial Y} \right) = \frac{\partial P}{\partial Y} + \mathrm{Pr} \left(\frac{\mu_b}{\mu_f} \right) \left(\frac{\partial^2}{\partial X^2} + \mathrm{RaPr} \frac{(\rho\beta)_b}{(\rho\beta)_f} \right) \Theta \tag{18}$$

$$\mathrm{Cr}\left(U\frac{\partial\Theta}{\partial X}+V\frac{\partial\Theta}{\partial Y}\right)=\frac{k_b}{k_f}\left(\frac{\partial^2\Theta}{\partial X^2}+\frac{\partial^2\Theta}{\partial Y^2}\right)-\frac{\sigma_{hnf}}{\sigma}Ha^2V \tag{19}$$

The dimensionless boundary conditions are

$U=V=0,\Theta=1$ *on the trapezoidal fin at the bottom*

$U=V=0,\Theta=0$ *on the side walls*

$U=V=0,\frac{\partial\Theta}{\partial Y}=0$ *on the rest adiabatic walls,*

and Ra and Pr are no dimensional quantities of Rayleigh and Prandtl numbers, respectively:

$$\mathrm{Ra}=\frac{g\rho_f\beta_f\Delta Te^3}{\alpha_f\mu_f} \tag{20}$$

$$\mathrm{Pr}=\frac{\mu_f}{\rho_f\alpha_f} \tag{21}$$

also,

$$\left(\frac{\rho_b}{\rho_f}\right)=(1-\phi)+\phi\left(\frac{\rho_p}{\rho_f}\right) \tag{22}$$

$$\left(\frac{\beta_b}{\beta_f}\right)=(1-\phi)+\phi\left(\frac{\beta_p}{\beta_f}\right) \tag{23}$$

Given that it is presumed that the thermal expansion of water is equivalent to that of NEPCMs, $\left(\beta_b/\beta_f\right)\sim 1$, Cr describes the ratio of heat capacity of the suspension over the water heat capacity: $Cr=\frac{(\rho C_p)_b}{(\rho C_p)_f}=(1-\phi)+\phi\lambda+\frac{\phi}{\delta\,\mathrm{Ste}}f$. (Ste) is the Stefan number, and it is defined as follows:

$$\lambda=\frac{\left(C_{p,co}+uC_{p,sh}\right)\rho_{cof}\rho_{\mathrm{sh}}}{(\rho C_p)_f(\rho_{\mathrm{sh}}+\rho_{\mathrm{co}})} \tag{24}$$

$$\varepsilon=\frac{T_{Mr}}{\Delta T} \tag{25}$$

$$\mathrm{Ste}=\frac{(\rho C_p)_f\Delta T(\rho_{\mathrm{sh}}+\rho_{\mathrm{co}})}{\alpha_f\left(h_{sf}\rho_{\mathrm{co}}\rho_{\mathrm{sh}}\right)} \tag{26}$$

Additionally, the nondimensional fusion expression, f, is given as

$$f=\frac{\pi}{2}\sin\left(\frac{\pi}{\varepsilon}\left(\Theta-\Theta_{fu}+\frac{\varepsilon}{2}\right)\right)\sigma \tag{27}$$

where

$$\sigma=\begin{cases}0, & \Theta<\Theta_{\mathrm{fu}}-\frac{\varepsilon}{2},\\ 1, & \Theta_{\mathrm{fu}}-\frac{\varepsilon}{2}<\Theta<\Theta_{\mathrm{fu}}+\frac{\varepsilon}{2}\\ 0, & \Theta>\Theta_{\mathrm{fu}}+\frac{\varepsilon}{2}.\end{cases} \tag{28}$$

Here, Θ_{fu}, the nondimensional fusion temperature is

$$\Theta_{fu}=\frac{T_{fu}-T_c}{\varepsilon T}. \tag{29}$$

The local Nusselt number of the heated side is obtained:

$$Nu = -(1 + Nc\phi)\frac{\partial\Theta}{\partial Y} \tag{30}$$

Additionally, the averaged form of the Nusselt number of the heated side is given as

$$\overline{Nu} = \int_{-0.5}^{0.5} Nu dY \tag{31}$$

3. Numerical Methodology

Various numerical methods can be exploited in order to find solutions of the differential equations whether in parallel or in sequential (see, e.g., [45–48]). Here, the main equations and accompanying initial boundary conditions are addressed using the Galerkin finite element method. The non-linear partial differential equations are transformed into linear equations using the weighted residual technique [42].

For the validation, the (Nu) on the hot surface at (Re = 500, Ha = 0, ϕ = 4% and N = 4) is utilized. Table 2 displays the results of the mesh independence study. The findings illustrate that the grid size of 21,999 is the ideal option.

Table 2. Grid independence test for Re = 100, Ha = 0, ϕ = 4% and N = 4.

No. of Elements	1141	2318	5400	21,999	81,359
Ψ_{max}	33.072	33.175	33.236	33.225	33.221
Nu_a	10.602	10.624	10.622	10.624	10.624

By using numerical research, the existing results were validated. The results from the model used in the current study are compared to those presented in Ghalambaz et al. [44], as seen in Figure 2.

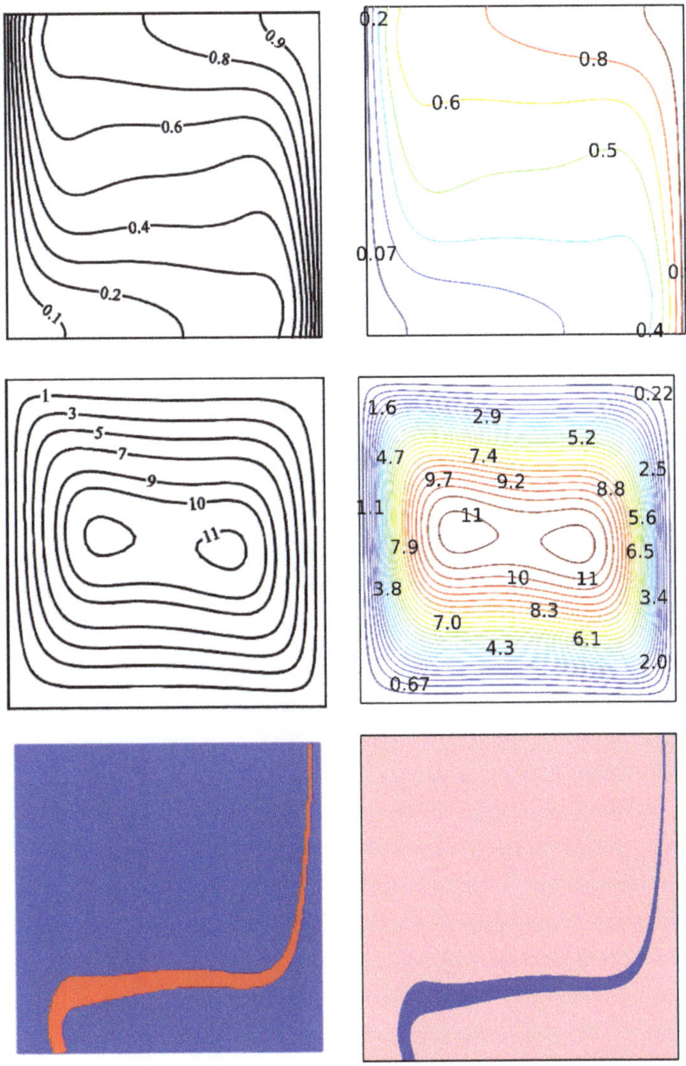

Figure 2. Comparison of current work with that of Mohammad Ghalambaz et al. reprinted/adapted with permission from Ref. [44]. 2022, Elsevier.

4. Results and Discussion

In this important part of the research, we present the results that show the impacts of Darcy's number (Da), Hartmann's number (Ha), Rayleigh's number (Ra) and the concentration of PCM nanoparticles on thermal activity and how the suspension (PCM nanoparticles + water) moves inside the room.

We note a scientific fact that occurs in a fluid: when the fluid absorbs a quantity of thermal energy, its weight decreases and, therefore, begins to move upward as it shifts downward. In contrast, the fluid moves downward as it loses heat energy. This kind of heat transfer accompanied by mass transfer is called Buoyancy-driven flow.

Figure 3 is devoted to understanding the effect of the Ra (between 10^3 and 10^6) number on the studied system for Ha = 0, Da = 10^{-2} and φ = 4%. This understanding is achieved by analyzing the contours of isotherms (dimensionless temperature), streamlines (line paths) and phase-changing zone (heat capacity Cr). The dimensionless temperature (isotherms) shows a gradual temperature distribution from the hot fin toward the cold walls of the container. We note that the fluid layers adjacent to the fin have a high temperature. Then this distribution is centered over the fin in an upward direction. In contrast, cold fluid layers are centered near the lateral walls. It is noted that the intensity of this distribution increases in terms of the number Ra. The streamlines are exactly equivalent to the isotherm patterns, meaning that the fluid layers above the trapezoidal fin move toward the top, while the cold layers on the lateral edges of the chamber move downwards. Finally, we note the formation of a circular motion of the suspension flow. This movement is divided into symmetric parts, one on the right side with the same clockwise direction and the second on the left side in the anticlockwise direction. We also note that the intensity of the movement of the two vortices increases in terms of the number Ra, whereas the center of this recirculation zone moves upward as the value of Ra increases.

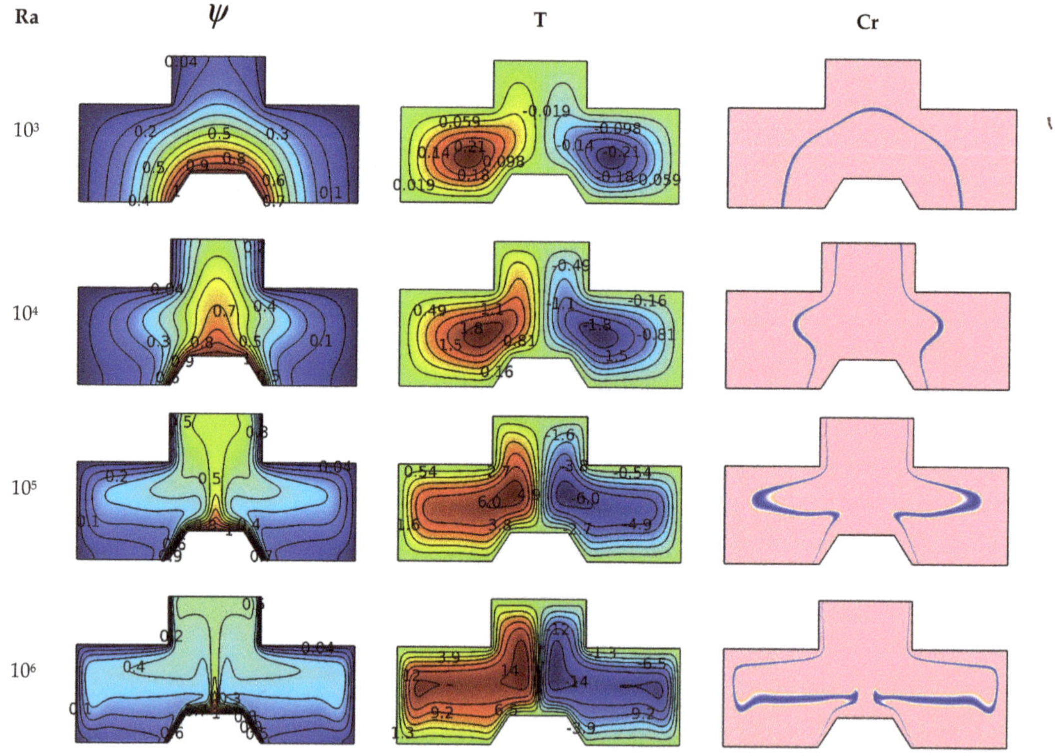

Figure 3. Ra number influence on streamlines, isotherms, and Cr for Da = 10^{-2}, Ha = 0, and φ = 4%.

When Ra = 10^6, the intensity of thermal buoyancy becomes very high, and we notice that the center of the vortex is divided into two parts, the first near the cold wall and the second near the hot fin. For the heat capacity contours (Cr), we notice an effect of the number Ra on the heat capacity (Cr). We notice that there is a line in the form of a semicircle around the trapezoidal fin for Ra = 10^3, then it is divided into two symmetric parts as the number Ra increases. The blue lines indicate the regions where the change of physical state occurs for PCM nanoparticles. We notice that the lines expand in terms of Ra because the movement of suspension particles increases in terms of Ra. In addition to this, it is noticed that the two heat capacity lines are thinner, close to both cold and hot walls, while they are thicker in the middle of the room. The greater the temperature gradient, the thinner the thickness of the equivalent heat capacity line for the zones where the change of physical state occurs.

In order to understand the impact of the values of Hartmann number (between the values 0 to 100) on the studied system, Figure 4 illustrates this effect of Da = 10^{-2}, Ra = 10^5 and φ = 4%. The scientific explanation for the influences of the Hartmann number is found by analyzing the contours of isotherms, streamlines and heat capacity (Cr). It is known that the magnetic field creates an electromagnetic force (Lorentz force) that hinders the movement of fluid particles, so we notice a gradual decline in isotherm distribution in terms of the number Ha. The streamlines also show a decrease in the intensity of vorticity by increasing the value of Ha. The contours of heat capacity (Cr) show that the blue lines of physical state change converge as the value of Ha increases because the velocity of the suspension particles decreases in terms of Ha.

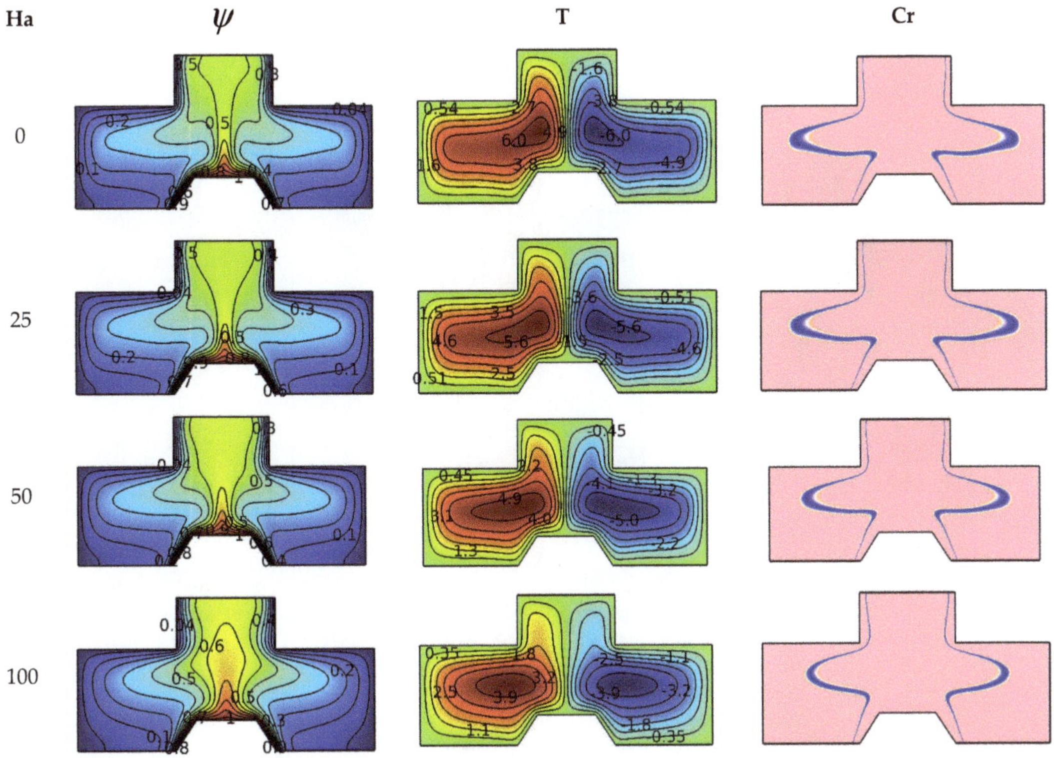

Figure 4. Ha number influence on streamlines, isotherms, and Cr for Ra = 10^5, Da = 10^{-2}, and φ = 4%.

The results of this work also include the effect of the medium permeability on the studied system, so Figure 5 illustrates the influences of the Darcy number (between Da = 10^{-2} and Da = 10^{-2}) on each of the isotherms, heat capacity and streamlines for Ra = 10^5, Ha = 0, and φ = 4%. It is known that the permeability of the space is defined in terms of the Darcy number; that is, the higher the value of this number, the greater the permeability. Based on this, the isotherms show an expansion of the dimensionless temperature in terms of Darcy's number because the suspension movement becomes easier. On the other hand, the streamlines depict an increase in the intensity of the vortices and an increase in their sizes as the value of Da increases. Heat capacity contours are also influenced by the Darcy number because the increase in the speed of the flow expands the blue lines of the regions with the change in the physical state of nano-encapsulated PCM.

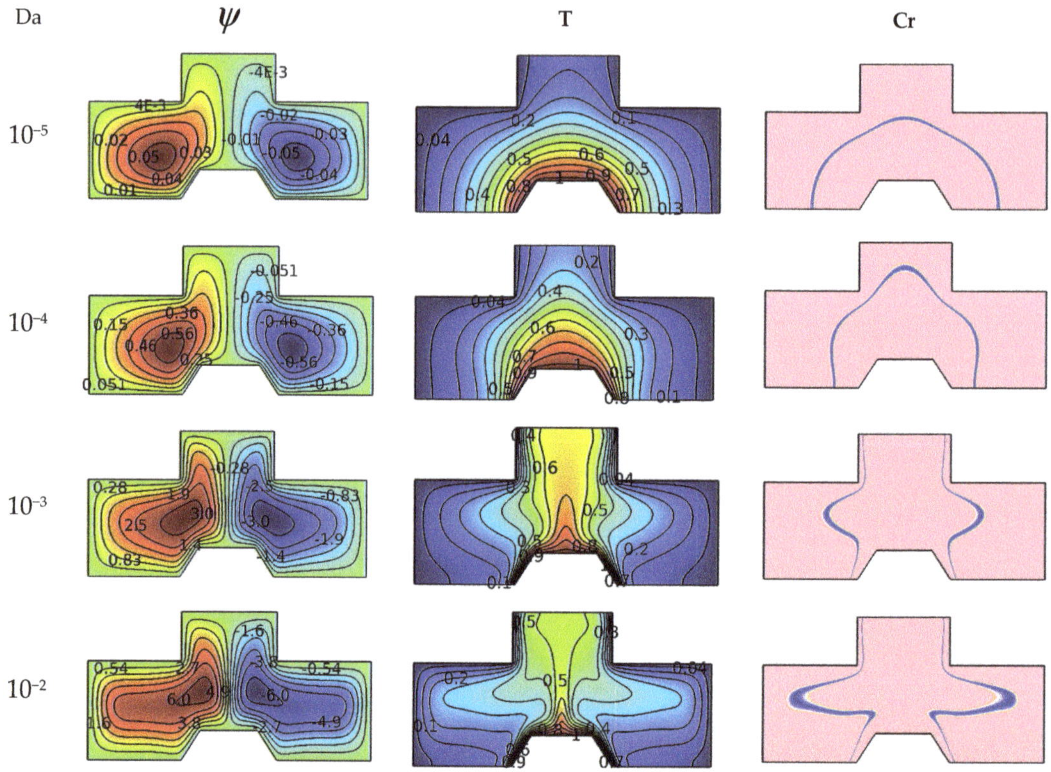

Figure 5. Da number influence on streamlines, isotherms, and Cr for Ra = 10^5, Ha = 0, and φ = 4%.

Figure 6 presents the developments of the mean values of the Nusselt number (Nu) of the hot trapezoidal fin in terms of Da, Ha, Ra and φ. Figure 6a is intended to show the effect of numbers Ha and Ra on Nu. For φ = 4% and Da = 10^{-2} we notice that the values of Nu increase with increasing values of Ra and decrease as the values of Ha increase, just as expected. That is, the higher the speed of the suspension particles, the better the convective heat transfer, and accordingly, the values of the number Nu increase. In addition to this, we notice that the increase in the values of Nu in terms of Ra gradually changes with the increase in the value of Ha. That is, as the value of Ha increases from 0 to 100, the evolution becomes nonlinear. Figure 6b is intended to present the influence of the nano-encapsulated PCM concentration and Ra on Nu. For Ha = 0 and Da = 10^{-2} we notice that in this studied space, there is no strong influence of the nano-encapsulated PCM concentration on Nu. The negligible effect of the ratio of NEPCM on Nu can be explained by the finite motion of the flow within the chamber. Figure 6c presents the effect of the Darcy number, i.e., the porosity of the medium and Ra number on Nu of the hot surfaces of trapezoidal fin for Ha = 0 and φ = 4%. We note an effective effect of the medium permeability (Da number) on Nu values. That is, the values of Nu increase with augmenting Da. The latter can be explained by the following: the expansion of the medium's permeability means a decrease in the flow obstruction, which allows for enhanced convective heat transfer.

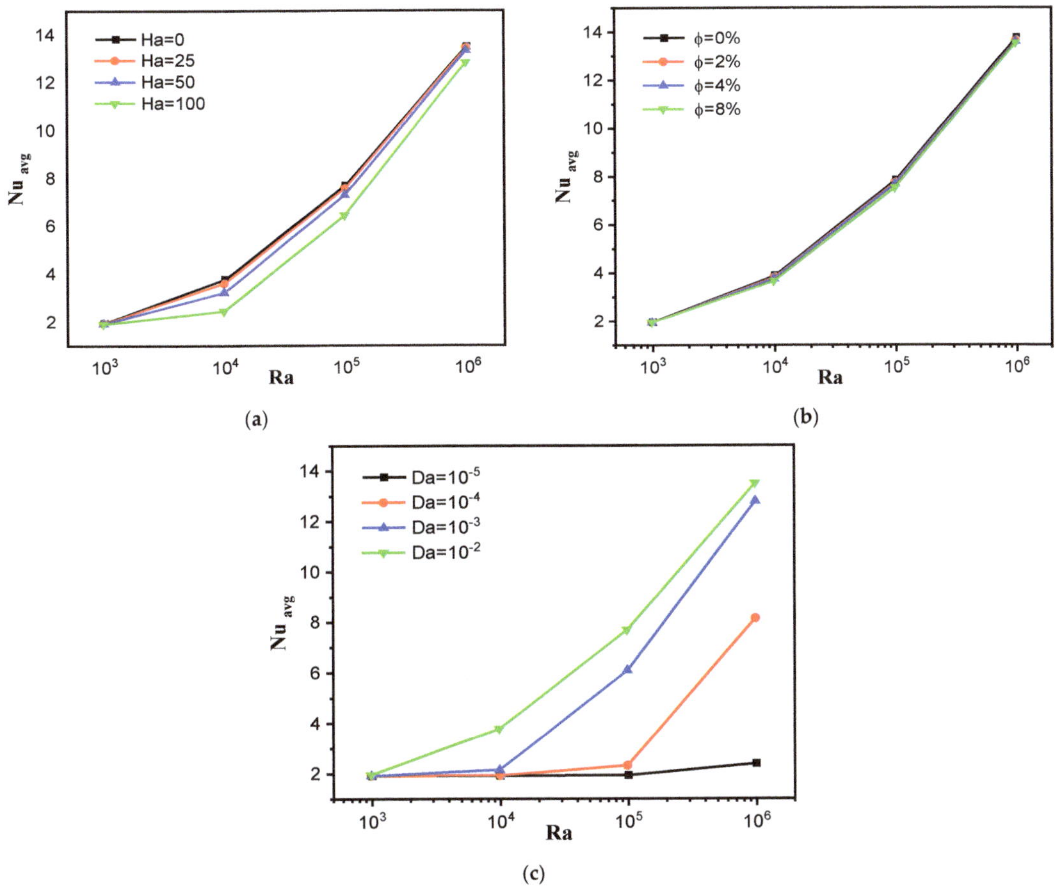

Figure 6. The effect of (**a**) Ha number, (**b**) nanoparticle volume fraction and (**c**) Da number on the Nu_{avg} for different values of Ra.

In order to know the quality of the convective heat transfer along the hot surfaces of trapezoidal fin, Figure 7 presents the distribution of the local values of Nu in terms of Da, Ra, and Ha and φ. Figure 7a illustrates the effect of Da on the local distribution of Nu for Ra = 10^5, φ = 4% and Ha = 0. We note that the local values of Nu on both sides of the trapezoidal fin are greater than the values in the middle of the fin because the velocity of suspension particles on both sides of the fin is greater. In addition, raising the number Da increases the local values of Nu number along the fin. Figure 7b is intended to display the local distribution of Nu number in terms of Ha for Ra = 10^5, φ = 4% and Da = 10^{-2}. All local values of Nu decrease with increasing Ha along the heated surfaces of the fin. Figure 7c is devoted to presenting the impact of the Ra number on local values of Nu along the heated fin for Ha = 0, φ = 4% and Da = 10^{-2}. It is clearly demonstrated that raising Ra increases all local values of Nu along the heated surfaces. Figure 7d is devoted to showing the concentration's impact on the local Nu distribution for Ha = 0, Ra = 10^5 and Da = 10^{-2}. We note that there is a very small effect of this element on the local values of Nu.

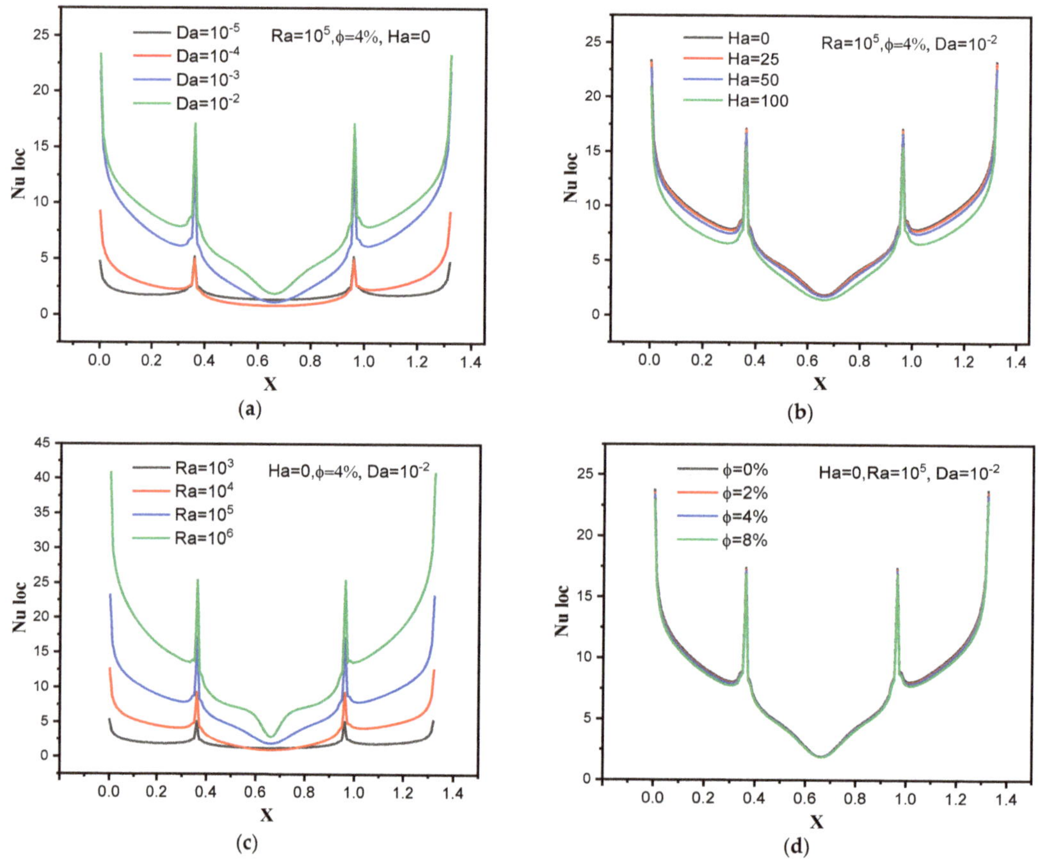

Figure 7. Effects of (**a**) Da number, (**b**) Ha number, (**c**) Ra number and (**d**) nanoparticle volume fraction on the local Nusselt number.

5. Conclusions

Through this work, we were able to create a digital simulation for suspension inside a room in the form of an inversed T shape. The suspension consists of water and the elements of nano-encapsulated PCM. The chamber is permeable and has a compound bottom with a hot trapezoidal fin, while the lateral ends are cold. We sought, through the simulation results, to highlight the heat transfer of free convection form between hot and cold elements by using the suspension as a thermal conductor. The study was also carried out under the effect of the intensity of the magnetic field. Analyzing the results of this research, we concluded the following:

- Increasing the porosity of the container makes the flow velocity better, increasing heat transmission.
- Rayleigh number controls the thermal buoyancy force; it was found that increasing this force moves the suspension flow better and has a positive effect on heat transfer.
- The magnetic field creates a force that opposes the movement of the flow. Therefore, the stronger the magnetic field, the lower the flow speed, and consequently, increased thermal activity.
- The movement of the flow within the space is characterized by the development of two identical vortices, as the higher the speed of the flow, the more the development of the vortices is served.

- The bar indicating the location of the change in the physical state of PCM elopements is characterized by two states: the first is the presence of a single band hovering around the heated surface for the low speed of the flow; the second case is characterized by the presence of two opposite bands, one on the right and the other on the left when the flow speed is high.

Author Contributions: Conceptualization, A.A. and A.M.; methodology, O.Y. and M.A.-K.; software, A.A.; validation, O.Y. and Z.D.; formal analysis, H.L.; investigation, H.L. and K.G.; resources, K.G. and O.Y.; data curation, A.M.; writing—original draft preparation, A.M. and H.L.; writing—review and editing, A.A., M.A.-K. and R.M.; visualization, A.M. and M.A.-K.; supervision, A.A.; project administration, A.A.; funding acquisition, M.A.-K. All authors have read and agreed to the published version of the manuscript.

Funding: Not applicable.

Institutional Review Board Statement: Not applicable.

Informed Consent Statement: Not applicable.

Data Availability Statement: Not applicable.

Acknowledgments: The authors extend their appreciation to the Deanship of Scientific Research at King Khalid University for funding this work through General Research Project under grant number (R.G.P.2/224/43). The authors would like to thank the Deanship of Scientific Research at Umm Al-Qura University for supporting this work by Grant Code: (22UQU4331317DSR75). Authors also extend their appreciation to Mathematics Department at Khalifa University for supporting this work.

Conflicts of Interest: The authors declare no conflict of interest.

References

1. Rashidi, M.M.; Mahariq, I.; Alhuyi Nazari, M.; Accouche, O.; Bhatti, M.M. Comprehensive review on exergy analysis of shell and tube heat exchangers. *J. Therm. Anal. Calorim.* **2022**, 1–11. [CrossRef]
2. Shahid, A.; Bhatti, M.M.; Ellahi, R.; Mekheimer, K.S. Numerical experiment to examine activation energy and bi-convection Carreau nanofluid flow on an upper paraboloid porous surface: Application in solar energy. *Sustain. Energy Technol. Assess.* **2022**, *52*, 102029. [CrossRef]
3. Molana, M.; Zarrinderafsh, V.; Chamkha, A.J.; Izadi, S.; Rafizadeh, S. Magnetohydrodynamics convection in nanofluids-filled cavities: A review. *Heat Transf.* **2020**, *49*, 1418–1443. [CrossRef]
4. Giwa, S.O.; Sharifpur, M.; Ahmadi, M.H.; Meyer, J.P. A review of magnetic field influence on natural convection heat transfer performance of nanofluids in square cavities. *J. Therm. Anal. Calorim.* **2021**, *145*, 2581–2623. [CrossRef]
5. Hemmat Esfe, M.; Afrand, M.; Esfandeh, S. Investigation of the effects of various parameters on the natural convection of nanofluids in various cavities exposed to magnetic fields: A comprehensive review. *J. Therm. Anal. Calorim.* **2020**, *140*, 2055–2075. [CrossRef]
6. Mourad, A.; Aissa, A.; Said, Z.; Younis, O.; Iqbal, M.; Alazzam, A. Recent advances on the applications of phase change materials for solar collectors, practical limitations, and challenges: A critical review. *J. Energy Storage* **2022**, *49*, 104186. [CrossRef]
7. Bhatti, M.M.; Jun, S.; Khalique, C.M.; Shahid, A.; Fasheng, L.; Mohamed, M.S. Lie group analysis and robust computational approach to examine mass transport process using Jeffrey fluid model. *Appl. Math. Comput.* **2022**, *421*, 126936. [CrossRef]
8. Raizah, Z.A.S.; Aly, A.M.; Ahmed, S.E. Natural convection flow of a nanofluid-filled V-shaped cavity saturated with a heterogeneous porous medium: Incompressible smoothed particle hydrodynamics analysis. *Ain Shams Eng. J.* **2021**, *12*, 2033–2046. [CrossRef]
9. Khan, Z.H.; Khan, W.A.; Elbaz, A.M.R.; Qasim, M.; Alharbi, S.O.; Sun, L. Natural convection in triangular fin-shaped cavity with partially heated base using nanofluid. *ZAMM J. Appl. Math. Mech.* **2021**, *101*, e202000306. [CrossRef]
10. Ghalambaz, M.; Hashem Zadeh, S.M.; Veismoradi, A.; Sheremet, M.A.; Pop, I. Free Convection Heat Transfer and Entropy Generation in an Odd-Shaped Cavity Filled with a Cu-Al_2O_3 Hybrid Nanofluid. *Symmetry* **2021**, *13*, 122. [CrossRef]
11. Asmadi, M.S.; Md Kasmani, R.; Siri, Z.; Saleh, H. Thermal performance analysis for moderate Rayleigh numbers of Newtonian hybrid nanofluid-filled U-shaped cavity with various thermal profiles. *Phys. Fluids* **2021**, *33*, 32006. [CrossRef]
12. Nabwey, H.A.; Rashad, A.M.; Khan, W.A.; Alshber, S.I. Effectiveness of magnetize flow on nanofluid via unsteady natural convection inside an inclined U-shaped cavity with discrete heating. *Alex. Eng. J.* **2022**, *61*, 8653–8666. [CrossRef]
13. Geridonmez, B.P.; Oztop, H.F. Natural convection of hybrid nanofluid flow in the presence of multiple vertical partial magnetic fields in a trapezoidal shaped cavity. *Eur. Phys. J. Spec. Top.* **2022**. [CrossRef]
14. Ahmed, S.E.; Abderrahmane, A.; Alotaibi, S.; Younis, O.; Almasri, R.A.; Hussam, W.K. Enhanced Heat Transfer for NePCM-Melting-Based Thermal Energy of Finned Heat Pipe. *Nanomaterials* **2021**, *12*, 129. [CrossRef]

15. Al-Kouz, W.; Aissa, A.; Devi, S.S.U.; Prakash, M.; Kolsi, L.; Moria, H.; Jamshed, W.; Younis, O. Effect of a rotating cylinder on the 3D MHD mixed convection in a phase change material filled cubic enclosure. *Sustain. Energy Technol. Assess.* **2022**, *51*, 101879. [CrossRef]
16. Bhatti, M.M.; Bég, O.A.; Ellahi, R.; Abbas, T. Natural Convection Non-Newtonian EMHD Dissipative Flow Through a Microchannel Containing a Non-Darcy Porous Medium: Homotopy Perturbation Method Study. *Qual. Theory Dyn. Syst.* **2022**, *21*, 97. [CrossRef]
17. Bhatti, M.M.; Ellahi, R.; Hossein Doranehgard, M. Numerical study on the hybrid nanofluid (Co_3O_4-Go/H_2O) flow over a circular elastic surface with non-Darcy medium: Application in solar energy. *J. Mol. Liq.* **2022**, *361*, 119655. [CrossRef]
18. NematpourKeshteli, A.; Iasiello, M.; Langella, G.; Bianco, N. Enhancing PCMs thermal conductivity: A comparison among porous metal foams, nanoparticles and finned surfaces in triplex tube heat exchangers. *Appl. Therm. Eng.* **2022**, *212*, 118623. [CrossRef]
19. Li, Z.-R.; Fu, G.-T.; Fan, L.-W. Synergistic effects of nano-enhanced phase change material (NePCM) and fin shape on heat storage performance of a finned shell-and-tube unit: An experimental study. *J. Energy Storage* **2022**, *45*, 103772. [CrossRef]
20. Cao, Y.; Mansir, I.B.; Mouldi, A.; Aouaini, F.; Bouzgarrou, S.M.; Marzouki, R.; Dahari, M.; Wae-hayee, M.; Mohamed, A. Designing a system for battery thermal management: Cooling LIBs by nano-encapsulated phase change material. *Case Stud. Therm. Eng.* **2022**, *33*, 101943. [CrossRef]
21. Mohammadpour, J.; Salehi, F.; Lee, A. Performance of nano encapsulated phase change material slurry heat transfer in a microchannel heat sink with dual-circular synthetic jets. *Int. J. Heat Mass Transf.* **2022**, *184*, 122265. [CrossRef]
22. Iachachene, F.; Haddad, Z.; Arıcı, M.; Abu-Nada, E.; Sheremet, M.A. The effect of nano encapsulated phase change materials and nanoparticles on turbulent heat transport: A conical diffuser scenario. *J. Energy Storage* **2022**, *52*, 104703. [CrossRef]
23. Abderrahmane, A.; Hatami, M.; Younis, O.; Mourad, A. Effect of double rotating cylinders on the MHD mixed convection and entropy generation of a 3D cubic enclosure filled by nano-PCM. *Eur. Phys. J. Spec. Top.* **2022**, 1–13.
24. Doshi, S.; Kashyap, G.; Tiwari, N. Thermo-hydraulic and entropy generation investigation of nano-encapsulated phase change material (NEPCM) slurry in hybrid wavy microchannel. *Int. J. Numer. Methods Heat Fluid Flow* **2022**. [CrossRef]
25. Bhatti, M.M.; Bég, O.A.; Abdelsalam, S.I. Computational Framework of Magnetized MgO–Ni/Water-Based Stagnation Nanoflow Past an Elastic Stretching Surface: Application in Solar Energy Coatings. *Nanomater.* **2022**, *12*, 1049. [CrossRef]
26. Bhatti, M.M.; Öztop, H.F.; Ellahi, R.; Sarris, I.E.; Doranehgard, M.H. Insight into the investigation of diamond (C) and Silica (SiO2) nanoparticles suspended in water-based hybrid nanofluid with application in solar collector. *J. Mol. Liq.* **2022**, *357*, 119134. [CrossRef]
27. Alehosseini, E.; Jafari, S.M. Nanoencapsulation of phase change materials (PCMs) and their applications in various fields for energy storage and management. *Adv. Colloid Interface Sci.* **2020**, *283*, 102226. [CrossRef]
28. Peng, H.; Wang, J.; Zhang, X.; Ma, J.; Shen, T.; Li, S.; Dong, B. A review on synthesis, characterization and application of nanoencapsulated phase change materials for thermal energy storage systems. *Appl. Therm. Eng.* **2021**, *185*, 116326.
29. Rodriguez-Cumplido, F.; Pabon-Gelves, E.; Chejne-Jana, F. Recent developments in the synthesis of microencapsulated and nanoencapsulated phase change materials. *J. Energy Storage* **2019**, *24*, 100821. [CrossRef]
30. Liu, C.; Cheng, Q.; Liu, X.; Cao, H.; Jin, S.; Liu, Y.; Rao, Z. Theoretical prediction and experimental investigation on nanoencapsulated phase change material with improved thermal energy storage performance. *Sol. Energy Mater. Sol. Cells* **2022**, *241*, 111741. [CrossRef]
31. Yuan, H.; Liu, S.; Hao, S.; Zhang, Z.; An, H.; Tian, W.; Chan, M.; Bai, H. Fabrication and properties of nanoencapsulated stearic acid phase change materials with Ag shell for solar energy storage. *Sol. Energy Mater. Sol. Cells* **2022**, *239*, 111653. [CrossRef]
32. Khanna, S.; Paneliya, S.; Marathey, P.; Shah, K.; Prajapati, P.; Chaudhari, R.; Vora, J. *Investigation of Thermophysical Properties of Synthesized N-Hexacosane-Encapsulated Titania Phase Change Material for Enhanced Thermal Storage Application BT—Recent Advances in Mechanical Infrastructure*; Parwani, A.K., Ramkumar, P.L., Abhishek, K., Yadav, S.K., Eds.; Springer Nature Singapore: Singapore, 2022; pp. 107–118.
33. Khanna, S.; Paneliya, S.; Prajapati, P.; Mukhopadhyay, I.; Jouhara, H. Ultra-stable silica/exfoliated graphite encapsulated n-hexacosane phase change nanocomposite: A promising material for thermal energy storage applications. *Energy* **2022**, *250*, 123729. [CrossRef]
34. Cao, Y.; Farouk, N.; Ayed, H.; Aly, A.A.; Jarad, F.; Dahari, M.; Wae-hayee, M.; Saleh, B. Heat transfer improvement between a pair of heater and cooler inside an energy storage by using nano-encapsulated phase change material/water: A numerical modeling. *Case Stud. Therm. Eng.* **2022**, *30*, 101770. [CrossRef]
35. Alhashash, A.; Saleh, H. Free convection flow of a heterogeneous mixture of water and nano-encapsulated phase change particle (NEPCP) in enclosure subject to rotation. *J. Energy Storage* **2022**, *51*, 104168. [CrossRef]
36. Koulali, A.; Abderrahmane, A.; Jamshed, W.; Hussain, S.M.; Nisar, K.S.; Abdel-Aty, A.H.; Yahia, I.S.; Eid, M.R. Comparative study on effects of thermal gradient direction on heat exchange between a pure fluid and a nanofluid: Employing finite volume method. *Coatings* **2021**, *11*, 1481. [CrossRef]

37. Zidan, A.M.; Nayak, M.K.; Karimi, N.; Sattar Dogonchi, A.; Chamkha, A.J.; Ben Hamida, M.B.; Galal, A.M. Thermal management and natural convection flow of nano encapsulated phase change material (NEPCM)-water suspension in a reverse T-shaped porous cavity enshrining two hot corrugated baffles: A boost to renewable energy storage. *J. Build. Eng.* **2022**, *53*, 104550. [CrossRef]
38. Al-Kouz, W.; Abderrahmane, A.; Shamshuddin, M.D.; Younis, O.; Mohammed, S.; Bég, O.A.; Toghraie, D. Heat transfer and entropy generation analysis of water-Fe_3O_4/CNT hybrid magnetic nanofluid flow in a trapezoidal wavy enclosure containing porous media with the Galerkin finite element method. *Eur. Phys. J. Plus* **2021**, *136*, 1184. [CrossRef]
39. Alshare, A.; Abderrahmane, A.; Guedri, K.; Younis, O.; Fayz-Al-Asad, M.; Ali, H.M.; Al-Kouz, W. Hydrothermal and Entropy Investigation of Nanofluid Natural Convection in a Lid-Driven Cavity Concentric with an Elliptical Cavity with a Wavy Boundary Heated from Below. *Nanomaterials* **2022**, *12*, 1392. [CrossRef]
40. Heydarian, R.; Shafii, M.B.; Rezaee Shirin-Abadi, A.; Ghasempour, R.; Alhuyi Nazari, M. Experimental investigation of paraffin nano-encapsulated phase change material on heat transfer enhancement of pulsating heat pipe. *J. Therm. Anal. Calorim.* **2019**, *137*, 1603–1613. [CrossRef]
41. Ho, C.J.; Liu, Y.-C.; Ghalambaz, M.; Yan, W.-M. Forced convection heat transfer of Nano-Encapsulated Phase Change Material (NEPCM) suspension in a mini-channel heatsink. *Int. J. Heat Mass Transf.* **2020**, *155*, 119858. [CrossRef]
42. Mehryan, S.A.M.; Ghalambaz, M.; Sasani Gargari, L.; Hajjar, A.; Sheremet, M. Natural convection flow of a suspension containing nano-encapsulated phase change particles in an eccentric annulus. *J. Energy Storage* **2020**, *28*, 101236. [CrossRef]
43. Sadr, A.N.; Shekaramiz, M.; Zarinfar, M.; Esmaily, A.; Khoshtarash, H.; Toghraie, D. Simulation of mixed-convection of water and nano-encapsulated phase change material inside a square cavity with a rotating hot cylinder. *J. Energy Storage* **2022**, *47*, 103606. [CrossRef]
44. Ghalambaz, M.; Chamkha, A.J.; Wen, D. Natural convective flow and heat transfer of nano-encapsulated phase change materials (NEPCMs) in a cavity. *Int. J. Heat Mass Transf.* **2019**, *138*, 738–749. [CrossRef]
45. Al-Khaleel, M.D.; Gander, M.J.; Ruehli, A.E. Optimized waveform relaxation solution of RLCG transmission line type circuits. In Proceedings of the 2013 9th International Conference on Innovations in Information Technology, IIT 2013, Al Ain, United Arab Emirates, 17–19 March 2013; pp. 136–140.
46. Wu, S.-L.; Al-Khaleel, M.D. Semi-discrete Schwarz waveform relaxation algorithms for reaction diffusion equations. *BIT Numer. Math.* **2014**, *54*, 831–866. [CrossRef]
47. Wu, S.-L.; Al-Khaleel, M. Convergence analysis of the Neumann–Neumann waveform relaxation method for time-fractional RC circuits. *Simul. Model. Pract. Theory* **2016**, *64*, 43–56. [CrossRef]
48. Wu, S.-L.; Al-Khaleel, M.D. Optimized waveform relaxation methods for RC circuits: Discrete case. *ESAIM Math. Model. Numer. Anal.* **2017**, *51*, 209–223. [CrossRef]

nanomaterials

Review

The Pool-Boiling-Induced Deposition of Nanoparticles as the Transient Game Changer—A Review

José Pereira [1], Ana Moita [1,2,*] and António Moreira [1]

[1] IN+, Center for Innovation, Technology and Policy Research, Instituto Superior Técnico, Universidade de Lisboa, Av. Rovisco Pais, 1049-001 Lisboa, Portugal
[2] CINAMIL—Centro de Investigação Desenvolvimento e Inovação da Academia Militar, Academia Militar, Instituto Universitário Militar, Rua Gomes Freire, 1169-203 Lisboa, Portugal
* Correspondence: anamoita@tecnico.ulisboa.pt

Abstract: It is widely known by the scientific community that the suspended nanoparticles of nanofluids can enhance the thermophysical properties of base fluids and maximize pool-boiling heat transfer. However, the nanoparticles may undergo extended boiling times and deposit onto the heating surfaces under pool-boiling conditions, thus altering their intrinsic characteristics such as wettability and roughness over time. The present study reviews the fundamental mechanisms and characteristics of nanoparticle deposition, and its impact on surface roughness and wettability, density of vaporized core points, and thermal resistance, among other factors. Moreover, the effect of the nanoparticle layer in long-term thermal boiling performance parameters such as the heat transfer coefficient and critical heat flux is also discussed. This work attempts to highlight, in a comprehensive manner, the pros and cons of nanoparticle deposition after extended pool-boiling periods, leading the scientific community toward further investigation studies of pool-boiling heat-transfer enhancement using nanofluids. This review also attempts to clarify the inconsistent results of studies on heat transfer parameters using nanofluids.

Keywords: nanofluids; fouling resistance; re-suspension; hydrodynamic instability; thermal resistance; roughness-to-particle size ratio

Citation: Pereira, J.; Moita, A.; Moreira, A. The Pool-Boiling-Induced Deposition of Nanoparticles as the Transient Game Changer—A Review. *Nanomaterials* **2022**, *12*, 4270. https://doi.org/10.3390/nano12234270

Academic Editor: Henrich Frielinghaus

Received: 10 October 2022
Accepted: 24 November 2022
Published: 1 December 2022

1. Introduction

The boiling process has been used to dissipate energy in the form of heat generated by macroscale operating equipment. As miniaturized, faster and more powerful electronic devices and systems are developed and implemented on a day-to-day basis, the operating thermal fluids need to be improved to augment the heat removal capability. Pool-boiling heat transfer depends on factors closely linked with the fluid and solid surface properties such as thermal conductivity, surface tension, viscosity, enthalpy, specific heat, and roughness, structure, and homogeneity of the heating surface. It also depends on the hydrodynamic state near the heating surface determined by the bubble departure frequency and diameter and hot/dry spots dynamics [1], among other factors. The suspension of nanoparticles into the base fluid, or in other words using a nanofluid, is one of the most followed routes to enhance the pool-boiling heat-transfer coefficient (HTC) and critical heat flux (CHF) and can modify many of the aforementioned factors and properties. The use of nanoparticles may have considerable influence on the thermophysical properties of the working thermal fluids and on the behavior of the triple solid–liquid–vapor contact line. The alteration of the liquid/vapor and solid surface tensions changes the acting force balance and length of the contact line [2], wettability, and number of active nucleation sites. The focus of the research studies on the field should be to achieve an optimized combination of the base fluid, concentration, shape, and dimensions of the nanoparticles and solid surface substrate to increase, at the same time, the CHF and HTC of the pool-boiling scenarios. Nevertheless, there are still features of vital importance that should be further studied such as the impact and control of the transient boiling-induced nanoparticle deposition [1]. In the course of the nucleate boiling process with nanofluids, the nanoparticles deposit onto the heating surface over time and can modify its properties including the wettability, water angle hysteresis, roughness, capillary wicking [3,4], and nucleation site density. Particularly, the surface roughness after the nanoparticle deposition is influenced by the fraction and intrinsic thermal properties of the nanoparticles, heating surface roughness, and on the thickness and morphology of the nanoparticle deposition layer. Moreover, the active nucleation site's density has a relevant role in promoting superior boiling heat-transfer performance, and can be controlled by factors such as the surface wettability and roughness, surface superheat value, and thermal properties of the fluid [5]. There is an intricate connection between the number of nucleation sites and the roughness and wettability of the heating surface: as the wettability improves, the likelihood of the liquid filling the cavities increases and, consequently, the number of active nucleation points decreases. In the work carried out by the authors Shoghl et al. [6], after the boiling process, using a alumina nanofluid as operating fluid, the deposited nanoparticle layer decreased the heating surface roughness in the case where the size of the nanoparticles was smaller than the roughness of the heat transfer surface. Moreover, Das et al. [7] reported that the smaller nanoparticles fill the cavities of the surface and reduce its original roughness. Since the nanoparticles agglomerate and deposit, the heating surface roughness decreases when the size of the clusters is smaller than the size of the surface cavities. Moreover, in the study conducted by Chopkar et al. [8], it was reported that after pool-boiling experiments using a 0.005 vol. % zirconia nanofluid, the heating surface roughness decreased. Moreover, the researchers revealed that the surface roughness decreased with the increasing number of successive pool-boiling experiments. Additionally, the authors Narayan et al. [9] employed machine vision to determine the nucleation site density after the completion of pool-boiling experiments with alumina nanofluids. The authors found that the number of active nucleation core points increased when the size of the nanoparticles of 47 nm was smaller than the roughness of the heating surface of 524 nm. In this case, the alumina nanoparticles penetrated into large surface cavities and enlarged the number of sites by splitting one active nucleation point into many points. On the other hand, if the nanoparticle size was similar to the average surface roughness of 48 nm, the number of nucleation sites was found to decrease considerably. Furthermore, the atomic force microscopy (AFM) technique was employed by White et al. [10] to measure the heating surface roughness after multiple pool-boiling tests with deionized water and with a 40 nm zinc oxide nanofluid. The research team observed that the surface roughness increased to 440 nm after seven nanofluid pool-boiling experiments. In addition, the heating surface roughness was reported by Bang and Chang [11] to increase after the nanofluid pool-boiling process if the alumina nanoparticles were larger than the surface roughness. Figure 1 shows the intertwined nanoparticle deposition time-dependent features at play in the pool-boiling heat transfer. Furthermore, the published results in the literature showed that the nanoparticle deposition layer affected the wettability more than the surface roughness [12]. In this direction, Buongiorno et al. [1] conducted pool-boiling experiments with nanofluids on smooth stainless-steel surfaces and found that the contact angle was modified because of the alterations in the solid surface tension provoked by the different chemistry and morphology of the deposited layer of nanoparticles.

It was previously reported in the literature that the CHF is usually enhanced by the alteration of the heat transfer surface characteristics [13,14]. Nevertheless, the nucleate boiling heat transfer of nanofluids is a complex phenomenon that depends on many factors such as the operating conditions [15,16]. Moreover, the experimental results so far published on the boiling heat transfer parameters of the nanofluids are not consistent with each other, even under similar experimental conditions [17]. In addition, the demanding time variation of the boiling heat transfer of nanofluids in which the most representative example is the nanoparticle deposition would be one of the main reasons behind such inconsistency of results [18]. To make matters worse, there is no available systematic experimental information or database regarding the transient nanoparticle deposition and its effects on pool-boiling heat transfer using nanofluids. Hence, further in-depth research works are recommended to achieve a better understanding of the mechanisms responsible for the conflicting trends. The main objective of the current work is to provide, given the circumstances, the most possible complete and accurate experimental and theoretical information about the nanoparticle deposition onto the heat transfer surfaces, in an effort to minimize the severe shortage of understanding of such deposition during pool boiling of nanofluids. Table 1 summarizes some of the main recent experimental works on the boiling-induced nanoparticle deposition and its fundamental effects on the poll boiling heat transfer characteristics.

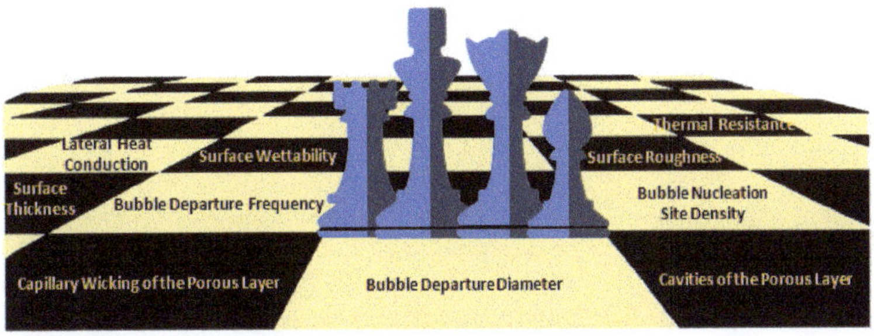

Figure 1. Nanoparticle deposition time-dependent features at play.

Table 1. Main recent experimental works on boiling-induced nanoparticle deposition.

Reference	Authors/Year	Nanoparticles Material	Substrate Material	Effects on Heat Transfer	Other Effects
[19]	Raveshi et al. /2013	Aluminum oxide	Copper	An HTC enhancement of up to 64% for the 0.75% vol. nanofluid was verified. Except for the increment of the base fluid properties, the heating surface alteration was the main factor influencing the HTC. Because the surface particle interaction parameter was more than the unity, only the increment of HTC was observed. At low concentration, the deposited layer was very thin, which altered the surface by multiplying the nucleate site and creating active cavities, and finally leading to enhancement of the heat transfer.	The thicker layer, which was recorded at the end of the boiling duration, was caused by the higher concentration of the nanoparticles.
[5]	Sarafraz and Hormozi/2014	Aluminum oxide	Stainless-steel	With increasing the concentration of nanofluids, due to the deposition of nanoparticles on the surface and equal size of nanoparticles and roughness of the heating surface, the average roughness of the surface decreased and, subsequently, the number of nucleation sites reduced which led to an HTC decrease.	An asymptotic behavior was reported for the particulate fouling at the heat transfer area, whereas a rectilinear behavior was found for the free convection region. The thickness and rate of fouling depended on the boiling duration and the fouling resistance because nucleate boiling was higher than those measured in the convection area.

Table 1. *Cont.*

Reference	Authors/Year	Nanoparticles Material	Substrate Material	Effects on Heat Transfer	Other Effects
[20]	Tang et al. /2014	Aluminum oxide		The aluminum oxide nanoparticles enhanced the heat transfer at concentrations of 0.001 vol. %, 0.01 vol. %, and 0.1 vol. % with the SDBS surfactant. The aluminum oxide nanoparticles deteriorate the heat transfer at 0.1 vol. % without SDBS due to the larger number of deposited nanoparticles. The SDBS deteriorated the heat transfer, and the deterioration was more than the enhancement by the nanoparticles at 0.001 vol. %.When the fractions were between 0.01 vol. % and 0.1 vol. %, the addition of the SDBS improved the heat transfer by the nanoparticles because the SDBS reduced the deposition of nanoparticles.	The impact of the change in the surface angle by the surface deposition of the nanoparticles can be negligible for R141b.
[21]	Manetti et al. /2017	Aluminum oxide	Copper	A decrease in the wall superheating up to 32% and 12% for a smooth and rough surfaces, respectively, was verified for the same heat flux in comparison with that of DI water. For low concentration nanofluids subjected to moderate heat flux, appreciable enhancement of the HTC was observed for the smooth and rough surfaces as compared with the DI water. It is argued that this phenomenon is related to the increase in the radius of the cavities due to changes in the morphology of the surface. For the rough surface, the HTC decreased appreciably with increasing heat flux due to the intensification of the nanoparticle deposition rate, and higher thermal resistance by the filling of the cavities of the heating surface with the nanoparticles. The surface modification due to the nanoparticle deposition increased the HTC only for low nanoparticle concentrations and when the particle interaction parameter (SIP) was greater than unity.	A higher nanoparticle deposition rate occurred for heat fluxes greater than 400 kW/m^2.
[22]	Nunes et al. /2020	Aluminum oxide	Copper	The coating layer formed on the heating surface increased the surface wettability; moreover, it provided a barrier to the heat transfer by increasing the thermal resistance on the heating surface, degrading the HTC for unconfined and confined boiling. For the latter, the wettability enhancement promoted a delay in the dryout incipience phenomenon.	The coating process delayed the dryout occurrence under confined conditions due to the influence of the nanostructures on the surface–fluid interaction mechanisms, e.g., the surface wettability, which is a pronounced effect for non-wetting fluids, such as the DI water.

Table 1. *Cont.*

Reference	Authors/Year	Nanoparticles Material	Substrate Material	Effects on Heat Transfer	Other Effects
[23]	Xing et al. /2016	Carbon	Copper	The multi-walled nanotubes (MWNTs) with CTAB nanofluid presents poor HTC, which decrease with increasing concentration for the deposition of nanoparticles. The deposition of nanoparticles onto the heating surface was not verified using covalent functionalization MWNTs nanofluids. Thus, the covalent functionalization MWNTs nanofluids show a higher HTC than the base fluid, and they increased as the MWNTs concentration increased. The maximum HTC enhancements are 34.2% and 53.4% for MWNTs-COOH and MWNTs-OH nanofluids, respectively.	…..
[24]	Li et al./2020	Copper oxide	Copper	The HTC was improved due to the partial fouling of the nanoparticles which increased the number of nucleation sites on the surface. After 1000 min of operation, the fouling layer changed the surface by decreasing the number of nucleation sites, inducing a thermal resistance to the surface and decreasing the bubble departure time.	…..
[25]	Cao et al./2019	Copper–zinc	Copper	The superheat value on the fully deposited surfaces was around 20 K lower than that on the smooth surface and the fully deposited surface had the highest HTC, around 100% enhancement than the smooth surface. The CHF was not enhanced on the fully deposited surface, but increased by 33% on the channel-pattern-deposited surfaces.	The experimental CHF on the channel pattern deposited surfaces agree well with the predicted model derived from hydrodynamic instability
[26]	Kiyomura et al. /2017	Iron oxide	Copper	The coated layer formed on the rough surfaces provided a barrier to the heat transfer and reduced the bubble nucleation, which led to the reduction in the number of microcavities and an increase in the thermal resistance, therefore degrading the HTC. For smooth surfaces, the deposition of nanoparticles tends to increase the nucleation site's density, increasing the boiling heat transfer. An increment in the HTC occurred only for low nanofluid concentrations, for which the thermal conductivity of the nanofluids was dominant as compared with the thermal resistance of the nanolayer formed on the heating surface.	The C coefficient that correlates to the HTC and the heat flux was used. Different Csf (surface–fluid parameter) and Cs (heating surface parameter) behaviors were found for the surfaces covered with nanoparticles. The Csf and Cs are influenced by the additional thermal resistance resulting from the nanoparticle deposition. The Cs underestimated the effects of wettability and surface roughness for the surfaces covered with nanoparticles

Table 1. *Cont.*

Reference	Authors/Year	Nanoparticles Material	Substrate Material	Effects on Heat Transfer	Other Effects
[27]	Stutz et al. /2011	Maghemite	Platinum	The coating made of nanoparticles reduced the HTC by introducing a thermal resistance that increased with layer thickness. The CHF enhancement depended on the covering rate of the heating surface by the nanoparticles, and evolved with boiling time. It reached a maximum when the heater was entirely covered with nanoparticles and then decreased slowly when the thickness of the coating increased. The observed increase in the CHF was due to the increase in the heat transfer area when the nanoporous layer was formed.	The effective thermal resistance of the layer appears to decrease substantially during boiling and seems to be coupled to the bubble dynamics. This may indicate that vapor generation occurs inside the layer, which reduces its effective thickness.
[28]	Souza et al. /2014	Maghemite	Copper	It was observed that the enhancement of the HTC is higher when the SIP was greater than unity. The HTC for the nanostructured surface with the deposition of nanoparticles of 10 nm diameter, corresponding to SIP = 16, was 55% higher than that for the bare surface. For the nanostructured surface with the deposition of nanoparticles having 80 nm of diameter, corresponding to a SIP equal to 2, the HTC decreased 29%.The HTC increased when the gap decreased, mainly for lower heat fluxes. For a gap length equal to 0.1 mm, a 145% HTC increase at heat fluxes lower than 45 kW/m^2 with the deposition of 10 nm sized nanoparticles was reported.	....
[29]	Heitich et al. /2014	Maghemite	Copper–nickel alloy	Nanostructured surfaces showed higher wettability as a consequence of the greater number of surface defects created by the nanoparticles. Surface defects affect the contact angle and may influence the heat transfer and CHF. The nanostructures have a greater number of these defects due to the small nanoparticle size. The nanostructures led to an increase in the CHF, especially with the maghemite deposition for which the value was around 139% higher than that of the smooth substrate. The CHF increased as the wettability increased. An increase in the CHF was observed as the contact angle decreased. The rough substrate samples showed an enhancement in the HTC of around 19%, while other samples showed an increase in the HTC values for high heat fluxes	The maghemite nanostructured surfaces showed greater porosity and roughness. These samples have a greater nanoparticle layer thickness and, consequently, a higher wettability compared with the molybdenum samples. The rough substrate showed a hydrophobic behavior, while the other samples with nanoparticle deposition showed a hydrophilic behavior. The small particle sizes in the nanostructures greatly promoted the wettability alteration.

Table 1. *Cont.*

Reference	Authors/Year	Nanoparticles Material	Substrate Material	Effects on Heat Transfer	Other Effects
[30]	Rostamian and Etesami/2018	Silicon oxide	Copper	Whatever the time of boiling on the heating surface increases, the differences between the boiling curves of nanofluid and deionized water becomes more due to the deposition of nanoparticles. In high concentrations of nanofluid, further deposition of nanoparticles on the surface causes a thickening of the layer made of nanoparticles. This thick layer enhances thermal resistance, so the HTC reduces. However, in low concentrations the surface roughness is less than that in high concentrations and during the nanofluid boiling on the surface, nucleation sites with sedimentation of nanoparticles became smaller in size and the number of nucleation sites increased; therefore, the HTC increased. The main cause for the CHF enhancement in nanofluid was the nanoparticle deposition which increased the surface wettability and CHF was delayed to a higher surface superheat value.	Whenever the concentration and the time of boiling increases, the surface roughness increases, too. Moreover, the increase in surface roughness happens more quickly for higher concentrations (0.01 vol. %).
[31]	Akbari et al. /2017	Silver	Copper	The deposition of nanoparticles was efficient in re-entrant inclined coated surfaces: up to 120% CHF increase compared with the smooth surface and up to 30% as compared with the uncoated inclined surface.	The bubbles generated on the coated re-entrant surface were larger in size.
[32]	Kumar et al. /2017	Titanium oxide	Nickel–chromium	The CHF was augmented up to a certain value of nanoparticle deposition, beyond which the rate of deposition was intangible. Approximately 80%, 88%, and 93% enhancements in the CHF were found for deposition up to 4 min, 8 min, and 16 min, respectively.	The larger the deposition or boiling time, the lower the contact angle leading to a higher CHF. The rate of deposition of nanoparticles was higher for boiling times up to 8 min and was comparatively lower beyond 8 min and above.

Table 1. *Cont.*

Reference	Authors/Year	Nanoparticles Material	Substrate Material	Effects on Heat Transfer	Other Effects
[33]	Hadzic et al. /2022	Titanium oxide	Copper	At a low nanoparticle concentration, the influence of nanofluid on boiling performance was minimal, with the HTC and CHF values comparable with those obtained using pure water on both the untreated and laser-textured surfaces. The boiling of a nanofluid with a high nanoparticle concentration resulted in a significant deposition of nanoparticles onto the boiling surface and CHF enhancement up to 2021 kW m^{-2}, representing double the value obtained on the untreated reference surface using water. However, very high surface superheat values (up to 100 K) were recorded, suggesting poor practical applicability. The decrease in heat transfer performance due to the boiling of nanofluids on laser textured surfaces can be explained through the deposition of nanoparticles into the laser-induced grooves and microcavities present on the surface, which decreased the number of active nucleation sites.	Thicker nanoparticle deposits resulted in added thermal resistance. While the surface porosity granted a delay in the CHF incipience due to the enhanced liquid replenishment, the surface superheat value was increased.
[34]	Kamel and Lezso-vits./2020	Tungsten oxide	Copper	The higher HTC enhancement ratio was 6.7% for a concentration of 0.01% vol. compared with deionized water. The HTC for nanofluids was degraded compared with the deionized water, and the maximum reduction ratio was about 15% for a concentration of 0.05% vol. relative to the baseline case. The reduction in the HTC was attributed to the deposition of tungsten oxide nanoflakes on the heating surface, which led to a decrease in the nucleation site's density.	.....
[35]	Gajghate et al. /2021	Zirconium oxide	Copper	The zirconia nanoparticle coating incremented the heat transfer and HTC. The peak enhancement in the HTC was of 31.52% obtained for 200 nm zirconium-oxide-coating thickness. The HTC increased with increasing coating thickness up to 200 nm, but a further increment in the thickness resulted in the reduction in HTC due to the rise in thermal resistance.	The increase in zirconium oxide nanoparticle concentration from 0.1 to 0.5 vol. % showed an increase in the coating thickness as well as surface roughness. The zirconium oxide layer gave hydrophilicity to the bare copper surface.

Table 1. *Cont.*

Reference	Authors/Year	Nanoparticles Material	Substrate Material	Effects on Heat Transfer	Other Effects
[36]	Minakov et al. /2017	Silicon, aluminum, iron oxides, and diamond	Nickel–chromium	Even at very small concentrations of nanoparticles, the CHF increased by more than 50% and continued growing with a further increase in the nanoparticle concentration. At high concentrations of nanoparticles, the growth rate of CHF slowed down and reached a constant value. Such behavior can be explained by the stabilization of the deposit size on the heating surface. With increasing boiling time, the CHF increased rapidly and reached a steady-state level. The correlation between CHF and the concentration, particle size and material, and boiling time confirmed the key role of the nanoparticle deposition.	At the same nanoparticle concentration, the desired height of the layer deposited on a smaller heating surface was formed much faster than that deposited on a larger heating surface.
[37]	Sulaiman et al. /2016	Titanium oxide, aluminum oxide and silicon oxide	Copper	The aluminum oxide nanofluids enhanced, whereas the silicon oxide nanofluids deteriorated, the heat transfer. The effect of the titanium oxide nanofluid on the heat transfer depended on the nanoparticle concentration. The maximum CHF was found for the most concentrated aluminum oxide nanofluid. Although significant detachment of the nanoparticle layer was found after the CHF measurement for the silicon oxide nanofluids, the value of CHF was not significantly different from those for the titanium oxide and aluminum oxide nanofluids.	Abnormal increase in the wall superheat value was observed for the titanium oxide and silicon oxide nanofluids when the heat flux was sufficiently high. It was considered that this phenomenon was related to the partial detachment of the nanoparticle layer formed on the heated surface since the defects of the nanoparticle layer were always detected when such a temperature rise took place.

2. Nanoparticle Deposition over Boiling Time

2.1. Causing Mechanisms

The causing mechanism for the nanoparticle deposition during pool boiling has been pointed out by Modi et al. [38] to be the evaporation in the microlayer beneath the vapor bubbles, where most of the heat and mass transfer is carried out. In this direction, the researchers Li et al. [39] also reported that the microlayer accumulates the nanoparticles that are deposited in the heating surface when the microlayer fully evaporates. The authors found that the microlayer evaporation was the key factor to promote the growing of the bubbles and that the nanoparticle deposition process would continue for as long as the duration of boiling. Figure 2 schematically represents the deposition of the nanoparticles in the microlayer of the vapor bubble. Moreover, Kim et al. reported in another study [40] that the growth of the deposition layer was promoted by the natural convection or by the gravity-driven sedimentation. Furthermore, as the heating surface temperatures increase beyond the operating fluid saturation point, the vapor bubbles are generated at the nucleation sites and start to grow because of the evaporation of the fluid at the contact line, nearby the zone of the bubbles, and inside the liquid microlayer underneath the vapor bubbles [41].

The buoyancy force is responsible for pushing the bubbles upwards until their detachment occurs. In addition, similar mechanisms generate bubbles within the nanofluids during pool boiling. Nevertheless, due to the evaporation rate of the microlayer, the fraction of the nanoparticles in the region around the heating substrate will be enhanced and the distance between the suspended nanoparticles will be shorter. With these conditions, more collisions are likely to happen between the nanoparticles, agglomeration, and sedimentation over the heat transfer surface.

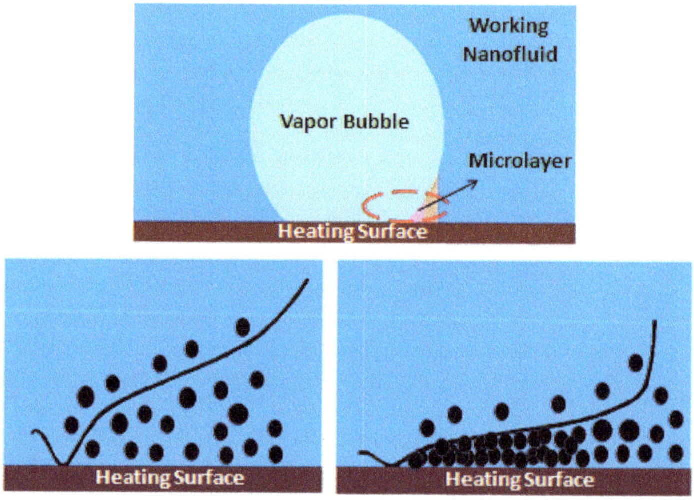

Figure 2. Schematic representation of the nanoparticle deposition in the vapor bubble microlayer. Adapted from [42].

2.2. Characteristics of the Deposition Layer

The first characteristic that can be highlighted for the deposited layer of nanoparticles during the pool boiling with nanofluids is the bonding strength with the heating surface. In this sense, the researchers Kwark et al. [43] reported that even after 16 consecutive pool-boiling experiments, the deposition layer kept a noticeable bonding strength with the substrate. Moreover, the deposition layer of nanoparticles has a porous structure over the heating surface [12,17,44] in the nanofluid pool boiling. The thickness of the porous structure and the associated capillarity effect are also relevant characteristics of the deposited nanoparticle layer. The thickness accommodates the extra liquid microlayer and the capillary wicking drives the fluid toward the dry spots at the base of the growing vapor bubbles. Throughout the pool-boiling process, the characteristics of the heating surface covered by the nanoparticle layer are continuously modified as the thickness of the layer tends to increase over time. The authors Kim et al. [45] investigated the evolution of the CHF under pool-boiling conditions on a heating NiCr wire covered with porous layers of nanoparticles. The authors obtained different deposit structures by changing the heat flux during the boiling of 0.01 volume fraction of titanium oxide nanoparticles suspended in water. The surface properties of the testing wires were measured to identify the parameters closely linked with the appreciable CHF increase. The investigation team noted that the heat capacity of the surface of the wires was altered due to the nanoparticle deposition and, hence, the heat capacity was not the main factor in interpreting the observed CHF enhancement. This experimental work may lead to the conclusion that when the CHF occurs on a very small heat transfer surface, such as a thin wire, it is described as being the consequence of the liquid dryout beneath the vapor patch associated with the dramatic increment in the surface superheat value caused by the merging of the vapor bubble rather than the hydrodynamic instability. The researchers stated that the thickness of the surface porous structures holding an extra fluid macrolayer and capillary wicking effect to conduct the working fluid toward the dry regions underneath the vapor bubbles were the key parameters responsible for the CHF increment. The thickness and structure of the deposition layer is affected by the concentration, size, and shape of the nanoparticles, and by the temperature of the heating substrate, rate of evaporation, and finally, induced heat flux. For instance, in the cases when the surface presents a very high temperature and, hence, enhanced evaporation rate, the porous layer becomes thicker and more condensed having larger agglomerations. A higher fraction of nanoparticles dispersed in the base fluid usually conducts to a thicker and more condensed deposition layer. Moreover, the morphology and intrinsic characteristics of the nanoparticles affect the deposition pattern during boiling. Another important characteristic is the roughness of the deposited layer, given that its modification during the boiling process will affect the nucleation site density [15–17]. Furthermore, it was already found that not only were the nanoparticles constituents of the deposited layer, since this one also contains solvable salts sedimented in the course of the pool boiling using nanofluids [46] and the influence of these compounds needs further understanding. The formation of a deposition layer during boiling requires engineering to minimize, or even eliminate, the negative consequences on the boiling heat transfer of nanofluids. Nevertheless, there are positive impacts encountered in most cases in the nanofluids and the boiling-induced nanoparticle deposition relative to the conventional thermal fluids such as, for instance, water. Figure 3 presents some of these positive effects.

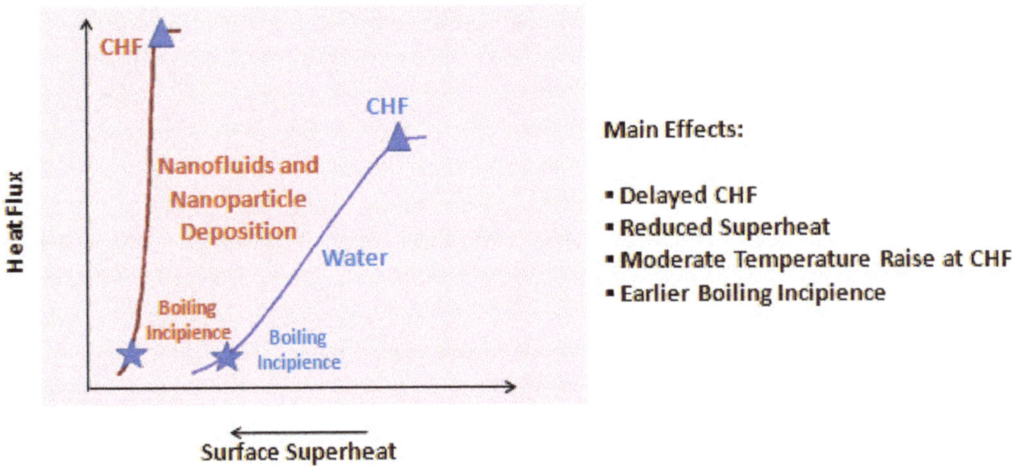

Figure 3. Positive effects of the nanofluids and nanoparticle deposition relative to water. Adapted from [47].

2.3. *Fundamental Time-Dependent Features*

2.3.1. Fouling Resistance

The fouling related with nanofluids is a type of particulate fouling phenomenon. The suspended nanoparticles lose their stability over time and adhere to the heating surface, mainly due to the interactions between the dispersed nanoparticles and the fluid and between the nanoparticles and the heating surface along with temperature gradients within the base fluid [48]. Figure 4 shows the typical time-dependent stages of the particulate fouling.

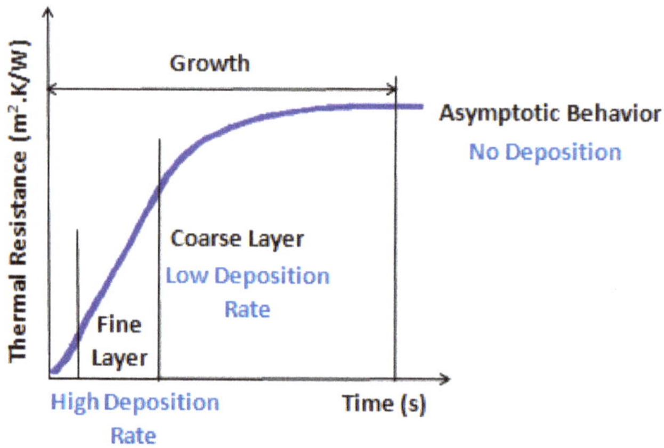

Figure 4. Thermal resistance vs. time and main stages of the particulate fouling. Adapted from [49].

Moreover, other parameters can be referred that determine the fouling occurrence when using nanofluids such as chemical composition, homogeneity, viscosity, diffusivity, density, interfacial properties, and compatibility of the nanoparticles with each other, the base fluid, and boiling surfaces. Regarding the nanoparticle concentration, an excessive amount of nanoparticles entails sedimentation concerns and, consequently, tends to promote particulate fouling over the surfaces that usually leads to a considerable decrement in the heat transfer performance. It should be stated that another vital concern of the particulate fouling over time is the clogging caused by the clustering of the nanoparticles. The fouling onto a boiling process surface has a considerable negative impact on the working of thermal management units, entailing negative consequences [50,51]. The latter are present in the operational damages caused by the shutdowns provoked by fouling, and in the maintenance costs that arise from cleaning the surfaces and equipment replacement [52]. The heat transfer performance decrement is the most observed effect of fouling [53,54]. This reduction is closely linked with the poor thermal conductivity of the fouling layer, pressure drop increase, clogged equipment, erosion and corrosion of the surfaces, and friction augmentation. To summarize briefly, the boiling fouling by nanoparticle deposition can be addressed through the interplay of the following phenomena [52]:

- Surface particulate deposition is controlled by the interactions between the thermal fluid and nanoparticles and between the heating surface and the nanoparticles.
- Re-suspension of the nanoparticles occurs after the deposition and is determined by the balance between the contact and hydrodynamic forces.
- Agglomeration occurs only in cases where the concentration of the dispersed nanoparticles in the thermal fluid is sufficient to promote consecutive interactions between the surfaces at play. The particle-to-particle collision rate is governed by the hydrodynamic forces associated with the motion of the nanoparticles, whereas the adhesion between the nanoparticles is determined by the short-range interactions between them. In the cases where the adhesion forces have a lower magnitude than the hydrodynamic counterparts, the break-up of aggregates of the nanoparticles may occur.

The hydrodynamic transport of the nanoparticles and attachment onto the heating surface of the nanoparticles are the involved phenomena in the two-step process of the structure of fouling [55,56]. In particulate fouling, the main thermophysical features involved are the motion of the nanoparticles by inertia, diffusion, and thermophoretic forces, linkage of the nanoparticles by the acting Van der Waals forces and superficial charges, and erosion. In pool-boiling scenarios, the fouling of nanoparticles resistance depends on [57]:

- Heat flux: given that with increasing heat flux, the fouling resistance also increases until a maximum value.
- The pH and ionic content of the nanofluids: the attachment and agglomeration of the nanoparticles are provoked by the action of the electric double layer and Van der Waals forces. The repulsion caused by the electric double-layer forces is mainly due to the accumulation of electric charges in the surface of the nanoparticles of hydroxides and metal oxides, and such electric charges are directly correlated with the pH of the suspension and its ionic content.
- Temperature of the surface: it was already demonstrated that the fouling resistance increased more rapidly with temperature at higher temperatures and more slowly at lower temperature values.
- Surface roughness: given that a smoother surface delays the fouling, and it is easier to clean. In contrast, considerable rough surfaces increased the nanoparticle deposition dramatically.
- Size of the nanoparticles: given that the fouling resistance depends on the particle size. However, a well-defined trend is not yet available in the published literature.

The fouling of the copper oxide nanoparticles onto a heat exchanger surface was investigated with a variety of experimental parameters and a correlation accounting with the flow of the nanoparticles was introduced by the authors Nikkhah et al. [58]. The researchers reported that with a larger amount of nanoparticles and increased heat flux, a more considerable surface fouling occurred. Moreover, a reduction in the fouling resistance was found with increasing surface temperature. In particular, the consecutive heating and cooling cycles or the high temperature of the nanofluids tended to promote the impact between the nanoparticles and, consequently, their aggregation trend.

2.3.2. Thermal Resistance

The boiling-induced nanoparticle deposition onto the interface of the solid and liquid phases seriously affects the interfacial thermal resistance (ITR) or Kapitsa resistance. The ITR can be modified by the wettability and morphology of the interface. For instance, the thermal conductance of different wettability interfaces functionalized with a self-assembled monolayer (SAM) has already been determined by the time-domain thermo-reflectance technique [59]. The authors reported that the ITR of a hydrophobic interface was near three-fold greater than that of a hydrophilic interface. Moreover, in the work performed by [5], the layer of carbon nanoparticles deposited onto the solid and liquid phases interface influenced the density of the working fluid, the heat transfer near the interface, and the ITR. The researchers numerically evaluated the ITR with changes in the intermolecular interactions between the fluid and the nanoparticles and those between the fluid and the surface. The researchers arrived at the following conclusions: (i) The nanoparticle layer deposited onto the solid–liquid interface decreased the ITR, which might become lower that the typical ITR value of a smooth heating surface, especially in the cases where the nanoparticles exhibit hydrophilicity. Additionally, it was common in the fullerene and amorphous nanoparticles. (ii) The considerable ITR decrease caused by the deposition layer enhanced the heat transfer from the heating surface to the nanoparticles, and, consequently, augmented the heat transfer of the nanoparticles. The key condition for the decrement in the ITR was the extraction, storing, and transport of the heat from the solid interface by the nanoparticles to the neighboring fluid outside the deposition layer. In addition to the heat flux increment, the temperature difference at the interface of the solid and liquid phases was decreased by the deposition of the nanoparticles. These alterations contributed to the ITR reduction. (iii) The deposit of nanoparticles onto the interface did not induce an appreciably extra thermal resistance as compared with the corresponding fluid layer, independently of the morphology of the nanoparticles. This was caused by the fact that the single layer of the nanoparticles was of such reduced dimensions that it could not be enough for producing the additional thermal resistance at the interface, although the thermal resistance of the used fullerene was high. (iv) The nanoparticle deposition changed the density of the thermal fluid and the temperature gradient at the interface. The ITR decreased with increasing fluid density in the deposition layer. The changes in the fluid density in the interface nearby region and in its temperature gradient by the nanoparticle deposition were two major conditions affecting the ITR.

2.3.3. Three-Phase Contact Line Behavior

During a nanofluid boiling situation, it is usually expected that the dispersed nanoparticles deposit uniformly onto the heating surface by the gravitational effect. Given that the nanofluids are a very dilute type of fluid, the quantity of nanoparticles deposited by gravity can be assumed to be negligible. As such, when the first vapor bubbles nucleate, the heat transfer surface is hydrophobic and the evaporation rate at the solid, liquid, and vapor phases contact line is the main mechanism associated with the growth stage of the initial vapor bubbles. The published transient imaging of the contact line evaporation in the literature has already demonstrated that the contact line of the bubbles presents a radial type of motion across the heating surface [60], provoking the expansion or, alternatively, the contraction of the dry patch. This dry patch can be characterized by a higher temperature area at the center of the active nucleation point. Given that the evaporative heat flux has its maximum value at the solid–liquid–vapor contact line, the suspended nanoparticles will deposit at the contact line as it evaporates and expands. It has been reported that a ring-like deposition pattern was observed at the contact line of a sessile droplet deposit onto a hydrophobic surface due to the delay in the depinning of the contact line. Moreover, it is expected that the radial motion of the contact line will be more constrained with the addition of nanoparticles. As a consequence, the initial deposition happens over only a narrow region under the form of a ring. Furthermore, the deposition at the contact triple line becomes very intense with the boiling time and, thus, modifies the surface wettability that, in turn, aids in the continuous shrinkage of the contact line radius corresponding to the dry patch dimensions. This particular mechanism conducts to the smearing out of the thinned ring to a wider ring structure, which, consequently, improves the wettability of an extended inner region of the nucleation sites. When the rewetting of this region is complete after the detachment of the bubbles, the growth stage of the following bubbles slowly changes from the dynamics of the contact line to the entrapment of the microlayer dynamics. The microlayer thickness augments in the radially outward direction and, consequently, evaporates faster near the center of the nucleation site, leaving behind a small dry patch. Because of the higher evaporation rate at the center of the front of the microlayer, a radially inward capillary flow is generated inside it, which conducts the nanoparticles toward the evaporative front of the microlayer. It should be stated that as the fraction of nanoparticles augments around the evaporative front because of such inward flow, the clustering rate of the nanoparticles raises up to produce larger particles. When the droplet evaporation happens on hydrophilic surfaces, it has been highlighted that the nanoparticles adhere and settle at the contact line region, which is caused by the conjugated action of the shape of the liquid and vapor phases interface and the balance between friction forces and capillarity acting on the nanoparticles. Taking into account these phenomena, it is believed that a similar mechanism is responsible for the settlement of the nanoparticles at the evaporative front of the microlayer considering that the bigger clustered particles come in contact with the evaporative front because of the presence of induced inward capillary flows. The clustering of the nanoparticles at the inner area of the deposition pattern can be easily observed using AFM imaging wherein the sub micro-scaled particles are deposited, and this fact can justify the high density and thickness of the deposit at the central regions of the nucleation point. Moving away from the central regions, although the microlayer possess a relatively large fraction of the working fluid to maintain the suspension of nanoparticles, the thickness of the deposit gradually decreases radially outward since the complete evaporation of the microlayer at these points does not occur because of its increasing thickness. The augmented microlayer thickness renders extra ITR to the heat transfer flow and the microlayer has lower rates of evaporation, which in turn, averts the complete dryout of the microlayer at the radially outward sites. The location of the dry patch can be estimated with the help of heat flux contour plots. If the maximum value of the dry patch radius and the deposited pattern radius is compared, it can be verified that the maximum dry patch radius is significantly smaller than the radius of the deposition pattern. Additionally, it has been found that the maximum dry patch radius is even smaller than the centrally deposited film radius. This fact that in the nanofluid boiling process, the microlayer does not evaporate totally in the radially outward regions and the dispersed nanoparticles settle primarily in such regions because of the gravity and attraction forces between the nanoparticles and surface like Van der Waals and electrostatic forces. Hence, the thickness of the deposit is radially decreased from the center of the nucleation points to their peripheral areas.

2.3.4. Capillary Wicking

Another important feature of the deposited layer of nanoparticles is the capillary wicking behavior. The formation of certain microstructures by the boiling-induced nanoparticle deposition may induce capillary wicking on the heating surface. For instance, in the work performed by Kim et al. [40], the wires covered by the nanoparticles of low fraction nanofluids do not induce a working fluid rise by capillary wicking since the formed microstructures on the heating surface at low concentrations are not sufficient to play the role of microflow pathways. The authors also confirmed that the material of the nanoparticles greatly influences the capillary wicking. In this sense, the oxide titanium nanoparticle-coated wire presented a maximum water capillary rise of 1.2 mm at maximum concentration, which then steeply increased for lower concentrations achieving 5.9 mm at the lowest concentration. Moreover, the alumina nanoparticle covered wire had a maximum water capillary rise value of 0.5 for all the tested concentrations. The fractal micro-scaled structures produced by the deposition of clustered titanium oxide nanoparticles induced fluid suction caused by the capillary wicking effect that increases the CHF of water and turns it considerably higher than that with nanofluids. Nevertheless, the much higher CHF of the heating surface covered by the nanoparticles is deteriorated by the use of nanofluids instead of water, given that the nanoparticles dispersed in the base fluids may clog the microflow pathways conducting the liquid bulk to the heat transfer surface through capillary wicking. Furthermore, the authors Kim and Kim [61] studied the influence of the surface wettability and capillarity of the boiling-induced nanoparticle deposition on the CHF. The authors reported that the deposition of nanoparticles during the boiling process induces capillary wicking on the porous deposition layer, whereby the supplied fluid delays the irreversible growth of dry patches. Moreover, the estimated heat flux gain based on the capillary liquid supply was of the same order of magnitude and consistent with the experimental CHF increase above the maximum CHF value of 1500 kW/m^2 obtained considering the wettability improvement. The researchers concluded that the appreciable CHF amelioration of nanofluid pool boiling is the result of not only improved surface wettability but also enhanced capillarity caused by the deposition of nanoparticles.

2.3.5. Roughness

With increasing concentration of the nanofluids, the deposited layer of nanoparticles grows on the heat transfer surface. However, the published results do not always confirm the direct relation between surface roughness and nanofluid concentration. This may be caused by the lack of uniformity of the deposited nanoparticle layer on the heating surface. Such a fact might conduct to higher heat transfer rates with the pool-boiling nanoparticle-deposited surfaces as compared with the ones obtained through the pool boiling of water alone. Nevertheless, it is possible to settle a well-defined correlation between the roughness of the surface and the deposited layer. Moreover, the authors Wen et al. [62] investigated the impact of surface roughness using rough and smooth surfaces. For the smooth surface, the increase in particle deposition on the heating surface led to an increased surface roughness. In the case of the rough surface, no considerable variation in the surface roughness was reported with increasing nanofluid concentrations. The researchers argued that the surface modification was performed by the features associated with the nanoparticles after the boiling process. Furthermore, the surface roughness was found to increase with an increasing wall superheat value for rough surfaces. In addition, the pool-boiling heat transfer of rough surface using nanofluids was around two-fold greater than that of the water alone. A reduction of around 30% in the heat transfer was reported for the pool boiling with nanofluids for a plain surface when compared with the base fluid itself. Although the surface roughness suffers a decrement from 0.167 to 0.099 μm after boiling completion, it achieved a CHF amelioration, which was likely due to alterations in the surface microstructure and topography. Moreover, Ham et al. [3] further investigated the boiling heat transfer of aluminum oxide nanoparticles dispersed in water on heating surfaces with different average roughness. The volume fraction of the nanoparticles ranged between 0 and 0.05 and the average surface roughness was of 177.5 nm and 292.8 nm. The researchers found that when the volume fraction was increased from 0 to 0.05, the CHF increased by 224.8% and 138.5% on the surfaces with roughness of 177.5 nm and 292.8 nm, respectively. The heat transfer capability of the aluminum oxide nanofluid with a 0.05 vol. % of concentration was lower than the one obtained with deionized water at R_a = 177.5 nm, but the maximum HTC value increased due to the increase in the CHF. Furthermore, Figure 5 schematically illustrates the boiling-induced nanoparticle deposition on two different rough heating surfaces.

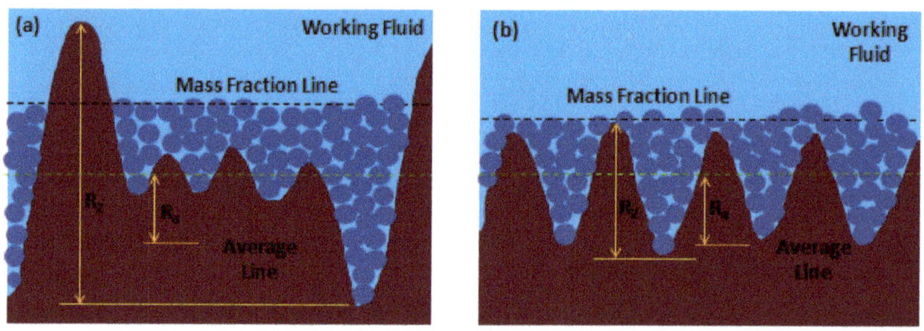

Figure 5. Schematic illustration of the boiling nanoparticle deposition on two different rough surfaces: (**a**) R_a and R_z with very different values and (**b**) R_a and R_z with approximate values.

The authors Ji et al. [63] made a qualitative analysis of the boiling-induced nanoparticle deposition on two different rough heating surfaces. The authors stated that it was easier to obtain a uniform deposition layer with a higher R_a and lower R_Z rough surface in the corresponding 5 (b) case. The mass concentration lines at the same height in the two different cases represent the equality of the concentration. The authors observed too that the heating surface with lower R_a and larger R_Z is not so easy to be uniformly coated by the nanoparticles. With lower weight fractions of the nanoparticles, it is possible to observe higher peaks outside the surface touching the water. Therefore, the more roughed surface was the one that exhibited improved heat transfer behavior. It can be stated that the number of deposited nanoparticles induces variations in the surface roughness of the heating surface. The continuous growth of the deposited layer produces greater thermal insulation, which leads to the decrement of the CHF. Further experimental studies are needed to establish the optimum value of the deposition layer thickness that maximizes the CHF.

2.3.6. Roughness-to-Particle Size Ratio

The ratio of the nanoparticle dimension to the average roughness of the heating surface can also be linked to the heat transfer behavior. With the use of extended ratios, the deposition of nanoparticles can further enhance the average roughness of the surface or, alternatively, reduce the average roughness. The authors Shoghl et al. [6] found a heat transfer decrement using zinc oxide and alumina nanoparticles dispersed in water. This deterioration was the consequence of the surface roughness decrease as compared with the heat transfer improvement that occurs with carbon nanotubes because of a notorious surface roughness enhancement. The authors reported that any improvement or any degradation of the boiling thermal performance was affected by the nature of the nanoparticles and the relative nanoparticle size ratio to the initial heating surface roughness. In addition, the researchers Narayan et al. [9], Shahmoradi et al. [64], and Wen et al. [65], among others, highlighted that the nucleate boiling enhancement generated by increasing the surface roughness factor was only viable with a well-defined range of size of the nanoparticle. For instance, the authors Narayan et al. [9] introduced the "surface interaction" factor, which represents the ratio of the average surface roughness to the average nanoparticle size, to address the capability of a given nanofluid to improve its nucleate boiling performance. In the cases where the nanoparticles are significantly smaller than the surface roughness features, having a "surface interaction" parameter considerably greater than one, the boiling heat transfer improved appreciably as the smaller particles deposited onto the initial nucleation sites and divided these initial single nucleation points into multiple ones. Nevertheless, in the cases where this interaction parameter had a value near one, the heat transfer was limited as most of the nanoparticles deposited in active nucleation points with approximated relative size and hindered the bubble nucleation. On the other hand, if the referred parameter presented a value considerably smaller than one, the heat transfer decrement was less intense than that verified with the parameter value around one, given that the larger nanoparticles decreased the number of the nucleation points being deactivated. Moreover, the "surface interaction" parameter is also of relevance in interpreting the boiling HTC degradation with augmented concentration of nanoparticles. In this direction, Shahmoradi et al. [64] found a decrement in the HTC for a water-based alumina nanofluid with the "surface interaction" parameter with values lower than one. Such decrement clearly worsened with enhanced nanoparticle volume fraction below 0.1. The researchers explained the worsening deterioration with the additional thermal resistance provided by the deposited layer of nanoparticles. Furthermore, Wen et al. [65] revealed that any augment in the boiling heat transfer was strongly influenced by the relative size of the particles dispersed in the working fluid and the initial dimensions of the superficial elements. Nevertheless, the researchers also highlighted that, since the surface modification by the nanoparticles was a cumulative phenomenon in time, the heat transfer trend was believed to continuously alter with successive pool-boiling experiments with only one heating surface. Moreover, Vafaei [66] reported an increment in the heating surface cavities' dimensions in the cases where the deposited nanoparticles were larger than the valleys of the surface roughness profile. The reported enhancement promoted the activation of nucleation points and improved the HTC at low heat fluxes. On the other hand, when the settled nanoparticles were smaller than the valleys of the surface roughness profile, the heating surface activation decreased, rendering a less effective boiling performance.

2.3.7. Thickness over Boiling Time

This subtopic has already been discussed throughout the current work, but nevertheless, and concerning the thickness of the deposition layer, it is worth mentioning the experimental work performed by the authors Park et al. [67] who evaluated the effects of the pool-boiling-induced nanoparticle deposition on the CHF improvement. With this purpose, boiling experiments were conducted to evaluate the impact of the thickness of the deposited nanoparticle layer on the CHF using a 0.01 vol. % alumina nanofluid with a Ni–Cr wire as the heating surface. The thickness of the nanoparticle-deposited layer was managed by different boiling times for pre-coating and the CHF evolution curve was acquired in function of the time of the pre-coating process. The authors found that the CHF remain unaltered regardless of the boiling time over a critical pre-coating time, whereas the CHF was dramatically increased during a relatively short pre-coating time. Moreover, the CHF occurrence was moderately decreased after the critical time region. It was also stated that the porosity decreased as the pre-boiling time increased since a denser layer was produced when the boiling time was extended. Moreover, the reduced number of pores affected the CHF decrement after that alteration. This fact denotes that the fluid fraction closely linked with the phase change diminished and this decrease would be promoted when employing a high concentration nanofluid during sufficient pool-boiling time. However, an excessive deposited layer thickness may result in a negative impact on the CHF. Overall, the authors concluded the following:

1. The CHF gradually decreased with long pre-coating times.
2. The CHF improvement may be interpreted by the enhanced surface wettability and hydrodynamic instability modifications with relatively short boiling durations.
3. The slow decrement in the CHF could be interpreted with the porosity decrease in the deposited layer with longer pre-coating durations.

2.3.8. Peripheral Regions of the Nucleation Sites and Deposition Points

During the initial part of the growing of the vapor bubbles in the nucleate boiling regime, the viscous effect can be large enough as compared withthe surface tension to hinder the working fluid motion and trap a very thin liquid layer underneath the vapor bubbles beneaththe growing bubbles. This layer is usually designated by a microlayer and has a thickness smaller than ten micrometers and a length up to one millimeter. Its lifetime is only a few milliseconds but it contributes to the major part of the heat transfer enhancement between the heating surface and the bubbles. For longervapor bubble growing times, the formation of a thin liquid layer designated by a macrolayer occurs, wherein the meniscus of the bubbles changes curvature suddenly. In terms of nanoparticle deposition, the particles deposited in the microlayer and macrolayer are usually nano-scaled particles [68]. Another aspect of the boiling-induced nanoparticle deposition process is related to what happens in the peripheral regions of the nucleate boiling points and microparticle deposition points. The periphery of the deposited micro-scaled particles corresponds to the maximum value of peripheral spreading of the area affected thermally by the nucleation sites, beyond which only negligible changes in the temperature are to be observed in the heating surface during bubble growth. This maximum peripheral region corresponds to the macrolayer of the bubbles. Large agglomerated microparticles up to 20 μm have been found in the peripheral region in the form of a ring on the deposited pattern, as it can be seen in Figure 6. Moreover, these micro-scaled particles are considerably larger than the particles deposited in high-density areas near the center of the deposited structure. Considering the associated complexities, it is challenging to interpret whether these large agglomerated microparticles migrated from some region of the liquid bulk to the periphery or, alternatively, whether these particles were formed at the periphery of the nucleation site. In this sense, the work carried out by Kangude and Srivastava [41] made a serious attempt to reveal the most likely phenomena responsible for the presence of the micro-sized particles at the periphery of the active nucleation sites. By AFM imaging observation of the nanoparticle deposition, the researchers confirmed that the nanoparticle deposition during the boiling process led to the formation of a porous deposited structure. They also concluded that the hydrophilicity of the porous deposited layer augmented the capillary flows by offering a chain of hierarchical nano/micro-scaled paths for the fluid to flow. The authors obtained a nanoparticle-deposited structure using the 0.005% nanofluid that was found to exhibit capillary flows.The presence of the capillary flows was ascertained through high-speed camera imaging by observing the motion of the contact line of a sessile droplet deposited onto the nanoparticle deposition layer. It was observed that the droplet achieved static equilibrium first and then the apparent contact line of the sessile droplet moved through the porous layer until it attained a very small contact angle as compared with that of the substrate. However, no such movement of the contact line was verified when the droplet was deposited onto regions of the substrate surface away from the nucleation sites. The authors concluded that the contact line motion confirmed the presence of capillary flows through the deposited porous layer of nanoparticles. Those flows had a strong impact on the microlayer evaporation by continuously replenishing the fluid to the evaporating front of the microlayer. As the microlayer starts evaporating at the central portion of the nucleation site, the capillary pressure forces the fluid to flow toward the contact line of the evaporating front of the microlayer from the periphery through the porous surface of the deposited layer. Furthermore, it should be stated that the thickness and density of the deposit increased from the periphery to the center of the nucleation site. When enough deposition occured, only the working fluid penetrated through the porous layer by capillary forces and the dispersed nanoparticles filtered out toward the peripheral region. The filtered-out nanoparticles agglomerated to form micrometer-sized particles at the periphery over time. Thus, the existence of the micrometer-sized particles in the peripheral region of the nanoparticle deposition points can be reasonably attributed to the presence ofcapillary-assisted radially-inward flows through the pores of the deposition layer during the microlayer evaporation.

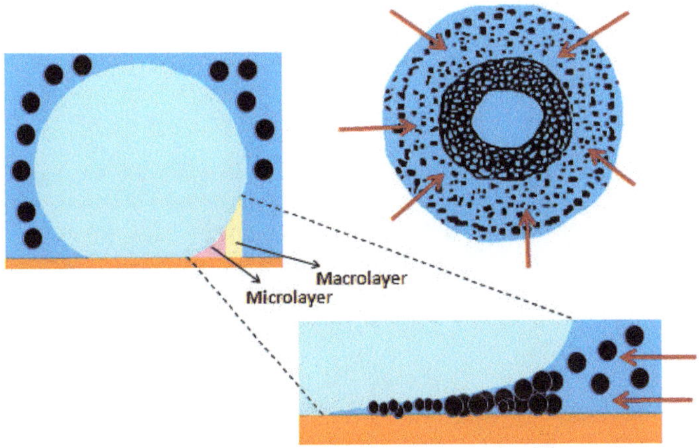

Figure 6. Schematic illustration of the nanoparticle deposition process and top view of the deposited structure showing the size distribution of the nanoparticles.

2.3.9. Vaporization Core Sites Density

The deterioration of the boiling HTC is caused mainly by the deposition of nanoparticles on the heating surface that enhances the heat transfer resistance. However, the wettability of the surface and the CHF were enhanced. For instance, welding metal foams onto a plain heating surface is a common technique to enhance the heat transfer area. Moreover, this procedure impacts the growth and detachment of vapor bubbles in the course of boiling. Moreover, to enhance the heat transfer capability of the metal foam structure, the researchers Xu et al. [69] studied the influences concerning the size and concentration of AlO and SiC nanoparticles deposited onto a copper foam of 7 mm thickness. The testing pore density was 5 PPI, 60 PPI, 100 PPI, and the corresponding porosity was of 0.9, 0.95, and 0.98. Some preliminary works reported that the boiling induced nanoparticle deposition onto the foamed copper might increase its capillary wicking and number of active vaporization core sites. Moreover, the thermal performance of nanofluids on the gradient hole surface was investigated by Xu and Zhao [70], since adding nanoparticles clogged the voids of the higher 100 PPI density copper foam, and, consequently, the vapor bubbles escape resistance became enhanced and the heat transfer capability degraded. One of the main effects of the heating surface modification is to increase the available heat transfer surface and, hence, enhance the number of available vaporization core sites. Moreover, the density of the nucleated bubbles will become higher in addition to the increasing density of gasification core sites. Therefore, during the pool-boiling process, the nanoparticles deposited onto a bigger heat transfer area, allowing high concentrations of the nanofluid to be employed, rapidly increased the thermal conductivity of the base fluid, and thus the nanoparticles did not deteriorate because of the porous deposition layer. Moreover, the vapor bubbles nucleated on a rough heating surface alter the capillary motion of the vapor bubbles, which are only impacted by the upward action of the buoyancy force.

2.3.10. Hydrodynamic Instability

Another relevant feature of the pool-boiling-induced nanoparticle deposition is the hydrodynamic instability alteration [71]. The vapor bubble dynamics during the pool-boiling process are affected by the deposition of nanoparticles onto the heating surface, which reduces the distance between the vapor bubble departure points. This distance is commonly designated by the Rayleigh–Taylor wavelength instability wavelength or hydrodynamic instability wavelength, which can be easily identified in the film boiling regime. The alteration of this wavelength is linked to the nanoparticle-deposited layer on the heating surface during boiling, given that the deposited layer will alter the spacing between nucleating bubbles (or vapor columns) and, consequently, the Rayleigh–Taylor instability wavelength. In addition, the hydrodynamic instability approach or hydrodynamic fluid-choking limit connects the characteristic wavelength of the hydrodynamic instability with the CHF enhancement of the nanoparticle-deposited porous layer. The hydrodynamic limit theory is based on the Rayleigh–Taylor instability wavelength and was introduced by Zuber [72] for a plain surface and can be broadened to a coated heating surface having capillary limit. Regarding the hydrodynamic limit, the CHF is generally caused by the vapor columns' instability. The deposited layer can modify the distance between vapor columns on the surface and, consequently, alter the critical instability wavelength. The researchers Liter and Kaviany [73] interpreted the impact of the modulated wavelength on the CHF of a modulated porous layer of nanoparticles (having periodic variations in the layer thickness) using the following expression:

$$Q''_{porous} = \frac{\pi}{8}\Delta h_{lv}\sqrt{\frac{\sigma \Delta \rho_{lv}}{\lambda_m}}$$

where λ_m is the modulated wavelength or the length scale that defines the vapor escape locations in the working pool-boiling fluid from the porous structure of the heating surface, Δh_{lv} is the enthalpy gradient between liquid and vapor phases, $\Delta \rho_{lv}$ is the density gradient between the liquid and vapor phases, and σ represents the fluid surface tension. For a smooth surface, the λ_m parameter is influenced by the balance between the buoyancy force and surface tension, being a function of the thermal characteristics of the working fluid. In the case of a surface coated with a porous layer, the parameter λ_m depends on the vapor escape pathways and, consequently, is a function of the porous structure of the deposition layer. In the experimental work performed by Park et al. [74], the wavelength alteration corresponded to the employed pool-boiling alumina and graphene/graphene oxide nanofluids CHF amelioration trend. As already stated by Liter and Kaviany, the wavelength can be taken as a well-defined geometrical parameter that depends on the surface conditions. As a consequence, the research team found that the change in the wavelength extends the wetting of the heating surface by enabling the working fluid to break through, resulting in the CHF improvement. Nevertheless, the researchers recognized that the recent published CHF enhancement and wavelengths are not consistent with the prediction of the Liter and Kaviany equation. Accordingly with the findings of the aforementioned authors, Park and Bang [71] stated that the onset of the CHF based on the hydrodynamic limit is due to the instability of vapor columns. The nanoparticle porous layer of the deposited nanoparticles during the boiling process could change the critical distance between vapor columns rising from the heating surface and, consequently, alter the critical instability wavelength. Moreover, a similar situation was observed in the droplet formation on the nanoparticle layer, which is closely linked with the detachment of the bubbles from the heating surface. The authors prepared different nanofluids and the used pool-boiling apparatus was designed to enable the direct observation of the Rayleigh–Taylor instability wavelengths. The authors reported that the distance between the bubbles was different for each nanoparticle-coated surface, which revealed a shorter distance between the bubbles than that of the bare heating surface. Moreover, the nanofluids that promoted a higher CHF enhancement exhibited shorter Rayleigh–Taylor instability wavelengths. A short wavelength allows the vapor to prevent the formation of a bulk of vapor by venting the vapor evenly across the heating surface. Furthermore, it was demonstrated that the shorter wavelengths also improved the wettability by allowing the liquid to break through the developing vapor film, which also increases the CHF.

2.3.11. Exfoliation

The deposited layer of nanoparticles on the heating surface may still suffer a peeling action throughout the pool-boiling process. The boiling HTC decreases with increasing fraction of the nanoparticles and the exfoliated area will be increased. This was confirmed by Watanabe et al. [75], who performed experiments to evaluate the force of adhesion of the deposition layer and the effect of its partial detachment on the boiling heat transfer behavior. The researchers reported that the amount of nanoparticles deposited on the heating surface and the force of adhesion of the nanoparticle layer were influenced by the constitutive material of the nanoparticles. With the used nanoparticles, the mass of the deposited nanoparticles could be ordered as titanium oxide > alumina > silica and the force of adhesion was ordered as silica > alumina > titanium oxide. The degradation of the boiling heat transfer was particularly considerable for the titanium oxide with the largest deposited mass. With the peeling of the deposition layer, the settled mass of nanoparticles decreased and the equilibrium contact angle was enhanced. This fact rendered that the degradation of the HTC and increment of the CHF became diminished with the gradually increasing deposition layer detachment. Nevertheless, a larger HTC and a smaller CHF were observed as compared with the bare heating surface for highly exfoliated nanoparticle layers. The authors also stated that the detachment of the deposited layer of nanoparticles happened through a non-homogeneous manner. Moreover, the research team found that the peculiar boiling process at the edge of the remaining deposited layer caused the enhancement of the HTC and the decrement of the CHF.

2.3.12. Re-Suspension of the Nanoparticles

The re-suspension of the deposited nanoparticles of the nanofluids can mitigate or, in some cases, even eliminate the potential deterioration of the nanofluids over time. Indeed, this re-suspension promotes the long-lasting heat transfer amelioration of the nanofluids. The particle re-suspension occurrence is usually observed in nature in, for instance, the re-suspension of the sediments of a riverbed that are carried away by water flow and the re-suspension of sand on the ground, which is blown away by the wind. In particular, the re-suspension of wall-surface-adhering particles by the action of turbulent flows has received great attention over the past few years from researchers. Nevertheless, most of the published scientific articles concerned with nanoparticle suspension comprise single-phase working fluids, and only a few deal with the two-phase fluid condition as it occurs, for example, in boiling. Since the boiling process may induce a very high fluid flow and disturbance of the vapor bubbles, the re-suspension of the nanoparticles deposited onto the heating surface is a very probable phenomenon. If those nanoparticles can be re-suspended, it generally denotes that the time-dependent deterioration of the nanofluids can be relieved without the help of any additional means or methods. This degradation mitigation of the nanofluids over time is a very promising way to achieve long-lasting heat transfer performance. Under pool-boiling scenarios, the movement of the working fluid is induced by convection and movement of the bubbles provoked by buoyancy action. A large part of the existing boiling research is derived from pool-boiling experiments. Consequently, pool boiling seems to be a suitable starting point for the in-depth study of the particle re-suspension behavior that can separate the effect of boiling from the action of the convection-driven flows on the nanoparticle re-suspension. In the work developed by [76], pool-boiling experiments were performed using surfaces with deposited nanoparticles produced by nanofluids. The researchers found the re-suspension ratio of these nanoparticles for different boiling times, densities of the deposition layers, and varying heat fluxes to infer on the impact of the experimental parameters on the re-suspension of the nanoparticles. The authors reported that a certain fraction of the deposited nanoparticles was re-suspended into the liquid bulk and, after that, migrated along with the liquid flow during the boiling process. Since the re-suspension does not take place before boiling, the disturbance suffered by the vapor bubble disturbance in the course of boiling must be an important factor for the re-suspension of the nanoparticles. Besides that, the main difference in the deposition surface between before and after boiling is the porosity of the surface. Hence, it can be stated that the porosity of the deposition layer is the major contributor for the re-suspension of the nanoparticles in the liquid bulk during the boiling process. For a better understanding, it can be noted that when a vapor bubble is in its nucleation state of evolution, the fluid near the bubble will be vaporized because of the superheat. The porous structure of the deposition layer enables the fluid to be vaporized in the layer to be enlarged from the outside of the layer, resulting in a fluid flow from outside toward the inside of the layer. This flow will steadily impinge on the deposited nanoparticles. In the cases where there is no fluid flowing through the deposition layer porous structure, the forces acting on the deposited nanoparticles are gravity, adhesive forces from the interacting neighboring nanoparticles, and brace force. In the cases where the outside-to-inside flow occurs through the pores, the operating fluid will create a dragging force on the nanoparticles around the pore. This force provokes the motion of the nanoparticles that tend to follow the fluid flow. As the generated vapor bubble continues to grow, the drag force increases. At the departure stage, when the vapor bubbles begin their raising from the surface, an extreme disturbing effect at the back region of the departing bubbles occurs. Under these conditions, the drag force overcomes the gravitational effect and forces of adhesion and provokes the detachment of the nanoparticles from the deposition layer, which moves together with the flowing fluid. Given that the fluid flow at the wake of the rising vapor bubbles is also upward driven, the detached nanoparticles will rise upwards together with the liquid flow and are, by this way, re-suspended. Hence, it should be emphasized that two main factors can be assumed for the re-suspension of the nanoparticles: the first is the effect of the fluid flow through the porous structure of the deposition layer during the growth stage of the bubbles and the second is the disturbing effect at the departure of the bubbles. Apart from these factors, many other factors may impact on the nanoparticle re-suspension trend. That is, in the case of the growing and rising process of the vapor bubbles, the nanoparticles dragged from the deposition layer can be re-suspended in the working fluid as a consequence of the departure of the rising bubbles or the breaking-up of the bubbles at the liquid and vapor phases interface. Nevertheless, further experimental and numerical works should be carried out to provide a better understanding of the governing mechanisms of the nanoparticle re-suspension. In the above-mentioned work, the authors plotted the re-suspension ration versus the deposition area density using 20 KW/m^2, 50 KW/m^2, 80 KW/m^2, and 100 KW/m^2 of heat flux. Moreover, the authors arrived at the following conclusions:

1. During boiling, a fraction of the deposited nanoparticles may be re-suspended within the operating fluid and, after that, migrate along with the fluid flow.
2. The nanopores of the deposition layer have a cardinal part in the re-suspension phenomenon and fluid flow through the pores at the growing stage of the vapor bubbles and the disturbance at the rear zone of the departing bubbles should be taken as the main factors for the re-suspension.
3. The re-suspension ratio increased with increasing heat flux, given that the former increased near 300% when the heat flux was enhanced from 20 to 100 kW/m^2. Moreover, when low heat fluxes near 20 kW/m^2 were applied, the re-suspension ratio

increased with increasing density of the deposition zone. When there were imposed moderate heat fluxes ranging between 50 kW/m^2 and 80 kW/m^2, the re-suspension ratio initially increased and then decreased with increasing deposition area density. When high heat fluxes of around 100 kW/m^2 were applied, the re-suspension ratio decreased with the density of the deposition area.

Moreover, Chen et al. [77] studied the re-suspended nanofluid pool boiling under the action of an electric field. The investigation team also discussed the difference in the thermal behavior between the re-suspended nanofluid and the base fluid alone. Figure 7 schematically illustrates the distribution of the deposited nanoparticles before and after the re-suspension phenomenon. Figure 8 shows the fluid flow and heat transfer of the nanoparticle re-suspension.

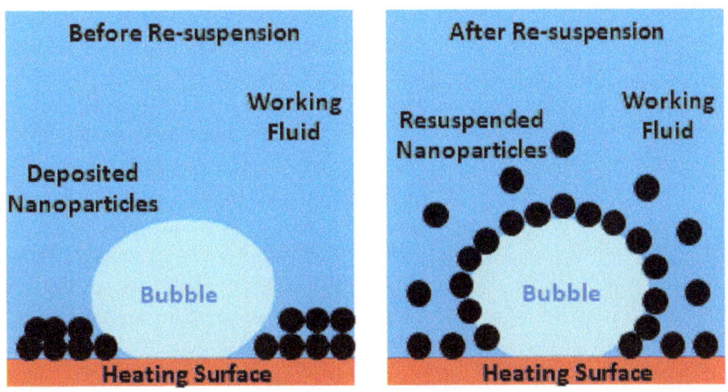

Figure 7. Boiling-induced nanoparticle deposition time-dependent features.

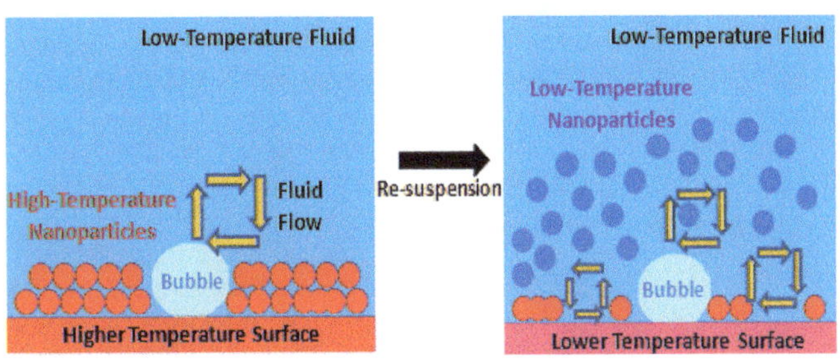

Figure 8. Fluid flow and heat transfer of the nanoparticle re-suspension.

The authors stated the following conclusions:

1. The nanoparticle re-suspension can be obtained by the action of an electric field and can be demonstrated by the thinned deposition layer of deposited nanoparticles and on the enhanced turbidity of the working fluid.
2. The HTC and CHF enhancement of the re-suspended nanofluid is generated because of the combined effect of the re-suspension and applied electric field. The latter can be enhanced with the voltage augmentation. Nevertheless, there is an optimum value of the concentration of the nanoparticles that maximizes the improving action of the electric field.
3. The mechanisms for the heat transfer improvement of the re-suspended nanofluid under the effect of an external electric field comprise the heat transport between the surface and the fluid promoted by the motion of the nanoparticles, further lowering the surface tension of the liquid and vapor phases interface and the thermal conductivity enhancement of the heat transfer media.

2.3.13. Sintering of the Nanoparticles

The authors Vafei et al. [78] introduced the approach of the nanofluid boiling as a process for the creation of a semiconductor titanium oxide nanoparticle film deposited onto a FTO (F-doped tin oxide) glass conductive substrate. A pool-boiling apparatus was employed to deposit the titanium oxide 20 nmsized nanoparticle nanofluid. The boiling of the nanofluid directly on the FTO glass substrate enables the deposition of the nanoparticles onto its surface. Using the as-deposited films, the crystal growth of the titanium oxide nanoparticles was controlled by altering the temperature, duration, and ramping rate of post-sintering. A densely packed titanium oxide layer was obtained for the as-deposited substrate through the pool-boiling process. For the maximum temperature at 550 °C, the titanium oxide grain sizes became larger and near 50 nm and more round-shaped titanium oxide nanostructures were observed. This work demonstared for the first time how the sintering of titanium oxide nanoparticles proceeds for the nanoporous totanium oxide films. It was observed that the titanium oxide nanoparticles fused with each other and crystal growth happened through the neighboring 2 to 4 nanoparticles at 550 °C. Hence, an extra beneficialfeature of the pool-boiling-induced nanoparticle deposition is that the heating surface will begin to sinter the thin deposited layer. Although this heat treatment is not enough to obtain the required properties, it is enough to increase the stability of the deposited layer. By taking advantage of the stability of the film and the sintering properties of the titanium oxide nanoparticles, a post-sintering treatment after the nanoparticle deposition considerably impacts on the prodictionof a uniform, robust, and dense film. In conclusion, this work demonstrated for the first time that sequential pool-boiling and sintering processes are alternative procedures to create uniform porous titanium oxide layers.

3. Effects of the Nanoparticle Deposition on the Boiling Heat Transfer Parameters

3.1. Heat Transfer Coefficient

The deposited nanoparticles together with the suspended remaining nanoparticles in the base fluid influence the boiling HTC. The conjunction of the substrate properties such as its base material, the vapor/liquid interface features, and the characteristics of the nanoparticles themselves play a relevant role in the pool-boiling HTC enhancement. This role is performed by the force balance changing and dynamics of the three-phase contact line, bubble growth stage, bubble frequency at departure, and wetting and rewetting. It was predictable that the poor thermal conductivity of the deposition layer increased the conduction thermal resistance and, hence, degraded the HTC. However, the experimental data denied it. In this sense, the experimental work conducted by White et al. [10] showed that the pool-boiling HTC increased with increasing thickness of the deposited layer of titanium oxide nanoparticles. The authors stated that no significant deterioration in the HTC of nanofluids was verified in the course of the deposition. Such a fact may reveal that the impact of the thermal resistance of the deposition layer is not significant when compared with that of the pool-boiling heat transfer by convection, even though the thermal conductivity of the titanium oxide layer is relatively low. Moreover, in the work conducted by Kathiravan et al. [60], the authors used high thermal conductivity nanoparticles of copper and observed that the deposited nanoparticle layer deteriorated the boiling HTC. The already published scientific articles proposed that the influence of the thermal resistance of the deposition layer is not significant and the deterioration of the HTC using nanofluids might be linked with the heating surface wettability and roughness alterations. Additionally, it can be stated that the heat flux directly affects the HTC since at low heat fluxes, the impact of the concentration of the nanoparticles on the pool-boiling HTC is negligible given that at low heat fluxes the larger heating surface cavities are the only ones that are active. Nevertheless, at high heat fluxes, the smaller cavities of heating surface are activated as well but the HTC is reduced with the increasing concentration of the nanoparticles [79]. This effect may be caused by the nanoparticle filling of the smaller surface cavities and by the decrease in the nucleation site's density.

Moreover, in specific studies concerning the impact on the pool-boiling HTC of the surface roughness alterations provoked by the deposition of nanoparticles, it has been proposed that the smaller nanoparticles fill the cavities of relatively rough surfaces and, consequently, decrease the surface roughness and number of active nucleation points, which in turn, demises the pool-boiling HTC and increases the wall superheat value [80,81]. Nevertheless, in the study carried out by Das et al. [7], pool-boiling tests were conducted on a considerably rough heating surface using a 0.005 vol. % zirconia nanofluid. The authors reported that in spite of the heating surface roughness decreasing, the HTC increased. On the other hand, the authors Narayan et al. [9] found that in the case of the average size of alumina nanoparticles being similar to the average surface roughness level, the HTC deteriorated with increasing nanoparticle fraction. In addition, Bang and Chang [11] verified that the pool-boiling HTC became lower when the average surface roughness was inferior to the size of the nanoparticles, and although the surface roughness became higher with the increasing concentration of the nanoparticles, the HTC deteriorated. Furthermore, many researchers reported the increment [79], decrement [41], and lack of reaction [81] of the deposited nanoparticles on the boiling HTC. In conclusion, it should be emphasized that the active nucleation site's density directly depends on the surface roughness and wettability, and on the average size of the nanoparticles. If the nanoparticles are small compared with the valleys of the surface roughness profile, the active nucleation site density decreases, and if the nanoparticles are not too small when compared with the average roughness of the surface, the nanoparticles will fill and split the surface cavities and the number of available nucleation points will increase. In the cases where the deposited nanoparticles are bigger than the valleys encountered in the surface roughness profile, the nucleation site density might change differently. The flooded cavities are not able to nucleate bubbles. In fact, cavities that are not completely flooded are the ones able to initiate the nucleation of the vapor bubbles and, hence, augment the HTC. The reduction in the surface wettability may avoid flooding its cavities and, in turn, create a higher number of nucleation sites. The authors Forrest et al. [18] studied the influence of the wettability of the heating surface on the HTC using hydrophilic, super hydrophilic, and hydrophobic heating-coated wires. The wires were coated with different surface-treated silica nanoparticles to generate different wettability effects and it must be stated that no visible change was found in the surface roughness after coating, which might indicate that the nanoparticles coated conformably to the micro-scaled surface deformities. The hydrophobic surface was found to possess a higher number of active nucleation sites and higher HTC, the super hydrophilic surface deteriorated the HTC using water, and in the hydrophilic surface, no alteration was observed in the HTC compared with that of the bare heating surface.

3.2. Critical Heat Flux

The CHF occurs when a layer of vapor is formed between the thermal fluid and the heat transfer surface. The continuous replacement of this vapor layer by the liquid keeps the heating surface temperature among a safe range. Moreover, the contributing mechanisms for the vapor layer removal from the surface or for the enhancement of the rewetting increased the CHF. Furthermore, the enhancement of the wettability caused by the deposited nanoparticle layer was considered to be a likely reason behind the enhancement of the nanofluid CHF. The effect of the wettability of the surface on the CHF was taken into account in the macrolayer dryout model proposed by Haramura and Katto [82] and in the hot/dry spot theory introduced by the authors Theofanous and Dinh [83]. Nevertheless, further experimental data revealed that all the hydrophobic, hydrophilic, and super hydrophilic surfaces improved the pool-boiling CHF [41]. Considering this fact, it is logical to assume that the wettability might not be the only possible characteristic for the CHF improvement. With the boiling-induced nanoparticle deposition phenomenon, the receding and advancing contact angle and consequent contact angle hysteresis, and the equilibrium contact angle are all subject to change [2]. In the experimental work performed by the researchers Forrest et al. [18], it was confirmed that heating surfaces having small receding contact angles enhanced the CHF. The values encountered for the equilibrium, receding, and advancing contact angles of the hydrophobic heating surface were found to be 140°, 20°, and 160°, respectively. The authors stated that this surface was found to, at the same time, improve the boiling HTC and CHF. Moreover, it should be noted that the lack of homogeneity of the fluids and surface are factors directly influencing the advancing and receding contact angles, and correspondent hysteresis. Furthermore, it is critical to infer on the conditions that increase the nucleate boiling HTC and the CHF of nanofluids simultaneously. Additionally, there are several theories for explaining the boiling mechanisms for the departure from nucleate boiling, which causes a sudden rise in the surface temperature and in the CHF. Such theories include, among others, the hydrodynamic instability, vapor bubble interaction, and the already mentioned macrolayer dryout and hot/dry spot theories. The hydrodynamic instability approach confirms that the hydrodynamic effect associated with the counter current flow of vapor and fluid in the nearby surface region is the fundamental reason for the departure from nucleate boiling. This theory also reclaims that the departure from nucleate boiling happens when the down flow of the fresh fluid to the heating surface is averted by the rising vapor [71]. The macrolayer approach defends that the bubbles are separated from the heating surface by the fluid macrolayer and the departure from nucleate boiling occurs when this macrolayer dries out [84]. The hot/dry spot theory fundamentally deals with the reversibility and irreversibility of hot/dry spots and the departure from nucleate boiling occurs in irreversible hotspots where the rewetting is no longer permitted [85]. In the bubble interaction theory, the departure from nucleate boiling is conducted when the density of the bubbles on the surface is high enough to achieve a complete covering of the surface with a vapor layer, preventing the access of the fluid to the surface [86].

3.3. Surface Superheat

The authors Gajghate et al. [35] studied the effect of the ZrO_2 layer settled during pool-boiling experiments on the heat transfer enhancement. The obtained boiling curves showed the conjugated influence of the nanoparticle layer thickness and surface roughness on the layer superheat value under different heat fluxes. The researchers observed a decrement in the wall superheat value with increasing heat flux and the maximum reduction was verified at 5.8 K with a heat flux of 109.8 kW/m^2 using a 200 nm ZrO_2-coated copper substrate having 227 nm of average roughness. The greatest reduction in the surface superheat value was of 31.52% compared with a smooth copper substrate. This effect is caused by the thickness increase in the ZrO_2 layer deposited onto the copper substrate during pool boiling, which incremented the effective heat transfer surface area and the nucleation core point density. The bubble dynamic at different heat fluxes was also studied showing the nucleation of an isolated vapor bubble on the copper surface at low heat fluxes. With the imposition of high heat fluxes, the bubble diameter was enhanced and lowered the bubble departure frequency. In addition, the authors [42] studied the superheat value of the deposited layer and concluded that it depended mainly on the diameter of the bubbles at the departure stage of evolution and on the boiling time. The layer superheat value was also found to decrease from using the 0.0025% nanofluid to the 0.005% nanofluid. It was observed that the bubble diameter at departure and the boiling time dramatically increased from 0.0025 to 0.005% nanofluid, while keeping almost the same growth time of the nucleated vapor bubbles. With these alterations, the magnitude of the evaporation superheat layer increased steadily with the third power of the diameter of the bubbles at departure and decreased almost linearly with increasing pool-boiling time. Therefore, the conjugated effect of the alteration in the departure diameter of the bubbles and ebullition time led to an increase in the deposition layer superheat value when using the 0.005% nanofluid rather than the 0.0025% nanoparticle concentration nanofluid. The researchers also stated that the HTC associated with each and every heat transfer mechanism was affected by the changes in the bubble dynamics and ITR. These factors are probably determined by the structure in the boiling-induced deposition layer at the active nucleation points. Furthermore, the researchers Hadzic et al. [33] investigated the influence of the nanoparticle dimension and concentration of titanium oxide nanofluids on the pool-boiling heat transport from laser-textured copper substrates. In their work, nanofluid pool boiling was evaluated with 4–8 nm titanium oxide nanoparticles with 0.001 wt. % and 0.1 wt. %. The boiling curve obtained for the 0.001 wt. % nanofluid was stable, while the boiling curve for the 0.1 wt. % nanofluid was shifted toward higher superheat values after the completion of each consecutive pool-boiling experiment. Such a shifting effect may be due to the decrement in the number of nucleation points through the deposition of titanium oxide nanoparticles. Employing the 0.1 wt. % nanofluids, the microcavities on the laser-textured surface were filled with nanoparticles during the initial experiment, which led to a decrease in the number of nucleation sites and an increase in the layer superheat value during the following experiments. With the increase in the number of consecutive experiments, the layer superheat value reached unpractical values and the boiling heat transfer degraded. The boiling experiments were repeated for large (490 nm) titanium oxide nanoparticles. At this time, the authors found with the 0.001 wt. % nanofluid, an appreciable shift in the boiling curve toward lower layer superheat values, while with the 0.1 wt. % nanofluid, the respective boiling curve was observed to be shifted toward higher superheat values. These findings were consistent with those using the smaller nanoparticles. Nevertheless, it was reported that after more than four hours of boiling, the more concentrated nanofluid boiling curve became unstable because of the thick layer settled on the surface that also locally flaked off. An additional study was conducted by mixing 0.05 wt. % small titanium oxide nanoparticles with 0.05 wt. % larger titanium oxide nanoparticles. At this time, the boiling curve was shifted toward a higher layer superheat value in accordance with the behavior observed in the case of the 0.1 wt. % nanofluid having either smaller or larger nanoparticles. It was also found that after the completion of the initial experiment, the layer superheat value was considerably enhanced, which is likely due to the filling of the surface microcavities and channels with the smaller sized nanoparticles that reduced the nucleation points. Moreover, as the latter decreased, the same happened with the boiling HTC. Overall, the authors arrived at the following conclusions:

1. The pool boiling of the nanofluids with higher nanoparticle concentration resulted in a considerable deposition of nanoparticles on the heating surface and a corresponding CHF improvement of up to 2021 kW/m^2. Nevertheless, were reported very high layer superheat values up to 100 K, which suggested poor practical applicability.
2. The heat transfer decrement of the pool boiling with nanofluids on laser textured surfaces may be explained by the penetration of the nanoparticles into the laser-made grooves and cavities, which decreased the active nucleation site density. Moreover, a thicker deposition resulted in extra thermal resistance, whereas the porosity of the surface assured an appreciable delay in the CHF incipience, and the surface superheat value in turn was dramatically enhanced.

4. Main Factors Impacting the Nanoparticle Deposition

4.1. Concentration of the Nanoparticles

The published results showed different characteristics of the natural thermo-convection of the nanofluids according with the concentration of the nanoparticles in the base fluid. One was the boiling heat transfer degradation for, in the majority of cases, volume fractions greater than 0.1. Another was the occurrence of optimum heat transfer rate and coefficient at a certain concentration level, beyond which a decrement was reported. Hence, each and every nanofluid may enhance theboiling heat transfer in an exact volume fraction of included nanoparticles for an exact case depending on the stability of the nanofluid, nature of the base fluid, thermal condition and cavities morphology, and density of the heating surface [87]. As was previouslymentioned throughout the present work, nanoparticle deposition on the pool-boiling heating surface is enhancedwith afraction of the nanoparticles suspended in the base fluid. An excessive nanoparticle deposition may cause the thickening of the deposition layer, which in turn, may lead to an increase in the thermal resistance and, consequently, to the pool-boiling heat-transfer deterioration. This fact was confirmed through the experimental work conducted by the researchers Mukherjee et al. [88] using silica nanofluids having different volume fractions. The authors reported that with lower concentrations of 0.0001 vol. % and 0.001 vol. %, the used nanofluids exhibited limited nanoparticle deposition, inducing a great number of nucleation sites and improving the heat transfer performance. Additionally, when concentrations of 0.01 vol. % and 0.1 vol. % were employed, a greater nanoparticle deposition occurred. The development of a thick deposit further impeded the pool-boiling heat transfer. The results confirmed an increment in the HTC and CHF for 0.0001 vol. % and 0.001 vol. % and a decrement with 0.01 vol. and 0.1 vol. fractions. Furthermore, and according tothe authors Kim et al. [40], the deposition layer produces a continuous modification in the surface morphology that directly influences the boiling heat transfer. Such surface morphology continuous alteration depends strongly on the nanoparticle concentration. In the cases where low nanoparticle volume fractions of 0.0001 and 0.001 were employed, only a slight modification of the heating surface was observed. However, when the concentration was increased, the nanoparticle deposit thickened, and more micro-scaled structures developed on the surface caused by the clustering of the silica nanoparticles. Figure 9 schematically represents the nanoparticle deposition process using different weight fractions of nanofluids. Mukherjee et al. [89] reported that the deposited nanoparticles fill the surface cavities, producing a smoother final surface and also concluded that a grater nanoparticle deposition led to a smoother heating surface. The authors noted that the surface modification is less pronounced at lower concentrations of 0.01 vol. % and 0.1 vol. %, due to the smaller amount of available nanoparticles and the higher stability of the used nanofluids that avert any further settlement of the nanoparticles. Nevertheless, such a tendency changes when the volume fraction is high at 1%. At this concentration, the amount of suspended nanoparticles is considerable and their deposition rate is greater and sufficient to fill up the cavities and alter the texture of the heating surface. Owing to a lesser change inthe surface and improved stability, the nanofluids exhibited enhancements in the HTC and CHF and the opposite trend was found at higher nanoparticle fractions.The effect of nanoparticle concentration in the base fluid was also investigated by Kole and Dey [90]. Two diverse concentrations of zinc oxide nanoparticles having a size of 30–40 nm in ethylene glycol were evaluated. It was observed that, after pool boiling, the nanoparticles were deposited over the heating surface, which prevented the active nucleation sites, and, therefore, the HTC decreased. By measuring the CHF values through the use of a thin copper–nickel alloy, the authors reported a considerable increase in the CHF by increasing the concentration of zinc oxide. The investigation team verified a maximum CHF enhancement of 117% for zinc oxide nanoparticle volume fractions of 2.6%. Moreover, the authors Minakov et al. [36] demonstrated that even at the 0.25 vol. % concentration of nanoparticles, the CHF increased by more than 50% and continued to grow with further increases in the nanoparticle fraction. It was stated that at high concentrations of nanoparticles, the growth rate of CHF slowed down and reached a constant value. Such behavior was due to the stabilization of the deposit size on the heat transfer surface. Moreover, the researchers Ahmed and Hamed [91] studied the pool-boiling heat transfer on smooth copper surfaces using nanofluid and water. Alumina nanoparticles of 40–50 nm were employed to prepare 0.01 vol. %, 0.1 vol. %, and 1 vol. % nanoparticle concentrations, and after performing the boiling experiments, the nanoparticle-coated surfaces were employed for pool-boiling experiments using pure water. The authors found that the concentration of the nanoparticles had great impact on the heat transfer behavior. At lower concentrations, the rate of deposition of the nanoparticles was lower, resulting in a greater improvement of the heat transfer, which can be attributed to the fact that, at low fractions of nanoparticles, the superior thermal conductivity of the nanofluids prevailed over the effect of the nanoparticle deposition onto the heating surface. In addition, the boiling of water on the surface coated with nanoparticles demonstrated that high deposition rates lead to an improvement in the heat transfer, which may be due to the less uniform deposited layer on the surface. Moreover, the authors Coursey and Kim [92] examined the potential of dispersed alumina nanoparticles in water and ethanol. The researchers demonstrated that at low concentrations of alumina, the CHF value remained unchanged, whileat higher nanoparticle concentrations, the CHF value was improved by 37%. In addition, the published results revealed that the carbon nanotube-based nanofluids exhibited appreciably higher thermal features, including thermal conductivity, boiling convective HTC, and CHF compared with the base fluids themselves, as well as other types of nanofluids. These enhanced properties further increased with increasing carbon nanotube concentration and temperature [93].

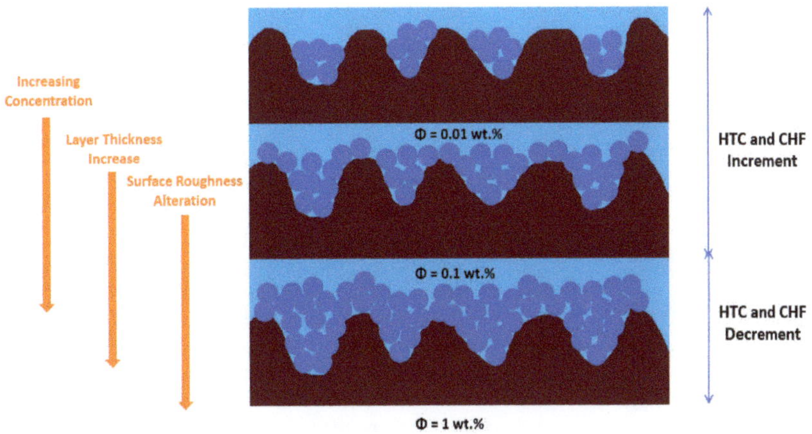

Figure 9. Schematic representation of the pool-boiling-induced nanoparticle deposition for different concentrations and effects on the heat transfer parameters.

4.2. Size and Shape of the Nanoparticles

The morphology of the nanoparticles has a considerable impact on the thermal conductivity enhancement. For instance, in the experimental work developed by [94], nanofluids were prepared by dispersing titanium oxidenanoparticles in rodshapes measuring 10 nm diameter and 40 nm length, and in spherical shapes measuring15 nm diameter in deionized water.The results revealed that the particle size and shape have effects on the thermal conductivity improvement. At 5 vol. %, with the titanium oxiderodsand spheres, the enhancement was found to be around 33% and 30%, respectively, compared with that of the base fluid itself. Moreover, for instance, in the case of ethylene-glycol-based nanofluids [95], the nanoparticle size effect on the thermal conductivity of the nanofluids was also not conclusive as some experimental works reported that the nanofluids with larger-sized nanoparticles exhibitedgreater enhancements in the thermal conductivity than the ones having smaller nanoparticles, whereas others found that the smaller the nanoparticles, the greater the improvement in the thermal conductivity. Hence, the size and shape of the added nanoparticles can also impact on the HTC and CHF. This is related to the fact that different parameters, including post-sintering pore shape, permeability, surface roughness, and thermal conductivity and diffusivity of the deposited layer are influenced by the morphology of the nanoparticles. Moreover, the authors Peng et al. [96] performed an empirical study on the impact of the size of copper nanoparticles on the nucleate pool-boiling heat transfer of a R113/oil thermal fluid. The researchers reported that the maximum increase of 23.8% was achieved for the HTC by reducing the size of the nanoparticles from 80 to 20 nm. Moreover, the smaller copper nanoparticles led to an increased pool-boiling HTC. The impact of the dimensions of the nanoparticles on the boiling behavior was also studied by Hu et al. [97] for a silica nanofluid. The authors observed an increasing tendency of the HTC by decreasing the nanoparticle size from 120 to 84 nm. Furthermore, it can be stated that the nanoparticle size strongly contributes to the amelioration of the pool-boiling CHF, and as the size of nanoparticles increases, better boiling heat transfer performance is achieved for the nanofluids. Additionally, the researchers Minakov et al. [36] established that the CHF using nanofluids depends on the nanoparticle size and that the CHF increased with increasing nanoparticle size. This fact can be explained by assuming that the particle deposition on the heating surface plays the key role in the CHF enhancement. The larger the size of deposited particles, the larger the scale of the final roughness on the surface, which promotes the formation of a deposit thickness enough for enhanced boiling. Shogl et al. [6] showed that the heating surface characteristics depended on the nanoparticles size and surface roughness. Hence, larger nanoparticle sizes improved the boiling performance of the nanofluid. Thus, the carbon nanotubes with water-based nanofluids enhanced the performance of the system and can be considered as the best heat removal method among the examined carbon nanotubes, alumina, and zinc oxide nanofluids, because both surface characteristics and boiling performance were improved with the carbon nanotubes.

4.3. Type of Nanoparticles

The authors Shogl et al. [6] performed the evaluation of the boiling performance of zinc oxide, alumina, and carbon nanotube nanofluids under heat fluxes up to 300,000 W/m^2. The results showed that, using nanoparticles may deteriorate or improve the HTC. To better understand the mechanism of the nanofluid pool boiling, it should be clarified that the decrement or enhancement in the HTC is an intrinsic characteristic of the nanofluids, or is the direct consequence of the heating surface modification, or a combination of both. The deterioration or enhancement depends on the type, size, and surface roughness of the nanoparticles. Inthis experimental work, the zinc oxide and alumina nanofluids formed smoother surfaces and the carbon nanotube CNT formed a rougher surface than the bare one. The overall effect of the zinc oxide/water and alumina/water nanofluids was the worsening of boiling heat transfer, whereas the effect of carbon nanotubes/water nanofluids was the improvement of boiling heat transfer. The type of the nanomaterials can also influence the thermal performance of the boiling systems, given that the different nanoparticle materials result in different thickness and surface relief of the deposits on the heating surface. It was also shown that with decreasing the available heat transfer area of the heating surface along with the boiling-induced nanoparticle deposition, the CHF using nanofluids increased significantly.

4.4. Wettability of the Nanoparticles

The nanoparticle wettability can affect the deposited nanoparticle layer morphology. On the one hand, as it can be seen in Figure 10, the suspended nanoparticles with moderate hydrophilicity are adsorbed to the liquid and vapor phases interface avoiding the fluid drainage among the vapor bubbles. This phenomenon hinders the coalescence concern of the bubbles and decreases their diameter at departure, resulting in a nucleate pool-boiling HTC and CHF amelioration. In contrast, the nanoparticles with high hydrophilicity will not adsorb onto the heating surface and, consequently, the bubble coalescence remains unchanged. As illustrated in Figure 10, the wettability of the nanoparticles influences the deposition layer morphology in which the highly hydrophilic layers are relatively smooth having a uniform nanoparticle dispersion onto the heat transfer interface, while the deposited layers produced with moderate to medium hydrophilicity possess higher roughness and more irregularities.

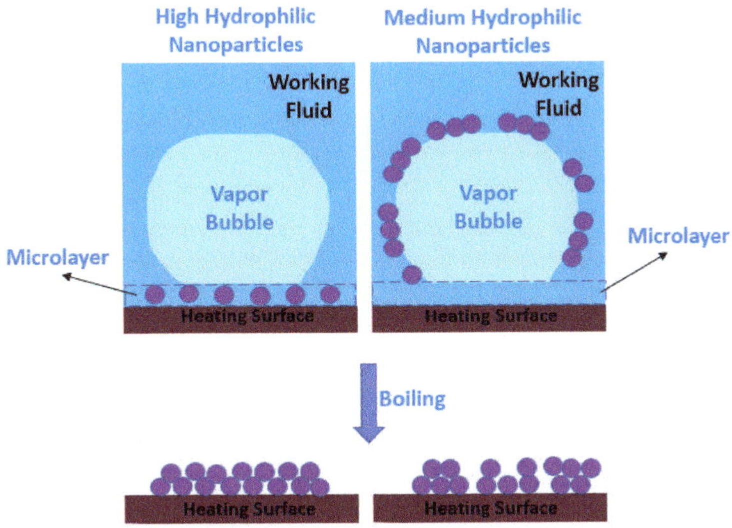

Figure 10. Nanoparticle hydrophilicity effect on the deposited layer morphology.

4.5. Base Fluid

The differences in the base fluid nature lead to great differences in the thermophysical characteristics of the nanofluid. For instance, if the viscosity of the base fluid is considerable, the viscosity of the nanofluid results higher. Moreover, the disturbance due to the nucleation and departure of the vapor bubbles is the main rationale behind the intense heat transfer. Moreover, the varying viscosity of the nanofluid is critical for the impact on the boiling bubbles because of the differences in the base fluid. It can be found in the published scientific articles that the inclusion of nanoparticles into a base fluid with relatively high viscosity strongly improves the boiling heat transfer. The main features arising from this fact are that the nanoparticles are suspended in a high viscosity fluid, which assures good stability over time. Moreover, the addition of nanoparticles has a negligible impact on the viscosity of the nanofluid that results in only a small alteration in the bubble growth and departure. In summary, the stability of the nanofluid is improved, the nanoparticle deposition is decreased, the number of gasification points together with the surface wettability is enhanced, and, hence, the pool-boiling heat transfer is improved.

4.6. Surfactants

The effect of the addition of a surfactant and clustering on the thermal conductivity of titanium oxide and alumina dispersed in water was experimentally studied by the authors [98]. The obtained results showed that the cluster size increased with increasing concentration of nanoparticles, while the thermal conductivity of the nanofluids decreased with increasing cluster size. The CTAB surfactant was effective in improving the dispersion of the nanoparticles and stability of the nanofluids. The added surfactant also contributed to the thermal conductivity of the nanofluids' improvement. Although the surfactant proved to be a benefit in the stability and thermal conductivity, the effect of the nanoparticle clustering on the same features was found to be negative. The authors Zhou et al. [99] studied the influence of the nanoparticle deposition and interfacial characteristics on the pool boiling using nanofluids and n-butanol as the surfactant over a platinum microwire. The researchers stated that the inclusion of n-butanol altered the liquid/vapor interface properties and it intensified the nanoparticle deposition at low heat fluxes. The obtained results confirmed that the CHF of the nanofluid became enhanced when the n-butanol was added to the nanofluid. The experimental data showed that the hindered bubble growth and increased nanoparticle agglomeration in the fluid wedge region were the reasons behind the degradation of the heat transport deterioration when the amount of the added surfactant was increased. Additionally, it was already proven that the impact of the surfactant on the heat transfer and CHF was greater than the unstable settlement of the nanoparticles onto the heating surface. The deposition profiles were significantly influenced by the n-butanol addition since the low concentration of the self-rewetting fluid produced higher surface tension and, consequently, attracted a greater amount of nanoparticles onto the microwire surface. However, the unstable settlement of the nanoparticles generated a less-uniform deposit as compared with that of the nanofluid alone, which is likely due to the greater disturbance derived from the surfactant addition. In conclusion, when the n-butanol was added to the nanofluid, this surfactant promoted the clustering of the nanoparticles at low heat fluxes. The nanoparticle deposition pattern indicated that this one modified the surface roughness and, thus, accelerated the emergence of the sweeping mechanism of the vapor bubbles. Moreover, it was found that the nanoparticle deposition is more intense with the further addition of a surfactant. The addition of the surfactant will increase the CHF and such an increase is higher for high-nanoparticle concentrations. Moreover, the CHF enhancement ratio decreased when more surfactant was added. In addition, the overall mechanism of the heat transfer enhancement in which the Marangoni flows imposed by the surfactant pushes the nanoparticles to the heating surface to produce the deposition layer and conducts the nanoparticles to penetrate the confined wedge of the vapor bubbles. Furthermore, in the work conducted by the authors Jung et al. [100], the addition of nitric acid as an ionic surfactant promoted the formation of self-assembled layers and structures of nanoparticles on the heat transfer surface. This alternative method builds a more uniform and smoother surface structure, reducing the CHF improvement.

5. Limitations and Challenges

Several questions and issues require investigation studies to further understand the main features of the nanoparticle deposition during the boiling of nanofluids. Hence, the following suggestions and challenges are stated for future work:

1. Despite the many published nanofluid-related studies, issues such as long-term stability, erosion, agglomeration, deposition, and maintenance procedures are still obstacles to large-scale commercialization of nanofluids. Hence, studies to extend the actual predictive correlations or innovative numerical simulation tools are highly recommended.
2. The methods for an effective nanoparticle deposition should be reviewed to find a replacement for the deposition rather than the nanofluid boiling; for instance, pre-boiling deposition of nanocoatings on the surface by physical or chemical vapor deposition. However, these depositing techniques require further in-depth investigation studies regarding the optimal thickness of the coatings to observe the delay of the CHF occurrence.
3. Concerning the depositing trend of the nanoparticles in the course of pool boiling, the coating of the surface with particles has been intensively explored to improve the HTC and CHF. In this context, the probable detachment or failure of the layer of particles over boiling time should be further addressed in a laboratory environment. Additionally, the complexity of correlations between the main properties of the heating surface should be addressed to further clarify the mechanisms of amelioration of the heat transfer parameters through the deposition of the nanoparticles.
4. Future experimental works should include the exploration of the possible boiling-induced nanoparticle deposition influencing parameters, including nanoparticle size, shape, type, and substrate material, and base fluid type to the wettability modification and its impact on the pool-boiling heat-transfer characteristics.
5. Regarding the effect of the deposition layer on the boiling surface heat transfer, the thickness of the layer should be optimized to induce the maximum value of latent heat at CHF occurrence.
6. The durability of the nanoparticle deposition layer and its effects on the CHF should be further systematically investigated. The initial studies indicated that the porous nanopar-

ticle deposited layer is affected by the dilatation effect and, under certain conditions, by the formation of a pocket of vapor between the heater and the deposited layer.

7. The particle sorting effect of the boiling-induced nanoparticle deposition requires further in-depth studies. The particle sorting effect has already been observed over the deposited layer of nanoparticles wherein larger micrometer-sized particles, compared with the ones observed close to the center of the deposited layer, have been found in the peripheral regions of the deposited layer. It should be also made clear if this phenomenon is indeed a particle sorting effect or, alternatively, if it is only the result of the boiling-induced capillary wicking flows through the deposition layer toward the center of the nucleation sites.
8. The conjugated effect of the wettability and capillary wicking should be further studied to reveal the heat transfer enhancement for different sizes and fractions of nanoparticles.
9. The bubble dynamics of a single bubble should be further investigated in experimental works and numerical simulations to determine the fraction of the deposited layer and of the suspended nanoparticles that contribute to the pool-boiling heat-transfer amelioration.
10. Efforts should be made to mitigate the initial heating surface differences and, also, to diminish the surface differences occurred during the boiling nanoparticle deposition, given that the modification of the boiling surface geometries is the main factor responsible for the contradictory literature reports on boiling heat transfer with nanofluids. The enhancement or deterioration of boiling heat transfer is dependent upon the relative size between the nanoparticles suspended in the fluid and the heating surface geometry, and respective interactions. Published experimental works already showed that for an initial smooth surface, the deposition of particles increases the surface roughness contributing to the improvement in the nucleate boiling heat transfer, whereas for a starting rough surface, no obvious change in the surface geometry is observed that results in a similar boiling curve.
11. The use of only one heating surface should be avoided, since this procedure will make the quantitative comparison of results more difficult. The experimental evaluation of the nanoparticle concentration effect will also be difficult. The heat transfer surface modification by the deposition of nanoparticles is an intrinsic feature of the use of nanofluids that occurs each time after boiling. Hence, the experimental results are affected by the number and frequency of the usage of the same heating surface.
12. Regarding the effect of the pressure of the pool-boiling system, the growth of the dry patches should be addressed under different pressure levels to elucidate its influence on the heat transfer enhancement from the deposition layer.
13. The possible melting characteristic at some sites on the coated heating surface after confined pool-boiling experiments should be further analyzed. In these cases, the bubble dynamics in the inner confined region and the enhanced residence time of the bubbles on the heating surface underneath could appreciably increase the temperature on the coated layer, melting the nanoparticles through a regressive melting process similar to liquid phase sintering.
14. Further attention must be paid to the possible residual chemical elements dissolved in the working fluid, given that these elements may interact with the deposited nanoparticles and, consequently, modify the morphology and chemical structure of the heating surface and, hence, impair the pool-boiling heat transfer effectiveness. Moreover, this phenomenon tends to take place more under confined boiling in which the hot spots are more sensitive because of the dryout phenomenon.
15. Future studies on the reusability of formerly boiled nanofluids and correspondent stability may be an adequate pathway to better understand the practical implications of the boiling-induced nanoparticle deposition process using nanofluids.
16. To better understand the underlying mechanism of the suspended nanoparticles deposited underneath the vapor bubbles, techniques such as infrared thermography

should be further explored to obtain the temperature and heat flux distributions of the active nucleation sites. Additionally, more microscopic measurements of the nucleation site should be carried out with the aid of optical profilometry and AFM to determine the shape of the nanoparticle deposited layer underneath the bubbles.

17. It is suggested that further studies be conducted on repetitive quenching with concentrated nanofluids by monitoring the heating surface wettability alteration and concentration changes after each individual quenching run to better understand the mechanisms of CHF enhancement and achieve a quantitative characterization of the impact of the suspended nanoparticles and its cumulative effect.
18. To better understand how the motion at the microscale of the nanoparticles influences the perturbation suffered by the bubbles in the heating surface, the motion path of the nanoparticles during boiling should be further examined with the aid of a component analyzer and by labeling the nanoparticles. Moreover, it is also advisable to observe and evaluate the generation of bubbles using a high-speed camera.
19. The impact of a multi-component solution on the nucleate pool-boiling heat transfer should be verified through the external condensation of the pool-boiling apparatus, and the relative motion between the different solutions should be observed and discussed. Furthermore, the movement and disturbance of the bubbles should be observed and analyzed by this procedure.
20. It is highly recommended to produce a database that will include the heat transport characteristics together with specific information about the deposited nanoparticle morphology and amount, and dispersion stability with or without the addition of surfactants, in which the enhanced pool-boiling thermal performance of promising nanofluids are prioritized.

6. Conclusions

The current work can be characterized as an overview of the transient characteristics of the pool boiling of nanofluids, namely, the nanoparticle deposition advantages and disadvantages during boiling time. The following conclusions should be highlighted:

1. The nanofluids already demonstrated the improvement of the CHF derived from the enhanced wettability of the heating surface after the deposition of the nanoparticles onto the heating surfaces. Nevertheless, the published data concerning the effect of nanofluids on the nucleate boiling HTC are still contradictory. This can be caused by the involvement of intricate concerns such as the nature of the thermal fluid, roughness of the heating surface, and imposed heat flux conjugated with the type, morphology, volume fraction, and preparation and functionalization methods of the nanoparticles. All these factors can significantly alter the thermophysical properties of the nanofluid and certain surface characteristics such as the wettability, surface roughness, number of active nucleation sites, and alterations in the three-phase contact line. Such limiting and complexing issues can strongly restrict the accurate modeling of the nanofluid pool boiling.
2. It was already confirmed that the thickness and density of the deposition layer decreases outward radially when the layer has a spherical shape. Taking into account the morphology and evaporation dynamics of the microlayer, the most verified thickness trend of the deposition layer has been suggested to be caused by the microlayer evaporation phenomenon. Nevertheless, the initiation of the deposition at the active nucleation sites may be caused by the contact line evaporation since such a mechanism is expected to be present in the nanofluid pool boiling.
3. Although the nanofluids have demonstrated great potential in improving boiling heat transfer, there are specific practical concerns that must be considered prior to any usage of the nanofluids in thermal management purposes including the agglomeration, sedimentation, and precipitation of the nanoparticles, equipment and systems clogging, boiling surface erosion, evolution in time of the heat transfer parameters, and inherent overall cost.

4. The nanoparticle deposition onto the heating surface alters its wettability and number of active gasification sites, which affects the heat transfer capability. Moreover, the nanoparticle deposited layer changes the generation of the bubbles and their departure frequency.
5. The morphology of the surface settlement of the various nanoparticles increases the capillary effect, thus enhancing the liquid replenishing after the detachment of the bubbles, which in turn increases the CHF.
6. The development of a closed porous deposit of nanoparticles increases the heat transport resistance of the heating surface and reduces the HTC. Moreover, the deposition causes the CHF to increase more rapidly than the layer superheat value, which results in an increment of the maximum HTC value. It is also common for a significant increase in the thermal conductivity of the base fluid to occur so that the HTC is enhanced. The perturbation of the heating surface caused by the nanoparticles turns the liquid microlayer thinner and enhances the disturbance of the bubbles, resulting in the amelioration of the HTC.
7. The prevalence of larger microparticles in the peripheral regions of the deposition layer has been observed to assume a well-defined circular shape. Nevertheless, the identification of the underlying mechanism still remains unclear, given that it has already been attributed to the eventual nanoparticle sorting or, alternatively, to the pool-boiling capillary wicking through the deposition layer toward the center of the active nucleation sites.
8. It was already observed that the suspension of the nanoparticles in a base fluid with higher viscosity brings benefits to the boiling heat transfer since the deposition of the nanoparticles is smaller; hence, the enhancement of the thermal conductivity of the fluid is more intense than the microscopic motion of the nanoparticles on the surface and the heat transport becomes enhanced.
9. In the cases where the fluid is a multi-component solution, the relative motion between the different composing solutions increases the movement of the nanoparticles due to the different evaporation rates. Therefore, the disturbance of the bubble on the heating surface increases and the boiling heat transfer performance is enhanced.
10. When compared with the thermal conductivity of the heat transfer surface, the deposition of nanoparticles having poor thermal conductivity on the heating surface decreases the heat dissipation and enhances the surface superheat value. Hence, the enhancement of the HTC should be attributed to the thermal conductivity of the nanoparticles and the effect of their movement in the disturbance of the surface vapor bubbles.
11. The different natures of the nanoparticles result in different thermal conductivity and the effects on the pool-boiling heat transfer strongly depend on the deposition pattern. Moreover, only small amounts of the deposited nanoparticles enhance the number of available active nucleation sites and, by this method, the pool-boiling heat transfer capability is improved.

Author Contributions: Conceptualization, J.P.; methodology, J.P. and A.M. (Ana Moita); software, A.M. (Ana Moita); validation, A.M. (Ana Moita) and A.M. (António Moreira); formal analysis, A.M. (Ana Moita); investigation, J.P.; resources, A.M. (António Moreira); data curation, J.P.; writing—original draft preparation, J.P.; writing—review and editing, J.P.; supervision, A.M. (Ana Moita) and A.M. (António Moreira); project administration, A.M. (António Moreira); funding acquisition, A.M. (António Moreira). All authors have read and agreed to the published version of the manuscript.

Funding: The authors are grateful to the Fundação para a Ciência e a Tecnologia (FCT), Portugal, for partially financing the Project "Estratégias interfaciais de arrefecimento para tecnologias de conversão com elevadas potências de dissipação", Ref. PTDC/EME-TED/7801/2020, António Luís Nobre Moreira, Associação do Instituto Superior Técnico para a Investigação e o Desenvolvimento (IST-ID). The author José Pereira also acknowledges FCT for his PhD Fellowship Ref. 2021.05830.BD.

Informed Consent Statement: Not applicable.

Data Availability Statement: Not applicable.

Conflicts of Interest: The authors declare no conflict of interest.

References

1. Buongiorno, J.; Hu, L.; Bang, I.C. Towards an Explanation of the Mechanism of Boiling Critical Heat Flux Enhancement in Nanofluids. In Proceedings of the ASME 2007 5th International Conference on Nanochannels, Microchannels, and Minichannels. ASME 5th International Conference on Nanochannels, Microchannels, and Minichannels, Puebla, Mexico, 18–20 June 2007; pp. 989–995. [CrossRef]
2. Vafei, S.; Wen, D. Spreading of triple line and dynamics of bubble growth inside nanoparticle dispersions on top of a substrate plate. *J. Colloid Interface Sci.* **2011**, *362*, 285–291. [CrossRef] [PubMed]
3. Ham, J.; Kim, H.; Shin, Y.; Cho, H. Experimental investigation of pool boiling characteristics in Al_2O_3nanofluid according to surface roughness and concentration. *Int. J. Therm. Sci.* **2017**, *14*, 86–97. [CrossRef]
4. Umesh, V.; Raja, B. A study on nucleate boiling heat transfer characteristics of pentane and CuO-pentane nanofluid on smooth and milled surfaces. *Exp. Therm. Fluid Sci.* **2015**, *64*, 23–29. [CrossRef]
5. Sarafraz, M.M.; Hormozi, F. Comparatively experimental study on the boiling thermal performance of metal oxide and multi-walled carbon nano tube nanofluids. *Powder Technol.* **2016**, *287*, 412–430. [CrossRef]
6. Shoghl, S.N.; Bahrami, M.; Jamialahmadi, M. The boiling performance of ZnO,α-Al_2O_3 and MWCNTs/water nanofluids: An experimental study. *Exp. Therm. Fluid Sci.* **2017**, *80*, 27–39. [CrossRef]
7. Das, S.K.; Putra, N.; Roetzel, W. Pool boiling characteristics of nano-fluids. *Int. J. Heat Mass Transf.* **2003**, *46*, 851–862. [CrossRef]
8. Chopkar, M.K.; Das, A.K.; Manna, I. Pool boiling heat transfer characteristics ofZrO2-water nanofluids from a flat surface in a pool. *Heat Mass Transf.* **2008**, *44*, 999–1004. [CrossRef]
9. Narayan, G.P.; Anoop, K.B.; Das, S.K. Mechanism of enhancement/degradation of boiling heat transfer using stable nanoparticle suspensions over vertical tubes. *J. Appl. Phys.* **2007**, *102*, 074317. [CrossRef]
10. White, S.B.; Shih, A.J.; Pipe, K.P. Boiling surface enhancement by electrophoretic deposition of particles from a nanofluid. *Int. J. Heat Mass Transf.* **2011**, *54*, 4370–4375. [CrossRef]
11. Bang, I.C.; Chang, S.H. Boiling heat transfer performance and phenomena ofAl_2O_3-water nano-fluids from a palin surface in a pool. *Int. J. Heat Mass Transf.* **2005**, *48*, 2407–2419. [CrossRef]
12. Abdollahi, A.; Salimpour, M.R.; Etesami, N. Experimental analysis of magnetic field effect on the pool boiling heat transfer of a ferrofluid. *Appl. Therm. Eng.* **2017**, *111*, 1101–1110. [CrossRef]
13. Sarafraz, M.M.; Hormozi, F. Pool boiling heat transfer to dilute copper oxide aqueous nanofluids. *Int. J. Therm. Sci.* **2015**, *90*, 224–237. [CrossRef]
14. Diao, Y.H.; Liu, Y.; Wang, R.; Zhao, Y.H.; Guo, L.; Tang, X. Effects of nanofluids and nanocoatings on the thermal performance of an evaporator with rectangular microchannels. *Int. J. Heat Mass Transf.* **2013**, *67*, 183–193. [CrossRef]
15. Liang, G.; Mudawar, I. Review of pool boiling enhancement with additives and nanofluids. *Int. J. Heat Mass Transf.* **2018**, *124*, 423–453. [CrossRef]
16. Malavasi, I.; Teodori, E.; Moita, A.S.; Moreira, A.L.N.; Marengo, M. Wettability Effect on Pool Boiling: A Review. In *Encyclopedia of Two-Phase Heat Transfer and Flow III*; World Scientific Publishing Co. Pte Ltd.: Singapore, 2018; pp. 1–61.
17. Liang, G.; Mudawar, I. Review of pool boiling enhancement by surface modification. *Int. J. Heat Mass Transf.* **2019**, *128*, 892–933. [CrossRef]
18. Forrest, E.; Williamson, E.; Buongiorno, J.; Hu, L.-W.; Rubner, M.; Cohen, R. Augmentation of nucleate boiling heat transfer and critical heat flux using nanoparticle thin-film coatings. *Int. J. Heat Mass Transf.* **2010**, *53*, 58–67. [CrossRef]
19. Raveshi, M.R.; Keshavarz, A.; Mojarrad, M.S.; Amiri, S. Experimental Investigation of pool boiling heat transfer enhancement of alumina-water-ethylene glycol nanofluids. *Exp. Therm. Fluid Sci.* **2013**, *44*, 805–814. [CrossRef]
20. Tang, X.; Zhao, Y.-H.; Diao, Y.-H. Experimental investigation of the nucleate boiling heat transfer characteristics of the δ-Al_2O_3-R141b nanofluids on a horizontal plate. *Exp. Therm. Fluid Sci.* **2014**, *52*, 88–96. [CrossRef]
21. Manetti, L.L.; Stephen, M.T.; Beck, P.A.; Cardoso, E.M. Evaluation of the heat transfer enhancement during pool boiling using low concentrations of Al_2O_3-water based nanofluid. *Exp. Therm. Fluid Sci.* **2017**, *87*, 191–200. [CrossRef]
22. Nunes, J.M.; Souza, R.R.; Rodrigues, A.R.; Safaei, M.R.; Cardoso, E.M. Influenceofcoatedsurfacesandgapsizeonboilingheattrans-ferofdeionizedwater. *J. Braz. Soc. Mech. Sci. Eng.* **2020**, *42*, 127. [CrossRef]
23. Xing, M.; Yu, J.; Wang, R. Effects of surface modification on the pool boiling heat transfer of MWNTs/water nanofluids. *Int. J. Heat Mass Transf.* **2016**, *103*, 914–919. [CrossRef]
24. Li, Z.; Mazinani, A.; Hayat, T.; Al-Rashed, A.A.A.A.; Alsulami, H.; Goodarzi, M.; Sarafraz, M.M. Transient pool boiling and particulate deposition of copper oxide nano-suspensions. *Int. J. Heat Mass Transf.* **2020**, *155*, 119743. [CrossRef]
25. Cao, Z.; Wu, Z.; Abbood, S.; Sundén, B. An analysis of pool boiling heat transfer on nanoparticle-coated surfaces. *Energy Procedia* **2019**, *158*, 5880–5887. [CrossRef]
26. Kiyomura, I.S.; Manetti, L.L.; Cunha, A.P.; Ribatski, G.; Cardoso, E.M. An analysis of the effects of nanoparticles deposition on characteristics of the heating surface and ON pool boiling of water. *Int. J. Heat Mass Transf.* **2017**, *106*, 666–674. [CrossRef]

27. Stutz, B.; Morcelli, C.H.S.; Silva, M.F.; Cioulachtjian, S.; Bonjour, J. Influence of nanoparticle surface coating on pool boiling. *Exp. Therm. Fluid Sci.* **2011**, *35*, 1239–1249. [CrossRef]
28. Souza, R.R.; Passo, J.C.; Cardoso, E.M. Influence of nanoparticle size and gap size on nucleate boiling using HFE7100. *Exp. Therm. Fluid Sci.* **2014**, *59*, 195–201. [CrossRef]
29. Heitich, L.V.; Passos, J.C.; Cardoso, E.M.; Silva, M.F.; Klein, A.N. Nucleate boiling of water using nanostructured surfaces. *J. Braz. Soc. Mech. Sci. Eng.* **2014**, *36*, 181–192. [CrossRef]
30. Rostamian, F.; Etesami, N. Pool boiling characteristics of silica/water nanofluid and variation of heater surface roughness in domain of time. *Int. Commun. Heat Mass Transf.* **2018**, *95*, 98–105. [CrossRef]
31. Akbari, E.; Gheitaghy, A.M.; Saffari, H.; Hosseinalipour, S.M. Effect of silver nanoparticle deposition in re-entrant inclined minichannel on bubble dynamics for pool boiling enhancement. *Exp. Therm. Fluid Sci.* **2017**, *82*, 390–401. [CrossRef]
32. Kumar, N.; Jothi, T.J.S.; Selvaraju, N. Effect of nanoparticle deposition rate on critical heat flux in pool boiling. *J. Eng. Res.* **2017**, *5*, 209–222.
33. Hadzic, A.; Moze, M.; Arhar, K.; Zupancic, M.; Golobic, I. Effect of Nanoparticle Size and Concentration on Pool Boiling Heat Transfer with TiO_2 Nanofluids on Laser-Textured Copper Surfaces. *Nanomaterials* **2022**, *12*, 2611. [CrossRef] [PubMed]
34. Kamel, M.S.; Lezsovits, F. Experimental Investigation on Pool Boiling Heat Transfer Performance Using Tungsten Oxide WO Nanomaterial-Based Water Nanofluids. *Materials* **2020**, *13*, 1922. [CrossRef] [PubMed]
35. Gajghate, S.S.; Bandurkar, A.V.; Das, S.; Saha, B.B.; Bhaumik, S. Effect of ZrO Nanoparticle Deposited Layer on Pool Boiling Heat Transfer Enhancement. *Heat Transf. Eng.* **2021**, *42*, 1184–1202. [CrossRef]
36. Minakov, A.V.; Pryazhnikov, M.I.; Guzei, D.V.; Zeer, G.M.; Rudyak, V.Y. The experimental study of nanofluid boiling crisis on cylindrical heaters. *Int. J. Therm. Sci.* **2017**, *116*, 214–223. [CrossRef]
37. Sulaiman, M.Z.; Matsuo, D.; Enoki, K.; Okawa, T. Systematic measurements of heat transfer characteristics in saturated pool boiling of water-based nanofluids. *Int. J. Heat Mass Transf.* **2016**, *102*, 264–276. [CrossRef]
38. Modi, M.; Kangude, P.; Srivastava, A. Performance evaluation of alumina nanofluids and nanoparticles-deposited surface on nucleate pool boiling phenomena. *Int. J. Heat Mass Transf.* **2020**, *146*, 118833. [CrossRef]
39. Li, X.; Cheung, S.C.P.; Tu, J. Nucleate boiling of dilute nanofluids— Mechanism exploring and modeling. *Int. J. Therm. Sci.* **2014**, *84*, 323–334. [CrossRef]
40. Kim, H.-D.; Kim, J.-B.; Kim, M.-H. Experimental Study on CHF Characteristics of Water-$Ti0_2$Nano-Fluids. *Nucl. Eng. Technol.* **2006**, *38*, 61–68.
41. Kangude, P.; Srivastava, A. On the mechanisms leading to ordered nanoparticle deposition during single bubble nucleate pool boiling regime. *Phys. Fluids* **2021**, *33*, 113306. [CrossRef]
42. Mori, S.; Utaka, Y. Critical heat flux enhancement by surface modification in a saturated pool boiling: A review. *Int. J. Heat Mass Transf.* **2017**, *108*, 2534–2557. [CrossRef]
43. Kwark, S.M.; Kumar, R.; Moreno, G.; Yoo, J.; You, S.M. Pool boiling characteristics of low concentration nanofluids. *Int. J. Heat Mass Transf.* **2010**, *53*, 972–981. [CrossRef]
44. Phan, H.T.; Caney, N.; Marty, P.; Colasson, S.; Gavillet, J. Surface Coating with Nanofluids: The Effects on Pool Boiling Heat Transfer. *Nanoscale Microscale Thermophys. Eng.* **2010**, *14*, 229–244. [CrossRef]
45. Kim, H.; Kim, E.; Kim, M.H. Effect of nanoparticle deposit layer properties on pool boiling critical heat flux of water from a thin wire. *Int. J. Heat Mass Transf.* **2014**, *69*, 164–172. [CrossRef]
46. Kwark, S.M.; Amaya, M.; Moon, H.; You, S.M. Effect of soluble additives, boric acid(H_3BO_3) and salt(NaCl) in pool boiling heat transfer. *Nucl. Eng. Technol.* **2011**, *43*, 195–204. [CrossRef]
47. Sezer, N.; Khan, S.A.; Koç, M. Boiling Heat Transfer Enhancement by Self-Assembled Graphene/Silver Hybrid Film for the Thermal Management of Concentrated Photovoltaics. *Energy Technol.* **2020**, *8*, 2000532. [CrossRef]
48. Neto, A.R.; Oliveira, J.L.G.; Passos, J.C. Heat transfer coefficient and critical heat flux during nucleate pool boiling of water in the presence of nanoparticles of alumina, maghemite and CNTs. *Appl. Therm. Eng.* **2017**, *111*, 1493–1506. [CrossRef]
49. Anuar, F.S.; Malayeri, M.R.; Hooman, K. Particulate Foulingand Challengesof Metal Foam Heat Exchangers. *Heat Transf. Eng.* **2017**, *38*, 730–742. [CrossRef]
50. Master, B.I.; Chunangad, K.S.; Pushpanathan, V. Fouling mitigation using helix changer heat exchangers. In Proceedings of the Heat Exchanger Fouling and Cleaning: Fundamentals and Applications, Santa Fe, NM, USA, 18–22 May 2003.
51. Mueller, S.H.; Malayeri, M.R.; Watkinson, A.P. Fouling of heat exchanger—New approaches to solve old problem. *Heat Transf. Eng.* **2005**, *26*, 1–4. [CrossRef]
52. Epstein, N. Thinking about Heat Transfer fouling: A 5 $\times$ 5 Matrix. *Heat Transf. Eng.* **1983**, *4*, 43–56. [CrossRef]
53. Awad, M.M. Fouling of Heat Transfer Surfaces, Heat Transfer. In *Theoretical Analysis, Experimental Investigations and Industrial Systems*; Belmiloudi, A., Ed.; In Tech: Bondi Junction, NSW, Australia, 2011.
54. Roy, U. Fouling and Its Effect on Heat Exchangers. In *Advanced Analytic and Control Techniques for Thermal Systems with Heat Exchangers*; Elsevier: Amsterdam, The Netherlands, 2020; pp. 233–246.
55. Elimelech, M.; Gregory, J.; Jia, X.; Williams, R. *Particle Deposition and Aggregation: Measurement, Modellingand Simulation*, 1st ed.; Butterworth-Heinemann: Woburn, MA, USA, 1995.
56. Hunter, R.J. *Foundations of Colloid Science*, 2nd ed.; Oxford University Press: Oxford, UK; NewYork, NY, USA, 2001.
57. Müller-Steinhagen, H. Cooling-Water Fouling in Heat Exchangers. *Adv. Heat Transf.* **1999**, *33*, 415–496. [CrossRef]

58. Nikkhah, V.; Sarafraz, M.M.; Hormozi, F. Application of Spherical Copper Oxide(II) Water Nano-fluid as a Potential Coolant in a Boiling Annular Heat Exchanger. *Chem. Biochem. Eng. Q.* **2015**, *29*, 405–415. [CrossRef]
59. Ge, Z.; Cahill, D.G.; Braun, P.V. Thermal Conductance of Hydrophilic and Hydrophobic Interfaces. *Phys. Rev. Lett.* **2006**, *96*, 186101. [CrossRef] [PubMed]
60. Ajaev, V.S.; Kabov, O.A. Heat and mass transfer near contact lines on heated surfaces. *Int. J. Heat Mass Transf.* **2017**, *108*, 918–932. [CrossRef]
61. Kim, H.; Kim, M. Experimental study of the characteristics and mechanism of pool boiling CHF enhancement using nanofluids. *Heat Mass Transf.* **2009**, *45*, 991–998. [CrossRef]
62. Wen, D.; Corr, M.; Hu, X.; Lin, G. Boilingheattransferofnanofluids:Theeffectofheatingsurfacemodification. *Int. J. Therm. Sci.* **2011**, *50*, 480–485. [CrossRef]
63. Ji, W.-T.; Zhao, P.-F.; Zhao, C.-Y.; Ding, J.; Tao, W.-Q. Pool boiling heat transfer of water and nanofluid outside the surface with higher roughness and different wettability. *Nanoscale Microscale Thermophys. Eng.* **2018**, *22*, 296–323. [CrossRef]
64. Shahmoradi, Z.; Etesami, N.; Esfahany, M.N. Pool boiling characteristics of nanofluid on flat plate based on heater surface analysis. *Int. Commun. Heat Mass Transf.* **2013**, *47*, 113–120. [CrossRef]
65. Wen, D.; Ding, Y. Experimental investigation into the pool boiling heat transfer of aqueous based γ-alumina nanofluids. *J. Nanoparticle Res.* **2005**, *7*, 265–274. [CrossRef]
66. Vafei, S. Nanofluid pool boiling heat transfer phenomenon. *Powder Technol.* **2015**, *217*, 181–192. [CrossRef]
67. Park, S.D.; Moon, S.B.; Bang, I.C. Effects of thickness of boiling-induced nanoparticle deposition on the saturation of critical heat flux enhancement. *Int. J. Heat Mass Transf.* **2014**, *78*, 506–514. [CrossRef]
68. Das, A.K.; Das, P.K.; Saha, P. Heat transfer during pool boiling based on evaporation from micro and microlayer. *Int. J. Heat Mass Transf.* **2006**, *49*, 3487–3499. [CrossRef]
69. Xu, J.; Ji, X.; Zhang, W.; Liu, G. Pool boiling heat transfer of ultra-light copper foam with open cells. *Int. J. Multiph. Flow* **2008**, *34*, 1008–1022. [CrossRef]
70. Xu, Z.G.; Zhao, C.Y. Experimental study on pool boiling heat transfer in gradient metal foams. *Int. J. Heat Mass Transf.* **2015**, *85*, 824–829. [CrossRef]
71. Park, S.D.; Bang, I.C. Experimental study of a universal CHF enhancement mechanism in nanofluids using hydrodynamic instability. *Int. J. Heat Mass Transf.* **2014**, *73*, 844–850. [CrossRef]
72. Zuber, N. On the Stability of Boiling Heat Transfer. *Trans. Am. Soc. Mech. Eng.* **1958**, *80*. Available online: https://www.osti.gov/biblio/4326542 (accessed on 21 September 2022). [CrossRef]
73. Liter, S.G.; Kaviany, M. Pool-boiling CHF enhancement by modulated porous-layer coating: Theory and experiment. *Int. J. Heat Mass Transf.* **2001**, *44*, 4287–4311. [CrossRef]
74. Park, S.D.; Lee, S.W.; Kang, S.; Bang, I.C.; Kim, J.H.; Shin, H.S.; Lee, D.W.; Lee, D.W. Effects of nanofluids containing graphene/graphene oxide nanosheets on critical heat flux. *Appl. Phys. Lett.* **2010**, *97*, 023103. [CrossRef]
75. Watanabe, Y.; Enoki, K.; Okawa, T. Nanoparticle layer detachment and its influence on the heat transfer characteristics in saturated pool boiling of nanofluids. *Int. J. Heat Mass Transf.* **2018**, *125*, 171–178. [CrossRef]
76. Lin, L.; Chang, Z.; Ding, G. Resuspension of deposited nanoparticles during pool boiling. *Int. J. Heat Mass Transf.* **2019**, *130*, 230–239. [CrossRef]
77. Chen, Y.; Guo, J.; Liu, X.; He, D. Experimental and predicted model study of resuspended nanofluid pool boiling heat transfer under electric field. *Int. Commun. Heat Mass Transf.* **2022**, *131*, 105847. [CrossRef]
78. Vafei, S.; Holmes, I.; Errion, B.; Thukka, Z.; Narita, R.; Sugiura, T.; Manseki, K. Manufacturing a TiO-Based Semiconductor Film with Nanofluid Pool Boiling and Sintering Processes toward Solar-Cell Applications. *Nanomaterials* **2022**, *12*, 1165. [CrossRef]
79. Kathiravan, R.; Kumar, R.; Gupta, A.; Chandra, R. Characterization and Pool Boiling Heat Transfer of Nanofluids. *J. Heat Transf.* **2009**, *131*, 081902. [CrossRef]
80. Sarafraz, M.M.; Hormozi, F. Nucleate pool boiling heat transfer characteristics of dilute A l_2O_3-ethylene glycol nanofluids. *Int. Commun. Heat Mass Transf.* **2014**, *58*, 96–104. [CrossRef]
81. Karimzadehkhouei, M.; Shojaeian, M.; Sendur, K.; Menguç, M.P.; Kosar, A. The effect of nanoparticle type and nanoparticle mass fraction on heat transfer enhancement in pool boiling. *Int. J. Heat Mass Transf.* **2017**, *109*, 157–166. [CrossRef]
82. Haramura, Y.; Katto, Y. A new hydrodynamic model of critical heat flux, applicable widely to both pool and forced convection boiling on submerged bodies in saturated liquids. *Int. J. Heat Mass Transf.* **1983**, *26*, 389–399. [CrossRef]
83. Theofanous, T.G.; Dinh, T.N. High heat flux boiling and burnout as microphysical phenomena: Mounting evidence and opportunities. *Multiph. Sci. Technol.* **2006**, *18*, 1–26. [CrossRef]
84. Hu, Y.W.; Li, H.R.; He, Y.R.; Wang, L.D. Role of nanoparticles on boiling heat transfer performance of ethylene glycol aqueous solution based graphene nanosheets nanofluid. *Int. J. Heat Mass Transf.* **2016**, *96*, 565–572. [CrossRef]
85. Theofanous, T.G.; Dinh, T.N.; Tu, J.P.; Dinh, A.T. The boiling crisis phenomenon: Part II: Dryout Dynamics and burnout. *Exp. Therm. Fluid Sci.* **2002**, *26*, 793–810. [CrossRef]
86. Rohsenow, W.M.; Griffith, P. *Correlation of Maximum Heat Flux Data for Boiling of Saturated Liquids*; Massachusetts Institute of Technology, Division of Industrial Cooperation: Cambridge, MA, USA, 1955.
87. Murshed, S.M.S.; Sharifpur, M.; Giwa, S.; Meyer, J.P. Experimental Research and Development on the Natural Convection of Suspensions of Nanoparticles—A Comprehensive Review. *Nanomaterials* **2020**, *10*, 1855. [CrossRef]

88. Mukherjee, S.; Ali, N.; Aljuwayhel, N.F.; Mishra, P.C.; Sen, S.; Chaudhuri, P. Pool Boiling Amelioration by Aqueous Dispersion of Silica Nanoparticles. *Nanomaterials* **2021**, *11*, 2138. [CrossRef]
89. Mukherjee, S.; Mishra, P.C.; Chaudhuri, P. Pool boiling performance of aqueousAl_2O_3andTiO_2nanofluids on a horizontal lyp laced fat polished surface: An experimental investigation. *J. Therm. Anal. Calorim.* **2021**, *146*, 415–433. [CrossRef]
90. Kole, M.; Dey, T.K. Investigations on the pool boiling heat transfer and critical heat flux of ZnO-ethylene glycol nanofluids. *Appl. Therm. Eng.* **2012**, *37*, 112–119. [CrossRef]
91. Ahmed, O.; Hamed, M.S. Experimental investigation of the effect of particle deposition on pool boiling of nanofluids. *Int. J. Heat Mass Transf.* **2012**, *55*, 3423–3436. [CrossRef]
92. Coursey, J.S.; Kim, J. Nanofluid boiling: The effect of surface wettability. *Int. J. Heat Fluid Flow* **2008**, *29*, 1577–1585. [CrossRef]
93. Murshed, S.M.S.; de Castro, C.A.N. Superior thermal features of carbon nanotubes-based nanofluids—A Review. *Renew. Sustain. Energy Rev.* **2014**, *37*, 155–167. [CrossRef]
94. Murshed, S.M.S.; Leong, K.C.; Yang, C. Enhanced thermal conductivity of TiO_2-water based nanofluids. *Int. J. Therm. Sci.* **2005**, *44*, 367–373. [CrossRef]
95. Murshed, S.M.S.; de Castro, C.A.N. Conduction and convection heat transfer characteristics of ethylene glycol based nanofluids—A review. *Appl. Energy* **2016**, *184*, 681–695. [CrossRef]
96. Peng, H.; Ding, G.; Hu, H.; Jiang, W. Effect of nanoparticle size on nucleate pool boiling heat transfer of refrigerant/oil mixture with nanoparticles. *Int. J. Heat Mass Transf.* **2011**, *54*, 1839–1850. [CrossRef]
97. Hu, Y.; Li, H.; He, Y.; Liu, Z.; Zhao, Y. Effect of nanoparticle size and concentration on boiling performance of SiO_2 nanofluid. *Int. J. Heat Mass Transf.* **2017**, *107*, 820–828. [CrossRef]
98. Murshed, S.M.S.; de Castro, C.A.N.; Lourenço, M.J.V. Effect of Surfactant and Nanoparticle Clustering on Thermal Conductivity of Aqueous Nanofluids. *J. Nanofluids* **2012**, *1*, 175–179. [CrossRef]
99. Zhou, L.; Wei, L.; Du, X.; Yang, Y.; Jiang, P.; Wang, B. Effects of nanoparticle behaviors and interfacial characteristics on subcooled nucleate pool boiling over microwire. *Exp. Therm. Fluid Sci.* **2014**, *57*, 310–316. [CrossRef]
100. Jung, J.Y.; Kim, H.; Kim, M.H. Effect of ionic additive on pool boiling critical heat flux of titania/water nanofluids. *Heat Mass Transf.* **2013**, *49*, 1–10. [CrossRef]

nanomaterials

MDPI

Article

Effect of Nanoparticle Size and Concentration on Pool Boiling Heat Transfer with TiO_2 Nanofluids on Laser-Textured Copper Surfaces

Armin Hadžić, Matic Može *, Klara Arhar, Matevž Zupančič and Iztok Golobič

Faculty of Mechanical Engineering, University of Ljubljana, Aškerčeva 6, 1000 Ljubljana, Slovenia; armin.hadzic@fs.uni-lj.si (A.H.); klara.arhar@fs.uni-lj.si (K.A.); matevz.zupancic@fs.uni-lj.si (M.Z.); iztok.golobic@fs.uni-lj.si (I.G.)

* Correspondence: matic.moze@fs.uni-lj.si; Tel.: +386-1-4771-309

Abstract: The enhancement of boiling heat transfer has been extensively shown to be achievable through surface texturing or fluid property modification, yet few studies have investigated the possibility of coupling both enhancement approaches. The present work focuses on exploring the possibility of concomitant enhancement of pool boiling heat transfer by using TiO_2-water nanofluid in combination with laser-textured copper surfaces. Two mass concentrations of 0.001 wt.% and 0.1 wt.% are used, along with two nanoparticle sizes of 4–8 nm and 490 nm. Nanofluids are prepared using sonification and degassed distilled water, while the boiling experiments are performed at atmospheric pressure. The results demonstrate that the heat transfer coefficient (HTC) using nanofluids is deteriorated compared to using pure water on the reference and laser-textured surface. However, the critical heat flux (CHF) is significantly improved at 0.1 wt.% nanoparticle concentration. The buildup of a highly wettable TiO_2 layer on the surface is identified as the main reason for the observed performance. Multiple subsequent boiling experiments using nanofluids on the same surface exhibited a notable shift in boiling curves and their instability at higher concentrations, which is attributable to growth of the nanoparticle layer on the surface. Overall, the combination of nanofluids boiling on a laser-textured surface proved to enhance the CHF after prolonged exposure to highly concentrated nanofluid, while the HTC was universally and significantly decreased in all cases.

Keywords: nanoparticles; nanofluids; boiling; heat transfer; heat transfer enhancement; nanoparticle deposition

Citation: Hadžić, A.; Može, M.; Arhar, K.; Zupančič, M.; Golobič, I. Effect of Nanoparticle Size and Concentration on Pool Boiling Heat Transfer with TiO_2 Nanofluids on Laser-Textured Copper Surfaces. *Nanomaterials* **2022**, *12*, 2611. https://doi.org/10.3390/nano12152611

Academic Editor: S. M. Sohel Murshed

Received: 18 July 2022
Accepted: 27 July 2022
Published: 29 July 2022

1. Introduction

Pool boiling heat transfer has been widely used in numerous engineering systems, such as aircraft and spacecraft thermal management, high power electronics cooling, heat exchangers, nuclear reactors, air conditioning, thermal power generation, etc. [1–6]. Compared to natural and forced convection without phase change, the advantage of pool boiling is a higher heat removal rate from a surface while maintaining a low superheat (i.e., the temperature difference between the surface and the boiling liquid), which is an important advantage for its use in dissipating highly concentrated thermal loads.

Pool boiling is the process of vaporization at the solid–liquid interface. and it occurs when the temperature of the surface exceeds the saturation temperature of the liquid at the given pressure. The characteristics of the pool boiling heat transfer process can be described by the boiling curve, which was first reported by Nukiyama [7]. In the first phase of boiling heat transfer, all of the heat is dissipated through single-phase natural convection. When the surface superheat is high enough, vapor bubbles begin to form in the cavities on the surface (i.e., heterogeneous boiling takes place), which represents the onset of nucleate boiling (ONB) and the inception of the nucleate boiling phase. The nucleate boiling heat transfer regime is the most effective heat transfer region of the pool boiling process due to very high heat transfer coefficients at low surface superheat values, making it the most suitable method for various engineering applications. The heat transfer coefficient (HTC), representing the ratio between dissipated heat flux and the corresponding wall superheat, is the most common metric to describe the heat transfer intensity. With increasing heat flux, the number of active nucleation sites on the surface increases. When the population of bubbles becomes too high (at a high heat flux), neighboring bubbles coalesce extensively (i.e., merge on or above the surface), forming an insulating blanket of vapor covering the heating surface and thereby significantly decreasing the heat transfer intensity. This phenomenon is known as the boiling crisis, and the maximum heat flux associated with its incipience is the critical heat flux (CHF). After CHF onset, the boiling process transitions towards film boiling, which is characterized by a high increase in wall temperature and a large decrease in the HTC [8,9].

While boiling heat transfer represents an efficient cooling method, the HTC and CHF need to be enhanced to meet the specifications of certain applications, allowing for the safe and efficient operation of such systems. The enhancement of heat transfer in the nucleate boiling regime can be achieved in various ways, including by (i) changing the characteristics of the boiling surface, (ii) modifying the surface-fluid interaction, (iii) modifying the working fluid, or (iv) changing the operating conditions [10,11]. The aim of these methods is generally to lower the ONB and increase both the CHF and the HTC [11–13]. Most approaches to boiling enhancement only consider one technique, meaning that possible combinations and synergies between more techniques are largely unexplored. This is addressed in the present study, where surface modification via laser texturing is coupled with fluid modification through the addition of nanoparticles, with the aim of obtaining superior boiling performance to that using a single enhancement approach.

1.1. Methods for Intensifying Boiling Heat Transfer

In recent years, researchers have thoroughly studied different techniques for intensifying boiling heat transfer, such as adding nanoparticles to the base fluid or modifying the surface characteristics of the boiling surface, namely its roughness, porosity, wettability, wickability, etc. [13–18]. Surface modification may be performed on the macro-, micro- or nanoscale [11]. The techniques for the modification of surface morphology can generally be divided into physical and chemical categories [19]. Common physical techniques include electro-deposition [20] and pulsed laser deposition [21], while in most studies, the chemical techniques are chemical vapor deposition [22], the sol-gel technique, and electrochemical deposition [23]. Overall, surface modification aims to alter the surface wettability and microstructure, which have a profound impact on the boiling process and the associated heat transfer parameters [16,24,25]. Free surface energy, roughness of the surface, and its micro- and nanotexture have a strong influence on wettability [10,26,27]. Based on the contact angle of a liquid droplet on the surface, the latter can be classified as phobic or philic to the liquid. As water is the most common testing medium, the surfaces are typically classified as hydrophilic (contact angle < 90°) or hydrophobic (contact angle > 90°). Based on previous experimental findings, the ONB will occur at a lower superheat, and CHF will be decreased on hydrophobic surfaces, while hydrophilic surfaces will exhibit delayed ONB, but increased CHF [10,28,29].

1.2. Boiling of Nanofluids

A common method of modify the properties of the coolant in boiling applications is the addition of various types of nanoparticles to the base fluid [30–32]. This changes its thermophysical properties, but more importantly, the nanoparticles are deposited onto the boiling surface during boiling, creating micro- and nanostructures that can favorably affect the boiling process [33]. The available literature mostly suggests that the boiling of nanofluids leads to an improvement of the HTC and the CHF, while some few studies make the opposite conclusion [34]. The increase in HTC and CHF tends to be correlated with increasing the concentration of nanoparticles, up to a certain point. At excessive concentrations, the HTC begins to decrease, while the CHF will remain unaltered [35]. At high nanoparticle concentration, the deposited layer on the surface becomes thicker, which leads to greater thermal resistance and decreases the HTC [35]. Due to the deposition of nanoparticles on the surface and their large effect on heat transfer during boiling, Fang et al. [31] suggested that further experiments on nanofluid boiling should be combined with a comprehensive study of the influence of nanoparticle deposition time for different concentration ranges. Long-term experiments on the variation of nanoparticle deposition time should be performed, and correlations to predict the evolution of deposited nanoparticle layers should be proposed. In addition to the concentration, the material and size of the nanoparticles and the preparation of the nanofluid also affect the boiling performance. Dadhich et al. [36] and Fang et al. [31] pointed out that there is a significant need for a database of thermal properties for different materials in different size ranges of nanoparticles.

The deposition of nanoparticles from the nanofluid during boiling is the most common enhancement strategy involving modified fluids. Manetti et al. [37] conducted an experimental study of HTC in pool boiling of deionized water and Al_2O_3-water nanofluid with low (0.0007 vol.%) and high (0.007 vol.%) volume concentration on smooth and rough copper surfaces within a heat flux range of 100 to 800 kW m^{-2}. They observed that HTC increased at low concentrations at average heat flux values, which was related to increasing the radius of the cavities through deposition of nanoparticles on the surfaces due to boiling. Increasing the heat flux led to a decrease in the HTC caused by the increased deposition rate of nanoparticles on the rough surface and the filling of the cavities with nanoparticles. Ahmed et al. [38] investigated pool boiling heat transfer performance on horizontal flat copper surfaces using nanofluid and pure water. They used 40–50 nm alumina nanoparticles to prepare three different nanoparticle concentrations (0.01 vol.%, 0.1 vol.%, and 1 vol.%). After performing boiling experiments, the nanoparticle-coated surfaces were used for pool boiling experiments using pure water. During the nanofluid boiling experiments, the authors concluded that the concentration of nanoparticles has a major effect on the heat transfer performance. At lower concentrations, the rate of deposition of the particles is lower, resulting in a greater enhancement of heat transfer, which can be attributed to the fact that the increased thermal conductivity of nanofluids has a greater dominance than the effect of nanoparticle deposition on the surface. On the other hand, the boiling of pure water on the surface coated with nanoparticles showed that high deposition rates lead to an improvement in heat transfer, which was explained by a less uniform deposition layer on the surface. Huang et al. [39] studied the enhancement of boiling on nickel wires coated with TiO_2 nanoparticles in pure water. The coating was produced by electrical heating of the wire in nanofluids with concentrations from 0.01 to 0.1 wt.% and heat flux up to 1000 kW m^{-2}. Experimental results of pure water boiling showed an enhancement of the CHF of up to 82.7% for coated nickel wire prepared in 0.1 wt.% nanofluids. On all nickel-coated wires, the HTC deteriorated due to the higher thermal resistance caused by the deposition of nanoparticles. Kiyomura et al. [40] investigated boiling heat transfer performance of surfaces coated with Fe_2O_3 nanoparticles in Fe_2O_3 water-based nanofluid at a high (0.29 g/L) and a low concentrations (0.029 g/L). The results showed the highest HTC values on coated copper surfaces with low mass concentration, and with increasing the concentration, the roughness of the surfaces increased. Salimpour et al. [41] performed boiling experiments on smooth and rough copper surfaces using iron-oxide-water-based nanofluid at low and high heat fluxes. They found that at low heat fluxes on smooth surfaces, and at high heat fluxes on rough surfaces, the deposition of nanoparticles on the surface enhanced the heat transfer during boiling. According to other research conducted in this field, there is no clear understanding of how the boiling heat transfer performance is changed by the various effects of nanoparticle deposition during nanofluid boiling [42,43].

In addition to in situ nanoparticle deposition, dip coating and drop casting techniques were also proposed for boiling applications. An example is the study by Yim et al. [44], who investigated the surface wettability in nucleate pool boiling on aluminum surfaces coated with TiO_2 nanoparticles from 1 wt.% TiO_2 ethanol-based nanofluids, using a drop casting technique. The obtained results showed that the performance of nucleate pool boiling was 64.1% higher for TiO_2-coated surfaces than for the bare surfaces.

1.3. Boiling on Laser-Texured Surfaces

Laser texturing is a very effective method for locally or globally changing the morphology of the surface to intensify boiling heat transfer. Može et al. [45] demonstrated a strong enhancement of boiling performance of water on superhydrophobic and superhydrophilic surfaces textured with a nanosecond fiber laser. By combining laser texturing with a superhydrophobic coating, they showed significant improvement in HTC and found that the intensification of boiling heat transfer is possible with a suitable surface morphology where the Wenzel wetting regime is achieved. Kurse et al. [46] studied the pool boiling heat transfer of deionized water on laser-textured surfaces. Microstructures were fabricated on stainless steel using a femtosecond laser. It was found that CHF and the maximum heat transfer coefficient on laser-textured surfaces were improved compared to those on the polished reference surface. It was also found that the improvement of CHF is related to the wetting and wicking capability of the surface, which allows for the replenishing of the evaporating liquid and affects the delay of the CHF. The study of the nucleate pool boiling of water and water-ethanol mixtures on untreated and laser textured stainless steel foils showed a significant improvement in boiling performance on laser-textured surfaces compared to the untreated surfaces for pure liquids and binary mixtures. The improvement in boiling performance is directly related to the microcavities, which act as active nucleation sites [47]. Serdyukov et al. [48] investigated nucleate pool boiling of water on laser textured silicon surfaces. Their results showed an enhancement of the heat transfer coefficient by up to 49.5%, compared to a rough silicon sample, and by up to 234%, compared to a polished sample. Additionally, the study showed that laser texturing of the surfaces resulted in a remarkable increase in the frequency and density of nucleation sites. It was concluded that a decrease in the diameter of the departure bubble and a lower nucleation temperature are characteristics that are associated with the laser-modified silicon surfaces. The study of the stability of copper surfaces before and after functionalization by laser texturing performed with a nanosecond fiber laser was presented by Može et al. [49]. The study showed the enhancement of CHF by up to 90% and HTC by up to 115% on textured surfaces compared to the reference surface. The results also demonstrate the tendency of constant shifts of boiling curves in each experimental run, while the shifts on laser textured surfaces did not occur after the second run. This fact is confirmed by changes in the morphological and chemical structure of the surface after the first onset of CHF and is associated with the effects of low-temperature annealing. Finally, only one study, published by Karthikeyan et al. [50], was found to have previously combined surface laser texturing and the use of nanofluids to enhance boiling heat transfer. The latter authors reported a notable enhancement of boiling heat transfer, but the nanofluid used was rather unconventional (carbon nanotubes dispersed in ethanol or water).

1.4. Scope and Aim of this Study

There are few studies that investigate the boiling performance of nanofluids on pre-modified surface (e.g., laser-textured surfaces) [51,52], despite the seemingly great potential for concomitant and synergistic enhancement of boiling performance. To fill this knowledge gap, we investigated the pool boiling heat transfer performance of TiO_2-water nanofluids prepared with two mass concentrations (0.001 and 0.1 wt.%) and with two different sizes of nanoparticles (small: 4–8, and large: 490 nm), on laser textured copper surfaces. Five consecutive measurements were performed on each laser-textured surface under pool boiling conditions. The surface morphology of deposited nanoparticles was analyzed using scanning electron microscopy (SEM), and the wettability changes were recorded through water contact angle (WCA) measurements. The results were analyzed through multiple comparisons to elucidate the effect of nanoparticle size and concentration on possible additional enhancement or deterioration of boiling performance of the laser-textured copper surfaces.

2. Methods

2.1. Sample Preparation and Amalysis

Samples for boiling experiments were prepared on high purity copper (>99.9% Cu). Each sample was first sanded using P1200 and P2000 grit sandpaper to achieve a surface roughness of approx. 0.15 μm. Afterwards, the samples were cleaned using isopropanol and lint-free wipes. One sample was tested without any further treatment, and it is denoted as REF (i.e., untreated reference sample). All other samples underwent direct laser texturing immediately before the boiling experiments.

To perform the laser texturing, a nanosecond pulsed fiber laser was used (FL-mark-C with JPT Opto-electronics Co., Ltd. "M7 30 W" MOPA source, Shenzhen, China). The laser system is equipped with an OPEX F-Theta lens with a focal distance of 100 mm and working field of 70 × 70 mm^2. A pattern of equidistant parallel lines (Δx = 60 μm) was used to create a channel-like microstructure on each sample. The pulse frequency was set to 110 kHz, the pulse duration to 45 ns, and the full power of 30 W was applied. With the focal beam diameter of ~25 μm and laser beam quality parameter $M^2 \leq 1.3$, the average laser pulse fluence was calculated to be ~56 J cm^{-2}.

The morphology and elemental composition of the samples were analyzed using scanning electron microscopy (ThermoFisher Scientific Quattro S, Waltham, MA, USA) and energy-dispersive X-ray spectroscopy (Oxford Instruments Ultim Max 65, Abingdon, UK). SEM images of the reference sample REF and laser-textured sample (LT) before exposure to boiling are shown in Figure 1.

Figure 1. SEM images of the (**a**) untreated reference sample REF (blue frame) and (**b**) laser-textured sample LT (red frame) before exposure to boiling.

The analysis of sample wettability was performed using a goniometer to record the static contact angles with water. Cleaning of the samples was performed with a UV/ozone cleaner (Ossila) to remove hydrocarbon contaminants after boiling/storage. Contact angles were measured: (i) before boiling, (ii) immediately after boiling experiments, (iii) 7 days after the boiling process, (iv) after UV/ozone cleaning (7 days after the experiments), and finally (v) three days after UV/ozone cleaning. Measurements were conducted using a goniometer (Ossila, ±1°). Five drops were deposited onto different parts of the surface, and the contact angles were recorded and averaged.

2.2. Boiling Performance Evaluation

The boiling performance evaluation of nanofluids on laser-textured copper surfaces was performed using a previously developed experimental setup shown in Figure 2 [53]. A glass cylinder with an inner diameter of 60 mm was placed between two stainless steel flanges, thus forming the boiling chamber. The latter was filled with 200 mL of the working fluid during experiments. As shown in Figure 2b, the sample was mounted on a copper heating block, and inserted through the bottom stainless-steel flange into the boiling chamber. PEEK bushing, a ring of flexible epoxy glue, and a silicone O-ring were used to ensure sealing, limit heat loss, and prevent parasitic boiling. Before every experiment, the working fluid was preheated and degassed with an immersion heater, which was controlled by a variable transformer. An immersion heater was also used to maintain the saturated state of the working fluid during the boiling experiment. The cartridge heaters, positioned inside the heating block, were used for the generation of heat and were also controlled by a variable transformer. All measurements were performed at atmospheric pressure, and the stability was confirmed by the stable temperature of the saturation. Vapor produced during measurements went to the water-cooled glass condenser and returned to the boiling chamber.

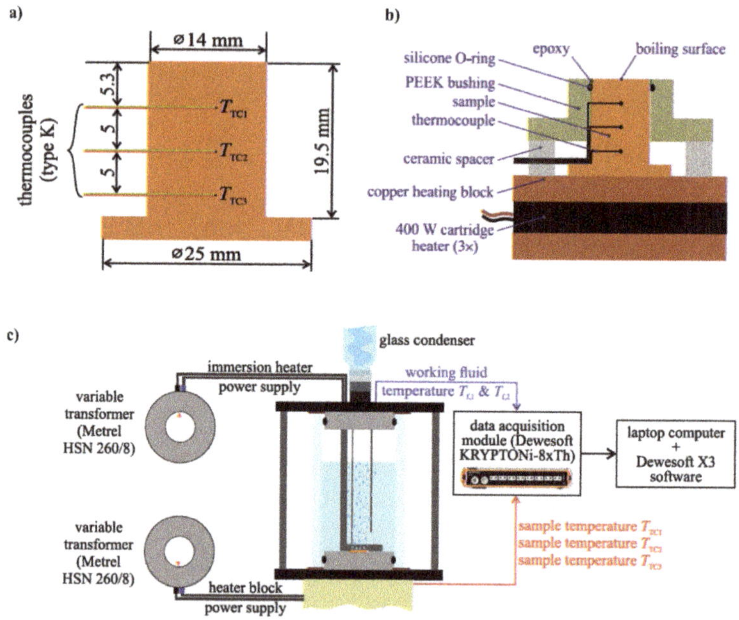

Figure 2. Dimensions of the sample including the position of thermocouples (**a**), cross section of the sample and the heating assembly (**b**), and experimental setup (**c**).

The temperatures inside the sample were measured by utilizing three K-type thermocouples spaced 5 mm apart from one another. The thermocouple closest to the boiling surface was positioned 5.3 mm from the top of the sample, as shown in Figure 2c. Two further K-type thermocouples were submerged into the boiling chamber at different heights to measure the temperature of the working fluid. A KRYPTONi-8xTH DAQ device was used for collecting all temperature signals as raw voltages. The temperature of each cold junction was recorded internally and used to offset the measurements to obtain correct temperature readings. The calculation of temperatures based on offset voltages was performed, utilizing NIST 9th degree polynomial. Data from the DAQ device were acquired using Dewesoft X3 software at a frequency of 10 Hz.

2.3. Data Reduction and Measurement Uncertainty

The calculation of relevant heat transfer parameters is based on measured temperatures. The parameters which are most relevant in boiling heat transfer systems are heat transfer coefficient, heat flux, and superheat of the surface. The methodology used for the calculation of the spatial temperature gradient between the highest positioned and the lowest positioned thermocouple in the sample utilizing the following linear interpolation (1) is suggested in [54]:

$$\frac{\Delta T}{\Delta x} = \frac{T_{TC3} - T_{TC1}}{2\Delta x_1} \tag{1}$$

Heat losses during the experiments are neglected because of favorable insulative properties achieved using a very low thermal conductivity material (PEEK). Consequently, a simplified case of 1D conduction through the sample towards the boiling surface is considered. The heat flux is calculated using Fourier's law of conduction:

$$\dot{q} = -k\frac{\Delta T}{\Delta x} \tag{2}$$

To accurately calculate the heat flux, thermal conductivity needs to be precisely evaluated. This is achieved by using a temperature-dependent value of thermal conductivity. Based on the average of all three temperatures measured within the sample, the thermal conductivity is calculated at the mean temperature (T) of all three temperatures in the sample according to the following equation:

$$k(T) = 0.000283T^2 - 0.1646T + 378.07, \tag{3}$$

The laser-flash measurement method of thermal diffusivity at various temperatures, and the determination of thermal conductivity based on temperature-dependent density and specific heat capacity are used to determine the latter expression. All temperature values are expressed in °C, and the thermal conductivity, as the result value, is returned in W m^{-1} K^{-1}. The temperature of the boiling surface is calculated by linear extrapolation using the previously determined heat flux and the temperature measured with the highest positioned thermocouple in the sample. The thermal conductivity is first determined at the highest positioned thermocouple in the sample, T_{TC1}, utilizing Equation (3) to obtain an estimate of surface temperature. The mean temperature of the top part of the sample is determined as the arithmetic average value between the estimated surface temperature and the temperature of the highest positioned thermocouple. The mean temperature of the top part of the sample is used to determine the average thermal conductivity, which is then used to evaluate the surface temperature with increased accuracy. The saturation temperature of the working fluid is calculated as the arithmetic average value of the temperatures measured with two immersed thermocouples. The difference between the saturation temperature of the working fluid and surface temperature is the surface superheat ($T_W - T_{sat}$). Dividing the heat flux by the corresponding surface superheat is used to determine the heat transfer coefficient:

$$h = \frac{\dot{q}}{T_w - T_{sat}}, \tag{4}$$

2.4. Peparation of Nanofluids

The nanofluids used were prepared by dispersing different amounts of nanoparticles in the base fluid (twice-distilled water). The nanofluids used in this study were prepared and stabilized using ultrasonic vibration, without using surfactants or adjusting the pH value. TiO_2 nanoparticles were chosen for the study because of their favorable chemical and physical stability and their hydrophilicity, which was previously shown to enhance the CHF value during boiling experiments. Two nanoparticle size ranges were used to examine the effect of their size, especially relative to the size of laser-induced surface structures. One type of nanofluid was prepared using very small nanoparticles (sized below 10 nm, much smaller than the majority of structures resulting from laser texturing), while the other type used nanoparticles that were two orders of magnitude larger, with diameters close to 500 nm (close to the scale of laser-induced surface structures). Furthermore, two nanofluids concentrations were tested for each type of nanofluid. Most studies typically report nanofluid concentrations in the range from 0.001 wt.% to 0.1 wt.%. Therefore, we opted to use the two extreme values, again separated by two orders of magnitude, in order to evaluate the effect of nanoparticle concentration on boiling performance.

One type of nanofluid was prepared using small nanoparticles, with a size of 4 to 8 nm (Carl Roth, Karlsruhe, Germany; ROTInanoMETIC $\geq$ 99.9%), and the second type was prepared using large nanoparticles, with a size of 490 nm (Nanografi Nano Technology, Ankara, 99.995+%). Both types of nanofluids were prepared at two different mass concentrations: 0.1 and 0.001 wt.%. The third type of nanofluid was prepared with 0.05 wt.% of the large-sized and small-sized nanoparticles, respectively. Twice-distilled water was degassed for 45 min via vigorous boiling before it was used to prepare the nanofluids to reduce the amount of entrapped gases and also reduce the need for further degassing in the boiling chamber, where deposition of nanoparticle on the immersion heater reduces the concentration of the nanofluid in an uncontrolled way. Nanofluids were sonicated for 1 h immediately after preparation in an ultrasonic bath (ASonic, Ultrasonic Cleaner-Pro 30, 40 kHz, 120 W). To stabilize the solution before use in the experiment, the nanofluid was again sonicated for 1 h. The performed experiments are summarized in Table 1.

Table 1. The characteristics of the performed experiments.

Name	Boiling Fluid	Surface	Size of Nanoparticles	Concentration (wt.%)
REF	water	Bare copper	/	/
LT	water	laser textured copper	/	/
TiO_2-S-0.001	TiO_2-water		small (4–8 nm)	0.001
TiO_2-S-0.1	TiO_2-water		large (490 nm)	0.1
TiO_2-L-0.001	TiO_2-water		small (4–8 nm)	0.001
TiO_2-L-0.1	TiO_2-water		large (490 nm)	0.1
TiO_2-Mix-0.05	TiO_2-water		small (4–8 nm) + large (490 nm)	0.05 small and 0.05 large

Overall, the used nanoparticle sizes and concentrations are based on the extreme values reported in the literature, and the corresponding extreme combinations, as used here, are expected to allow for the generalization of the results.

2.5. Measurement Protocol

After degassing the working fluid, five experimental runs were performed for each combination of the sample and working fluid. During each experimental run, the heat flux was continuously increased at a rate of 2 kW m^{-2} s^{-1}. The chosen methodology of a slow continuous increase in the heat flux can be considered as quasi-stationary, as shown by Može et al. [45]. The boiling curve was recorded until the CHF was reached or, when that was not possible, until a surface superheat was reached (~100 K, representing a safety limit of the experimental setup). When an experimental run was finished, cartridge heaters were turned off, and the sample was left to cool down on its own. The duration of each run was approximately 45 min, while all 5 runs were completed in a 5-hour period.

3. Results and Discussion

3.1. Boiling Heat Transfer with Water

To establish a baseline for experiments with nanofluids, the boiling performance of the bare copper surface and the laser textured surface was first evaluated utilizing twice-distilled water. The results are shown in Figure 3.

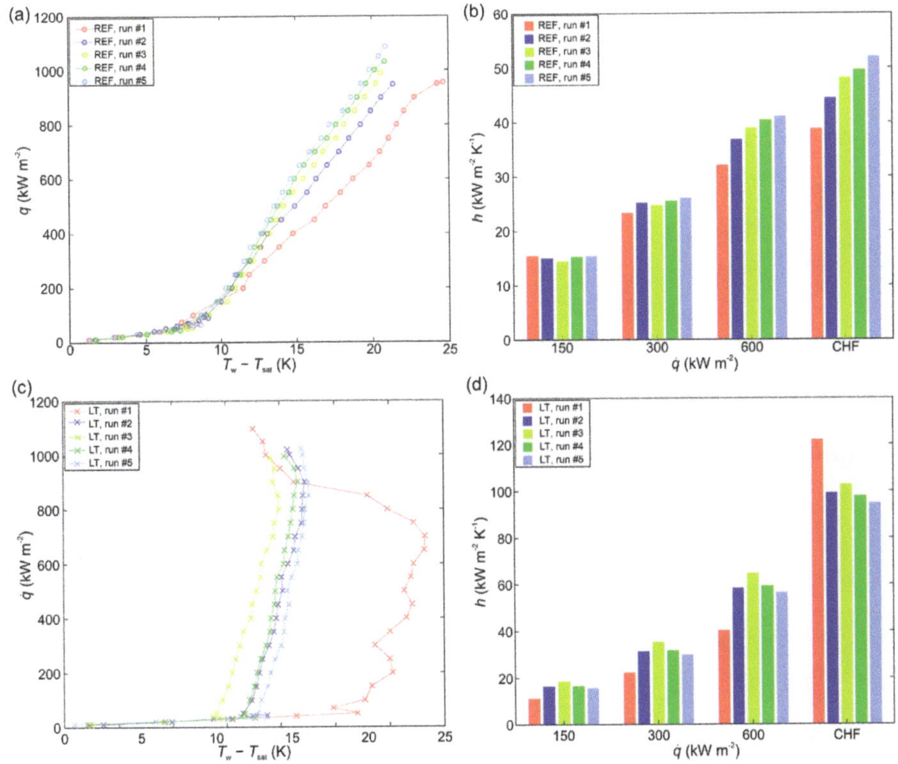

Figure 3. Boiling curves (**a**) and heat transfer coefficients (**b**) on the reference surface; boiling curves (**c**) and heat transfer coefficients (**d**) on the laser-textured surface.

Figure 3a,c shows the boiling curves (i.e., heat flux as a function of surface superheat) recorded on the bare surface (REF) and the laser textured surface (LT), respectively. In both cases, the boiling curves were recorded until the incipience of the CHF. It is evident that the boiling curves on the REF surface are slightly shifted towards lower superheats after each experimental run. The same shifting of the boiling curve also occurs on the LT surface during the first two runs, but then boiling curves stabilize, with a slight reversal of the trend. This is also consistent with the wettability of the examined surfaces. The REF surface is transitioning from a hydrophilic state (before boiling) towards the hydrophobic state (after boiling), and the LT surface is transitioning from a super hydrophilic state to a hydrophobic state. This is in correlation with the previous research, where it was shown that the shift in the boiling curves ensues due to changes in surface morphology and chemistry [49,53]. The HTC of the REF and LT surfaces at selected heat fluxes are shown in Figure 3b,d, respectively.

The highest HTC values were recorded at the point of CHF incipience, where the LT surface provides a significant improvement compared to the untreated reference surface. Secondary boiling effects were detected on the LT surface, leading to a significant decrease in wall superheat temperature near the CHF, which in turn leads to the remarkable enhancement of the HTC. The reduction in surface superheats is caused by the higher density of active nucleation sites at high heat flux, when boiling also starts to take place on the peaks of the surface morphology instead of just within the cavities [46,55].

SEM images of the laser-textured sample (LT) after exposure to the boiling of water are shown in Figure 4. It is evident that the microtopography remains the same, but changes are present on the submicron scale, where the transition of oxide species from copper(II) oxide to copper(I) oxide is evident and perfectly matches previous observations [49,53]. In [49], the appearance of copper(I) oxide in the form of sub-micron sized cubes or (truncated) octahedra was shown to take place after CHF incipience as a result of low temperature annealing, resulting in a transition of needle-shaped copper(II) oxide in the reduced form.

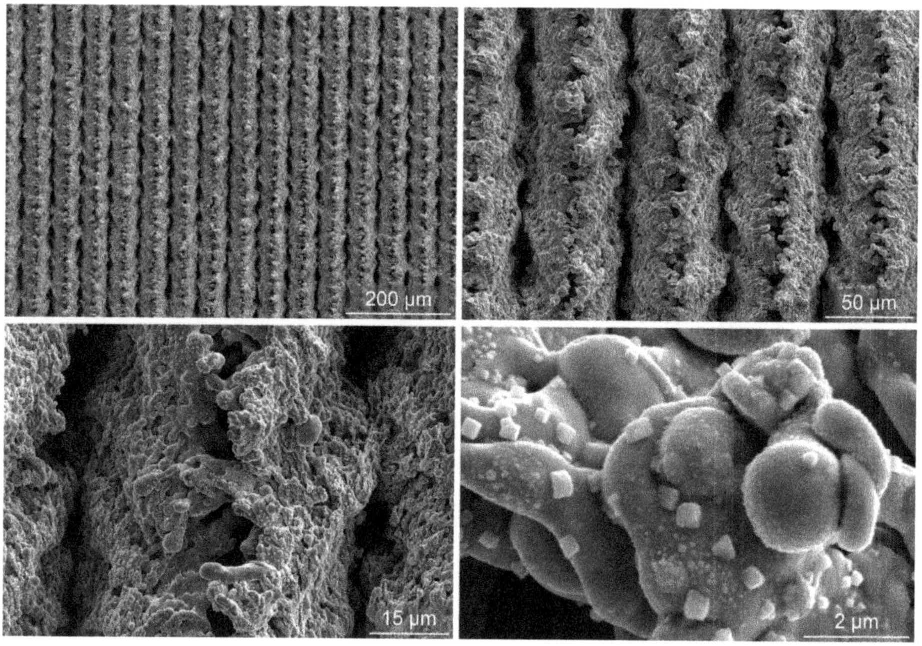

Figure 4. SEM images of the laser-textured sample LT after exposure to the boiling of water.

3.2. *The Effect of Concentration on Boiling of Nanofluid with Small and Large TiO_2 Nanoparticles*

The boiling of nanofluid with small (4–8 nm) TiO_2 nanoparticles on the copper laser textured surface was evaluated at two different mass concentrations of 0.001 and 0.1 wt.%. The corresponding boiling curves and HTCs are shown in Figure 5a,b for 0.001 wt.% nanofluid, while the data for the 0.1 wt.% nanofluid is shown in Figure 5c,d. The boiling curves obtained using 0.001 wt.% nanofluid are very stable, while the curves for the 0.1 wt.% nanofluid are shifted towards higher superheat temperatures with each consecutive run. This shifting can be explained through a decrease in the active nucleation site density [33,35] due to the deposition of nanoparticles onto the surface. At 0.1 wt.%, many microcavities on the laser textured surface become filled with nanoparticles during the first experimental run, which leads to a decrease in active nucleation site density and an increase in surface superheat during the next boiling runs. With increasing the number of runs, the surface superheat increases to unacceptable levels for practical use and the boiling performance is deteriorated. The highest HTC value, recorded during the first run (approx. 1 h of boiling on the surface), was 79,4 kW m^{-2} K^{-1} near CHF incipience. After the 5th run, the HTC value was 83% lower at same heat flux. During the second boiling run, the high surface temperature prevented the CHF incipience from being recorded, as the setup was turned off due to safety reasons. The highest CHF of 1457 kW m^{-2} was recorded at 0.1 wt.%, which represents an enhancement of 35% compared to the highest recorded CHF at 0.001 wt.% (1082 kWm^{-2}). The enhancement of CHF is attributed due to a reduction in static contact angle and an improvement in surface wettability.

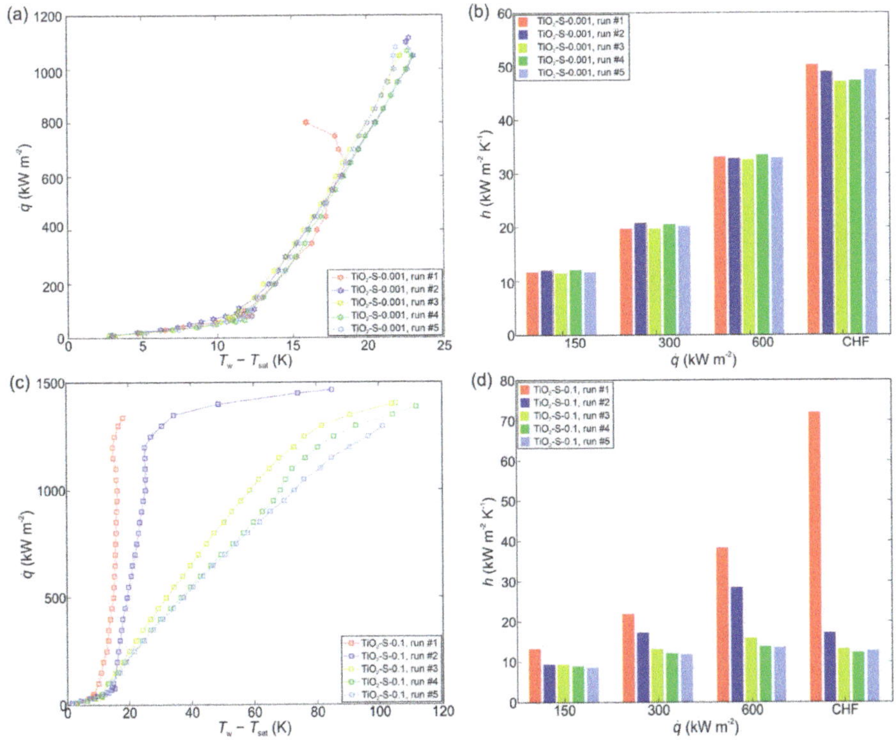

Figure 5. Boiling of TiO_2-water nanofluid with small size nanoparticles at 0.001 wt.%: boiling curves (**a**) and heat transfer coefficients (**b**). Boiling of TiO_2-water nanofluid with small size nanoparticles at 0.1 wt.%: boiling curves (**c**) and heat transfer coefficients (**d**).

SEM images of the laser-textured sample after exposure to boiling of 0.1 wt.% nanofluid with smaller nanoparticles are shown in Figure 6. It is observable that a thick nanoparticle deposit has formed on the surface, obscuring the shape of the laser-induced microstructure below. However, the deposited layer is not homogeneous and uniform, as evident from the missing patches, under which the laser-textured copper surface is visible. This was confirmed through EDS analysis, which detected a mixture of titanium and oxygen stemming from the TiO_2 deposits on most of the surface, while a lower percentage of these two elements was detected, along with a notable percentage of copper on the flaked-off patch.

The boiling of nanofluid with larger nanoparticles (490 nm) was performed at 0.001 and 0.1 wt.%. Boiling curves at 0.001 and 0.1 wt.% are shown in Figure 7a,c, respectively, while the HTCs are shown in Figure 7b,d, respectively.

Figure 6. SEM images of the laser-textured sample LT after exposure to boiling 0.1 wt.% nanofluid with small nanoparticles (i.e., S-0.1).

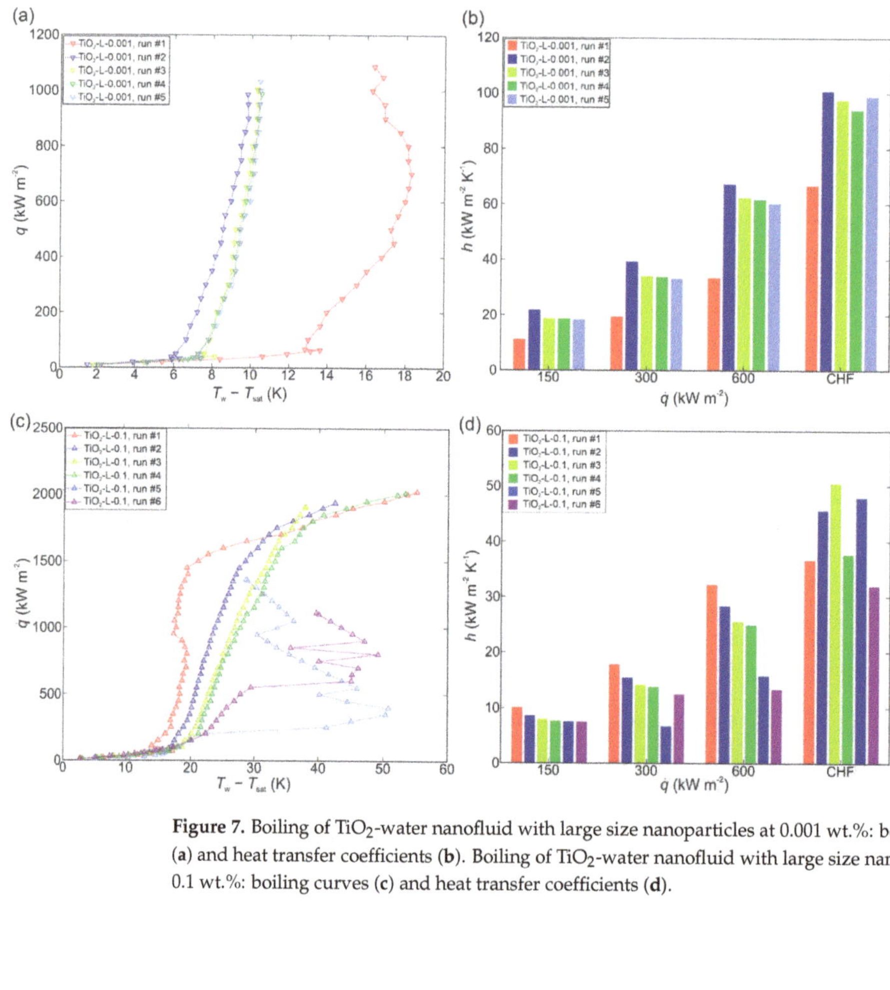

Figure 7. Boiling of TiO_2-water nanofluid with large size nanoparticles at 0.001 wt.%: boiling curves (**a**) and heat transfer coefficients (**b**). Boiling of TiO_2-water nanofluid with large size nanoparticles at 0.1 wt.%: boiling curves (**c**) and heat transfer coefficients (**d**).

At 0.001 wt.%, a significant shift in the boiling curves towards a lower surface superheat was observed, while at 0.1 wt.%, the boiling curves are shifted toward higher superheat values. The general observations match those made for the smaller nanoparticle size. However, it was observed that after prolonged boiling of highly concentrated nanofluid (0.1 wt.% for 4+ h), the boiling curves became very unstable (e.g., Figure 7c) due to the very thick deposited layer on the surface, which also flakes off locally. SEM images of the laser-textured sample after exposure to the boiling of 0.1 wt.% nanofluid with larger nanoparticles are shown in Figure 8. Here, the underlying laser-induced microstructure is evident, but the deposited layer is again heterogeneous. It is estimated that significant flaking of the deposited layer occurred during the boiling process and after removal of the sample due to the low adhesion strength. The EDS analysis confirmed a nearly perfect atomic ratio of oxygen to titanium (2:1) for the TiO_2 deposit.

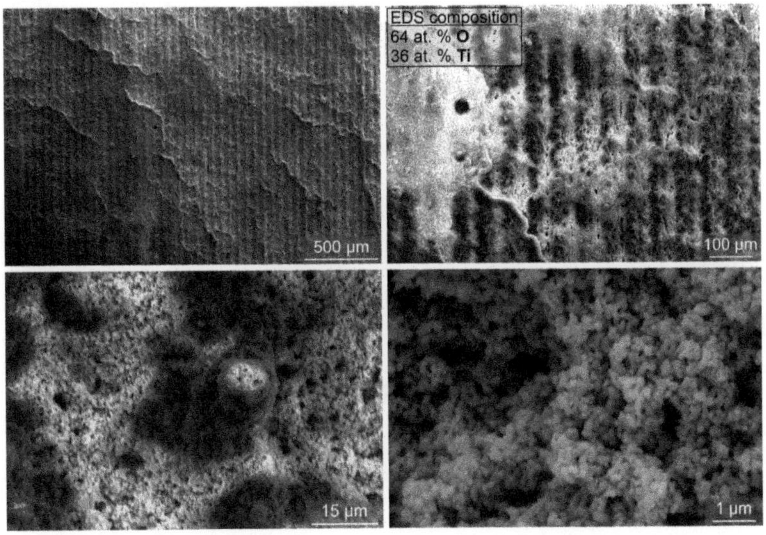

Figure 8. SEM images of the laser-textured sample LT after exposure to boiling 0.1 wt.% nanofluid with large nanoparticles (i.e., L-0.1).

Interestingly, the boiling performance increase detected with larger nanoparticles at 0.001 wt.% was not observed at the same concentration using the smaller nanoparticle size. It is likely that the small nanoparticles (two orders of magnitude smaller than the large nanoparticles used in this study) form a thin, compact deposited layer without additional nucleation sites (i.e., the HTC is not enhanced at comparable heat flux values), while the increased wettability is able to raise the CHF value. On the other hand, much larger nanoparticles (490 nm) form a more porous deposit, which offers additional cavities for bubble nucleation, thus enhancing the HTC, but the overall effect on increasing the surface wettability (and with that, the CHF value) is lower. A comparison of the SEM images taken after boiling of 0.001 wt.% nanofluid, with both small and large nanoparticle size, is shown in Figure 9. Only a thin layer of small nanoparticles remains on the laser-induced structured after boiling the 0.001 wt.% nanofluid and is only evident at higher magnifications. On the contrary, larger nanoparticles deposited at the same concentration are more clearly visible and seem to fill a part of the laser-induced surface channels.

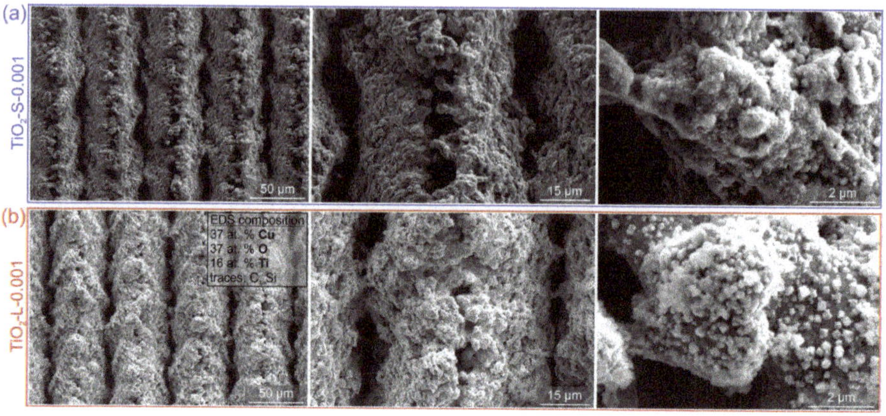

Figure 9. (**a**) SEM images of the laser-textured sample LT after exposure to boiling 0.001 wt.% nanofluid with small nanoparticles (i.e., S-0.001) and (**b**) after exposure to boiling 0.001 wt.% nanofluid with large nanoparticles (i.e., L-0.001).

The HTC is enhanced by 49% after 5 h of boiling at the concentration of 0.001 wt.%, and both the CHF and the HTC enhancements are stable. The CHF enhancement is more pronounced with larger nanoparticles at 0.1 wt.%, with the highest value reaching 2021 kW m^{-2}, which represents an 86% enhancement over the highest CHF recorded for the 0.001 wt.% Enhancement of the CHF was recorded for all performed runs, which was attributed to better wettability characteristics of the surface exposed to the boiling of the concentrated nanofluid. Increased wettability compared to the results for the boiling of pure water on the LT surface was confirmed by the measurement of the static contact angle after boiling. The contact angle after boiling with the 0.1 wt.% nanofluid was 13.5°, and 29.8° for the 0.001 wt.% nanofluid.

A further investigation was performed by mixing 0.05 wt.% of small and 0.05 wt.% of large size nanoparticles, respectively. The boiling curves and HTCs are shown in Figure 10.

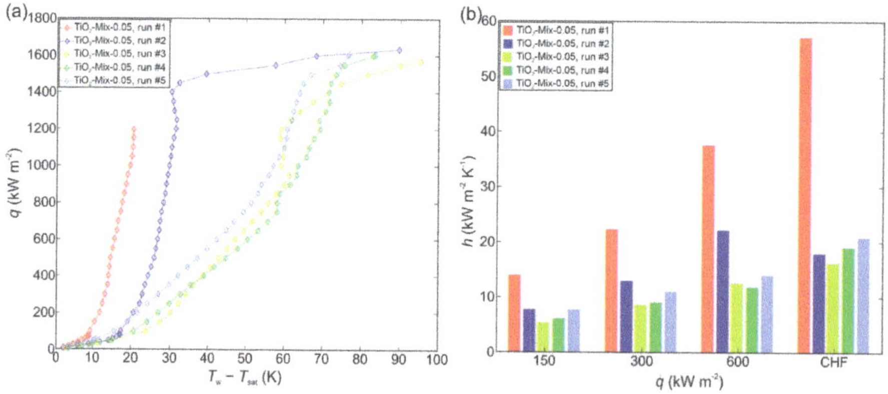

Figure 10. Boiling of TiO_2-water nanofluid with 0.05 wt.% of small size and 0.05 wt.% large size nanoparticles, respectively: boiling curves (**a**) and heat transfer coefficients (**b**).

The boiling curves are again shifted toward a higher surface superheat, which matches the behavior when 0.1 wt.% nanofluid was boiled with either small or large nanoparticles. After the first experimental run, the surface superheat increased significantly, which was likely caused by the presence of small size nanoparticles that filled the microcavities and surface channels, thus decreasing the active nucleation site density. As the latter decreases, so does the HTC. CHF was again not recorded after the first hour of boiling due to very high surface temperatures and the danger of damage to the setup, but the achieved maximal heat flux values were notably lower than for the 0.1 wt.% nanofluid with large particles, but higher than for the same concentration of nanofluid with small nanoparticles.

SEM images of the surface after exposure to the boiling of nanofluid with mixed particle size (i.e., MIX-0.05) are shown in Figure 11.

Figure 11. SEM images of the laser-textured sample LT after exposure to boiling of nanofluid with a mixture of 0.05 wt.% of small nanoparticles and 0.05 wt.% of large nanoparticles (i.e., MIX-0.05).

It is noticeable that the surface is rather uniformly covered with a deposited layer, and the laser-induced structures are not visible. However, the deposit is again inhomogeneous.

3.3. Contact Angle Measurments

The static contact angle of water was measured on each surface to determine its wettability and help explain the observed boiling behaviors, especially in terms of the effect of nanoparticle deposition onto the surface during the experiments. The recorded values, obtained through measurements at different points in time, as previously explained in Section 2.1, are depicted and compared in Figure 12.

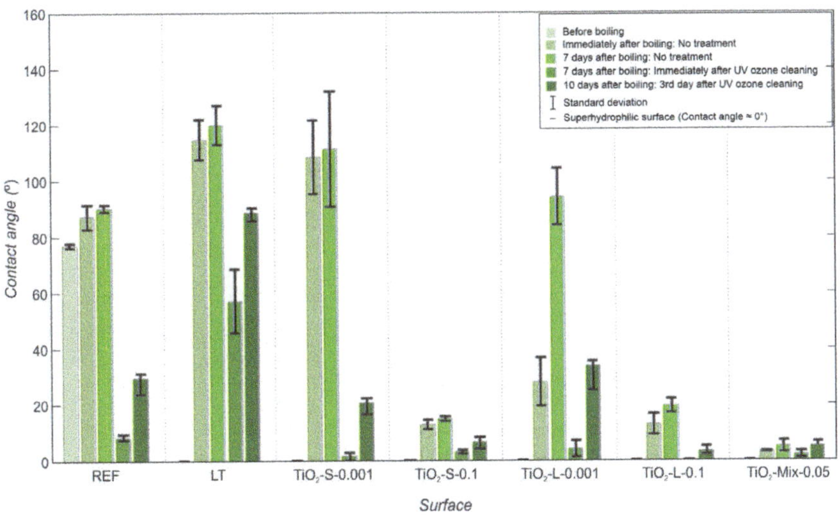

Figure 12. Static contact angle measurements on all surfaces before boiling, after boiling, and after storage.

Laser textured surfaces were initially superhydrophilic immediately after processing, and the contact angle of the bare copper surface was 82° before the boiling experiments. After the boiling experiments with water on the laser textured surface and the reference surface, the contact angles increased, as was shown in previous studies [51]. After the boiling experiments, the wettability of the surfaces changed due to the deposition of nanoparticles. The increase in the contact angles of surfaces is slightly higher after boiling with low-concentration nanofluids than for the contact angles measured after boiling with high-concentration nanofluids. This could be due to the higher deposition rate at higher concentrations, which causes a thicker layer of deposited TiO_2 nanoparticles on the surface. TiO_2 nanoparticles are (super)hydrophilic, and a thick layer of them on the surface leads to the surface staying in a hydrophilic state. At lower concentrations and for the small-sized nanoparticles, the deposited layer on the surface is thinner. On the other hand, the larger nanoparticles form more porous deposits, resulting in lower contact angles compared to those for the smaller nanoparticles used in this study.

Additionally, all surfaces were exposed to ambient conditions for several days after the boiling experiments were finished and then cleaned with a UV/ozone cleaner to remove volatile organic compounds (VOC) and other carbon-based impurities. Contact angle measurements were performed before and immediately after cleaning with the UV/ozone cleaner. Afterward, the surfaces were again exposed to ambient air for three days, and the contact angles were measured again. The results of the measurements show the contact angles of all surfaces gradually increased over time, which was observed in many previous studies [56,57], and this was confirmed by previous research [58,59]. The change in the wettability is attributed to the adsorption of hydrophobic contaminants from the air. The adsorption of polymeric organosilicon compounds was recently found to be important as the most probable reason for the wettability transition of such samples in the laboratory environment [60]. The air-exposed surfaces were then cleaned with a UV/ozone cleaner for 30 min. This removes most typical contaminants, such as oils and greases, and other contamination adsorbed during prolonged exposure to air [61]. The wettability of all surfaces increased dramatically after the UV/ozone cleaning, but after exposing them to the ambient air conditions for a further 3 days, the wettability again decreased.

3.4. Comparison of the Overall Boiling Performance

It was observed that regardless of the size of the nanoparticle, an increase in concentration had the same effect on the CHF and HTC, which agrees with the findings of previous research [33,35,62–64]. The research groups in this field concluded that an increase in nanoparticle concentration leads to the creation of a fouling layer on the surface, which deteriorates HTC due to a decrease in active nucleation site density. In our case, the laser-induced microcavities were filled with nanoparticles, leading to a decrease in contact angle and a reduction in the number of cavities suitable for bubble nucleation [65,66]. Additionally, the adhesion energy, which is defined as the horizontal component of the surface tension force acting against the bubble growth, significantly increased with nanoparticle deposition. This caused the bubble departure frequency to decrease and prolonged their growth times. On top of that, the fluid was wetting the entire deposited layer, which lead to deterioration of the active nucleation sites on the surface [67]. On the other hand, the layer formed on the surface improved the wettability, capillary wicking action, and constitution of inflow liquid inside the fouling layer [62,63,68], which effectively increased the achievable heat flux values without CHF incipience.

A comparison of the boiling performance of nanofluids with small nanoparticles with the performance of twice-distilled water on the reference and LT surface is shown in Figure 13. Nanofluids at both concentrations deteriorated the heat transfer parameters, in comparison with the boiling of pure water, on the LT surface. Additionally, at the low concentration, the boiling performance also decreased compared to the boiling performance of water on the untreated reference surface. The highest deterioration in HTC recorded at CHF was measured to be 60% at the low and 89.5% at the high nanoparticle concentration compared to the HTC recorded on the LT surface with the base fluid (water) at CHF.

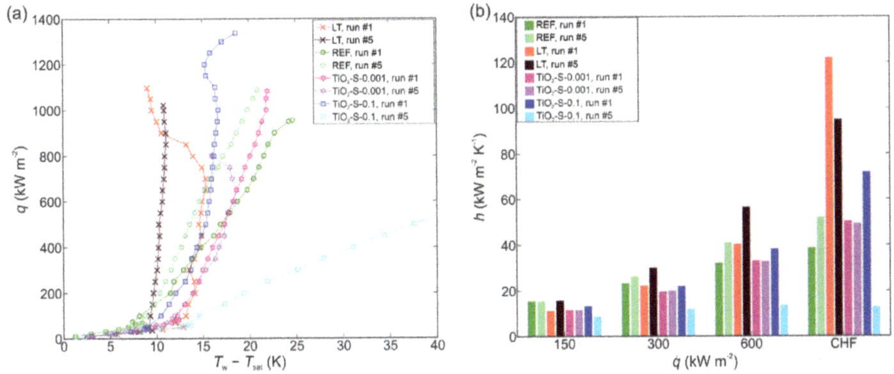

Figure 13. Comparison of boiling characteristics of TiO_2-water nanofluids with small size nanoparticles with the performance of pure water on reference and LT surface (1st and 5th run): boiling curves (**a**) and heat transfer coefficients (**b**).

Furthermore, Figure 14 shows a comparison of the boiling performance of nanofluids with the performance of twice-distilled water on the reference and LT surface. Here, a deterioration of the HTC when using the nanofluid instead of pure water is also recorded.

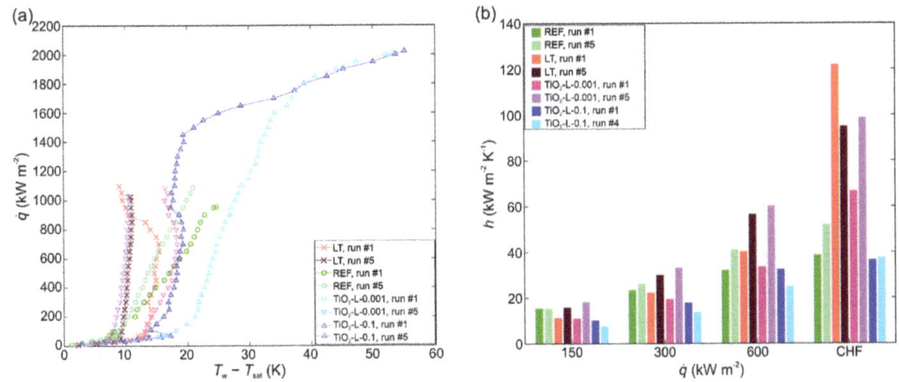

Figure 14. Comparison of boiling characteristics of TiO_2-water nanofluids-small size NPs with base fluid on the reference and LT surface (1st and 5th run): boiling curves (**a**) and HTCs (**b**).

Figure 15 shows a comparison of the performance of the laser-textured surface, when used either with water (denoted as LT), or with nanofluids, relative to the boiling performance of pure water on the untreated reference surface (REF). At low and medium heat flux values, the reference surface provides superior performance during the first experimental run and only at high heat fluxes, the laser-texturing and nanoparticle deposition enhance the HTC (Figure 15a). However, after 5 experimental runs were performed on each surface, the performance of the laser-textured surface used with pure water or 0.001 wt.% large nanoparticle nanofluid significantly exceeds the performance of the untreated surfaces. This is mainly attributed to the increased number of potential active nucleation sites on both surfaces, as shown in the SEM images and discussed previously. On the other hand, the deterioration of HTC is exacerbated on other surfaces, with its values being up to 75% lower compared to those of pure water boiling on the untreated surface. The main cause for the deterioration of boiling performance is the growth of a thick nanoparticle deposit layer, which raises the surface temperature due to increased thermal resistance of the deposited titanium dioxide layer. The obtained results are in accordance with the findings of other authors for large nanoparticles and mixed sizes of nanoparticles [69], and also for nanofluids with small nanoparticle diameters [70].

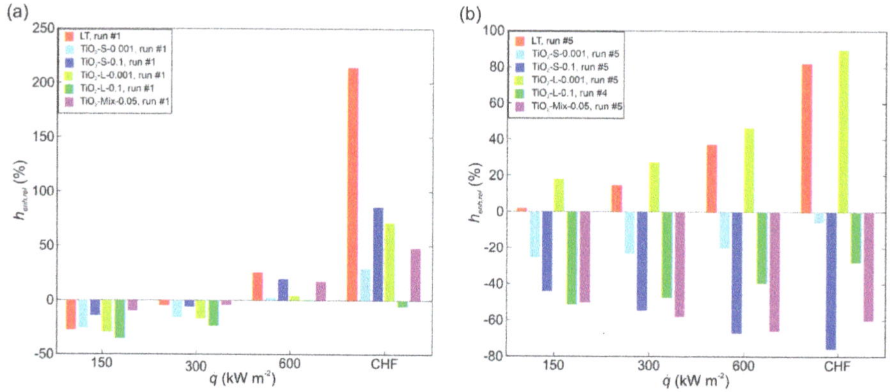

Figure 15. Comparison of HTC for different boiling media (pure water or nanofluids) relative to the performance of the untreated reference surface during the 1st run (**a**) and during the 5th run (**b**).

Significant enhancement of the CHF was not observed at 0.001 wt.% compared to the CHF for pure water on either the laser-textured or untreated surface, while the highly concentrated nanofluid evidently caused thicker deposits, resulting in significant increases in the CHF value. Enhanced CHF can be mainly be attributed to the increased wettability [69,71] of the surfaces and their porosity; this also agrees with the contact angle measurements after boiling, shown in Figure 12.

Finally, a comparison of all boiling curves for the first and last experimental run is made in Figure 16a,b, respectively, to elucidate the effect of both nanoparticle concentration and size on boiling performance. Additionally, the corresponding plots comparing HTC values at selected heat fluxes are shown in Figure 16c,d.

During the first experimental run, an enhancement of boiling performance in terms of increased HTC at medium and high heat flux values was observed for the laser-textured surface tested with pure water, for both mixed nanoparticle size nanofluid and for both low-concentration nanofluids (0.001 wt.%). A noticeable CHF increase was obtained with highly concentrated nanofluids (0.1 wt.%), but high surface superheat values were already recorded for the nanofluid with large nanoparticles, hinting at thicker deposits and problems during future testing. It can be concluded that large nanoparticles agglomerate much faster, which improves the deposition rate, leading to a reduction in the number of active nucleation sites. This observation is not in accordance with previous research, which concluded that if the size of nanoparticles is much smaller than the roughness of the surface, an enhancement of boiling performance will be achieved. While even the large nanoparticles are approximately two orders of magnitude smaller than the laser-induced roughness, a deterioration in boiling performance was observed, matching previous reports for larger ratios of nanoparticle size to surface roughness [70,72]. Here, we report that regardless of the sizes of nanoparticles, the HTC of nanofluids at the higher concentration on laser-textured surface decreases compared to the boiling of water on the same surface. The last experimental run on each surface revealed that highly concentrated nanofluids provided a notable CHF enhancement, but universally at the expense of notable surface superheat increase, which reached 100 K for nanofluid containing 0.1 wt.% of small nanoparticles. Overall, CHF enhancement due to the boiling of nanofluid on laser-textured surfaces was not observed at the low concentration.

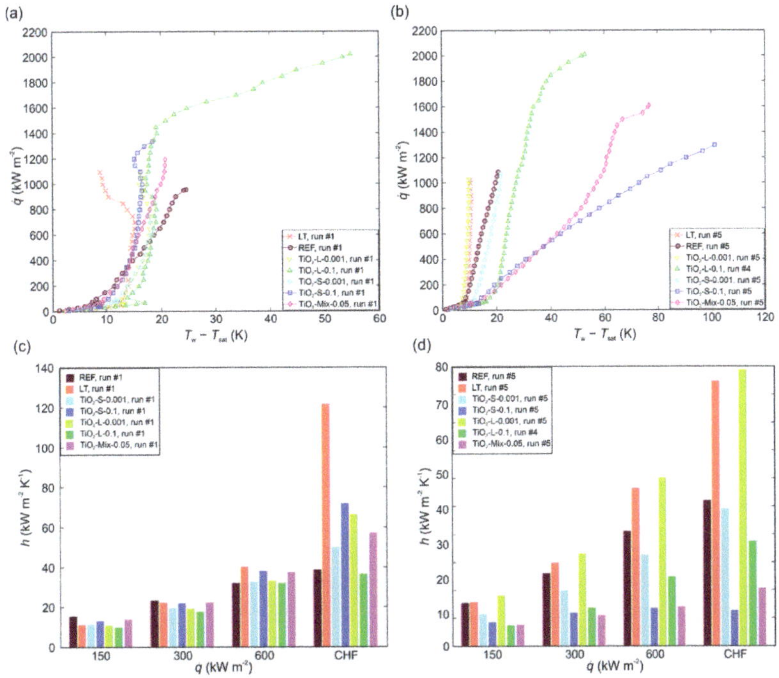

Figure 16. Summary of boiling performance for all tested cases: boiling curves during the first run (**a**) and during the last run (**b**) on each surface; HTC values during the first run (**c**) and during the last run (**d**) on each surface.

4. Conclusions

The pool boiling performance of TiO_2-water nanofluids with two different sizes of nanoparticles (4–8 nm and 490 nm) at two different concentrations (0.001 wt.% and 0.1 wt.%) was investigated on laser-textured copper surfaces. The following conclusions were made based on the results of the performed experiments:

1. The boiling of nanofluids on an already enhanced (i.e., laser-textured) surface failed to provide a notable (additional) enhancement of the heat transfer coefficient.
2. At a low nanoparticle concentration, the influence of nanofluid on boiling performance is minimal, with heat transfer coefficient and CHF values comparable to those obtained using pure water on both the untreated and laser-textured surface.
3. The boiling of a nanofluid with a high nanoparticle concentration resulted in significant deposition of nanoparticles onto the boiling surface and CHF enhancement up to 2021 kW m^{-2}, representing double the value obtained on the untreated reference surface using water. However, very high surface superheat values (up to 100 K) were recorded, suggesting poor practical applicability.
4. The decrease in heat transfer performance due to the boiling of nanofluids on laser-textured surfaces can be explained through the deposition of nanoparticles into the laser-induced grooves and microcavities present on the surface, which decreased the number of active nucleation sites. Furthermore, thicker nanoparticle deposits resulted in added thermal resistance. While the surface porosity granted a notable delay in CHF incipience due to enhanced liquid replenishment, the surface superheat was massively increased.

In summary, the use of nanofluids on surfaces previously functionalized with surface treatment methods (e.g., laser texturing) was generally proved to be unable to provide an additional viable heat transfer performance enhancement.

Author Contributions: Conceptualization, M.M., I.G. and A.H.; methodology, M.M., K.A., A.H. and M.Z.; software, M.M. and A.H.; validation, K.A. and I.G.; investigation, A.H.; resources, I.G.; data curation, A.H., M.M. and M.Z.; writing—original draft preparation, A.H., M.M., M.Z. and K.A.; writing—review and editing, A.H., M.M. and I.G.; visualization, M.M. and K.A.; supervision, M.Z. and I.G.; project administration, I.G.; funding acquisition, M.Z. and I.G. All authors have read and agreed to the published version of the manuscript.

Funding: The authors acknowledge the financial support from the state budget of the Slovenian Research Agency (Program No. P2-0223 and Project Nos. J2-2486 and N2-0251).

Data Availability Statement: All data is available from the corresponding author upon request.

Conflicts of Interest: The authors declare no conflict of interest.

References

1. Dhir, V.K. Nucleate and Transition Boiling Heat Transfer Under Pool and External Flow Conditions. *Int. J. Heat Fluid Flow* **1991**, *12*, 129–155. [CrossRef]
2. Liang, G.; Mudawar, I. Review of spray cooling—Part 1: Single-phase and nucleate boiling regimes, and critical heat flux. *Int. J. Heat Mass Transf.* **2017**, *115*, 1174–1205. [CrossRef]
3. Mori, S.; Utaka, Y. Critical heat flux enhancement by surface modification in a saturated pool boiling: A review. *Int. J. Heat Mass Transf.* **2017**, *108*, 2534–2557. [CrossRef]
4. Zhang, C.; Yu, F.; Li, X.; Chen, Y. Gravity–capillary evaporation regimes in microgrooves. *AIChE J.* **2019**, *65*, 1119–1125. [CrossRef]
5. Zhang, C.; Chen, Y.; Wu, R.; Shi, M. Flow boiling in constructal tree-shaped minichannel network. *Int. J. Heat Mass Transf.* **2011**, *54*, 202–209. [CrossRef]
6. Mudawar, I. Recent advances in high-flux, two-phase thermal management. *J. Therm. Sci. Eng. Appl.* **2013**, *5*, 021012. [CrossRef]
7. Nukiyama, S. The maximum and minimum values of the heat Q transmitted from metal to boiling water under atmospheric pressure. *Int. J. Heat Mass Transf.* **1966**, *9*, 1419–1433. [CrossRef]
8. Lahey, R.T., Jr. *Boiling Heat Transfer: Modern Developments and Advances*, 1st ed.; Elsevier Science: Amsterdam, The Netherlands, 1992.
9. Hewitt, G.F.; Tien, C.L. *Series in Chemical and Mechanical Engineering Tong and Tang, Boiling Heat Transfer and Two-Phase Flow*, 2nd ed.; Routledge: New York, NY, USA, 1997.
10. Malavasi, I.; Teodori, E.; Moita, A.S.; Moreira, A.L.N.; Marengo, M. Wettability Effect on Pool Boiling: A Review. In *Encyclopedia of Two-Phase Heat Transfer and Flow III*; World Scientific Publishing Co. Pte Ltd.: Singapore, 2018; pp. 1–61.
11. Liang, G.; Mudawar, I. Review of pool boiling enhancement by surface modification. *Int. J. Heat Mass Transf.* **2019**, *128*, 892–933. [CrossRef]
12. Pioro, I.L.; Rohsenow, W.; Doerffer, S.S. Nucleate pool-boiling heat transfer. I: Review of parametric effects of boiling surface. *Int. J. Heat Mass Transf.* **2004**, *47*, 5033–5044. [CrossRef]
13. Može, M.; Vajc, V.; Zupančič, M.; Golobič, I. Hydrophilic and hydrophobic nanostructured copper surfaces for efficient pool boiling heat transfer with water, water/butanol mixtures and Novec 649. *Nanomaterials* **2021**, *11*, 3216. [CrossRef]
14. Zupančič, M.; Može, M.; Gregorčič, P.; Golobič, I. Nanosecond laser texturing of uniformly and non-uniformly wettable micro structured metal surfaces for enhanced boiling heat transfer. *Appl. Surf. Sci.* **2017**, *399*, 480–490. [CrossRef]
15. Chu, K.H.; Enright, R.; Wang, E.N. Structured surfaces for enhanced pool boiling heat transfer. *Appl. Phys. Lett.* **2012**, *100*, 241603. [CrossRef]
16. Ji, X.; Xu, J.; Zhao, Z.; Yang, W. Pool boiling heat transfer on uniform and non-uniform porous coating surfaces. *Exp. Therm. Fluid Sci.* **2013**, *48*, 198–212. [CrossRef]
17. Li, J.Q.; Mou, L.W.; Zhang, J.Y.; Zhang, Y.H.; Fan, L.W. Enhanced pool boiling heat transfer during quenching of water on superhydrophilic porous surfaces: Effects of the surface wickability. *Int. J. Heat Mass Transf.* **2018**, *125*, 494–505. [CrossRef]
18. Lee, C.Y.; Zhang, B.J.; Kim, K.J. Morphological change of plain and nano-porous surfaces during boiling and its effect on nucleate pool boiling heat transfer. *Exp. Therm. Fluid Sci.* **2012**, *40*, 150–158. [CrossRef]
19. Shi, J.; Jia, X.; Feng, D.; Chen, Z.; Dang, C. Wettability effect on pool boiling heat transfer using a multiscale copper foam surface. *Int. J. Heat Mass Transf.* **2020**, *146*, 118726. [CrossRef]
20. Li, S.; Furberg, R.; Toprak, M.S.; Palm, B.; Muhammed, M. Nature-inspired boiling enhancement by novel nanostructured macroporous surfaces. *Adv. Funct. Mater.* **2008**, *18*, 2215–2220. [CrossRef]
21. Lee, S.W.; Park, S.D.; Bang, I.C. Critical heat flux for CuO nanofluid fabricated by pulsed laser ablation differentiating deposition characteristics. *Int. J. Heat Mass Transf.* **2012**, *55*, 6908–6915. [CrossRef]
22. Ahn, H.S.; Sathyamurthi, V.; Banerjee, D. Pool boiling experiments on a nano-structured surface. *IEEE Trans. Compon. Packag. Technol.* **2009**, *32*, 156–165. [CrossRef]
23. Gupta, S.K.; Misra, R.D. Experimental study of pool boiling heat transfer on copper surfaces with Cu-Al_2O_3 nanocomposite coatings. *Int. Commun. Heat Mass Transf.* **2018**, *97*, 47–55. [CrossRef]
24. O'Connor, J.P.; You, S.M.; Chang, J.Y. Gas-Saturated Pool Boiling Heat Transfer from Smooth and Microporous Surfaces in FC-72. *ASME J. Heat Transf.* **1996**, *118*, 662–667. [CrossRef]

25. Takata, Y.; Hidaka, S.; Masuda, M.; Ito, T. Pool boiling on a superhydrophilic surface. *Int. J. Energy Res.* **2003**, *27*, 111–119. [CrossRef]
26. O'Hanley, H.; Coyle, C.; Buongiorno, J.; McKrell, T.; Hu, L.W.; Rubner, M.; Cohen, R. Separate effects of surface roughness, wettability, and porosity on the boiling critical heat flux. *Appl. Phys. Lett.* **2013**, *103*, 024102. [CrossRef]
27. Kim, S.J.; Bang, I.C.; Buongiorno, J.; Hu, L.W. Effects of nanoparticle deposition on surface wettability influencing boiling heat transfer in nanofluids. *Appl. Phys. Lett.* **2006**, *89*, 153107. [CrossRef]
28. Jo, H.; Ahn, H.S.; Kang, S.; Kim, M.H. A study of nucleate boiling heat transfer on hydrophilic, hydrophobic and heterogeneous wetting surfaces. *Int. J. Heat Mass Transf.* **2011**, *54*, 5643–5652. [CrossRef]
29. Phan, H.T.; Caney, N.; Marty, P.; Colasson, S.; Gavillet, J. Surface wettability control by nanocoating: The effects on pool boiling heat transfer and nucleation mechanism. *Int. J. Heat Mass Transf.* **2009**, *52*, 5459–5471. [CrossRef]
30. Barber, J.; Brutin, D.; Tadrist, L. A review on boiling heat transfer enhancement with nanofluids. *Nanoscale Res. Lett.* **2011**, *6*, 280. [CrossRef]
31. Fang, X.; Chen, Y.; Zhang, H.; Chen, W.; Dong, A.; Wang, R. Heat transfer and critical heat flux of nanofluid boiling: A comprehensive review. *Renew. Sustain. Energy Rev.* **2016**, *62*, 924–940. [CrossRef]
32. Gerardi, C.; Buongiorno, J.; Hu, L.-W.; McKrell, T. Infrared thermometry study of nanofluid pool boiling phenomena. *Nanoscale Res. Lett.* **2011**, *6*, 232. [CrossRef]
33. Vafaei, S.; Borca-Tasciuc, T. Role of nanoparticles on nanofluid boiling phenomenon: Nanoparticle deposition. *Chem. Eng. Res. Des.* **2014**, *92*, 842–856. [CrossRef]
34. Murshed, S.M.S.; Nieto De Castro, C.A.; Loureno, M.J.V.; Lopes, M.L.M.; Santos, F.J.V. A review of boiling and convective heat transfer with nanofluids. *Renew. Sustain. Energy Rev.* **2011**, *15*, 2342–2354. [CrossRef]
35. Khan, A.; Ali, H.M. A comprehensive review pool boiling using nanofluids. *Therm. Sci.* **2019**, *23*, 3209–3237. [CrossRef]
36. Dadhich, M.; Prajapati, O.S. A brief review on factors affecting flow and pool boiling. *Renew. Sustain. Energy Rev.* **2019**, *112*, 607–625. [CrossRef]
37. Manetti, L.L.; Stephen, M.T.; Beck, P.A.; Cardoso, E.M. Evaluation of the heat transfer enhancement during pool boiling using low concentrations of Al2O3-water based nanofluid. *Exp. Therm. Fluid Sci.* **2017**, *87*, 191–200. [CrossRef]
38. Ahmed, O.; Hamed, M.S. Experimental investigation of the effect of particle deposition on pool boiling of nanofluids. *Int. J. Heat Mass Transf.* **2012**, *55*, 3423–3436. [CrossRef]
39. Huang, C.K.; Lee, C.W.; Wang, C.K. Boiling enhancement by TiO_2 nanoparticle deposition. *Int. J. Heat Mass Transf.* **2011**, *54*, 4895–4903. [CrossRef]
40. Kiyomura, I.S.; Manetti, L.L.; da Cunha, A.P.; Ribatski, G.; Cardoso, E.M. An analysis of the effects of nanoparticles deposition on characteristics of the heating surface and ON pool boiling of water. *Int. J. Heat Mass Transf.* **2017**, *106*, 666–674. [CrossRef]
41. Salimpour, M.R.; Abdollahi, A.; Afrand, M. An experimental study on deposited surfaces due to nanofluid pool boiling: Comparison between rough and smooth surfaces. *Exp. Therm. Fluid Sci.* **2017**, *88*, 288–300. [CrossRef]
42. Taylor, R.A.; Phelan, P.E. Pool boiling of nanofluids: Comprehensive review of existing data and limited new data. *Int. J. Heat Mass Transf.* **2009**, *52*, 5339–5347. [CrossRef]
43. Park, S.D.; Moon, S.B.; Bang, I.C. Effects of thickness of boiling-induced nanoparticle deposition on the saturation of critical heat flux enhancement. *Int. J. Heat Mass Transf.* **2014**, *78*, 506–514. [CrossRef]
44. Yim, K.; Lee, J.; Naccarato, B.; Kim, K.J. Surface wettability effect on nucleate pool boiling heat transfer with titanium oxide (TiO2) coated heating surface. *Int. J. Heat Mass Transf.* **2019**, *133*, 352–358. [CrossRef]
45. Može, M.; Senegačnik, M.; Gregorčič, P.; Hočevar, M.; Zupančič, M.; Golobič, I. Laser-Engineered Microcavity Surfaces with a Nanoscale Superhydrophobic Coating for Extreme Boiling Performance. *ACS Appl. Mater. Interfaces* **2020**, *12*, 24419–24431. [CrossRef]
46. Kruse, C.M.; Anderson, T.; Wilson, C.; Zuhlke, C.; Alexander, D.; Gogos, G.; Ndao, S. Enhanced pool-boiling heat transfer and critical heat flux on femtosecond laser processed stainless steel surfaces. *Int. J. Heat Mass Transf.* **2015**, *82*, 109–116. [CrossRef]
47. Zakšek, P.; Zupančič, M.; Gregorčič, P.; Golobič, I. Investigation of Nucleate Pool Boiling of Saturated Pure Liquids and Ethanol-Water Mixtures on Smooth and Laser-Textured Surfaces. *Nanoscale Microscale Thermophys. Eng.* **2020**, *24*, 29–42. [CrossRef]
48. Serdyukov, V.; Starinskiy, S.; Malakhov, I.; Safonov, A.; Surtaev, A. Laser texturing of silicon surface to enhance nucleate pool boiling heat transfer. *Appl. Therm. Eng.* **2021**, *194*, 117102. [CrossRef]
49. Može, M.; Zupančič, M.; Hočevar, M.; Golobič, I.; Gregorčič, P. Surface chemistry and morphology transition induced by critical heat flux incipience on laser-textured copper surfaces. *Appl. Surf. Sci.* **2019**, *490*, 220–230. [CrossRef]
50. Karthikeyan, A.; Coulombe, S.; Kietzig, A.M. Boiling heat transfer enhancement with stable nanofluids and laser textured copper surfaces. *Int. J. Heat Mass Transf.* **2018**, *126*, 287–296. [CrossRef]
51. Freitas, E.; Pontes, P.; Cautela, R.; Bahadur, V.; Miranda, J.; Ribeiro, A.P.C.; Souza, R.R.; Oliveira, J.D.; Copetti, J.B.; Lima, R.; et al. Article pool boiling of nanofluids on biphilic surfaces: An experimental and numerical study. *Nanomaterials* **2021**, *11*, 125. [CrossRef]
52. Pontes, P.; Freitas, E.; Fernandes, D.; Teixeira, A.; Ferreira, R.; Bellmann, S.; Cautela, R.; Moita, A.S.; Bahadur, V.; Miranda, J.; et al. Pool boiling of nanofluids on biphilic surfaces. In Proceedings of the 8th European Thermal Sciences Conference (EUROTHERMAL 2021), Virtual, 20–22 September 2021; Journal of Physics: Conference Series. 2021; Volume 2116. [CrossRef]

53. Može, M.; Vajc, V.; Zupančič, M.; Šulc, R.; Golobič, I. Pool boiling performance of water and self-rewetting fluids on hybrid functionalized aluminum surfaces. *Processes* **2021**, *9*, 1058. [CrossRef]
54. Može, M.; Zupančič, M.; Golobič, I. Investigation of the scatter in reported pool boiling CHF measurements including analysis of heat flux and measurement uncertainty evaluation methodology. *Appl. Therm. Eng.* **2020**, *169*, 114938. [CrossRef]
55. Može, M.; Zupančič, M.; Golobič, I. Pattern geometry optimization on superbiphilic aluminum surfaces for enhanced pool boiling heat transfer. *Int. J. Heat Mass Transf.* **2020**, *161*, 120265. [CrossRef]
56. Gregorčič, P.; Conradi, M.; Hribar, L.; Hočevar, M. Long-term influence of laser-processing parameters on (Super)hydrophobicity development and stability of stainless-steel surfaces. *Materials* **2018**, *11*, 2240. [CrossRef]
57. Trdan, U.; Hočevar, M.; Gregorčič, P. Transition from superhydrophilic to superhydrophobic state of laser textured stainless steel surface and its effect on corrosion resistance. *Corros. Sci.* **2017**, *123*, 21–26. [CrossRef]
58. Ta, D.V.; Dunn, A.; Wasley, T.J.; Kay, R.W.; Stringer, J.; Smith, P.J.; Connaughton, C.; Shephard, J.D. Nanosecond laser textured superhydrophobic metallic surfaces and their chemical sensing applications. *Appl. Surf. Sci.* **2015**, *357*, 248–254. [CrossRef]
59. Long, J.; Zhong, M.; Fan, P.; Gong, D.; Zhang, H. Wettability conversion of ultrafast laser structured copper surface. *J. Laser Appl.* **2015**, *27*, S29107. [CrossRef]
60. Gregorčič, P. Comment on "Bioinspired Reversible Switch between Underwater Superoleophobicity/Superaerophobicity and Oleophilicity/Aerophilicity and Improved Antireflective Property on the Nanosecond Laser-Ablated Superhydrophobic Titanium Surfaces". *ACS Appl. Mater. Interfaces* **2020**, *13*, 2117–2127. [CrossRef]
61. Vig, J.R.; Bus, J.W.L. UV/Ozone Cleaning of Surfaces. *IEEE Trans. Parts Hybrids Packag.* **1976**, *12*, 365–370. [CrossRef]
62. Vafaei, S. Nanofluid pool boiling heat transfer phenomenon. *Powder Technol.* **2015**, *277*, 181–192. [CrossRef]
63. Sarafraz, M.M.; Hormozi, F.; Peyghambarzadeh, S.M. Pool boiling heat transfer to aqueous alumina nano-fluids on the plain and concentric circular micro-structured (CCM) surfaces. *Exp. Therm. Fluid Sci.* **2016**, *72*, 125–139. [CrossRef]
64. Coursey, J.S.; Kim, J. Nanofluid boiling: The effect of surface wettability. *Int. J. Heat Fluid Flow* **2008**, *29*, 1577–1585. [CrossRef]
65. Park, K.J.; Jung, D.; Shim, S.E. Nucleate boiling heat transfer in aqueous solutions with carbon nanotubes up to critical heat fluxes. *Int. J. Multiph. Flow* **2009**, *35*, 525–532. [CrossRef]
66. Kathiravan, R.; Kumar, R.; Gupta, A.; Chandra, R. Characterization and pool boiling heat transfer studies of nanofluids. *J. Heat Transf.* **2009**, *131*, 081902. [CrossRef]
67. Ciloglu, D.; Bolukbasi, A. A comprehensive review on pool boiling of nanofluids. *Appl. Therm. Eng.* **2015**, *84*, 45–63. [CrossRef]
68. Kim, H.D.; Kim, M.H. CHF Enhancement in Pool Boiling of Nanofluid: Effect of Nanoparticle-Coating on Heating Surface. In Proceedings of the KNS spring meeting, Jeju, Korea, 26–27 May 2005.
69. Suriyawong, A.; Wongwises, S. Nucleate pool boiling heat transfer characteristics of TiO_2-water nanofluids at very low concentrations. *Exp. Therm. Fluid Sci.* **2010**, *34*, 992–999. [CrossRef]
70. Hu, Y.; Li, H.; He, Y.; Liu, Z.; Zhao, Y. Effect of nanoparticle size and concentration on boiling performance of SiO_2 nanofluid. *Int. J. Heat Mass Transf.* **2017**, *107*, 820–828. [CrossRef]
71. Soltani, S.; Etemad, S.G.; Thibault, J. Pool boiling heat transfer performance of Newtonian nanofluids. *Heat Mass Transf. Stoffuebertragung* **2009**, *45*, 1555–1560. [CrossRef]
72. Das, S.K.; Prakash Narayan, G.; Baby, A.K. Survey on nucleate pool boiling of nanofluids: The effect of particle size relative to roughness. *J. Nanoparticle Res.* **2008**, *10*, 1099–1108. [CrossRef]

nanomaterials

MDPI

Article

Appearance of a Solitary Wave Particle Concentration in Nanofluids under a Light Field

Abram I. Livashvili [1], Victor V. Krishtop [2,*], Polina V. Vinogradova [1], Yuriy M. Karpets [1], Vyacheslav G. Efremenko [1], Alexander V. Syuy [3], Evgenii N. Kuzmichev [4] and Pavel V. Igumnov [4]

1 Institute of Natural Sciences, Far Eastern State Transport University, 47, Seryshev St., 680021 Khabarovsk, Russia; livbru@mail.ru (A.I.L.); vpolina17@hotmail.com (P.V.V.); kjum1947@mail.ru (Y.M.K.); oblako3@yandex.ru (V.G.E.)
2 Department of General Physics, Perm National Research Polytechnic University, 29, Komsomolsky Prospekt, 614990 Perm, Russia
3 Department of General Physics, Moscow Institute of Physics and Technology, 9, Institutskiy Per., 141701 Dolgoprudny, Russia; alsyuy271@gmail.com
4 Institute of Materials Technology of Khabarovsk Centre of FEC The Russian Academy of Sciences, 153, Tihookeanskaya St., 680042 Khabarovsk, Russia; e_kuzmichev@mail.ru (E.N.K.); 407320@mail.ru (P.V.I.)
* Correspondence: krishtop@list.ru

Abstract: In this study, the nonlinear dynamics of nanoparticle concentration in a colloidal suspension (nanofluid) were theoretically studied under the action of a light field with constant intensity by considering concentration convection. The heat and nanoparticle transfer processes that occur in this case are associated with the phenomenon of thermal diffusion, which is considered to be positive in our work. Two exact analytical solutions of a nonlinear Burgers-Huxley-type equation were derived and investigated, one of which was presented in the form of a solitary concentration wave. These solutions were derived considering the dependence of the coefficients of thermal conductivity, viscosity, and absorption of radiation on the nanoparticle concentration in the nanofluid. Furthermore, an expression was obtained for the solitary wave velocity, which depends on the absorption coefficient and intensity of the light wave. Numerical estimates of the concentration wave velocity for a specific nanofluid—water/silver—are given. The results of this study can be useful in the creation of next-generation solar collectors.

Keywords: nanofluid; nonlinear dynamics; colloidal suspension; solitary wave; Burgers–Huxley equation

Citation: Livashvili, A.I.; Krishtop, V.V.; Vinogradova, P.V.; Karpets, Y.M.; Efremenko, V.G.; Syuy, A.V.; Kuzmichev, E.N.; Igumnov, P.V. Appearance of a Solitary Wave Particle Concentration in Nanofluids under a Light Field. *Nanomaterials* **2021**, *11*, 1291. https://doi.org/10.3390/nano11051291

Academic Editors: Patrice Estellé and S M Sohel Murshed

Received: 7 April 2021
Accepted: 11 May 2021
Published: 14 May 2021

1. Introduction

In recent years, the optical properties of colloidal suspensions (nanofluids) have been actively studied [1–6]. Researchers are particularly interested in nonlinear optical effects that are realized in such media. In particular, studies have focused on four-wave interactions and the self-action of light waves [7–10]. Without detailed knowledge of the optical properties of nanofluids, it is impossible to create next-generation solar collectors [11–15]. For example, [16] summarized the results of studies on the nanocolloids of ionic liquids (i.e., ionic liquids with nanoparticles in suspension), which can be directly applied to convective heat transfer. In [17], machine learning was used to develop Gaussian process regression models to describe the statistical correlations between the thermal conductivity and physical parameters of two-phase nanofluid components. For this purpose, approximately 300 samples of nanofluids, dispersions of metal, and ceramic nanoparticles in water, ethylenecol, and transformer oil have been investigated. The modeling approach demonstrates a high degree of accuracy and stability, facilitating efficient and inexpensive thermal conductivity estimates. Work [18] considered a liquid consisting of a stable colloidal suspension of magnetic maghemite nanoparticles in water. It has been found that these nanoparticles constitute an excellent absorber of solar radiation and simultaneously an amplifier of thermoelectric power output with a very small volume fraction when the liquid is heated from above. These results demonstrate that the investigated nanofluid has great potential as a coolant for the co-production of heat and energy in completely new hybrid flat solar thermal collectors, for which top heating geometry is required. The main mechanisms for optical nonlinearity in these cases are the phenomena of thermal diffusion and electrostriction of nanoparticles [19,20]. Despite the many studies on this problem [21–25], several questions still remain. In particular, the dynamics of the concentration of nanofluid particles are unknown in the presence of concentration dependences on the coefficients of the thermal conductivity, viscosity, and absorption of radiation of the medium. Providing a theoretical description of the processes of heat and mass transfer for the nanofluid and radiation system is fraught with serious mathematical difficulties that are associated with the search for analytical solutions of the corresponding nonlinear equations. In this study, we developed a theoretical model for the dynamics of the concentration of nanoparticles in a liquid-phase medium when subjected to constant-intensity laser irradiation. Further, the study considers the dependence of the coefficients of absorption of radiation, thermal conductivity, and viscosity of the medium on the concentration of nanoparticles. It should be noted that, in the works cited above, the dynamics of the concentration of nanoparticles were studied assuming constant values of these coefficients.

2. Theoretical Model

We consider that the particle sizes satisfy the following condition: $a_0 << \lambda$, where a_0 is the linear size; and λ is the wavelength of light. Thus, we do not consider diffraction and light scattering processes. We also exclude the processes associated with particle sedimentation.

Let us consider a liquid-phase medium with nanoparticles irradiated by a light beam of intensity I_0 that is uniformly distributed over a region (Figure 1).

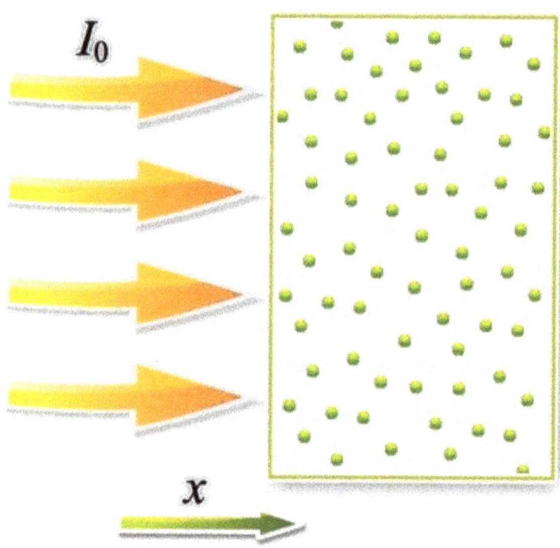

Figure 1. Geometry of the problem.

Temperature and concentration gradients arise as a result of the action of the light field in the medium, and are then used to determine the heat- and mass-transfer processes (Soret effect). These phenomena are described by a system of balanced equations for the temperature and particles [26,27].

We define the system of balanced equations for heat conduction and the mass of nanoparticles transferred as follows:

$$C_p\rho\frac{\partial T}{\partial t} = div(\lambda(C)\ (\overrightarrow{grad}T)) + \alpha(C)I_0, \tag{1}$$

$$\frac{\partial C}{\partial t} = div(D\overrightarrow{grad}C) + D_T div(C(1-C)\overrightarrow{grad}T) - \overrightarrow{V}\cdot\overrightarrow{grad}C, \tag{2}$$

where T is the temperature of the medium; C is the volume concentration of the medium; $\lambda(C)$ is the thermal conductivity of the medium; $\alpha(C)$ is the absorption coefficient of the light wave; D is the diffusion coefficient of nanoparticles; D_T Tis the thermal diffusion coefficient; V is the concentration convection velocity; and C_p and ρ are known thermophysical constants. It should be noted that, in Equation (2), we take into account the incompressibility of the nanofluid: $div\ \overrightarrow{V} = 0$ [27].

We now consider the one-dimensional case, neglecting the Dufour effect owing to its small contribution. We do not consider flows caused by the forces of pressure on the particles from the side of the light field. In further calculations, we assume:

$$div\left(\lambda(C)\frac{\partial T}{\partial x}\right) \approx \lambda(C)\frac{\partial^2 T}{\partial x^2},\ div\left(D\frac{\partial C}{\partial x}\right) \approx D\frac{\partial C}{\partial x}, \tag{3}$$

$$div\left(C(1-C)\overrightarrow{grad}T\right) \approx C(1-C)\frac{\partial^2 T}{\partial x^2}, \tag{4}$$

The validity of these approximations can be verified by direct calculations. We study the dynamics of nanoparticles against the background of the stationary temperature of the medium, i.e., $\partial T/\partial t = 0$ (thermal processes are assumed to be 2–3 orders of magnitude faster than diffusion). We focus on processes with $C \ll 1$; this inequality ensures that the coagulation (coalescence) of nanoparticles can be disregarded.

According to theoretical and experimental studies [28,29], the concentration dependence of the thermal conductivity of a medium at low concentrations can be considered to be linear, as follows:

$$\lambda(C) = \lambda_0(1 + pC), \tag{5}$$

where λ_0 is the value of the thermal conductivity coefficient of the fluid (without nanoparticles), and p is a linear coefficient. We consider the concentration dependence of the light absorption coefficient to be of the form: $\alpha = \beta C$ (where β exceeds zero). Given the stationary temperature regime, the approximations (Equations (4) and (5)), and low concentration, we obtain the following from the heat equation:

$$\frac{\partial^2 T}{\partial x^2} = \frac{-\beta C}{\lambda_0(1+pC)}I_0 \approx -\frac{\beta C}{\lambda_0}I_0(1-pC)(pC<1), \tag{6}$$

Using the approximations in Equations (3), (4) and (6), Equation (2) can be rewritten as follows:

$$\frac{\partial C}{\partial t} = D\frac{\partial^2 C}{\partial x^2} - \frac{D_T\beta I_0}{\lambda_0}(1-pC)C^2 - V\frac{\partial C}{\partial x}, \tag{7}$$

For a complete description of the transport processes in the system under consideration, Equation (7) must be supplemented by the Navier-Stokes equation (to determine the velocity, V). In this case, the formulated problem can be solved numerically [25]. However, here, we use a different approach to derive the analytical solution. In particular, we represent the convective velocity in the following form:

$$V(C) = \frac{\eta(C)}{\rho(C)l}, \tag{8}$$

where $\eta(C)$ is the dynamic viscosity coefficient of the nanofluid; $\rho(C)$ is its density; and l is the characteristic length of the system, the value of which is determined later.

We consider the dependence of the viscosity coefficient on concentration to be linear, such that:

$$\eta(C) = \eta_0(1+\gamma C), \tag{9}$$

where η_0 is the value of the viscosity coefficient of the base fluid devoid of nanoparticles. A similar dependence was obtained theoretically and experimentally, as confirmed in previous studies [28–30]. As for $\rho(C)$, a linear dependence on concentration is also permissible here [31,32]:

$$\rho = \rho_0(1+\chi C), \tag{10}$$

where ρ_0 is the average density of the medium, and χ is the coefficient of the concentration expansion. As $\gamma \gg \chi$ is real, we consider the density dependence on concentration to be insignificant.

Therefore, the expression for the velocity (Equation (8)) can be represented using Equations (9) and (10):

$$V(C) = \frac{\eta_0(1+\gamma C)}{\rho_0 l(1+\chi C)} \approx \frac{\eta_0}{\rho_0 l}(1+\gamma C), \tag{11}$$

As a result, the diffusion equation (Equation (7)) can be rewritten as follows:

$$\frac{\partial C}{\partial t} = D\frac{\partial^2 C}{\partial x^2} - \frac{\eta_0}{l\rho_0}(1+\gamma C)\frac{\partial C}{\partial x} - \frac{D_T\beta I_0}{\lambda_0}C^2(1-pC), \tag{12}$$

We now introduce the dimensionless variables and parameterize Equation (12). As a result, we obtain:

$$\frac{\partial C}{\partial \tau} = \frac{\partial^2 C}{\partial y^2} - \delta\frac{\partial C}{\partial y} - \delta\gamma C\frac{\partial C}{\partial y} - C^2(1-pC), \tag{13a}$$

The following notation is accepted here:

$$\tau = \frac{S_T D\beta I_0}{\lambda_0}t, y = \frac{1}{\sqrt{b}}x, b = \frac{\lambda_0}{S_T\beta I_0}, \sqrt{b} = l, \delta = \frac{\eta_0}{\rho_0 D}, \tag{13b}$$

Thus, we demonstrate that light-induced thermal diffusion in nanofluids, in the low-particle-concentration approximation, against the background of a steady temperature, and taking into account concentration convection, can be described by nonlinear Equation (13a), which differs from the Burgers–Huxley equation [33] owing to the derivative in the last linear term.

First, we consider the two spatially homogeneous stationary states derived from $C^2(1-pC) = 0$, which correspond to the roots of the equation, namely, $C_1 = C_2 = 0$, $C_3 = 1/p(p>1)$. The kinetics of a dissipative system strongly depend on the stabilities of the stationary states. In our case, the states $C = C_{1,2}$ are twofold degenerate and unstable (they contain derivatives from the source $F'(C) > 0$), whereas state $C = C_{13}$ is stable. Thus, the medium studied herein is not bistable, unlike that studied by Ognev et al. [34].

We note that similar parabolic equations with cubic nonlinearities have been considered in previously published studies, in which they were applied to a model dissipative medium with arbitrary parameters [34], and to a nanofluid + radiation system [35,36]. We look for particular solutions in the form of the Cole-Hopf transform [36]:

$$C(y,\tau) = \frac{W'_y}{W}\cdot\mu, W = W(y,\tau), \tag{14}$$

where μ is a parameter, and $'$ denotes the derivative.

By substituting Equation (14) into (13a) and equating the coefficients for the various powers of W to zero, we obtain an overdetermined system of equations for function $W(y,\mu)$:

$$W''_{y\tau} = W'''_{yyy} - \delta W''_{yy}, \tag{15}$$

$$W'_{\tau} = 3W''_{yy} + \delta\gamma\mu W''_{yy} + (\mu - \delta)W'_{y}, \tag{16}$$

$$p\mu^2 + \delta\mu + 2 = 0, \tag{17}$$

From the last equation of this system, we obtain the values of parameter μ:

$$\mu_{1,2} = \frac{1}{2p}, \tag{18}$$

The estimates of parameters γ and δ, which are provided below, show that roots $\mu_{1,2}$ are real. Furthermore, by integrating Equation (15) with respect to variable y, we obtain:

$$W'_{\tau} = W''_{yy} - \delta W'_{y} + C_1(\tau), \tag{19}$$

Using Equation (16), we obtain:

$$(2 + \delta\gamma)W''_{yy} + \mu W'_{y} + C_1(\tau) = 0, \tag{20}$$

The solution for this equation can be represented as:

$$W(y,\tau) = \frac{C_1(\tau)}{\mu}y + C_2(\tau) + C_3(\tau)exp(-\omega y), \tag{21}$$

where $\omega = \mu/(2 + \delta\gamma\mu)$.

$C_i(\tau)$ can be determined using Equations (20) and (21), and it can be used to express the solution for function $W(y,\tau)$ as follows:

$$W(y,\tau) = \widetilde{C_1}\left(\left(1 - \frac{\delta}{\mu}\right) + \frac{1}{\mu}y\right)\tau + \widetilde{C_2} + \widetilde{C_3}exp(\omega(\omega + \delta)\tau - \omega y), \tag{22}$$

where $\widetilde{C_i}$ are constants.

According to Equations (14) and (22), the desired concentration can be represented as:

$$C(y,\tau) = \mu\frac{\widetilde{C_1} - \omega\widetilde{C_3}exp(\omega(\omega + \delta)\tau - \omega y)}{\acute{C}_1((\mu - \delta)\tau + y) + \widetilde{C_2} + \widetilde{C_3}exp(\omega(\omega + \delta)\tau - \omega y)}, \tag{23}$$

3. Solution Analysis

In this section, we examine the dependence of the determined exact solution (23) on parameters μ, ω, and δ. Parameter δ is estimated to be $\approx 10^5$ by assuming that $\eta_0 = 10^{-3}$ kg/(m·s), $\rho_0 = 10^3$ kg/m^3, and $D = 10^{-11}$ m^2/s in Equation (13a). Furthermore, from the findings of previous studies [24–26], it follows that $p \approx 1 \div 1.5$. Then, we must consider that $4p \ll \delta^2\gamma^2$ in Equation (18). Therefore, for sufficiently accurate roots $\mu_{1,2}$, we obtain $\mu_1 = -2/\delta, \omega_1 = -2, \mu_2 = -\delta/p, \omega_2 = -1/p$.

Furthermore, by substituting the corresponding expressions for parameters and δ (their estimates) into Equation (23), we obtain two solutions:

$$C_1(y,\tau,\mu_1) = \frac{2}{\delta\gamma}\frac{1 + 2c_3exp(-2(\delta\tau - y))}{(\delta\tau - y) + c_2 - c_3exp}, \tag{24}$$

$$C_2(y,\tau,\mu_2) = \delta\gamma, \tag{25}$$

Here, c_2 and c_3 are the newly redefined constants.

Let us consider Equation (24), the graph of which is shown in Figure 2; we see that it is a function of the variable traveling wave, $z = (\delta\tau - y)$. On the graph, the solution is presented in the form of soliton-like pulses moving to the right (with increasing time).

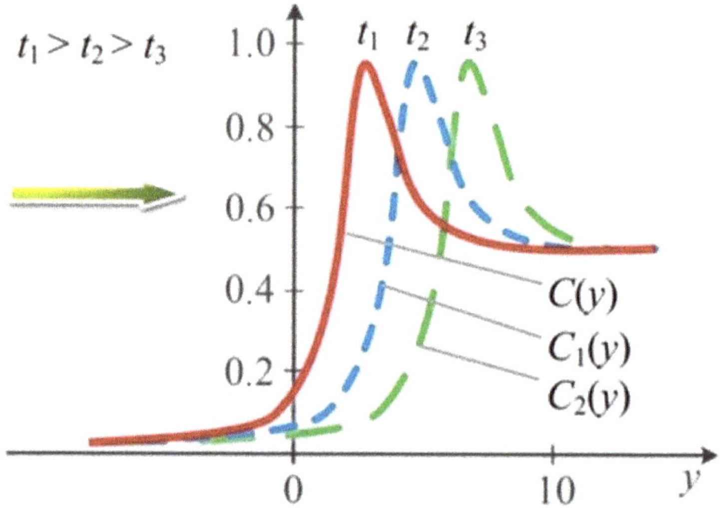

Figure 2. Solution of Equation (24).

A distinctive feature of Equation (25) is the presence in the denominator of the $(\delta(\gamma/p - 1)\tau - y)$ term. Clearly, the nature of curve $C(y)$ in Equation (25) strongly depends on the γ/p ratio. When plotting the function, we set $\gamma/p = 3$ (see Figure 3). Note that the wave-pulse profiles are not similar in this case.

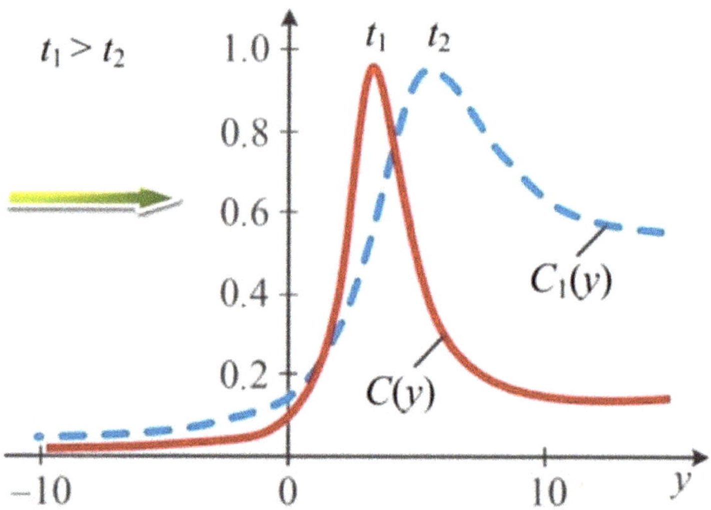

Figure 3. Solution profiles of the concentration wave obtained from Equation (25) with increasing time.

The velocity of the wave front of Equation (24) can be determined using Equation (13b). As a result, we obtain:

$$v = \frac{\eta_0}{\rho_0 \sqrt{b}}. \quad (26)$$

The speed depends on the thermodynamic, hydrodynamic, and optical characteristics of the nanofluid + radiation system. It follows from Equations (26) and (13b) that, in the case of anomalous thermal diffusion ($S_T < 0$—nanoparticles move in a higher-temperature region), the velocity acquires an imaginary term, which has no physical meaning. We believe that this case requires a separate consideration, which we plan to carry out in the future.

Wenumerically estimate the wavefront propagation velocity according to Equation (26) and consider water/silver as a nanofluid. Because the absorption coefficient β is present in Equation (26) through parameter b (see Equation (13b)), its evaluation requires the following equation [37]:

$$\beta = \frac{12\pi}{\lambda} C Im(\frac{m^2 - 1}{m^2 + 2}),$$

where $m = m_{particles}/m_{fluid}, m = n + ik$.

Here, by assuming $m_p = 0.15 + 3.5i$, $m_f = 1.33 + 0.2i$, $\lambda = 6.5 \cdot 10^{-7}$ m, and $C = 1 \cdot 10^{-4}$, we obtain $\beta \approx 2 \cdot 10^3$. Furthermore, by substituting $\eta_0 = 0.003$ kg/m·s, $\rho_0 = 1 \cdot 10^3$, $I_0 = 1 \cdot 10^5$ W/m^2, and $\lambda_0 = 0.5$ W/m·K into Equations (26) and (13b), we obtain a velocity estimate: $v \approx 2 \cdot 10^{-3}$ m/s. The estimated velocity value depends on the initial concentration distribution, which can be obtained from Equation (24) for $\tau = 0$. It should be noted that our approach cannot solve the problem analytically under arbitrary initial conditions.

4. Conclusions

1. Two exact analytical solutions of a nonlinear one-dimensional Burgers–Huxley-type equation were obtained. These solutions describe the dynamics of the concentration of nanoparticles in a liquid-phase medium by taking into account concentration convection. In this case, the coefficients of thermal conductivity, viscosity, and absorption of radiation by particles were found to be concentration dependent.
2. One of the solutions found was represented as a solitary wave. Both solutions were expressed in the form of traveling single-phase waves.
3. In the framework of the formulated approximations, the nanofluid + radiation system under study exhibited one stable (doubly degenerate) state and one unstable state. It should be noted that convection does not affect the nature of the stability.
4. Within the framework of these approximations, it is possible to obtain the spatiotemporal dependence of the particle absorption coefficient, which exhibits the same wave characteristics as those of Equations (24) and (25).

Invariably, we did not consider some issues. In particular, based on the fact that the equation under consideration is autonomous, studying the equation on the phase plane is of significant interest and will be the subject of our further research.

Author Contributions: Conceptualization, A.I.L. and P.V.V.; data curation, A.V.S. and P.V.I.; formal analysis, A.I.L. and P.V.V.; investigation, Y.M.K. and P.V.I.; methodology, A.I.L. and Y.M.K.; project administration, V.V.K., V.G.E. and A.V.S.; resources, V.G.E. and A.V.S.; software, E.N.K.; supervision, V.V.K.; visualization, E.N.K.; writing—original draft, A.I.L.; writing—review and editing, V.V.K. and V.G.E. All authors have read and agreed to the published version of the manuscript.

Funding: This research was funded by the Ministry of Education and Science of the Russian Federation within the framework of the program of activities of the world-class scientific and educational center "Rational Subsoil Use".

Conflicts of Interest: The authors declare no conflict of interest.

References

1. Ali, H.M. *Hybrid Nanofluids for Convection Heat Transfer*, 1st ed.; Academic Press: London, UK, 2020.
2. Rudyak, V.Y.; Minakov, A.V. *Modern Problems of Micro- and Nanofluidics*; Nauka: Novosibirsk, Russia, 2016.
3. Lee, G.J.; Attri, P.; Choi, E.H.; Kwon, Y.-W.; Krasnikov, I.; Seteikin, A. Optical and structural properties of nanobiomaterials. *J. Nanosci. Nanotechnol.* **2014**, *14*, 221–249. [CrossRef]
4. Handelman, A.; Apter, B.; Lapshina, N.; Rosenman, G. Bioinspired Peptide-Based Photonic Integrated Devices. In *Advanced Photonics 2018 (BGPP, IPR, NP, NOMA, Sensors, Networks, SPPCom, SOF), Proceedings of the Novel Optical Materials and Applications, Zurich, Switzerland, 2–5 July 2018*; OSA Technical Digest (online); Optical Society of America: Washington, DC, USA, 2018.
5. Omrani, E.; Siddaiah, A.; Moghadam, A.; Garg, U.; Rohatgi, P.; Menezes, P. Ball Milled Graphene Nano Additives for Enhancing Sliding Contact in Vegetable Oil. *Nanomaterials* **2021**, *11*, 610. [CrossRef]
6. Aboalhamayie, A.; Festa, L.; Ghamari, M. Evaporation Rate of Colloidal Droplets of Jet Fuel and Carbon-Based Nanoparticles: Effect of Thermal Conductivity. *Nanomaterials* **2019**, *9*, 1297. [CrossRef] [PubMed]
7. Livashvili, A.I.; Krishtop, V.V.; Yakunina, M.I. Electrostrictive self-action mechanism of radiation in nanofluids. *Adv. Condens. Matter. Phys.* **2013**, *5*, 591087. [CrossRef]
8. El-Ganainy, E.; Christodoulides, D.N.; Rotschild, C.; Segev, M. Soliton dynamics and self-induced transparency in nonlinear nanosuspensions. *Opt. Express* **2007**, *15*, 10207–10218. [CrossRef]
9. Vorobyeva, E.V.; Ivakhnik, V.V.; Savelyev, M.V. Spatial and temporal characteristics of a four-wave radiation converter in a transparent medium based on electrostriction and Dufour effect. *Comput. Opt.* **2014**, *38*, 223–228. [CrossRef]
10. Ivanov, V.I. *Thermo-Induced Mechanisms of Dynamic Holograms Recording*; FESTU: Khabarovsk, Russia, 2006.
11. Ivanov, V.I.; Livashvili, A.I. Electrostriction mechanism of self-radiation in a liquid with nanoparticles. *Bull. Novosib. State Univ. Ser. Phys.* **2009**, *4*, 58–60.
12. Sani, E.; Papi, N.; Mercatelli, L.; Barison, S.; Agresti, F.; Rossi, S.; Dell'Oro, A. Optical Limiting of Carbon Nanohorn-Based Aqueous Nanofluids: A Systematic Study. *Nanomaterials* **2020**, *10*, 2160. [CrossRef] [PubMed]
13. Khan, M.S.; Yan, M.; Ali, H.M.; Amber, K.P.; Bashir, M.A.; Akbar, B.; Javed, S. Comparative performance assessment of dif-ferent absorber tube geometries for parabolic trough solar collector using nanofluid. *J. Therm. Anal. Calorim.* **2020**, *142*, 2227–2241. [CrossRef]
14. Sreekumar, S.; Joseph, A.; Kumar, C.S.; Thomas, S. Investigation on influence of antimony tin oxide/silver nanofluid on direct absorption parabolic solar collector. *J. Clean. Prod.* **2020**, *249*, 119378. [CrossRef]
15. Kaya, H.; Alkasem, M.; Arslan, K. Effect of nanoparticle shape of Al_2O_3/Pure water nanofluid on evacuated U-Tube solar collector efficiency. *Renew. Energy* **2020**, *162*, 267–284. [CrossRef]
16. Minea, A.A.; Murshed, S.S. Ionic Liquids-Based Nanocolloids—A Review of Progress and Prospects in Convective Heat Transfer Applications. *Nanomaterials* **2021**, *11*, 1039. [CrossRef] [PubMed]
17. Zhang, Y.; Xu, X. Predicting the thermal conductivity enhancement of nanofluids using computational intelligence. *Phys. Lett. A* **2020**, *384*, 126500. [CrossRef]
18. Sani, E.; Martina, M.R.; Salez, T.J.; Nakamae, S.; Dubois, E.; Peyre, V. Multifunctional Magnetic Nanocolloids for Hybrid Solar-Thermoelectric Energy Harvesting. *Nanomaterials* **2021**, *11*, 1031. [CrossRef]
19. Livashvili, A.I.; Kostina, G.V.; Yakunina, M.I. Temperature dynamics of a transparent nanoliquid acted on by a periodic light field. *J. Opt. Technol.* **2013**, *80*, 124–126. [CrossRef]
20. Rudyak, V.Y.; Belkin, A.A. Modeling of transport coefficients of nanofluids. *Nanosistemy Fiz. Khim.* **2010**, *1*, 156–177.
21. Ashkin, A.; Dziedzic, J.M.; Bjorkholm, J.E.; Chu, S. Observation of a single-beam gradient force optical trap for dielectric par-ticles. *Opt. Lett.* **1986**, *11*, 288–290. [CrossRef]
22. Jesacher, A.; Fürhapter, S.; Maurer, C.; Bernet, S.; Ritsch-Marte, M. Holographic optical tweezers for object manipulations at an air-liquid surface. *Opt. Express* **2006**, *14*, 6342–6352. [CrossRef]
23. Livashvili, A.I.; Krishtop, V.V.; Bryukhanova, T.N.; Kostina, G.V. Concentration Dynamics of Nanoparticles under a Periodic Light Field. *Phys. Procedia* **2015**, *73*, 156–158. [CrossRef]
24. Smorodin, B.L.; Cherepanov, I.N. Convection in a colloidal suspension in a closed horizontal cell. *J. Exp. Theor. Phys.* **2015**, *120*, 319–326. [CrossRef]
25. de Groot, S.R.; Mazur, P. *Non-Equilibrium Thermodynamics*; Courier Corporation: North Chelmsford, MA, USA, 2013.
26. Landau, L.D.; Lifshitz, E.M. *Fluid Mechanics*, 2nd ed.; Pergamon: London, UK, 1987.
27. Rudyak, V.Y.; Minakov, A.V.; Pryazhnikov, M.I. Thermal properties of nanofluids and their similarity criteria. *Tech. Phys. Lett.* **2017**, *43*, 23–26. [CrossRef]
28. Choi, S.U.; Eastman, J.A. *Enhancing Thermal Conductivity of Fluids with Nanoparticles (No. ANL/MSD/CP-84938; CONF-951135-29)*; Argonne National Lab.: Lemont, IL, USA, 1995.
29. Venerus, D.C.; Buongiorno, J.; Christianson, R.; Townsend, J.; Bang, I.C.; Chen, G.; Dubois, F. Viscosity measurements on colloidal dispersions (nanofluids) for heat transfer applications. *Appl. Rheol.* **2010**, *20*, 11–17.
30. Rudyak, V.; Belkin, A. Molecular dynamics simulation of fluid viscosity in nanochannels. *Nanosyst. Physics Chem. Math.* **2018**, *9*, 349–355. [CrossRef]
31. Smorodin, B.L.; Cherepanov, I.N.; Myznikova, B.I.; Shliomis, M.I. Traveling-wave convection in colloids stratified by gravity. *Phys. Rev.* **2011**, *84*, 026305. [CrossRef]

32. Yefimova, O.Y.; Kudryashov, N.A. Exact solutions of the Burgers-Huxley equation. *J. Appl. Math. Mech.* **2004**, *3*, 413–420. [CrossRef]
33. Livashvili, I.A.; Krishtop, V.V.; Karpets, Y.M.; Bryuhanova, T.N.; Kireeva, N.M. Laser beam-induced bistability of concentration in nanofluids. *J. Phys. Conf. Ser.* **2016**, *737*, 12011. [CrossRef]
34. Ognev, M.V.; Petrovskii, S.V.; Prostokishin, V.M. Dynamics of formation of a switching wave in a dissipative bistable medi-um. *J. Tech. Phys.* **1995**, *40*, 521–524.
35. Livashvili, A.; Krishtop, V.; Kostina, G.; Vinogradova, P.; Kireeva, N. Dynamics of Switching Waves in a Nanofluid in a Light Field. *KnE Energy* **2018**, *3*, 165–172. [CrossRef]
36. Kasaeian, A.; Eshghi, A.T.; Sameti, M. A review on the applications of nanofluids in solar energy systems. *Renew. Sustain. Energy Rev.* **2015**, *43*, 584–598. [CrossRef]
37. Bohren, C.F.; Huffman, D.R. *Absorption and Scattering of Light by Small Particles*; John Wiley & Sons: Hoboken, NJ, USA, 2008; p. 544.

nanomaterials

Article

Molecular Dynamics Simulation on Behaviors of Water Nanodroplets Impinging on Moving Surfaces

Hao Zhang, Ling Pan * and Xuqing Xie

School of Mechanical Engineering and Automation, Fuzhou University, Fuzhou 350108, China; hao_capable@126.com (H.Z.); n190227107@fzu.edu.cn (X.X.)
* Correspondence: panling@fzu.edu.cn

Abstract: Droplets impinging on solid surfaces is a common phenomenon. However, the motion of surfaces remarkably influences the dynamical behaviors of droplets, and related research is scarce. Dynamical behaviors of water nanodroplets impinging on translation and vibrating solid copper surfaces were investigated via molecular dynamics (MD) simulation. The dynamical characteristics of water nanodroplets with various Weber numbers were studied at five translation velocities, four vibration amplitudes, and five vibration periods of the surface. The results show that when water nanodroplets impinge on translation surfaces, water molecules not only move along the surfaces but also rotate around the centroid of the water nanodroplet at the relative sliding stage. Water nanodroplets spread twice in the direction perpendicular to the relative sliding under a higher surface translation velocity. Additionally, a formula for water nanodroplets velocity in the translation direction was developed. Water nanodroplets with a larger Weber number experience a heavier friction force. For cases wherein water nanodroplets impinge on vibration surfaces, the increase in amplitudes impedes the spread of water nanodroplets, while the vibration periods promote it. Moreover, the short-period vibration makes water nanodroplets bounce off the surface.

Keywords: water nanodroplets; molecular dynamics; dynamical behaviors; translation surface; vibration surface

Citation: Zhang, H.; Pan, L.; Xie, X. Molecular Dynamics Simulation on Behaviors of Water Nanodroplets Impinging on Moving Surfaces. *Nanomaterials* **2022**, *12*, 247. https://doi.org/10.3390/nano12020247

Academic Editors: Antonio Gloria and S M Sohel Murshed

Received: 4 December 2021
Accepted: 7 January 2022
Published: 13 January 2022

1. Introduction

Droplets impinging on solid surfaces are pervasive in nature and industrial production, such as pesticide spraying [1], spray cooling, and inkjet printing [2,3]. A further understanding of this phenomenon is of great importance with the increasing demand for droplet manipulation, such as self-cleaning [4], anti-icing [5,6], and energy harvesting [7]. Although many efforts have been made in this field by researchers, the mechanisms of dynamical behaviors for impinging droplets are not fully clarified.

It is important to recognize that impinging droplets experience a process of spreading, retraction, and eventually bouncing off or depositing on surfaces [8–10]. Many experimental investigations have analyzed the dynamical behaviors of droplets. Wang et al. investigated water droplets impinging on pillar-arrayed polydimethylsiloxane (PDMS) surfaces with different solid fractions and suggested that the lower and upper limits of the Weber number (*We*) for the complete rebound of droplets decreased with solid fractions [11]. Hu et al. revealed the asymmetric spreading and retraction processes in the axial and spanwise directions when droplets impinged on ridged superhydrophobic surfaces. They proposed two theoretical models to quantitatively describe the spreading of droplets for size ratios (the ratio of the ridge diameter to the droplet diameter) as smaller and larger than the critical size ratio [12]. Work by Refael et al. showed the dynamical behaviors of droplets impinging on moving liquid surfaces. Droplets can splash, coalesce with, or bounce off liquid surfaces, which depends on the liquid properties and the velocity of droplets and surfaces [13]. Almohammadi et al. investigated the asymmetric spreading of droplets impinging on a moving surface and developed a model predicting the shape of the lamella. They also discussed the effect of surface wettability on droplets in the retraction process [14].

Recently, the exploration of nanodroplets has become a highly debated topic due to the development of micro and nano technology, such as nanoprinting and nanocoating. However, many important details cannot be displayed in experiments. Molecular dynamics (MD) simulations provide a powerful tool to probe the dynamical behaviors of nanodroplets on an atomic scale. The mechanisms of spreading, break-up, bouncing, and other behaviors of nanodroplets are discussed via MD [15–19]. Chen et al. simulated polymer nanodroplets impinging on a solid surface and found that the viscous dissipation of water nanodroplets stemmed from the velocity gradients in both the impinging and spreading directions [20]. Song et al. investigated the deformation behaviors of water nanodroplets in an electric field and observed the deformation hysteresis phenomenon. The distribution of average dipole orientations in water molecules also showed hysteresis [21]. Kwon et al. presented the dynamical behaviors of a water nanodroplet impinging on a stepped surface with different wetting gradients and step heights [22]. They found three phenomena: fully climbing the step, partially climbing the step, and being blocked by the step, which relied on the normalized step height. MD simulations were also adopted to investigate the influences of surfaces vibrating with high frequency on the wetting mechanism, evaporation, and transportation of nanodroplets [23–26].

However, studies on water nanodroplets impinging on moving solid surfaces are still scarce, especially on high-frequency vibrating surfaces. Due to the rapid development of piezoelectric materials, the frequency of surface acoustic waves (SAW), based on the principle of the inverse piezoelectric effect [27], is up to the level of GHz [28–33]. Therefore, high vibrations of surfaces with a frequency on the order of GHz can be realized. In this paper, the dynamical behaviors of water nanodroplets impinging on translation and vibration surfaces are investigated through MD simulations, respectively. The effects of parameters related to translation and vibration on the dynamical behaviors are explored.

2. Simulation Model and Methodology

The initial configuration of water nanodroplets impinging on a moving surface is schematically shown in Figure 1. The flat substrate is composed of copper atoms, which are arranged in the face-centered-cube (FCC) structure with a lattice constant of 3.615 Å [34]. A TIP4P model of water is adopted because it can present more accurate dynamical characteristics [35,36]. The parameters corresponding to the TIP4P model are listed in Table 1 (q_O: the charge of oxygen atom; q_H: the charge of hydrogen atom; e: electron charge; r_{OH}: the bond length between oxygen and hydrogen atoms; θ_{HOH}: the bond angle of hydrogen-oxygen-hydrogen). The water nanodroplet is constituted by 5991 water molecules, with 35 Å in radius. To prevent atoms in the initial configuration from overlapping, which leads to the infinite interaction force among atoms, water molecules are arranged in the body-centered-cube (BCC) structure, whose lattice constant is determined by the density of water ρ at 298 K.

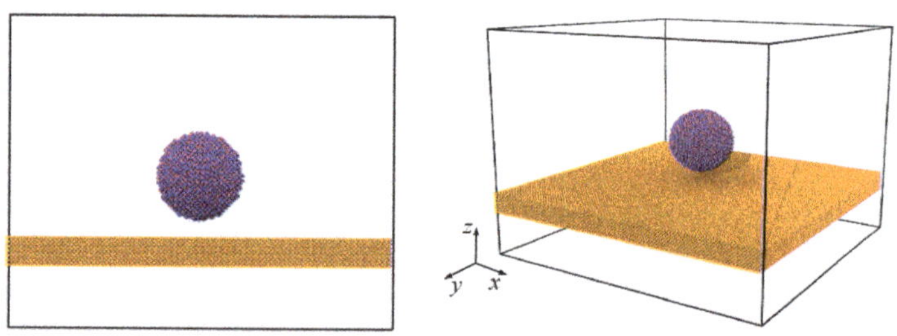

(a) (b)

Figure 1. The initial configuration of a water nanodroplet impinging on a moving surface: (**a**) front view; (**b**) perspective view.

Table 1. Parameters of TIP4P model [35].

q_O (e)	q_H (e)	r_{OH} (Å)	θ_{HOH} (°)
−1.0484	+0.5242	0.9572	104.52

The whole simulation is programmed and conducted with LAMMPS (large-scale atomic/molecular massively parallel simulator) [37]. Periodical boundary conditions are applied in all three directions. A SHAKE algorithm is adopted to fix the bond length and angle of water molecules [38]. Copper atoms are constrained to initial equilibrium position during simulation [39–41]. The interaction between water molecules [42], containing a 12-6 Lennard-Jones (LJ) term and a long-range Coulomb term, is shown as follows:

$$U = 4\varepsilon_{OO}\left[\left(\frac{\sigma_{OO}}{r_{OO}}\right)^{12} - \left(\frac{\sigma_{OO}}{r_{OO}}\right)^{6}\right] + \sum_{a=1}^{3}\sum_{b=1}^{3}\frac{q_{ia}q_{jb}}{4\pi\varepsilon_0 r_{ia,jb}} \quad (1)$$

where r_{OO} denotes the distance between oxygen atoms in water molecules i and j. ε_{OO} and σ_{OO} represent the well depth and equilibrium distances, respectively. $r_{ia,jb}$ stands for the distance between charge site a of molecule i and charge site b of molecule j. q_{ia} and q_{jb} are charges of sites a and b, respectively. ε_0 denotes the vacuum permittivity. Although Lennard-Jones potential was presented for noble gas at first, it essentially describes the interaction between electronic neutral atoms [43]. Hendrik et al. justified the reliability of this potential for copper and showed that 12-6 LJ potential is more suitable for large simulation systems (~10^6 atoms) compared with embedded atom model (EAM) and density functional method [44]. In this paper, a large-sized solid surface consisting of 201,344 copper atoms is adopted to avoid the influences of adjacent mirror images during the spreading of the droplet. Therefore, both the interactions between copper atoms and those between oxygen atoms and copper atoms are implemented by using the 12-6 LJ potential [45–47]. The interaction parameters for oxygen and copper are presented in Table 2. Since the solid surface is frozen in the simulation, the parameters of copper atoms are mainly used to calculate the interaction between copper and oxygen through Lorentz–Berthelot mix rules that are shown in Equations (2) and (3).

$$\sigma_{mn} = \frac{1}{2}(\sigma_m + \sigma_n) \quad (2)$$

$$\varepsilon_{mn} = \sqrt{\varepsilon_m \varepsilon_n} \quad (3)$$

where σ_m, ε_m, σ_n, and ε_n are parameters of 12-6 LJ potential for atoms m and n; σ_{mn} and ε_{mn} are parameters of 12-6 LJ potential between atoms m and n.

Table 2. Interaction parameters for oxygen and copper atoms [48].

Atom Type	O	Cu
ε (kcal/mol)	ε_O = 0.1628	ε_{Cu} = 0.2379
σ (Å)	σ_O = 3.1644	σ_{Cu} = 2.3400

The cutoff distances of 12-6 LJ potential and Coulomb potential are 10 Å and 12 Å [34], respectively. Velocity–Verlet algorithm is used to integrate the Newton equation of motion and update the velocities and coordinates of atoms with a time step of 1.0 fs. The long-range Columbic interaction is calculated by PPPM (particle–particle–particle–mesh) method in K space [49].

Water nanodroplets experience minimization and relaxation lasting for 0.5 ns in NVT ensemble, leading them equilibrated and the potential energy being the local minimum. During the relaxation process, a Nose–Hoover thermostat is adopted to realize temperature control and the initial equilibrium velocity of water molecules is in accord with Maxwell–Boltzmann distribution [50,51]. After relaxation, the simulation system achieves equilibrium state, and there are some water molecules in gas phase randomly distributed in simulation domain. Then, the NVT ensemble is removed, eliminating the temperature control, and the system is simulated under NVE ensemble. Meanwhile, a vertical initial velocity V_z is applied to the water nanodroplet. V_z is in the negative z direction. The moment water nanodroplets touch the solid surface is considered 0 ps.

3. Results and Discussion

3.1. Calculation of the Contact Angle and Verification of Water Nanodroplet Size

The contact angle is an essential parameter indicating the wettability of solid surfaces. The density contour of a water nanodroplet on the smooth copper surface illustrated in Figure 2 is based on the time-average density of every single chunk the water nanodroplet is being divided into. Considering the symmetrical projection of the water nanodroplet on x–y plane, only the right-sided water nanodroplet is investigated for contact angle calculation. The curve where the density is half that of bulk water is selected as the boundary of the liquid and gas phase (the black line in Figure 2) [52]. The density profile is considered as a part of a circle and is fitted by Equation (4). The contact angle can be obtained through Equation (5) [48]:

$$(x-a)^2 + (z-b)^2 = R^2 \tag{4}$$

$$\theta = arcsin\left(\frac{b - z_{\text{sub}}}{R}\right) + 90^o \tag{5}$$

where a and b are x and z coordinates of the centroid and R denotes the radius of the fitted circle. Only chunks that are more than 5 Å above the solid surface are considered to avoid the effects of density fluctuations at the liquid–solid interface. Therefore, z_{sub} is the z coordinate of a virtual surface 5 Å above the solid surface. According to the calculation, the equilibrium contact angle θ of water droplets on the smooth copper surface is 108.7°, and the corresponding experimental result is 102° [53]. The slight difference between them is attributed to the surface used in the experiment being not as smooth as that used in the simulation.

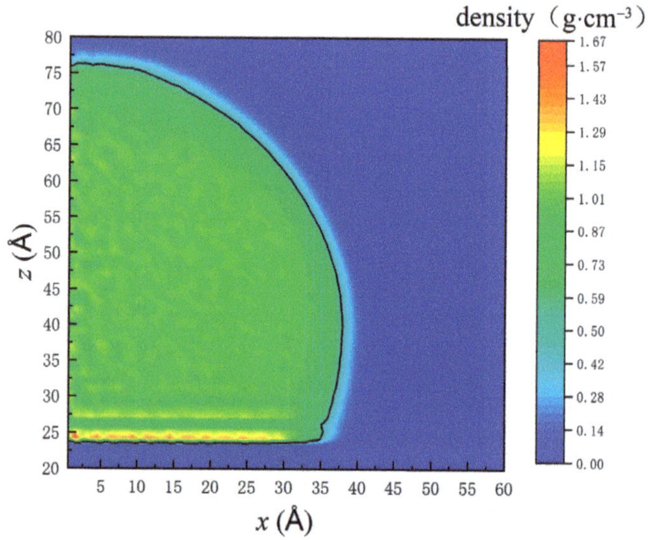

Figure 2. Density profile of the right-sided water nanodroplet on the smooth copper surface.

To explore the effect of the number of molecules on the dynamical behaviors, nanodroplets with different radii impinging on the immobile surface are simulated. Radii and the corresponding number of water molecules are listed in Table 3. The dimensionless number (R/R_0, R is the spreading radius, and R_0 represents the initial radius) of those nanodroplets is shown in Figure 3.

Table 3. Nanodroplets with different radii and the number of water molecules.

Nanodroplet Radius (Å)	Number of Molecules
35	5991
40	8953
45	12,721

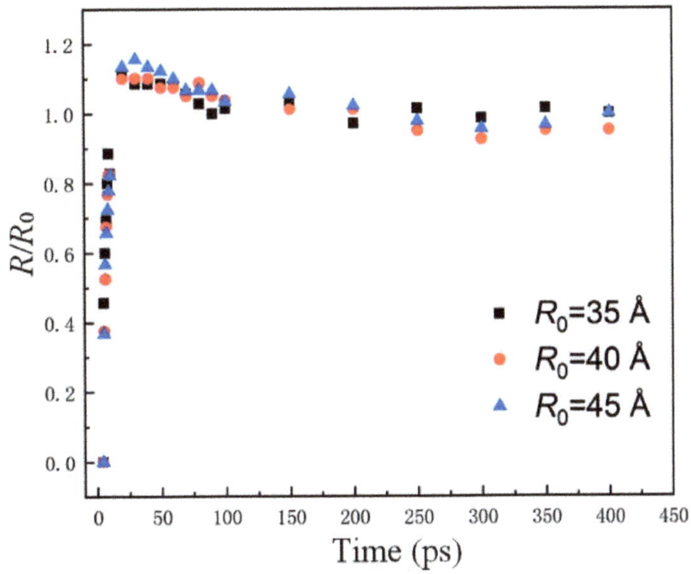

Figure 3. Effects of droplets size on spreading time.

It can be seen that although the number of water molecules varies, the R/R_0 of different nanodroplets are similar to each other, and the maximum relative error is 6.9%. Therefore, the effect of the number of molecules arranged in the droplet can be neglected. Taking the computation accuracy and efficiency into consideration, nanodroplets with R_0 = 35 Å are adopted.

3.2. *Effects of Translation Velocity When Water Nanodroplets Impinge on Translation Surfaces*

Considering the inertia force plays an important role in the impinging process, the Weber number (*We*) is chosen to represent the characteristics of water nanodroplets [54]. The effects of translation velocity of surfaces V_s on the dynamical behaviors of water nanodroplets with *We* = 7.41 are investigated. *We* is defined as

$$We = \frac{\rho R V_z^2}{\gamma} \tag{6}$$

where R is the radius of the droplet, V_z represents the velocity of droplets, and the surface tension γ = 72.75 mN/m [52]. The morphological evolution of water nanodroplets is illustrated in Figure 4. Due to the sufficient simulation period, water nanodroplets can exceed the simulation box boundary, and they re-enter the box from the opposite side because of the periodical boundary conditions. For the sake of clarity, the dimensions of boxes and surfaces in x direction are assumed to be infinite.

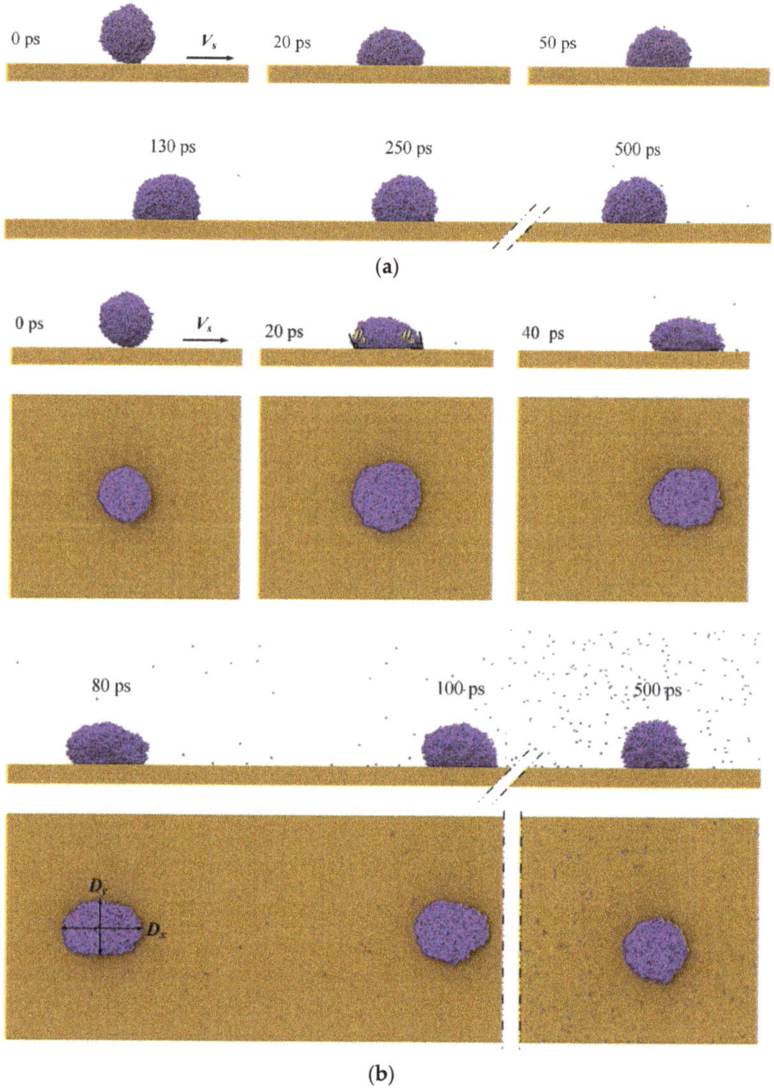

Figure 4. Morphological evolution of droplets impinging on surfaces with different translation velocities: (**a**) V_s = 0.5 Å/ps; (**b**) V_s = 9 Å/ps.

Although the surface is smooth, Figure 4 shows that both deformation and displacement of water nanodroplets can be observed, which is due to microscopic forces such as the Van der Waals force and capillary force taking a leading role in the interaction on a nanoscale. The component of F, interaction force between the water nanodroplet and surface, in the direction of relative sliding, is defined as friction force F_x, and the component in the direction perpendicular to the contact area is defined as normal pressure F_z. Therefore, the deformation and displacement of water nanodroplets result from the F_x applied by moving surfaces.

As seen in Figure 4a, the morphology of the water nanodroplet remains nearly symmetrical in the spreading and retraction processes. With V_s increasing to 9 Å/ps, contact angle hysteresis is observed during the process, so that the water nanodroplet moves along the V_s direction and spreads. Contact angle hysteresis refers to the difference between the advancing contact angle θ_a and the receding contact angle θ_r. The extent of hysteresis experiences a gradual drop to zero as time passes, which means the contact angle hysteresis eventually disappears. The water nanodroplet spreads asymmetrically in x and y directions at time t = 40 ps, and the asymmetry becomes more significant as time elapses. In particular, the water nanodroplet evolves into an ellipsoid at t = 80 ps. Afterwards, the water nanodroplet gradually recovers the sphere shape under the action of surface tension and then moves alongside the surface.

The evolution of velocity of the water nanodroplet in the x direction, V_x, is shown in Figure 5. It can be seen that the variation of V_x can be divided into two stages: the relative sliding stage and the stable stage. At the relative sliding stage, the acceleration of V_x increases with the surface velocity. Especially, during 0~50 ps, V_x almost linearly increases with t. This is because the deformation of water nanodroplets becomes more severe with the increase in surface velocity, making the spreading area enlarged, and thus the interaction between droplets and surfaces is enhanced. Once the water nanodroplet reaches the maximum spreading state, it begins to retract, and the interaction becomes weakened, resulting in a slowdown in the acceleration of V_x. At a stable stage, the translation velocity of the water nanodroplet is roughly equal to that of the surface, and there is no relative sliding between them.

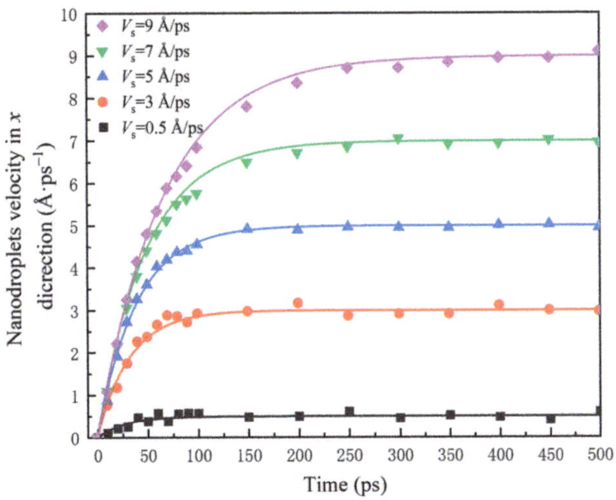

Figure 5. Time evolution of velocity of the water nanodroplet in the x direction.

The mathematical expression, presented in Equation (7), quantitatively describes the change in V_x in the whole dynamical process. Water nanodroplets with *We* ranging from 7.41 to 66.67 impinging on the translation surface are also simulated, and this equation can provide good approximations.

$$V_x = V_s - V_s\left(0.962 + 0.019\frac{V_s}{We}\right)^t \quad 7.41 \leq We \leq 66.67 \tag{7}$$

Taking the surface with V_s = 9 Å/ps as an example, the motion mechanisms of water molecules at different stages are explored. According to Figure 5, the water nanodroplet is at a relative sliding stage during 0~250 ps and at a stable stage during 250~500 ps. As illustrated in Figure 6a, the top and bottom sections of the water nanodroplet are colored blue and red at t = 0 ps, respectively. The color property is fixed during 0~250 ps. According to the snap shots, water molecules in the blue and red regions rotate counterclockwise around the center of the water nanodroplet, accompanying the diffusion of water molecules into the entire water nanodroplet. Then, the water nanodroplet is colored again at 250 ps, as shown in Figure 6b. During the period of 250~500 ps, water molecules in blue and red regions only move downward or upward due to the diffusion, and no rotation is observed. Therefore, diffusion aside, water molecules in the water nanodroplets rotate around the centroid while they move along the surface in the relative sliding stage, and the rotation of water molecules disappears in the stable stage.

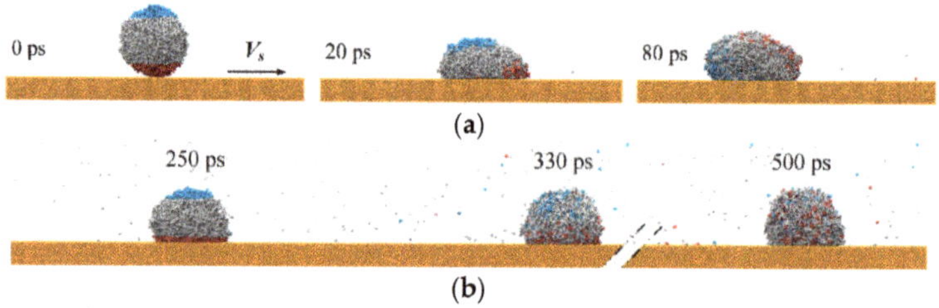

Figure 6. Different motion mechanisms of water molecules inside the water nanodroplet at two stages: (**a**) relative sliding stage; (**b**) stable stage.

Friction force F_x applied to water nanodroplets is shown in Figure 7. When the surfaces translate at high speed (e.g., V_s = 9 Å/ps), there are continuous positive values of F_x in a long period, which is about 80 ps, as shown in Figure 7c. Additionally, that period shrinks as V_s decreases, as illustrated in Figure 7a,b. As V_s increases to 3 and 9 Å/ps from 0.5 Å/ps, the maximum friction force climbs to 20.5 and 36 Kcal·mole^{-1}·Å^{-1}, which means that F_x increases with the V_s. In the stable stage, F_x fluctuates around 0. The fluctuation results from the random thermal motion that leads to a slight change in the spreading area.

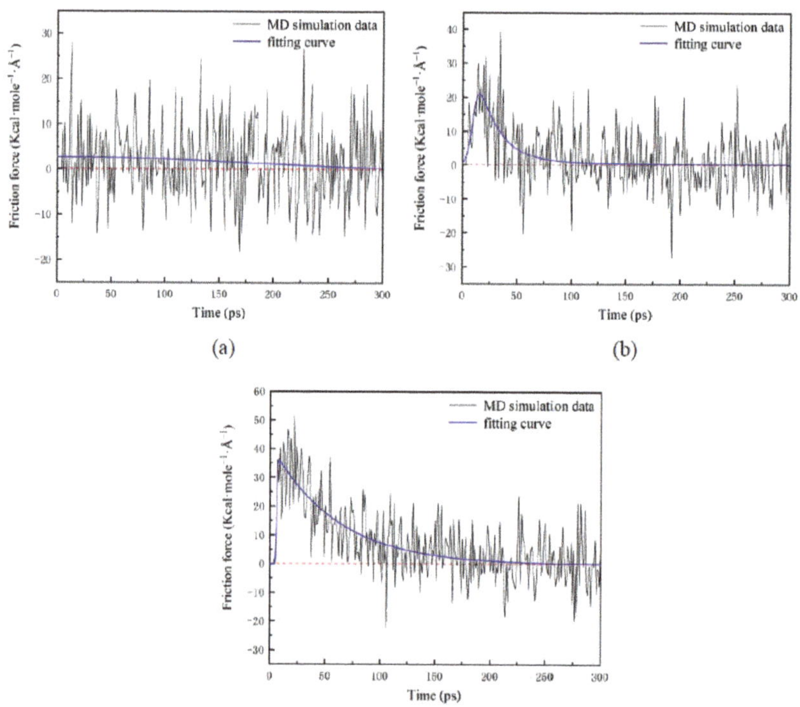

Figure 7. Time evolution of the friction force applied to droplets: (**a**) V_s = 0.5 Å/ps; (**b**) V_s = 3 Å/ps; (**c**) V_s = 9 Å/ps.

Figure 8 shows the spreading factors β_x and β_y. β_x (β_y) denotes the ratio spreading length D_x (D_y) of water nanodroplets in the x(y) direction to the initial diameter D_0. $\beta_{x\max}$ ($\beta_{y\max}$) is the maximum of β_x (β_y).

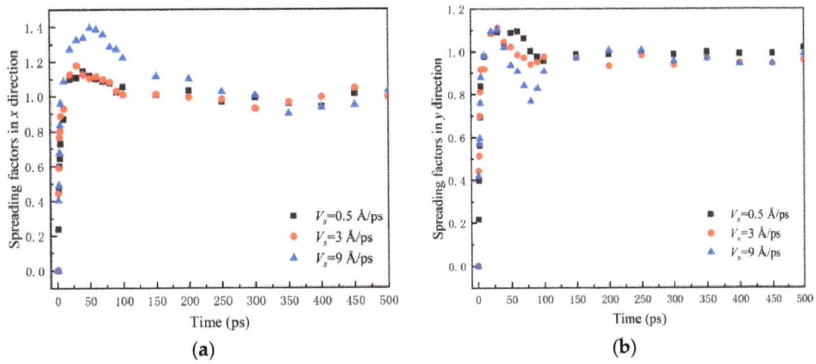

Figure 8. Time evolution of spreading factors in x and y directions: (**a**) β_x; (**b**) β_y.

For the case V_s = 9 Å/ps, it can be observed from Figure 8b that β_y reaches its maximum at 30 ps and then descends to 0.76 rapidly at 80 ps. After, it climbs again and finally obtains stability. This inflection point indicates the water nanodroplet spreads twice in the y direction, which is perpendicular to the direction of relative sliding. This is due to the fact that the water nanodroplet suffers reactive force when it impinges on solid surfaces, and that makes the water nanodroplet spread at first. The $\beta_{y\max}$ is realized soon, and then the water nanodroplet begins to retract in the y direction. Meanwhile, the tensile deformation of water nanodroplets in the x direction becomes severe with the increase in V_x and thus accelerates the retraction in the y direction since the volume of the water nanodroplet is constant. Once the $\beta_{x\max}$ is realized, the water nanodroplet begins to retract in the x direction, which results in the water nanodroplet spreading again in the y direction.

3.3. *Effects of the Weber Number When Water Nanodroplets Impinge on Translation Surfaces*

To investigate the influence of *We* on the behavioral evolution, the water nanodroplet with *We* ranging from 0.82 to 66.67 hitting the translation surfaces is simulated. The solid surface translates at the speed of V_s = 7 Å/ps along the positive direction of x axis.

The performance of water nanodroplets is shown in Figure 9. It can be observed that water nanodroplets with different *We* lie on various positions of the surfaces despite the identical V_s, which suggests the dynamical behaviors of water nanodroplets are determined by the joint actions of V_s and *We*.

The contact hysteresis can be seen for water nanodroplets with *We* = 0.82~20.58, as shown in Figure 9a,b. However, the degree of hysteresis decreases with increasing *We*, and the hysteresis cannot be observed for *We* = 66.67, as shown in Figure 9c. This can be explained by the fact that when *We* is relatively low, the spreading velocity after impinging is also low due to inertia, which makes the influence of friction force more remarkable, and thus the tensile deformation becomes more obvious. Therefore, the contact angle hysteresis is more remarkable. As *We* increases, the spreading velocity after impinging increases, and the friction force plays a less important role in the dynamical behaviors. As a consequence, the difference between θ_a and θ_r decreases gradually. It is important to note that a hole appears near the center of the water nanodroplet during 20~30 ps, as shown in Figure 9c, because the higher *We* facilitates the spreading of water nanodroplets, and the water molecules move towards the edge of the water nanodroplet constantly during the spreading process. However, the hole disappears under the action of surface tension during the retraction process.

Figure 10 illustrates the friction force F_x applied to water nanodroplets with various *We*. F_x increases with *We* at the relative sliding stage, and thus, the water nanodroplet comes into the stable stage early, where the water nanodroplet and surface have the same velocity. This is because the increased spreading area leads to the enhancement of the interaction between water and solid surfaces.

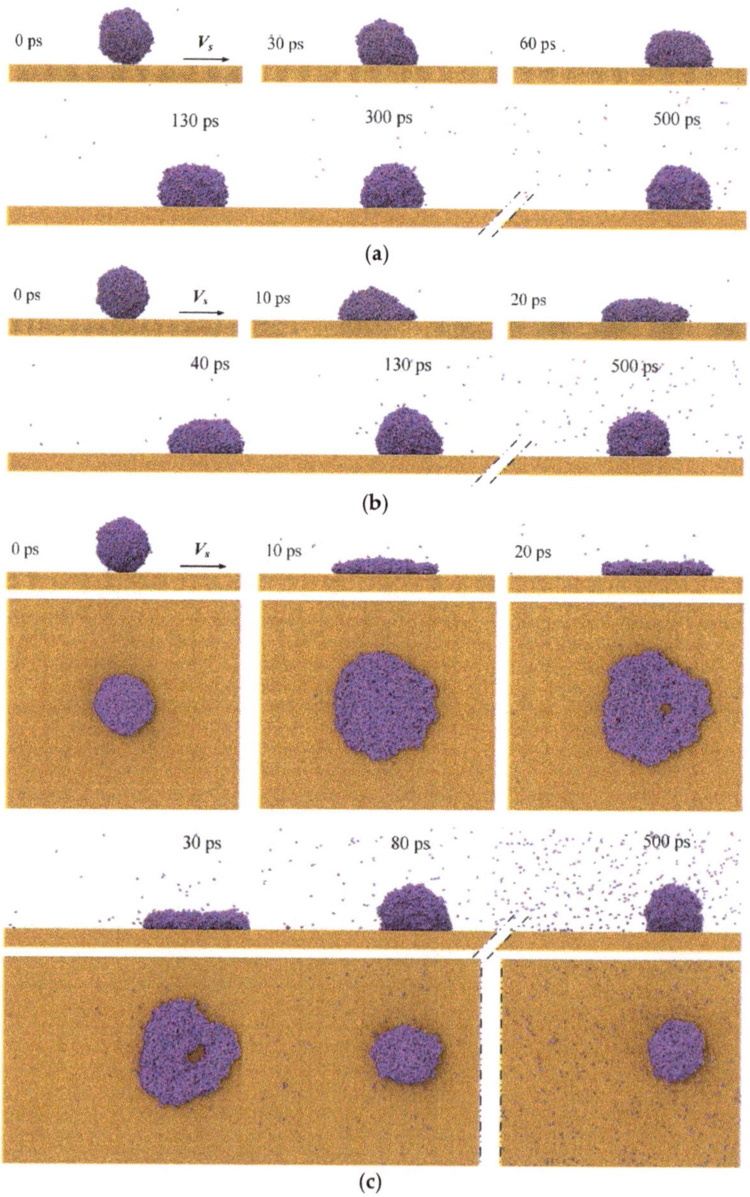

Figure 9. Morphological evolution of water nanodroplets with various *We* impinging on the translation surfaces: (**a**) *We* = 0.82; (**b**) *We* = 20.58; (**c**) *We* = 66.67.

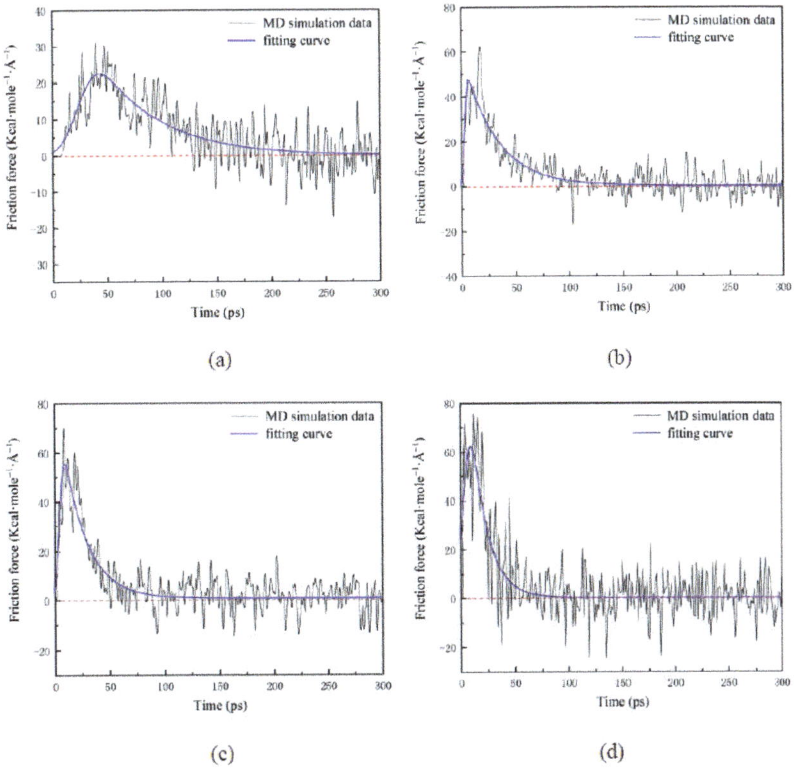

Figure 10. Time evolution of the friction force applied to water nanodroplets with various *We*: (**a**) *We* = 0.82; (**b**) *We* = 20.58; (**c**) *We* = 40.33; (**d**) *We* = 66.67.

From the spreading factors shown in Figure 11, it can be seen that both β_x and β_y are approximately 1.0 after 150 ps despite the different *We*, which indicates that *We* has little influence on the steady state of water nanodroplets and mainly takes a role in the spreading and retraction processes. In addition, $\beta_{x\max}$ and $\beta_{y\max}$ increase with the *We*. It should be noted that in the dashed line region shown in Figure 11b, besides *We* = 0.82, β_y decreases first and then gradually increases to a stable level. Therefore, the water nanodroplet experiences a secondary spreading in the *y* direction.

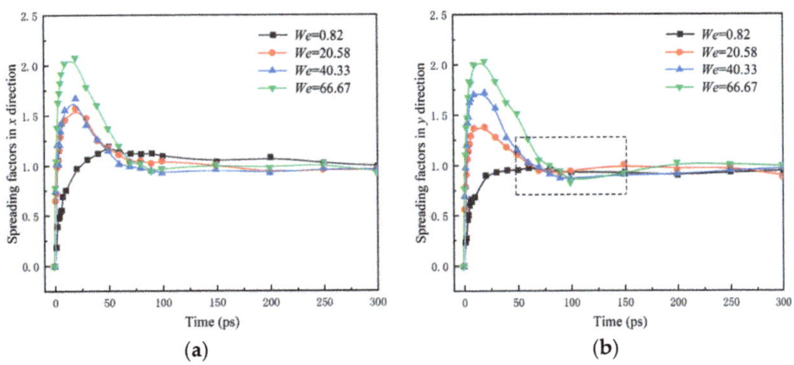

Figure 11. Time evolution of spreading factors in *x* and *y* directions: (**a**) β_x; (**b**) β_y.

3.4. Effects of Vibration Amplitudes on Dynamical Behaviors When Water Nanodroplets Impinge on Vibration Surfaces

To explore the influence of vibration amplitudes A on the dynamical behaviors, water nanodroplets impinging on surfaces with vibration amplitudes ranging from 1 to 4 Å, and the vibration period T = 12 ps is simulated. Vibration is achieved by applying periodic displacement up and down along the z axis on the solid surface, as shown in Figure 12.

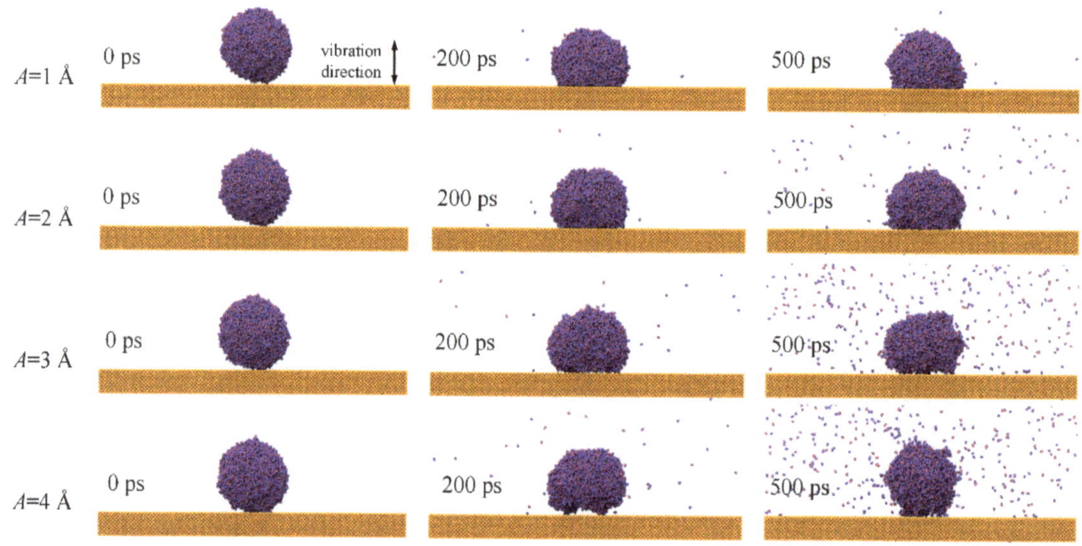

Figure 12. Morphological evolution of water nanodroplets impinging on vibration surfaces with various amplitudes.

The spreading area reduces with the increasing A, as shown in Figure 12. In particular, when A = 3 and 4 Å, there are some gaps between the water nanodroplets and surfaces. Hence, the adhesion between the water nanodroplet and surface is weakened, and more water molecules escape from the bulk of water nanodroplets with the increase in A.

Figure 13 shows the centroid height h of water nanodroplets. In general, h increases, and water nanodroplets have a tendency to depart from surfaces as A increases. During 0~20 ps, h decreases rapidly. That is because in this period, inertia is the dominant factor determining the dynamical behaviors, and the water nanodroplet spreads at high speed, and as a result, h almost decreases linearly. Once h decreases to the minimum, indicating the maximum spreading state being realized, the water nanodroplet starts to retract due to surface tension. However, energy dissipation in the spreading process leads to a low retraction velocity. Hence, the vibration of surfaces gradually comes into play, and the saw-tooth fluctuation appears in a single vibration period. After 70 ps, the retraction almost finishes, and the dynamical behaviors of water nanodroplets are solely affected by the vibration surfaces. The amplitudes of saw-tooth fluctuations decrease, and the shape of the saw tooth becomes uniform.

The variation of spreading factors β with time is recorded in Figure 14. It also shows apparent fluctuation due to the vibration surfaces. The maximum spreading factor β_{max} and the maximum spreading area reduce with the increasing A, suggesting a larger vibration amplitude not conducive to the spreading of water nanodroplets.

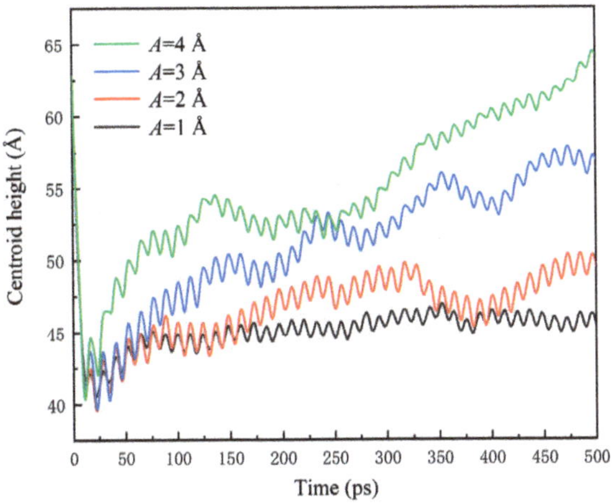

Figure 13. Centroid height of water nanodroplets impinging on vibration surfaces with various amplitudes.

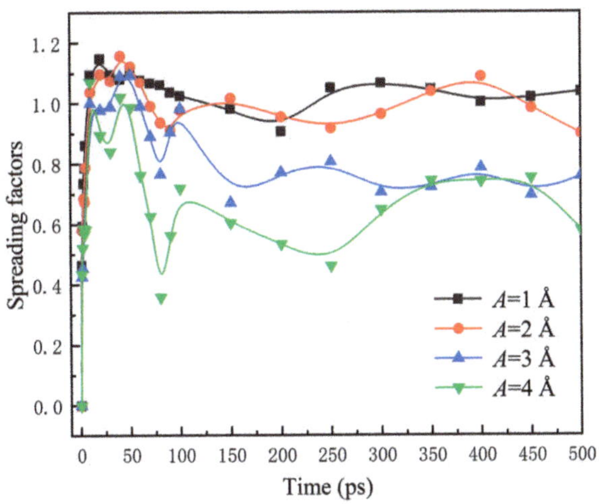

Figure 14. Time evolution of spreading factors of droplets impinging on surfaces with different vibration amplitudes.

3.5. *Effects of Vibration Periods on Dynamical Behaviors When Water Nanodroplets Impinge on Vibration Surfaces*

A water nanodroplet with $We = 7.41$ impinging on vibration surfaces with periods T ranging from 2 to 16 ps and $A = 2$ Å is simulated to analyze the influence of T on dynamical behaviors.

Figure 15 presents the snapshots in that process. Water nanodroplets eventually deposit on the solid surfaces when T = 4~16 ps. It can be seen that a longer T makes the connection between water nanodroplets and solid surfaces tighter, e.g., T = 16 ps in Figure 15c. It should be noted that the water nanodroplet finally bounces off the solid surface when T = 2 ps. This provides a method to facilitate water nanodroplets bouncing, except when designing special rough structures on the surfaces mentioned in the previous study [55]. Moreover, the water nanodroplet has a shorter contact time with the solid surface than the alternative.

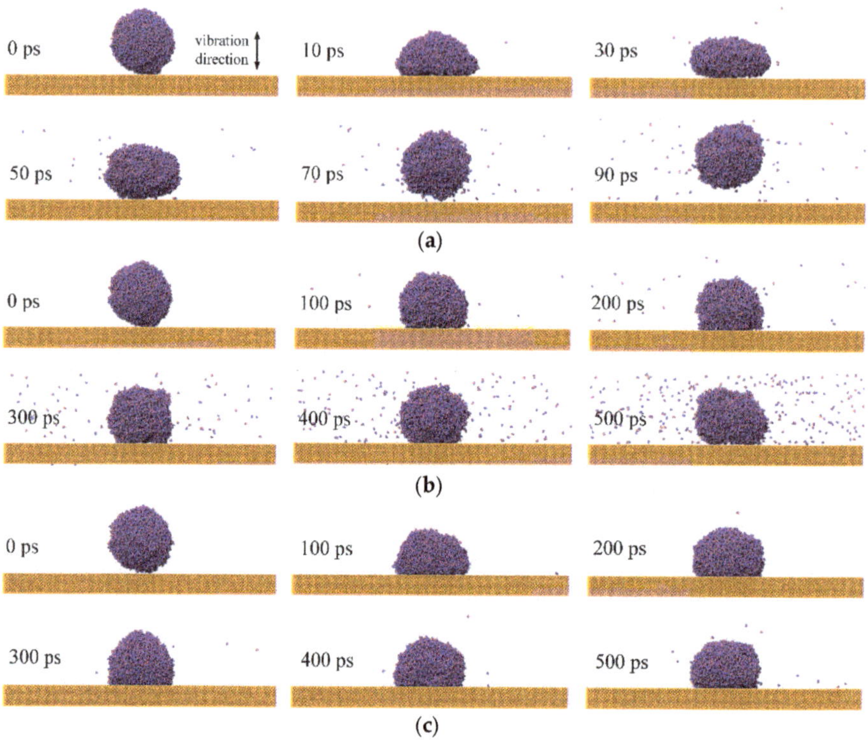

Figure 15. Dynamical behaviors of water nanodroplets impinging on surfaces with different vibration periods: (**a**) T = 2 ps; (**b**) T = 8 ps; (**c**) T = 16 ps.

The variation in centroid height h for water nanodroplets is shown in Figure 16. h nearly remains stable after 25 ps, when T = 12 and 16 ps. As T decreases, h increases. Especially, the water nanodroplet bounces off the solid surfaces at 55 ps when T = 2 ps.

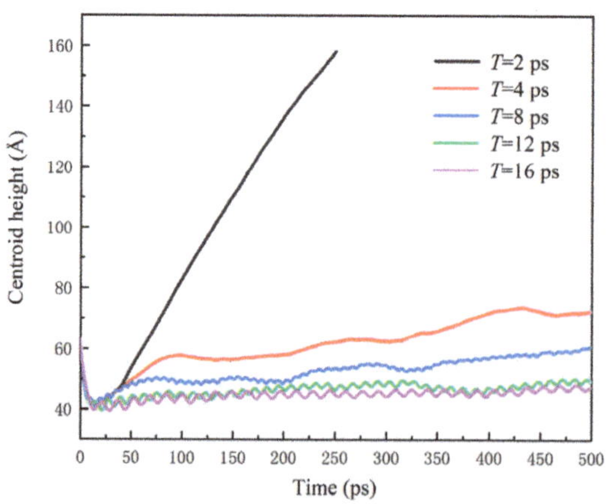

Figure 16. Time evolution of water nanodroplets impinging on surfaces with different vibration periods.

According to Figure 17, as a whole, β decreases with the decrese in the vibration period, and even it finally decreases to 0 for T = 2 ps. As a result, the increasing T induces the further spread of the water nanodroplet.

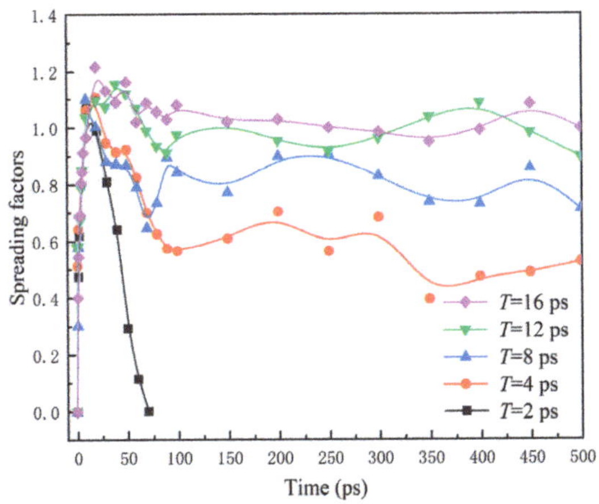

Figure 17. Spreading factors of water nanodroplets impinging on solid surfaces with various vibration periods.

To quantify the bounce of water nanodroplets, the bounce domain, determined by T and A, of water nanodroplets impinging on vibration surfaces, is showcased in Figure 18. This provides the theoretical basis for practical applications.

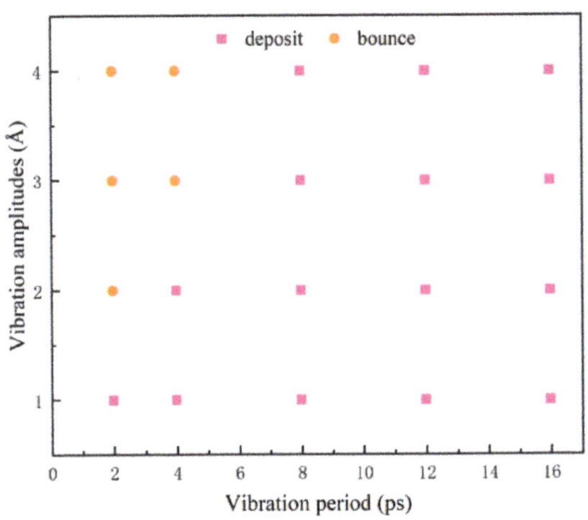

Figure 18. The bounce domain as water nanodroplets impinge on vibration surfaces.

4. Conclusions

In this work, molecular dynamics simulation was used to investigate the dynamical behaviors of water nanodroplets impinging on the translation surfaces and vibration surfaces. The influences of the translation velocity of the surface, the *We* of water nanodroplets, the vibration amplitudes, and the vibration periods on the dynamical behaviors were investigated. The main conclusions are as follows:

(1) The expression of velocity for water nanodroplets in the direction of translation was developed, which quantitively describes the velocity evolution of water nanodroplets as they impinge on translation surfaces.

(2) At the relative sliding stage, water molecules rotate around the centroid of the water nanodroplet and move along the translation direction. At the stable stage, the rotation of water molecules disappears.

(3) The higher the translation velocity is, the larger the friction force applied to the water nanodroplets is, and thus, the asymmetric spreading is more apparent. A higher translation velocity results in the water nanodroplet spreading twice in the direction perpendicular to the relative sliding.

(4) The increase in vibration amplitudes impedes the spreading of water nanodroplets, while the increase in vibration periods facilitates it. Additionally, the bounce domain of water nanodroplets was mapped.

Author Contributions: Conceptualization, H.Z. and L.P.; methodology, H.Z. and L.P.; software, H.Z. and X.X.; validation, L.P.; formal analysis, H.Z.; investigation, H.Z.; resources, H.Z.; data curation, H.Z.; writing—original draft preparation, H.Z.; writing—review and editing, H.Z. and L.P.; visualization, H.Z. and X.X.; supervision, L.P.; project administration, L.P.; funding acquisition, L.P. All authors have read and agreed to the published version of the manuscript.

Funding: This research was funded by the National Natural Science Foundation of China (grant number 51875105), the Industry–Academy Cooperation Project of Fujian Province (grant number 2020H6025), and the Scientific Research Program of the Jinjiang Science and Education Park Development Center Fuzhou University, China (grant number 2019-JJFDKY-54).

Institutional Review Board Statement: Not applicable.

Informed Consent Statement: Not applicable.

Data Availability Statement: The data related to this investigation are available on reasonable request.

Acknowledgments: The authors would like to acknowledge the support from the National Natural Science Foundation of China, the Industry–Academy Cooperation Project of Fujian Province, and the Scientific Research Program of the Jinjiang Science and Education Park Development Center, Fuzhou University, China. The authors also express gratitude to the supercomputing center of Fujian Province, China.

Conflicts of Interest: The authors declare no conflict of interest.

References

1. Dorr, G.J.; Wang, S.; Mayo, L.C.; McCue, S.W.; Forster, W.A.; Hanan, J.; He, X. Impaction of spray droplets on leaves: Influence of formulation and leaf character on shatter, bounce and adhesion. *Exp. Fluids* **2015**, *56*, 143. [CrossRef]
2. Roisman, I.V.; Breitenbach, J.; Tropea, C. Thermal atomisation of a liquid drop after impact onto a hot substrate. *J. Fluid Mech.* **2018**, *842*, 87–101. [CrossRef]
3. Zhang, Y.; Ren, T.; He, J. Inkjet Printing Enabled Controllable Paper Superhydrophobization and Its Applications. *ACS Appl. Mater. Interfaces* **2018**, *10*, 11343–11349. [CrossRef]
4. Shi, S.; Lv, C.; Zheng, Q. Drop Impact on Two-Tier Monostable Superrepellent Surfaces. *ACS Appl. Mater. Interfaces* **2019**, *11*, 43698–43707. [CrossRef]
5. Pan, S.; Wang, N.; Xiong, D.; Deng, Y.; Shi, Y. Fabrication of superhydrophobic coating via spraying method and its applications in anti-icing and anti-corrosion. *Appl. Surf. Sci.* **2016**, *389*, 547–553. [CrossRef]
6. Zhang, X.; Liu, X.; Wu, X.; Min, J. Impacting-freezing dynamics of a supercooled water droplet on a cold surface: Rebound and adhesion. *Int. J. Heat Mass Transf.* **2020**, *158*, 119997. [CrossRef]
7. Xu, W.; Zheng, H.; Liu, Y.; Zhou, X.; Zhang, C.; Song, Y.; Deng, X.; Leung, M.; Yang, Z.; Xu, R.X.; et al. A droplet-based electricity generator with high instantaneous power density. *Nature* **2020**, *578*, 392–396. [CrossRef]
8. Lee, D.J.; Kim, H.M.; Song, Y.S.; Youn, J.R. Water Droplet Bouncing and Superhydrophobicity Induced by Multiscale Hierarchical Nanostructures. *ACS Nano* **2012**, *6*, 7656–7664. [CrossRef]
9. Shi, Q.M.; Jia, Z.H.; Lin, Q.Y. Dynamic behavior of droplets impacting on microstructured hydrophobic surfaces. *Chem. Ind. Eng. Prog.* **2016**, *35*, 3818–3824. [CrossRef]

10. Jia-Lue, P.; Hao, G.; Tian-Ya, Y.; Xian-Bing, J.; Jin-Liang, X. Behavioral characteristics of droplet collision on Janus particle spheres. *Acta Phys. Sin.-Chin. Ed.* **2021**, *70*, 044701. [CrossRef]
11. Wang, L.-Z.; Zhou, A.; Zhou, J.-Z.; Chen, L.; Yu, Y.-S. Droplet impact on pillar-arrayed non-wetting surfaces. *Soft Matter* **2021**, *17*, 5932–5940. [CrossRef]
12. Hu, Z.; Zhang, X.; Gao, S.; Yuan, Z.; Lin, Y.; Chu, F.; Wu, X. Axial spreading of droplet impact on ridged superhydrophobic surfaces. *J. Colloid Interface Sci.* **2021**, *599*, 130–139. [CrossRef]
13. Castrejón-Pita, J.R.; Muñoz-Sánchez, B.N.; Hutchings, I.M.; Castrejón-Pita, A.A. Droplet impact onto moving liquids. *J. Fluid Mech.* **2016**, *809*, 716–725. [CrossRef]
14. Almohammadi, H.; Amirfazli, A. Asymmetric Spreading of a Drop upon Impact onto a Surface. *Langmuir* **2017**, *33*, 5957–5964. [CrossRef]
15. Xie, F.; Lv, S.; Yang, Y.; Wang, X. Contact Time of a Bouncing Nanodroplet. *J. Phys. Chem. Lett.* **2020**, *11*, 2818–2823. [CrossRef]
16. Hubao, A.; Yang, Z.; Hu, R.; Chen, Y.-F.; Yang, L. Effect of Solid–Liquid Interactions on Substrate Wettability and Dynamic Spreading of Nanodroplets: A Molecular Dynamics Study. *J. Phys. Chem. C* **2020**, *124*, 23260–23269. [CrossRef]
17. Liu, H.; Chu, F.; Zhang, J.; Wen, D. Nanodroplets impact on surfaces decorated with ridges. *Phys. Rev. Fluids* **2020**, *5*, 074201. [CrossRef]
18. Kumar, V.A.; Sajith, V.; Sathian, S.P. Influence of Nanoparticles on the Evaporation of a Nanodroplet from Solid Substrates: An Experimental and Molecular Dynamics Investigation. *Colloids Surf. A Physicochem. Eng. Asp.* **2021**, *615*, 126227. [CrossRef]
19. Drewienkiewicz, A.; Żydek, A.; Trybula, M.; Pstruś, J. Atomic Level Insight into Wetting and Structure of Ag Droplet on Graphene Coated Copper Substrate—Molecular Dynamics versus Experiment. *Nanomaterials* **2021**, *11*, 1465. [CrossRef]
20. Wang, Y.-B.; Wang, X.-D.; Yang, Y.-R.; Chen, M. The Maximum Spreading Factor for Polymer Nanodroplets Impacting a Hydrophobic Solid Surface. *J. Phys. Chem. C* **2019**, *123*, 12841–12850. [CrossRef]
21. Song, F.; Ju, D.; Fan, J.; Chen, Q.; Yang, Q. Deformation hysteresis of a water nano-droplet in an electric field. *Eur. Phys. J. E* **2019**, *42*, 120. [CrossRef]
22. Kwon, T.W.; Park, Y.G.; Park, S.H.; Ha, M.Y. Dynamic Behavior of a Nanosized Water Droplet on the Stepped Surface with a Wetting Gradient. *Langmuir* **2020**, *37*, 330–338. [CrossRef]
23. Pillai, R.; Borg, M.K.; Reese, J.M. Acoustothermal Atomization of Water Nanofilms. *Phys. Rev. Lett.* **2018**, *121*, 104502. [CrossRef]
24. Guo, W.; Ma, K.; Wang, Q.; Xue, H. The wetting of Pb droplet on the solid Al surface can be promoted by ultrasonic vibration—Molecular dynamics simulation. *Mater. Lett.* **2020**, *264*, 127118. [CrossRef]
25. Pillai, R.; Borg, M.K.; Reese, J.M. Dynamics of Nanodroplets on Vibrating Surfaces. *Langmuir* **2018**, *34*, 11898–11904. [CrossRef] [PubMed]
26. Russell, J.T.; Wang, B.; Král, P. Nanodroplet Transport on Vibrated Nanotubes. *J. Phys. Chem. Lett.* **2012**, *3*, 353–357. [CrossRef]
27. Sappati, K.K.; Bhadra, S. Piezoelectric Polymer and Paper Substrates: A Review. *Sensors* **2018**, *18*, 3605. [CrossRef]
28. Ding, X.; Li, P.; Lin, S.-C.S.; Stratton, Z.S.; Nama, N.; Guo, F.; Slotcavage, D.; Mao, X.; Shi, J.; Costanzo, F.; et al. Surface acoustic wave microfluidics. *Lab Chip* **2013**, *13*, 3626–3649. [CrossRef]
29. Li, R.; Li, G.; Hong, W.-C.; Reyes, P.I.; Tang, K.; Yang, K.; Wang, S.-Y.; Ye, H.; Li, Y.; Zhang, L.; et al. Tunable surface acoustic wave device using semiconducting MgZnO and piezoelectric NiZnO dual-layer structure on glass. *Smart Mater. Struct.* **2018**, *27*, 085025. [CrossRef]
30. Johnston, S.R.; Yang, Y.; Cui, Y.-T.; Ma, E.Y.; Kämpfe, T.; Eng, L.M.; Zhou, J.; Chen, Y.-F.; Lu, M.; Shen, Z.-X. Measurement of surface acoustic wave resonances in ferroelectric domains by microwave microscopy. *J. Appl. Phys.* **2017**, *122*, 074101. [CrossRef]
31. Su, R.; Fu, S.; Shen, J.; Chen, Z.; Lu, Z.; Yang, M.; Wang, R.; Zeng, F.; Wang, W.; Song, C.; et al. Enhanced Performance of $ZnO/SiO_2/Al_2O_3$ Surface Acoustic Wave Devices with Embedded Electrodes. *ACS Appl. Mater. Interfaces* **2020**, *12*, 42378–42385. [CrossRef] [PubMed]
32. Zheng, J.; Zhou, J.; Zeng, P.; Liu, Y.; Shen, Y.; Yao, W.; Chen, Z.; Wu, J.; Xiong, S.; Chen, Y.; et al. 30 GHz surface acoustic wave transducers with extremely high mass sensitivity. *Appl. Phys. Lett.* **2020**, *116*, 123502. [CrossRef]
33. Fu, S.; Wang, W.; Qian, L.; Li, Q.; Lu, Z.; Shen, J.; Song, C.; Zeng, F.; Pan, F. High-Frequency Surface Acoustic Wave Devices Based on ZnO/SiC Layered Structure. *IEEE Electron Device Lett.* **2019**, *40*, 103–106. [CrossRef]
34. Zhang, M.; Ma, L.; Wang, Q.; Hao, P.; Zheng, X. Wettability behavior of nanodroplets on copper surfaces with hierarchical nanostructures. *Colloids Surfaces A Physicochem. Eng. Asp.* **2020**, *604*, 125291. [CrossRef]
35. Jorgensen, W.L.; Chandrasekhar, J.; Madura, J.D.; Impey, R.W.; Klein, M.L. Comparison of simple potential functions for simulating liquid water. *J. Chem. Phys.* **1983**, *79*, 926–935. [CrossRef]
36. Fu, T.; Wu, N.; Lu, C.; Wang, J.; Wang, Q. Effect of nanostructure on wettability on copper surface: A molecular dynamic study. *Mol. Simul.* **2019**, *45*, 35–39. [CrossRef]
37. Plimpton, S. Fast Parallel Algorithms for Short-Range Molecular Dynamics. *J. Comput. Phys.* **1995**, *117*, 1–19. [CrossRef]
38. Ryckaert, J.-P.; Ciccotti, G.; Berendsen, H.J.C. Numerical integration of the cartesian equations of motion of a system with constraints: Molecular dynamics of n-alkanes. *J. Comput. Phys.* **1977**, *23*, 327–341. [CrossRef]
39. Gonzalez-Valle, C.U.; Kumar, S.; Ramos-Alvarado, B. Investigation on the Wetting Behavior of 3C-SiC Surfaces: Theory and Modeling. *J. Phys. Chem. C* **2018**, *122*, 7179–7186. [CrossRef]
40. Koishi, T.; Yasuoka, K.; Zeng, X.C. Molecular Dynamics Simulation of Water Nanodroplet Bounce Back from Flat and Nanopillared Surface. *Langmuir* **2017**, *33*, 10184–10192. [CrossRef]

41. Ren, H.; Yang, F.; Li, C.; Deng, C. Head-on collision of binary nanodroplets on rough surfaces: Impact velocity dependent spreading dynamics. *Appl. Surf. Sci.* **2021**, *541*, 148426. [CrossRef]
42. Yin, Z.-J.; Ding, Z.-L.; Zhang, W.-F.; Su, R.; Chai, F.-T.; Yu, P. Dynamic behaviors of nanoscale binary water droplets simultaneously impacting on flat surface. *Comput. Mater. Sci.* **2020**, *183*, 109814. [CrossRef]
43. Yu, Y.; Xu, X.; Liu, J.; Liu, Y.; Cai, W.; Chen, J. The study of water wettability on solid surfaces by molecular dynamics simulation. *Surf. Sci.* **2021**, *714*, 121916. [CrossRef]
44. Heinz, H.; Vaia, R.A.; Farmer, B.L.; Naik, R.R. Accurate Simulation of Surfaces and Interfaces of Face-Centered Cubic Metals Using 12−6 and 9−6 Lennard-Jones Potentials. *J. Phys. Chem. C* **2008**, *112*, 17281–17290. [CrossRef]
45. Wu, N.; Zeng, L.; Fu, T.; Wang, Z.; Deng, X. Mechanism of heat transfer enhancement by nanochannels copper plate interface wettability: A molecular dynamics study. *Int. J. Therm. Sci.* **2021**, *159*, 106589. [CrossRef]
46. Bai, P.; Zhou, L.; Du, X. Molecular dynamics simulation of the roles of roughness ratio and surface potential energy in explosive boiling. *J. Mol. Liq.* **2021**, *335*, 116169. [CrossRef]
47. Gao, S.; Liu, W.; Liu, Z. Tuning nanostructured surfaces with hybrid wettability areas to enhance condensation. *Nanoscale* **2019**, *11*, 459–466. [CrossRef]
48. Chen, L.; Wang, S.-Y.; Xiang, X.; Tao, W.-Q. Mechanism of surface nanostructure changing wettability: A molecular dynamics simulation. *Comput. Mater. Sci.* **2020**, *171*, 109223. [CrossRef]
49. Bekele, S.; Evans, O.G.; Tsige, M. Spreading Dynamics of Water Droplets on a Completely Wetting Surface. *J. Phys. Chem. C* **2020**, *124*, 20109–20115. [CrossRef]
50. Hoover, W.G. Canonical dynamics: Equilibrium phase-space distributions. *Phys. Rev. A* **1985**, *31*, 1695–1697. [CrossRef]
51. Song, F.; Ma, L.; Fan, J.; Chen, Q.; Zhang, L.; Li, B.Q. Wetting Behaviors of a Nano-Droplet on a Rough Solid Substrate under Perpendicular Electric Field. *Nanomaterials* **2018**, *8*, 340. [CrossRef]
52. Li, H.; Zhang, K. Dynamic behavior of water droplets impacting on the superhydrophobic surface: Both experimental study and molecular dynamics simulation study. *Appl. Surf. Sci.* **2019**, *498*, 143793. [CrossRef]
53. Cheng, Z.; Du, M.; Lai, H.; Zhang, N.; Sun, K. Super-hydrophobic Copper Surface with Controlled Adhesion Prepared via Ammonia Corrosion. *Chem. J. Chin. Univ.-Chin.* **2013**, *34*, 606–609. [CrossRef]
54. Josserand, C.; Thoroddsen, S.T. Drop Impact on a Solid Surface. *Annu. Rev. Fluid Mech.* **2016**, *48*, 365–391. [CrossRef]
55. Pan, L.; Zhang, H.; Lin, G. Molecular dynamics simulation on dynamic behaviors of nanodroplets impinging on solid surfaces decorated with nanopillars. *Acta Phys. Sin.-Chin. Ed.* **2021**, *70*, 307–316. [CrossRef]

MDPI
St. Alban-Anlage 66
4052 Basel
Switzerland
Tel. +41 61 683 77 34
Fax +41 61 302 89 18
www.mdpi.com

Nanomaterials Editorial Office
E-mail: nanomaterials@mdpi.com
www.mdpi.com/journal/nanomaterials

www.ingramcontent.com/pod-product-compliance
Lightning Source LLC
Chambersburg PA
CBHW061806210926
51850CB00006B/627
* 9 7 8 3 0 3 6 5 7 3 1 3 7 *